Complex Variables

Complex Variables

M. Ya. Antimirov
Riga Technical University
Riga, Latvia

A. A. Kolyshkin
Riga Technical University
Riga, Latvia

Rémi Vaillancourt
University of Ottawa
Ottawa, Canada

ACADEMIC PRESS
San Diego London Boston
New York Sydney Tokyo Toronto

Academic Press
525 B Street, Suite 1900, San Diego, CA, 92101-4495, USA
1300 Boylston St., Chestnut Hill, MA 02167, USA
http://www.apnet.com

ACADEMIC PRESS LIMITED
24–28 Oval Road, London NW1 7DX, UK
http://www.hbuk.cc.uk/ap/

Library of Congress Cataloging-in-Publication Data

Antimirov, M. Ya.,
Complex Variables / M. Ya. Antimirov, Andrei Kolyshkin, Rémi Vaillancourt.
p. cm.
Includes bibliographical references and index.
ISBN 0-12-059545-1 (alk. paper)
1. Functions of complex variables. I. Kolyshkin, A. A. II. Vaillancourt, Rémi. III. Title
QA331.7.A57 1998
515'.9—dc21 97–47110
CIP

Printed in the United States of America

97 98 99 00 01 IP 9 8 7 6 5 4 3 2 1

To our understanding wives,
Anna, Valentina, and Jeanne

Contents

Preface

This book is the fruit of many years of teaching complex variables to students in applied mathematics by the first author and research by the third author with the close collaboration of the second author, who translated a preliminary Russian version of the text and collected and solved all the exercises. It is an extended course in complex analysis and its applications, written in a style that is particulary well-suited for students in applied mathematics, science and engineering, and for users of complex analysis in the applications.

The first half of the book is a clear and rigorous introduction to the theory of functions of one complex variable. The second half contains the evaluation of many new integration formulae and the summation of new infinite series by the calculus of residue. The last chapter is concerned with the Fatou–Julia theory for meromorphic functions for finding selective roots of some transcendental equations as found in the applications.

A second book in preparation will cover Schwarz–Christoffel transformations, aplications of conformal mapping in heat transfer, electrostatics, hydrodynamics and aerodynamics, and applications of Laplace transform in engineering, etc.

Chapter 1 reviews the representation of complex numbers and introduces analytic (holomorphic) functions. In Chapter 2, both traditional and non-traditional problems in conformal mapping are solved in great detail.

Chapter 2 depends only on Chapter 1 and is independent of the other chapters; thus it can be taken any time, after the study of the first chapter. Chapters 3, 4, 5, 6 and 9 can be covered in that order.

Chapters 7, 8, 10 and 11 cover more specialized topics and are beyond a usual introduction to analytic functions.

The short bibliography lists common references in English and in Russian and a few research papers.

The exercises are elementary and aim at the understanding of the theory of analytic functions. Some of them can be easily solved with symbolic software on computers. Answers to almost all odd-numbered exercises are found at the end of the book.

In chapters 1 to 11 of this book, for the convenience of the reader, generally every reference to a theorem, formula, table, etc. in a book listed in the bibliography includes the page number (first and last page numbers if the reference is to more than one page) where the referenced theorem, formula, table, etc. is to be found in the referenced book. For example, "Table 5 on p. 757 in [14]" refers to Table 5 on p. 757 of the referenced book listed as No. 14 in the bibliography. However, in the Answers to Odd-Numbered Exercises, following Chapter 11, a reference of the form "Page 12 in the text" means page 12 of this book.

The text benefitted from the remarks made by generations of students at the Riga Technical University and at the University of Ottawa. Miss Ellen Yanqing Zheng has read a preliminary version of the book in the winter of 1994 and made many corrections.

The authors express their thanks to Dr. Thierry Giordano, who has used the manuscript as lecture notes since 1995. He has made many valuable suggestions for improving the first part of the book.

Mr. André Montpetit of the Centre de recherches mathématiques of the Université de Montréal has been generous in offering invaluable assistance for the composition of the text in *AMS − LaTeX*.

This book benefitted from the supports of the Natural Sciences and Engineering Council of Canada, the University of Ottawa, Riga Technical University, and the Centre de recherches mathématiques of the Université de Montréal.

The authors express their warmest thanks to the dynamic and collaborative editorial and production team of Academic Press.

M. Ya. Antimirov
A. A. Kolyshkin
Rémi Vaillancourt
Riga, Ottawa, 4 December 1997

CHAPTER 1

Functions of a Complex Variable

1.1. Complex numbers

1.1.1. Algebraic operations on complex numbers.

DEFINITION 1.1.1. A *complex number* z is an ordered pair, (x, y), of real numbers, x and y, where x is called the *real part* of z, written $x = \Re z$, and y is called the *imaginary part* of z, written $y = \Im z$. The set of complex numbers is denoted by $\mathbb{C}$.

For clarity, the expressions *z-plane* and *w-plane* will be used to mean $z \in \mathbb{C}$ and $w \in \mathbb{C}$, respectively, when referring to different copies of $\mathbb{C}$.

Two complex numbers, $z_1 = (x_1, y_1)$ and $z_2 = (x_2, y_2)$, are *equal*, written $z_1 = z_2$, if and only if their real and imaginary parts are equal; that is, if and only if $x_1 = x_2$ and $y_1 = y_2$.

DEFINITION 1.1.2. The *sum* of two complex numbers, $z_1 = (x_1, y_1)$ and $z_2 = (x_2, y_2)$, is defined to be the complex number

$$z = z_1 + z_2 = (x_1 + x_2, y_1 + y_2).$$

The *commutativity* and the *associativity* of the addition,

$$\begin{aligned} z_1 + z_2 &= z_2 + z_1, \\ z_1 + (z_2 + z_3) &= (z_1 + z_2) + z_3, \end{aligned}$$

follow from Definition 1.1.2. The complex number *zero*, $0 = (0, 0)$, such that $z + 0 = z$ for all $z \in \mathbb{C}$, is introduced in the same way as the real number 0 in the set of real numbers.

DEFINITION 1.1.3. The *product* of two complex numbers, $z_1 = (x_1, y_1)$ and $z_2 = (x_2, y_2)$, is defined to be the complex number

$$z = z_1 z_2 = (x_1 x_2 - y_1 y_2, x_1 y_2 + x_2 y_1).$$

The *commutativity*, the *associativity* and the *distributivity* of the multiplication,

$$\begin{aligned} z_1 z_2 &= z_2 z_1, \\ z_1(z_2 z_3) &= (z_1 z_2) z_3, \\ (z_1 + z_2) z_3 &= z_1 z_3 + z_2 z_3, \end{aligned}$$

follow from Definition 1.1.3.

The set $\mathbb{R}$ of real numbers becomes a subset of the set $\mathbb{C}$ of complex numbers if $a \in \mathbb{R}$ is identified with $a = (a, 0) \in \mathbb{C}$. It then follows, from Definitions 1.1.2 and 1.1.3 of addition and multiplication, respectively, that all the known properties of the addition and the multiplication of real numbers are also valid for complex numbers. Therefore the set $\mathbb{C}$ of complex numbers can be considered as an extension of the set $\mathbb{R}$ of real numbers.

Note that $(a, 0) \times (x, y) = (ax, ay)$.

The complex numbers are not ordered. Hence the order relations $<$ and $>$ cannot be applied to complex numbers; that is, given two distinct nonreal complex numbers, z_1 and z_2, it is impossible to write $z_1 > z_2$ or $z_1 < z_2$, without violating some properties of the real numbers.

DEFINITION 1.1.4. A complex number of the form $(0, y)$ is said to be a *pure imaginary* number.

The complex number $(0, 1)$ is called the *imaginary unit* and is denoted by a symbol i: $i = (0, 1)$. The number $(0, y)$ can be considered as the product of the real number $y = (y, 0)$ and the imaginary unit $(0, 1)$,

$$(y, 0) \times (0, 1) = (y \times 0 - 0 \times 1, y \times 1 + 0 \times 0) = (0, y).$$

Therefore we can write $(0, y) = iy$.

Squaring the imaginary unit, we have

$$i \times i = (0, 1) \times (0, 1) = (0 \times 0 - 1 \times 1, 0 \times 1 + 1 \times 0) = (-1, 0),$$

that is,

$$i^2 = -1. \tag{1.1.1}$$

1.1.2. Algebraic form of complex numbers. The previous relation (1.1.1) allows one to give a direct computationally convenient algebraic meaning to complex numbers.

DEFINITION 1.1.5. The *algebraic form* of the complex number

$$z = (x, y) = (x, 0) + (0, y)$$

is

$$z = x + iy. \tag{1.1.2}$$

NOTATION 1.1.1. Complex numbers in the algebraic form are usually denoted by $z = x + iy$, $\zeta = \xi + i\eta$, $w = u + iv$, and $a = \alpha + i\beta$. The letters c and d are also used.

To perform addition and multiplication of complex numbers, one simply uses the usual rules of the algebra of polynomials plus the rules $i^2 = -1$, $i^3 = -i$ and $i^4 = 1$.

DEFINITION 1.1.6. The complex number $\bar{z} = x - iy$ is called the *complex conjugate* of $z = x + iy$.

The *subtraction* of complex numbers is defined as the inverse of the addition. Given two complex numbers $z_1 = x_1 + iy_1$ and $z_2 = x_2 + iy_2$, the difference, $z_2 - z_1$, is the complex number z such that $z_1 + z = z_2$. Thus

$$z = z_2 - z_1 = x_2 - x_1 + i(y_2 - y_1).$$

The *division* of complex numbers is defined as the inverse of the multiplication. If $z_1 = x_1 + iy_1$ and $z_2 = x_2 + iy_2 \neq 0$, then $z = z_1/z_2$ if

$$z_2 z = z_1. \tag{1.1.3}$$

Letting $z = x+iy$ in (1.1.3), performing the multiplication and equating the real and imaginary parts on the right- and left-hand sides of (1.1.3), respectively, we obtain a system of equations for $x = \Re z$ and $y = \Im z$. Solving this system, we get

$$x + iy = \frac{x_1 + iy_1}{x_2 + iy_2} = \frac{x_1x_2 + y_1y_2}{x_2^2 + y_2^2} + i\,\frac{x_2y_1 - x_1y_2}{x_2^2 + y_2^2}. \tag{1.1.4}$$

It is easy to check that the same result can be found by multiplying the numerator and the denominator of the fraction z_1/z_2 by $\bar{z}_2 = x_2 - iy_2$.

1.1.3. Geometric representation of complex numbers. We shall represent the complex number $z = x + iy$ by the point A in the plane with coordinates (x, y) referred to the Cartesian coordinate system $x0y$. Such a plane is called the *complex plane*, the x-axis being called the *real axis* and the y-axis being called the *imaginary axis*. There is a one-to-one correspondence between the points of the complex plane and the set of complex numbers. Therefore in the sequel we shall not distinguish between a complex number and its corresponding point in the complex plane, so that we shall say, for example, the "point $3 + 2i$," the "triangle with vertices z_1, z_2 and z_3," etc.

In Fig 1.1, the vector $\overrightarrow{OA} = (x, y)$ is identified with the complex number $z = x + iy$. The angle θ formed by $\overrightarrow{OA}$ and the positive x-axis is called the

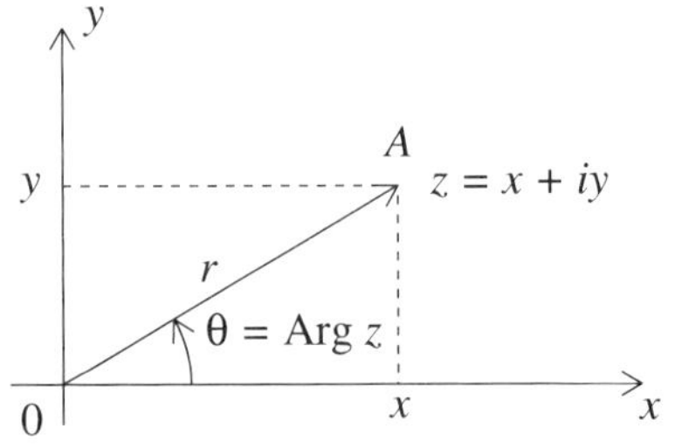

FIGURE 1.1. The vector $\overrightarrow{OA} = (x, y)$ identified with the complex number $z = x + iy$.

argument of z and is denoted by $\arg z$:

$$\theta = \arg z, \quad \text{if} \quad \tan(\arg z) = \frac{y}{x}. \tag{1.1.5}$$

The length r of the vector $\overrightarrow{OA}$ is called the *modulus* of the complex number z and is denoted by $|z|$,

$$|z| = |\overrightarrow{OA}| = r = \sqrt{x^2 + y^2} \geq 0. \tag{1.1.6}$$

The angle $\arg z$ is usually taken in one of the half-open intervals,

$$(2k - 1)\pi < \arg z \leq (2k + 1)\pi, \qquad k = 0, \pm 1, \pm 2, \dots, \tag{1.1.7}$$

or

$$2k\pi \leq \arg z < 2(k + 1)\pi, \qquad k = 0, \pm 1, \pm 2, \dots. \tag{1.1.8}$$

The *principal value* of the argument of z is defined to be the angle $\operatorname{Arg} z$ such that

$$\tan(\operatorname{Arg} z) = \frac{y}{x}, \qquad -\pi < \operatorname{Arg} z \leq \pi, \tag{1.1.9}$$

by taking $k = 0$ in (1.1.7), or

$$\tan(\operatorname{Arg} z) = \frac{y}{x}, \qquad 0 \leq \operatorname{Arg} z < 2\pi, \tag{1.1.10}$$

by taking $k = 0$ in (1.1.8).

In this book, the choice of (1.1.9) or (1.1.10) will be dictated by each problem in hand and should be clear from the context. Generally, (1.1.9) is used in Chapters 1, 3, 4 and 5, and (1.1.10) is used in Chapters 2, 6, 7 and 8. Most computers use the the principal value given by (1.1.9).

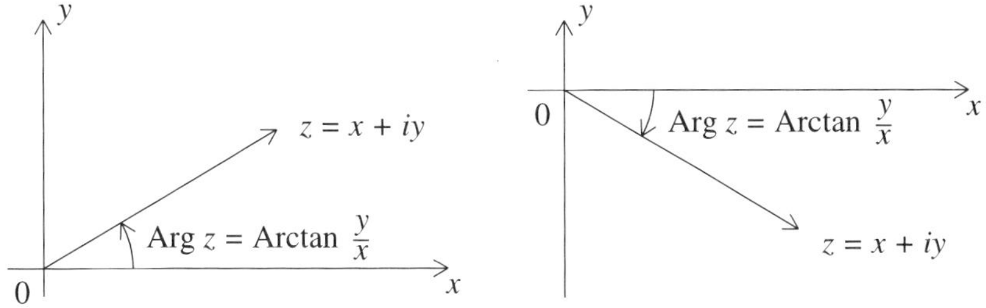

FIGURE 1.2. The principal value of arg z for $x > 0$.

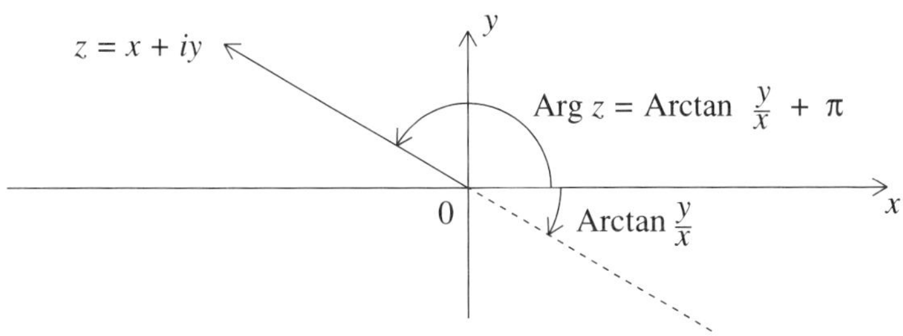

FIGURE 1.3. The principal value of arg z for $x < 0$, $y > 0$.

With the choice (1.1.9), there are three cases to be considered for Arg z:
(a) If $x > 0$ (see Fig 1.2), $\operatorname{Arg} z = \operatorname{Arctan} \dfrac{y}{x}$.
(b) If $x < 0$ and $y > 0$ (see Fig 1.3), $\operatorname{Arg} z = \operatorname{Arctan} \dfrac{y}{x} + \pi$.
(c) If $x < 0$ and $y < 0$ (see Fig 1.4), $\operatorname{Arg} z = \operatorname{Arctan} \dfrac{y}{x} - \pi$.
Hence, for (1.1.9),

$$\operatorname{Arg} z = \begin{cases} \operatorname{Arctan}\left(\frac{y}{x}\right), & x > 0, \\ \operatorname{Arctan}\left(\frac{y}{x}\right) + \pi, & x < 0,\ y > 0, \\ \operatorname{Arctan}\left(\frac{y}{x}\right) - \pi, & x < 0,\ y < 0. \end{cases} \tag{1.1.11}$$

In any case, one sees that $\arg z = \operatorname{Arg} z + 2k\pi$ for $k \in \mathbb{Z}$, that is, arg is periodic of period 2π.

Note 1.1.1. The definitions (1.1.9) or (1.1.10) of Arg z mean that a cut is made along the negative or positive real axis, respectively. In general terms, a *cut* is a double line that is not allowed to be crossed when angles are measured. Therefore, with (1.1.9) $\operatorname{Arg} z = \pi$ on the upper part of the cut and $\operatorname{Arg} z = -\pi$ on the lower part of the cut. Such a cut can be taken

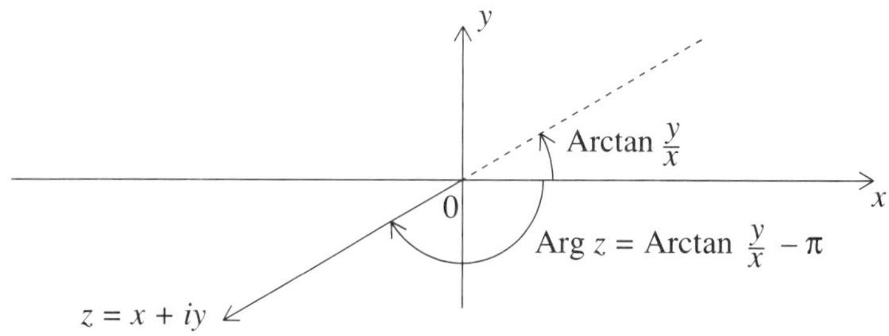

FIGURE 1.4. The principal value of arg z for $x < 0$, $y < 0$.

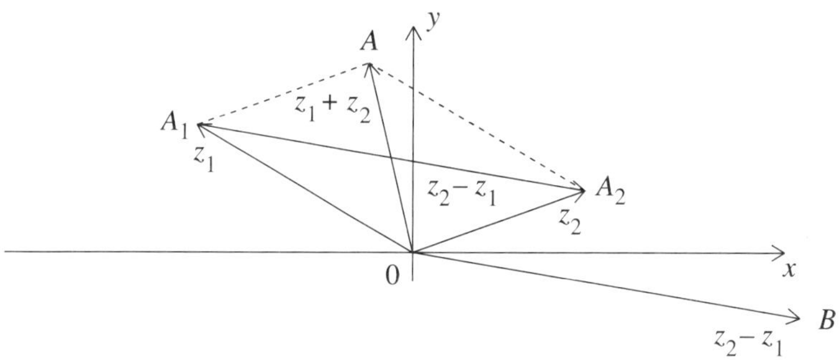

FIGURE 1.5. Geometric representation of the sum, $\overrightarrow{OA}$, and difference, $\overrightarrow{OB} = \overrightarrow{A_1A_2}$, of two complex numbers.

along an arbitrary direction, but formula (1.1.11) differs from cut to cut. Most computers and calculators take the cut along the negative real axis so that the principal value, Arg z, of the argument of z is given by (1.1.9) so that (1.1.11) holds. In the Russian mathematical literature, the roles of arg and Arg are interchanged.

Let us consider the geometric meaning of the sum and difference of the two complex numbers $z_1 = x_1 + iy_1$ and $z_2 = x_2 + iy_2$.

In Fig 1.5 the vectors $\overrightarrow{OA_1} = (x_1, y_1)$ and $\overrightarrow{OA_2} = (x_2, y_2)$ correspond to z_1 and z_2, respectively.

Since $z_1 + z_2 = (x_1 + x_2) + i(y_1 + y_2)$, then the vector

$$\overrightarrow{OA} = (x_1 + x_2, y_1 + y_2)$$

corresponds to the complex number $z_1 + z_2$. Thus, the sum of the vectors

$\overrightarrow{OA_1}$ and $\overrightarrow{OA_2}$,

$$\overrightarrow{OA} = \overrightarrow{OA_1} + \overrightarrow{OA_2}, \tag{1.1.12}$$

corresponds to the sum of the complex numbers z_1 and z_2. Similarly, the vector

$$\overrightarrow{OA} = \overrightarrow{OA_1} + \overrightarrow{OA_2} + \cdots + \overrightarrow{OA_n} \tag{1.1.13}$$

corresponds to the sum $z_1+z_2+\cdots+z_n$ of the complex numbers $z_1, z_2, \cdots, z_n$ represented by the vectors $\overrightarrow{OA_1}, \overrightarrow{OA_2}, \ldots, \overrightarrow{OA_n}$, respectively. The vector $\overrightarrow{OA}$ joins the beginning and the end of the polygonal line $OA_1A_2\cdots A_n$. It follows from Fig 1.5 and formulae (1.1.12) and (1.1.13) that

$$|\overrightarrow{OA}| \leq |\overrightarrow{OA_1}| + |\overrightarrow{OA_2}|, \qquad |\overrightarrow{OA}| \leq |\overrightarrow{OA_1}| + |\overrightarrow{OA_2}| + \cdots + |\overrightarrow{OA_n}|,$$

that is, we have the *triangle inequality*,

$$|z_1 + z_2| \leq |z_1| + |z_2|, \tag{1.1.14}$$

and its generalization to n numbers,

$$|z_1 + z_2 + \cdots + z_n| \leq |z_1| + |z_2| + \cdots + |z_n|.$$

These inequalities can be written in the short form

$$\left|\sum_{k=1}^{n} z_k\right| \leq \sum_{k=1}^{n} |z_k|. \tag{1.1.15}$$

Equality in (1.1.14) and (1.1.15) holds only if all the complex numbers z_k lie on the same straight line in the complex plane.

Inequality (1.1.15) is basic for estimating the moduli of sums of complex numbers and integrals of functions of a complex variable.

On the other hand, since $z_2 - z_1 = (x_2 - x_1) + i(y_2 - y_1)$, then the vector $\overrightarrow{OB} = (x_2 - x_1, y_2 - y_1)$ corresponds to the complex number $z_2 - z_1$. In this case,

$$\overrightarrow{OB} = \overrightarrow{A_1A_2} = \overrightarrow{OA_2} - \overrightarrow{OA_1}, \tag{1.1.16}$$

that is, the vector $\overrightarrow{OB}$ corresponds to the difference of the given complex numbers and is represented by a difference of the vectors $\overrightarrow{OA_2}$ and $\overrightarrow{OA_1}$.

It follows from Fig 1.5 and formula (1.1.16) that

$$|z_2 - z_1| = |\overrightarrow{A_1A_2}| = \sqrt{(x_2 - x_1)^2 + (y_2 - y_1)^2}, \tag{1.1.17}$$

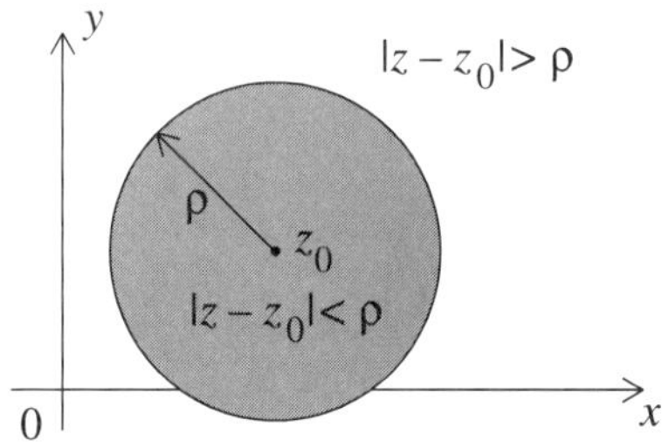

FIGURE 1.6. (Shaded) interior and (unshaded) exterior of a disk.

that is, the modulus of the difference, $z_2 - z_1$ of two complex numbers is equal to the distance between the points z_1 and z_2 in the complex plane. Since the distance in $\mathbb{C}$ and $\mathbb{R}^2$ is given by the same formula, it will be seen in the next subsections that the definition of a neighborhood of a point, the set of interior or exterior points of a disk in $\mathbb{C}$, etc., will be the same as in $\mathbb{R}^2$. Hence $\mathbb{C}$ and $\mathbb{R}^2$ have the same notions of continuity and limit, that is, the same topology.

For example, if $z_0 = x_0 + iy_0 =$ constant and $\rho =$ constant > 0, then the formula

$$|z - z_0| = \rho \tag{1.1.18}$$

represents the geometric locus of all the points z which are at distance ρ from the point z_0. Thus (1.1.18) is the equation of a circle centered at z_0 and of radius ρ (see Fig 1.6). If $z = x + iy$ and $z_0 = x_0 + iy_0$, it follows from (1.1.17) and (1.1.18) that

$$(x - x_0)^2 + (y - y_0)^2 = \rho^2, \tag{1.1.19}$$

which is the Cartesian equation of the circle of radius ρ, centered at (x_0, y_0). In Fig 1.6, the inequality $|z - z_0| < \rho$ represents the (shaded) set of points inside the disk whereas the inequality $|z - z_0| > \rho$ represents the (unshaded) set of points outside the same disk.

1.1.4. Trigonometric form of complex numbers. One easily sees from Fig 1.1 that if $z = x + iy$, then $x = r\cos\theta$ and $y = r\sin\theta$ with $r = |z|$. Thus, we have the following definition.

DEFINITION 1.1.7. The *trigonometric form* of the complex number $z = x + iy$ is

$$z = r(\cos\theta + i\sin\theta), \tag{1.1.20}$$

where $x = r\cos\theta$, $y = r\sin\theta$ and $r = |z|$.

The trigonometric form (1.1.20) of complex numbers allows one to give a simple geometric meaning to the product and quotient of two complex numbers. Given

$$z_1 = r_1(\cos\theta_1 + i\sin\theta_1), \qquad z_2 = r_2(\cos\theta_2 + i\sin\theta_2),$$

by the usual rules of algebra the product of z_1 and z_2 is

$$z_1 z_2 = r_1 r_2\big[\cos\theta_1\cos\theta_2 - \sin\theta_1\sin\theta_2 + i(\cos\theta_1\sin\theta_2 + \sin\theta_1\cos\theta_2)\big], \quad (1.1.21)$$

which, upon using trigonometric identities for sums of angles, reduces to

$$z_1 z_2 = r_1 r_2\big[\cos(\theta_1+\theta_2) + i\sin(\theta_1+\theta_2)\big]. \quad (1.1.22)$$

It follows from (1.1.22) that

$$|z_1 z_2| = |z_1|\,|z_2|, \qquad \arg(z_1 z_2) = \arg z_1 + \arg z_2, \quad (1.1.23)$$

that is, the modulus of the product of two complex numbers is equal to the product of their moduli, while the argument of the product is equal to the sum of their arguments. It can easily be proved by mathematical induction that relations similar to (1.1.22) and (1.1.23) hold for any finite number of complex numbers:

$$z_1 z_2 \cdots z_n = r_1 r_2 \cdots r_n\left[\cos\left(\sum_{k=1}^{n}\theta_k\right) + i\sin\left(\sum_{k=1}^{n}\theta_k\right)\right]; \quad (1.1.24)$$

thus

$$|z_1 z_2 \cdots z_n| = |z_1|\,|z_2|\cdots|z_n|, \qquad \arg(z_1 z_2\cdots z_n) = \sum_{k=1}^{n}\arg z_k. \quad (1.1.25)$$

Similarly, if $z_2 \neq 0$,

$$\frac{z_1}{z_2} = \frac{r_1}{r_2}\frac{\cos\theta_1 + i\sin\theta_1}{\cos\theta_2 + i\sin\theta_2} = \frac{r_1}{r_2}\frac{(\cos\theta_1 + i\sin\theta_1)(\cos\theta_2 - i\sin\theta_2)}{\cos^2\theta_2 + \sin^2\theta_2}.$$

Multiplying the numerator out and applying trigonometric identities for difference of angles, we have

$$\frac{z_1}{z_2} = \frac{r_1}{r_2}\left[\cos(\theta_1-\theta_2) + i\sin(\theta_1-\theta_2)\right]. \quad (1.1.26)$$

It follows from (1.1.26) that

$$\left|\frac{z_1}{z_2}\right| = \frac{|z_1|}{|z_2|}, \qquad \arg\frac{z_1}{z_2} = \arg z_1 - \arg z_2, \tag{1.1.27}$$

that is, the modulus of the ratio of two complex numbers is equal to the ratio of their moduli, and the argument of the ratio is equal to the difference of their arguments.

Letting $z_1 = z_2 = \cdots = z_n = z = r(\cos\theta + i\sin\theta)$ in (1.1.24), we obtain

$$z^n = r^n(\cos n\theta + i\sin n\theta); \tag{1.1.28}$$

thus

$$|z^n| = |z|^n, \qquad \arg z^n = n\arg z. \tag{1.1.29}$$

1.1.5. Exponential form of complex numbers. We introduce at this point a third form of complex numbers, called the exponential form, even though the exponential function for a complex variable will be defined later in Section 1.5.

Thus, to avoid breaking the logical order of presentation, we introduce *Euler's formula*,

$$e^{i\theta} = \cos\theta + i\sin\theta, \tag{1.1.30}$$

and postpone its derivation, as (1.5.9), until Subsection 1.1.5. We shall also assume the *law of exponents* (1.5.11) in the form

$$e^{i\theta_1}e^{i\theta_2} = e^{i(\theta_1+\theta_2)},$$

which will be proved later.

Substituting (1.1.30) in (1.1.20), we have the following definition.

DEFINITION 1.1.8. The *exponential form* of the complex number

$$z = r(\cos\theta + i\sin\theta)$$

is

$$z = r\,e^{i\theta} \qquad \text{or} \qquad z = |z|\,e^{i\arg z}. \tag{1.1.31}$$

Relations (1.1.22)–(1.1.29) can be easily obtained by means of (1.1.31). For example, if $z_1 = |z_1|\,e^{i\arg z_1}$ and $z_2 = |z_2|\,e^{i\arg z_2}$, then

$$z_1z_2 = |z_1|\,|z_2|\,e^{i(\arg z_1+\arg z_2)}. \tag{1.1.32}$$

Formulae (1.1.22) and (1.1.23) follow from (1.1.32).

1.1.6. Powers and roots of complex numbers.

DEFINITION 1.1.9. Given $n \in \mathbb{N}$, the complex number $w = z^{1/n}$ is called an *nth root* of the complex number z if $w^n = z$.

We have the following theorem.

THEOREM 1.1.1. *A nonzero complex number $z = r(\cos\theta + i\sin\theta)$ has exactly n distinct nth roots given by the formula*

$$z^{1/n} = |z|^{1/n}\left(\cos\frac{\operatorname{Arg} z + 2k\pi}{n} + i\sin\frac{\operatorname{Arg} z + 2k\pi}{n}\right),$$
$$k = 0, 1, \dots, n-1. \quad (1.1.33)$$

PROOF. Given $z = r(\cos\theta + i\sin\theta) \neq 0$, we determine the real numbers $\rho \geq 0$ and φ such that

$$w = \rho(\cos\varphi + i\sin\varphi) = z^{1/n}. \quad (1.1.34)$$

It follows from the relation $w^n = z$ and (1.1.28) that

$$\rho^n(\cos n\varphi + i\sin n\varphi) = r(\cos\theta + i\sin\theta). \quad (1.1.35)$$

Thus $\rho^n = r$ and

$$\rho = r^{1/n}, \quad (1.1.36)$$

where it is understood that the positive real value of $r^{1/n}$ taken. Moreover,

$$\cos n\varphi = \cos\theta \quad \Longrightarrow \quad n\varphi = \theta + 2k\pi, \qquad k = 0, 1, \dots .$$

Thus,

$$\varphi = \arg w = \frac{\theta + 2k\pi}{n}, \qquad k = 0, 1, \dots, n-1, \quad (1.1.37)$$

where the largest value of k in (1.1.37) is $k = n-1$ because, upon setting $k = n, n+1, \dots, 2n-1$ in (1.1.37), the n points with arguments

$$\frac{\theta}{n} + 2\pi, \quad \frac{\theta + 2\pi}{n} + 2\pi, \quad \dots, \quad \frac{\theta + 2(n-1)\pi}{n} + 2\pi$$

correspond to the n points with arguments

$$\frac{\theta}{n}, \quad \frac{\theta + 2\pi}{n}, \quad \dots, \quad \frac{\theta + 2(n-1)\pi}{n},$$

respectively. Hence by induction we see that, for $k = n, n+1, \dots,$ there are no new values of $z^{1/n}$ in the complex plane.

Substituting (1.1.36) and (1.1.37) into (1.1.34) we obtain formula (1.1.33), which, by Euler's formula, becomes

$$z^{1/n} = |z|^{1/n}\, e^{i(\operatorname{Arg} z + 2k\pi)/n}, \qquad k = 0, 1, \dots, n-1, \quad (1.1.38)$$

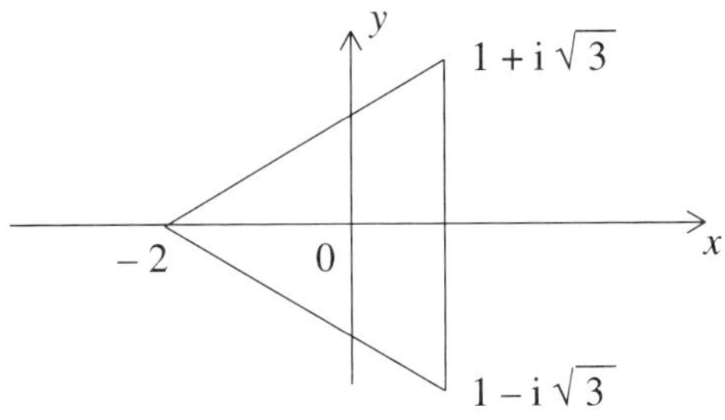

FIGURE 1.7. The three values of $(-8)^{1/3}$.

where, as always, $|z|^{1/n}$ denotes the positive real nth root of $|z|$. □

We see from formula (1.1.33) that the radii of the nth roots of $z \neq 0$ are equal to $|z|^{1/n}$, but their arguments differ by $2\pi/n$. These roots lie at the n vertices of a regular polygon in the complex plane, except in the case $z = 0$ where they are all zero.

EXAMPLE 1.1.1. *Find the three third roots, $(-8)^{1/3}$, of -8.*

SOLUTION. Since $-8 = 8\,e^{i\pi}$, then

$$(-8)^{1/3} = (8e^{i\pi})^{1/3} = 2\exp\left(i\,\frac{\pi + 2k\pi}{3}\right), \qquad k = 0, 1, 2,$$

that is,

$$(-8)^{1/3} = \begin{cases} 2e^{i\pi/3} & = & 2(\cos\frac{\pi}{3} + i\sin\frac{\pi}{3}) & = & 1 + i\sqrt{3}, \\ 2e^{i\pi} & = & 2(\cos\pi + i\sin\pi) & = & -2, \\ 2e^{-i\pi/3} & = & 2(\cos\frac{\pi}{3} - i\sin\frac{\pi}{3}) & = & 1 - i\sqrt{3}. \end{cases}$$

The three values of $(-8)^{1/3}$ are shown in Fig 1.7. □

Exercises for Section 1.1

If $z_1 = -1 + i$, $z_2 = 3 + 2i$ and $z_3 = -4 - 3i$, evaluate the following expressions.

1. $z_1 z_2 - z_3^2$.
2. $\left|\dfrac{z_1}{z_2}\right|$.
3. $\Re\left[\dfrac{z_3}{z_1 + z_2} + z_1^3\right]$.
4. $\Im[\bar{z}_1(z_2 + \bar{z}_3)]$.

5. $\operatorname{Arg}(z_1\bar{z}_2)$.

6. $\arg(z_1\overline{\bar{z}_2 z_3})$.

If $z_1 = 2 + i$, $z_2 = -1 + 3i$ and $z_3 = 4i$, evaluate the following expressions.

7. $(z_1 + z_2)^2 - \bar{z}_3$.

8. $|z_1\bar{z}_2 + z_2\bar{z}_3|$.

9. $\arg\left(\bar{z}_3^3\right)$.

10. $\operatorname{Arg}\left(\dfrac{\bar{z}_1}{z_2}\right)$.

11. $\Re[z_1 - \bar{z}_2 z_3^2]$.

12. $\Im\left[\dfrac{z_1}{z_2} + \dfrac{z_3}{z_1}\right]$.

13. Find real numbers x and y such that

$$2x + 3iy - 4ix + 5y + 1 = -(2y + 9x) + (x + 5y + 4)i.$$

Solve the following equations.

14. $z(4 - 3i) = 1 + 8i$.

15. $(1 + 2i)z = 2 - 4i$.

16. $|z|^2 - 2z = 3 + 4i$.

If $z = (-1+i\sqrt{3})/2$ and $-\pi < \operatorname{Arg} z \leq \pi$, find all the values of the following expressions.

17. $|z|$, $\arg z$, $\operatorname{Arg} z$.

18. $\arg(-z)$, $\operatorname{Arg}(-z)$, $\arg(\bar{z})$, $\operatorname{Arg}(\bar{z})$.

Find the real and imaginary parts of the following numbers.

19. $\dfrac{3 - i}{4 + 2i}$.

20. $\left(\dfrac{\sqrt{2} - i\sqrt{2}}{2}\right)^3$.

21. $\dfrac{2}{1 - i} + \dfrac{3 - i}{1 + 2i}$.

22. $\left(\dfrac{1 - i}{1 + i}\right)^3$.

23. Show that $\arg(\bar{z}) = -\arg(z)$, where $z \neq 0$.

24. Find the values of z for which $\operatorname{Arg}(\bar{z}) = -\operatorname{Arg}(z)$.

Find the complex numbers which are complex conjugates of

25. their own squares.
26. their own cubes.

Prove the following identities.

27. $|\bar{z}| = |z|$.
28. $\overline{z_1 + z_2} = \bar{z}_1 + \bar{z}_2$.
29. $\overline{z_1 z_2} = \bar{z}_1 \bar{z}_2$.
30. $\overline{\left(\dfrac{z_1}{z_2}\right)} = \dfrac{\bar{z}_1}{\bar{z}_2}$.

31. When do three points, z_1, z_2 and z_3, lie on a straight line?

32. Let σ be the line segment joining the points z_1 and z_2. Find the point z which divides σ in the ratio $\lambda_1{:}\lambda_2$.

33. Show that $|z_1 - z_2|^2 = |z_1|^2 + |z_2|^2 - 2\Re(z_1 \bar{z}_2)$.

34. Prove the parallelogram law: $|z_1 - z_2|^2 + |z_1 + z_2|^2 = 2(|z_1|^2 + |z_2|^2)$.

35. Let z_1, z_2 and z_3 be consecutive vertices of a parallelogram. Find the fourth vertex z_4 (opposite to z_2).

36. Find the point in the complex plane which is symmetric to $x + iy$ with respect to the line $y = x$.

37. By which angle should the vector $3\sqrt{2} + i2\sqrt{2}$ be rotated in order to obtain the vector $-5 + i$?

38. Prove the Cauchy–Schwarz inequality

$$|z_1 w_1 + z_2 w_2| \leq \sqrt{|z_1|^2 + |z_2|^2}\,\sqrt{|w_1|^2 + |w_2|^2},$$

and generalize it to n terms, that is, $|z_1 w_1 + z_2 w_2 + \cdots + z_n w_n|$.

39. Prove that

$$|1 - \bar{z}w|^2 - |z - w|^2 = (1 - |z|^2)(1 - |w|^2).$$

40. Prove that

$$|z| \leq |\Re z| + |\Im z| \leq \sqrt{2}\,|z|,$$

and give examples to show that either inequality may be an equality.

41. Show that if $|z| = 1$ and $z \neq a$, then $z/(z-a) = 1/(1-a\bar{z})$.

42. Prove that $|(1+i)z^3 + iz| < 3/4$ if $|z| < 1/2$.

43. Prove that $|z_1 + z_2| \geq \big| |z_1| - |z_2| \big|$. When does equality hold?

Represent the following numbers in trigonometric form.

44. $-7i$.

45. $-1 + i\sqrt{3}$.

46. $2 - 4i$.

47. $\dfrac{\sqrt{2} - i\sqrt{2}}{(\sqrt{3}+i)^2}$.

48. $\dfrac{(1-i)^6}{(1+i)^4}$.

49. $\dfrac{(1+i)^5}{(\sqrt{3}-i)^6}$.

50. $(i^3 + i^6)^{10}$.

51. $\left(\dfrac{i}{\sqrt{2}+i\sqrt{2}}\right)^8$.

52. Show that

$$\left(\frac{1+i\tan\alpha}{1-i\tan\alpha}\right)^n = \frac{1+i\tan n\alpha}{1-i\tan n\alpha}, \qquad \alpha \in \mathbb{R}.$$

53. Show that

$$(\cos\alpha + i\sin\alpha)^n = 1 \implies (\cos\alpha - i\sin\alpha)^n = 1, \qquad \alpha \in \mathbb{R}.$$

Find all complex numbers z for which the ratio $\dfrac{2-z}{2+z}$

54. is real.

55. is pure imaginary.

Find all the values of the following roots and plot them in the complex plane.

56. $\sqrt[4]{-1}$.

57. $\sqrt[3]{27i}$.

58. $\sqrt[4]{\dfrac{-\sqrt{3}+i}{1-i}}$.

59. $\sqrt[6]{\dfrac{\sqrt{2}-i\sqrt{2}}{1+i\sqrt{3}}}$.

60. Prove that if $z_1+z_2+z_3=0$ and $|z_1|=|z_2|=|z_3|=1$, then the points z_1, z_2, z_3 are the vertices of an equilateral triangle inscribed in the unit circle $|z|=1$.

61. Let a be any nth root of unity other than 1, where $n>1$. Prove that

$$1+2a+3a^2+\cdots+na^{n-1}=\frac{n}{a-1}.$$

(Hint. Multiply by $1-a$.)

62. Prove that the sum of all distinct nth roots of unity is zero, and interpret this fact geometrically.

1.2. Continuity in the complex plane

1.2.1. Domains, regions and boundaries.

DEFINITION 1.2.1. Given a positive real number, $\delta>0$, the set of all complex numbers z, which satisfy the inequality

$$|z-z_0|<\delta, \tag{1.2.1}$$

is called a *δ-neighborhood* of the point z_0.

The inequality (1.2.1) describes the set of points inside the open disk $D_{z_0}^{\delta}$ of radius δ centered at z_0.

DEFINITION 1.2.2. A set $U\subset\mathbb{C}$ is a *neighborhood* of $z_0\in\mathbb{C}$ if U contains a δ-neighborhood of z_0.

DEFINITION 1.2.3. Given a set $S\subset\mathbb{C}$, a point z_0 is

(a) an *interior point* of S, if there exists a $D_{z_0}^{\delta}$ such that $D_{z_0}^{\delta}\subset S$,
(b) an *exterior point* of S, if there exists a $D_{z_0}^{\delta}$ such that $D_{z_0}^{\delta}\cap S=\emptyset$,
(c) a *boundary point* of S, if every $D_{z_0}^{\delta}$ contains both interior and exterior points of S.

DEFINITION 1.2.4. A set S is *open* if all its points are interior points; it is closed if it contains all its interior and boundary points. The closure of S is denoted by $\overline{S}$.

For example, $|z|<1$ is an open set and $|z|\le 1$ is a closed set.

DEFINITION 1.2.5. A point set S is said to be *connected* if any two points of S can be joined by a polygonal line consisting entirely of points of S.

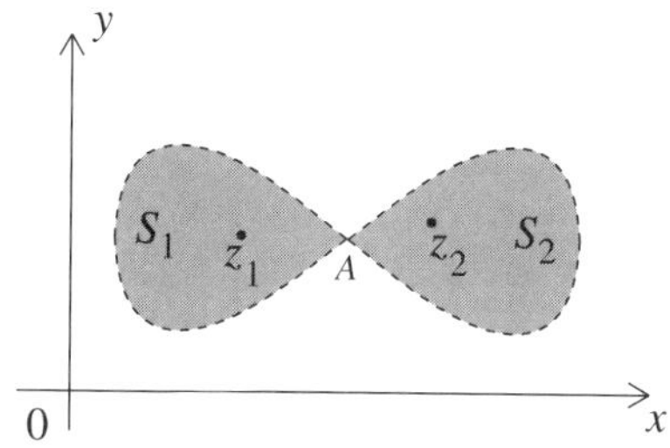

FIGURE 1.8. Open set $S = S_1 \cup S_2$ disconnected at point A.

DEFINITION 1.2.6. A *domain*, Ω, is an open connected set. A *region*, R, is a domain together with some, none or all of its boundary points.

It follows from the definition that "region" is more general than "domain."

For example, the open unit disk $|z| < 1$ is a domain while the closed unit disk $|z| \leq 1$ is not a domain, but a region.

The open set $S = S_1 \cup S_2$ shown in Fig 1.8 is neither a domain nor a region, because it is not connected at the point A. For example, the points z_1 and z_2 cannot be joined by a polygonal line that lies in the set.

DEFINITION 1.2.7. If the boundary of a domain Ω consists of a single closed non-self-intersecting (rectifiable) curve γ, then the domain is called *simply connected*; otherwise it is said to be *multiply connected.*

DEFINITION 1.2.8. The *positive direction* of the boundary γ of a domain Ω is that direction for which the points of Ω lie to the left of γ.

In Fig 1.9, simply, doubly and triply connected domains are shown, where the arrows indicate the positive direction along the boundary.

Consider a curve $y = f(x)$ in $\mathbb{R}^2$. The equation of this curve in the complex plane is $z = x + if(x)$, where $z = x + iy$.

For example, the equation of the parabola $y = x^2$ in $\mathbb{R}^2$ is written, in the complex plane, in the form $z = x + ix^2$.

If a curve $\gamma(t)$ is given by the *parametric equations*

$$x = x(t), \quad y = y(t), \qquad t_1 \leq t \leq t_2, \tag{1.2.2}$$

then its equation in the complex plane is

$$z(t) = x(t) + iy(t), \qquad t_1 \leq t \leq t_2. \tag{1.2.3}$$

For instance, the complex form of the equation of the circle

$$\begin{aligned} x - x_0 &= r\cos t, \\ y - y_0 &= r\sin t, \end{aligned} \qquad 0 \leq t \leq 2\pi, \tag{1.2.4}$$

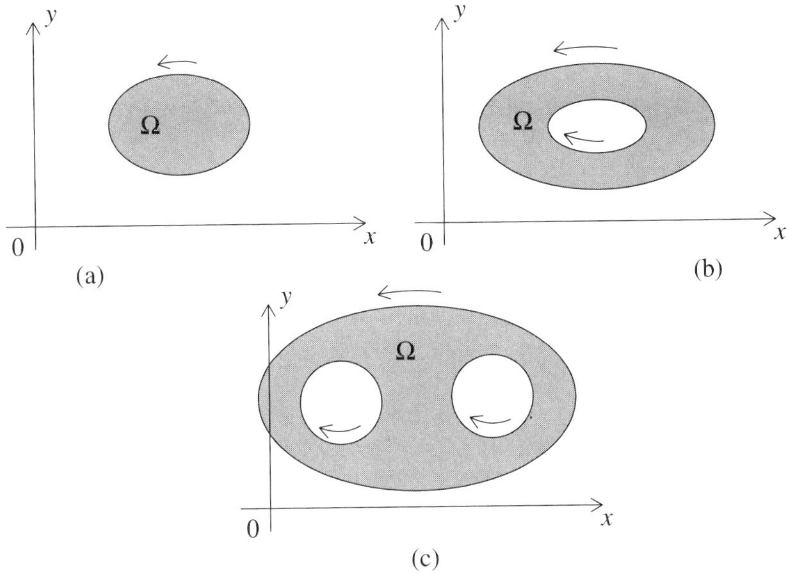

FIGURE 1.9. (a) Simply connected, (b) doubly connected and (c) triply connected domains. The arrows indicate the positive direction along the boundary.

is

$$z = x_0 + r\cos t + i(y_0 + r\sin t),$$

or, with $z_0 = x_0 + iy_0$,

$$z = z_0 + r(\cos t + i\sin t), \qquad 0 \le t \le 2\pi,$$

which, by Euler's formula (1.1.30), becomes

$$z = z_0 + r\,e^{it}, \qquad 0 \le t \le 2\pi. \tag{1.2.5}$$

Since $\left|e^{it}\right| = 1$, we see that (1.2.5) coincides with (1.1.18) with r changed to ρ.

EXAMPLE 1.2.1. *Give a geometric meaning to the following simple inequations:*

$$r_1 \le |z - z_0| \le r_2, \qquad 0 \le \operatorname{Arg} z \le \pi/4, \qquad \pi/6 \le \operatorname{Arg}(z - 2i) \le \pi/3,$$
$$2 \le \Re z \le 3, \qquad 1 \le \Im z \le 3.$$

SOLUTION. The respective geometric regions are as follows:

(a) $r_1 \le |z - z_0| \le r_2$ is an annulus centered at z_0 with radii r_1 and r_2, shown in Fig 1.10(a).
(b) $0 \le \operatorname{Arg} z \le \pi/4$ is a wedge with vertex at the origin of the coordinate system, shown in Fig 1.10(b).
(c) $\pi/6 \le \operatorname{Arg}(z - 2i) \le \pi/3$ is a wedge with vertex at the point $2i$, shown in Fig 1.10(c).
(d) $2 \le \Re z \le 3$ is a strip of unit width parallel to the imaginary axis, shown in Fig 1.10(d).
(e) $1 \le \Im z \le 3$ is a strip of width 2 parallel to the real axis, shown in Fig 1.10(e). □

To explain part (c) further, using the substitution $z - 2i = z_1$, we obtain the inequality $\pi/6 \le \operatorname{Arg} z_1 \le \pi/3$, which describes a wedge centered at $z_1 = 0$, that is, centered at the point $z = 2i$.

1.2.2. Limit of a sequence of complex numbers. As in real analysis, a sequence, $\{z_n\}$, of complex numbers is defined as the ordered set of values of a function, f, whose argument is a set of positive integers,

$$z_n = a_n + ib_n = f(n), \tag{1.2.6}$$

where $\{a_n\}$ and $\{b_n\}$ are sequences of real numbers.

DEFINITION 1.2.9. A complex number a is called the *limit of a sequence*, $\{z_n\}$, of complex numbers, as $n \to \infty$, if for every $\varepsilon > 0$ there exists $N_\varepsilon \in \mathbb{N}$ such that for all $n > N_\varepsilon$,

$$|z_n - a| < \varepsilon,$$

and we write

$$a = \lim_{n\to\infty} z_n.$$

Note 1.2.1. The inequality $|z_n - a| < \varepsilon$ means that, for $n > N_\varepsilon$, all the terms of the sequence are located in the open disk D_a^ε of center a and radius ε.

The limit of a sequence of complex numbers is equivalent to the limit of two sequences of real numbers as proved in the following theorem.

THEOREM 1.2.1. *Let $\{z_n\}$ be a sequence of complex numbers. A necessary and sufficient condition for the existence of a limit*

$$a + ib = \lim_{n\to\infty} z_n, \tag{1.2.7}$$

where $z_n = a_n + ib_n$, is the existence of the limits

$$a = \lim_{n\to\infty} a_n, \qquad b = \lim_{n\to\infty} b_n. \tag{1.2.8}$$

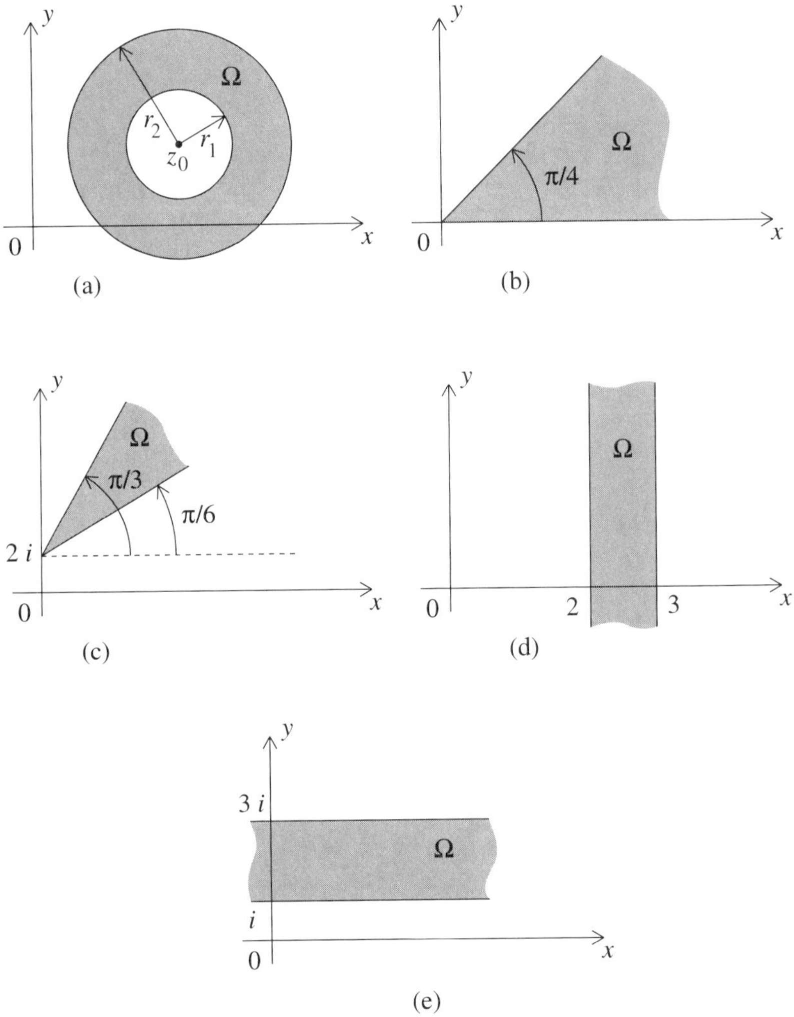

FIGURE 1.10. Geometric figures for Example 1.2.1 (a) to (e).

PROOF. **Necessity.** Suppose that the limit in (1.2.7) exists, that is,

$$\forall \varepsilon > 0 \quad \exists N_\varepsilon : \quad \forall n > N_\varepsilon, \quad |z_n - a| < \varepsilon,$$

which we write explicitly as

$$\sqrt{(a_n - a)^2 + (b_n - b)^2} < \varepsilon. \tag{1.2.9}$$

It follows from (1.2.9) that

$$\forall n > N_\varepsilon, \quad |a_n - a| < \varepsilon, \quad |b_n - b| < \varepsilon; \tag{1.2.10}$$

but (1.2.10) implies that the limits in (1.2.8) exist. Geometrically, (1.2.9) and (1.2.10) means that if the hypotenuse of a right-angle triangle is smaller than ε, then the adjacent sides must also be smaller than ε.

Sufficiency. Suppose that the limits in (1.2.8) exist, that is, for every $\varepsilon > 0$,

$$\exists N_1 : \quad \forall n > N_1 \quad |a_n - a| < \frac{\varepsilon}{\sqrt{2}}, \tag{1.2.11}$$

$$\exists N_2 : \quad \forall n > N_2 \quad |b_n - b| < \frac{\varepsilon}{\sqrt{2}}. \tag{1.2.12}$$

Then for all $n > N = \max\{N_1, N_2\}$ the inequalities (1.2.11) and (1.2.12) are satisfied simultaneously. But inequality (1.2.9) follows from (1.2.11) and (1.2.12) for all $n > N$ (if we square (1.2.11) and (1.2.12) and add the corresponding inequalities). The latter implies that the limit in (1.2.7) exists. □

It follows from the previous theorem that the study of the properties of sequences, $\{z_n\}$, of complex numbers can be reduced to study of the properties of pairs of sequences, $\{a_n\}$ and $\{b_n\}$, of real numbers.

1.2.3. The point at infinity. Let $\{z_n\}$ be a sequence of complex numbers such that for every $R > 0$ there exists N such that for all $n > N$, $|z_n| > R$. Such a sequence, $\{z_n\}$, is called an increasing sequence with no finite limit. Introducing the complex number $z = \infty$, called the *point at infinity*, we say that $\{z_n\}$ converges to infinity and write

$$\lim_{n \to \infty} z_n = \infty.$$

A region outside a disk of sufficiently large radius R ($|z| > R$) is called a *neighborhood* of the point $z = \infty$.

We use the so-called *stereographic projection* to illustrate this idea.

Suppose that a sphere of radius 1, called a *Riemann sphere*, is supported by the complex plane with the south pole, S, of the sphere located at the

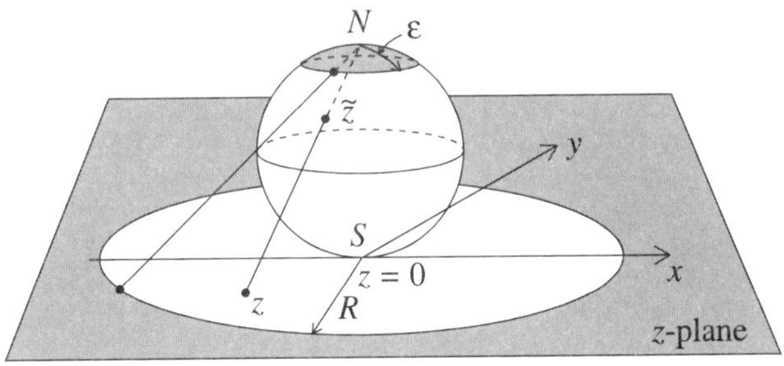

FIGURE 1.11. Stereographic projection from the Riemann sphere to the z-plane.

origin, $z = 0$, of the coordinate system (see Fig 1.11). The equation of the sphere is

$$x_1^2 + x_2^2 + (x_3 - 1)^2 = 1. \tag{1.2.13}$$

If we draw a ray from the north pole, N, to the point $z = x + iy$ of the complex z-plane and let $\tilde{z}(x_1, x_2, x_3)$ be the point of intersection of the ray with the Riemann sphere, then it is seen from Fig 1.11 that the three points $N(0, 0, 2)$, $\tilde{z}$ and $z(x, y, 0)$, lie on the straight line

$$\frac{x_1 - 0}{x - 0} = \frac{x_2 - 0}{y - 0} = \frac{x_3 - 2}{0 - 2}. \tag{1.2.14}$$

Expressing x_2 and x_3 in terms of x_1 from the equation of the line and substituting these values into the equation of the sphere, we obtain

$$x_1 = \frac{4x}{x^2 + y^2 + 4}.$$

Similarly,

$$x_2 = \frac{4y}{x^2 + y^2 + 4} \quad \text{and} \quad x_3 = \frac{2(x^2 + y^2)}{x^2 + y^2 + 4}.$$

Since $z = x + iy$, we have

$$x_1 = \frac{2(z + \bar{z})}{|z|^2 + 4}, \qquad x_2 = \frac{2(z - \bar{z})}{i(|z|^2 + 4)}, \qquad x_3 = \frac{2|z|^2}{|z|^2 + 4}. \tag{1.2.15}$$

It follows from these formulae that to each (finite) point $z = x + iy \in \mathbb{C}$ there corresponds a unique point $\tilde{z}(x_1, x_2, x_3)$ on the Riemann sphere.

Conversely, from equation (1.2.14) of the line, we have

$$x = \frac{2x_1}{2 - x_3}, \qquad y = \frac{2x_2}{2 - x_3}. \tag{1.2.16}$$

Hence, to each point $\tilde{z}(x_1, x_2, x_3)$ on the Riemann sphere there corresponds a unique point $z = x + iy \in \mathbb{C}$ (except for the north pole, N).

Therefore, there is a one-to-one correspondence between the points of the complex z-plane and the points of the sphere. The only point of the sphere to which there does not correspond any point in the finite part of the complex z-plane is the north pole. If we let the point $z = \infty$ correspond to N, then the exterior of a disk of radius R in $\mathbb{C}$ corresponds to an ε-neighborhood of N where ε decreases as R increases.

DEFINITION 1.2.10. The complex z-plane together with the point $z = \infty$ is called the *extended complex plane* and the z-plane without the point $z = \infty$ is called the *open plane.*

Exercises for Section 1.2

In Exercises 1 to 8, for each set S_i, $i = 1, \dots, 8$, draw S_i and show whether (a) S_i is open or closed, and (b) its interior is connected or not (if the interior is not empty).

1. $S_1 = \{z;\ 1 < \Im z < 2\}$.
2. $S_2 = \{z;\ |z| = 2\}$.
3. $S_3 = \{z;\ \Re(z^2) \ge 3\}$.
4. $S_4 = \{z;\ 0 \le \operatorname{Arg} z < \pi/4\}$.
5. $S_5 = \{z;\ |z+1| < 1\} \cup \{z;\ |z-5| \le 1\}$.
6. $S_6 = \{z;\ |z| > 2|z-1|\}$.
7. $S_7 = \{z = x + iy;\ x \le 2\} \cap \{z = x + iy;\ y \ge 3\}$.
8. $S_8 = \{z = x + iy;\ x = 4\} \cap \{z = x + iy;\ y > 0\}$.

In Exercises 9 to 16, describe geometrically each set S_i, $i = 9, \dots, 16$, and show whether it is open or closed.

9. $S_9 = \{z;\ |z - 2 + i| \le 2\}$.
10. $S_{10} = \{z;\ 1 < |z| < 2\} \cap \{z;\ 0 < \operatorname{Arg} z < \pi/4\}$.
11. $S_{11} = \{z;\ |z-1| < |z-i|\}$.
12. $S_{12} = \{z;\ |z-1| < 1\} \cap \{z;\ |z| = |z-2|\}$.
13. $S_{13} = \{z;\ \Im(z^2) < 1\}$.
14. $S_{14} = \{z;\ z^2 + \bar{z}^2 = 1\}$.
15. $S_{15} = \left\{z;\ \Im\left(\overline{z^2 - \bar{z}}\right) = 2 - \Im z\right\}$.

16. $S_{16} = \{z;\ |z|^2 > z + \bar{z}\}$.

What curves are represented by the following functions? Draw the curves.

17. $z(t) = \cos t - i \sin t, \qquad 0 \le t \le \pi.$

18. $z(t) = 3 + 2i + 4(\cos t + i \sin t), \qquad 0 \le t \le 2\pi.$

19. $z(t) = z_0 + r(\cos t + i \sin t), \qquad 0 \le t \le \pi$, where $z_0 \in \mathbb{C}$ and $r > 0$.

20. $z(t) = t + 2 + it^2, \qquad -2 \le t \le 1.$

21. $z(t) = \cosh t + i \sinh t, \qquad -1 \le t \le 1.$

22. $z(t) = t + i/t, \qquad 1 \le t \le 2.$

Represent the following curves in parametric form as $z = z(t)$.

23. $y = 2x + 1$, from $(0, 1)$ to $(1, 3)$.

24. $y = 5x^2 + 2$, from $(0, 2)$ to $(2, 22)$.

25. The semicircle in the left half-plane whose diameter joins the point $(0, -R)$ to the point $(0, R)$.

26. $x^2 + y^2 = 9$.

27. $9(x - 1)^2 + 16(y + 3)^2 = 144$.

28. $\frac{1}{4} x^2 - y^2 = 1$.

Find the limit, if any, as $n \to \infty$, of each of the following sequences.

29. $z_n = i^n$.

30. $z_n = \dfrac{i^n + (-1)^n}{n^2}$.

31. $z_n = \left(1 + \dfrac{1}{n}\right)\left(\cos \dfrac{\pi}{n} + i \sin \dfrac{\pi}{n}\right)$.

32. $z_n = \dfrac{(1 + i)^n}{n!}$.

33. $z_n = \dfrac{3^n}{n!} + \dfrac{i^n}{2^n}$.

34. $z_n = \cos\left(\dfrac{\pi}{2} + \dfrac{1}{3n}\right) + i \sin\left(\dfrac{\pi}{2} + \dfrac{1}{3n}\right)$.

35. $z_n = \sqrt[n]{2} + i \sin \dfrac{2}{n}$.

36. $z_n = \sqrt[n]{n} + \dfrac{in}{3^n}$.

37. Describe the relative positions of the images of z, $-z$ and $\bar{z}$ on the Riemann sphere.

38. Suppose $z_n \to \infty$ as $n \to \infty$. What are the implications on $\Re z_n$, $\Im z_n$, $|z_n|$ and $\arg z_n$?

39. Prove that if $z_n \to \alpha$ as $n \to \infty$, then $|z_n| \to |\alpha|$ as $n \to \infty$. Show that the converse is not true.

40. What curve on the Riemann sphere is the image under stereographic projection of a straight line in the extended plane?

41. What is the relation satisfied by two points, z_1 and z_2, which are the images under stereographic projection of a pair of diametrically opposite points of the Riemann sphere?

1.3. Functions of a complex variable

1.3.1. Definitions.

DEFINITION 1.3.1. A *function* f defined on a set $S \subset \mathbb{C}$ is a rule which assigns to each value of z in S a complex number w. The complex number w is called the *value* of f at z and is denoted by $f(z)$; that is,

$$w = f(z).$$

The set S is called the *domain of definition* of f.

It is to be remarked that the domain of definition of a function is an essential part of the definition of a function. When the domain is not specified it is taken to be as large as possible but still preserving the single-valuedness of the function.

EXAMPLE 1.3.1.

(a) *The expression* $w = \arg z = \operatorname{Arg} z + 2k\pi$, *for* $k \in \mathbb{Z}$, *defines infinitely many functions, one for each value of* k. *We say that each of these functions is a* branch *of* $w = \arg z$.

(b) *The expression*

$$w = z^{1/n} = |z|^{1/n} \exp\left(i\,\frac{\operatorname{Arg} z + 2k\pi}{n}\right), \qquad k = 0, 1, \dots, n-1,$$

defines n *functions, one for each value of* k. *We say that each of these functions is a branch of* $w = z^{1/n}$.

(c) *The expression* $w = z^2$ *is a function because only one value of* w *corresponds to each value of* z.

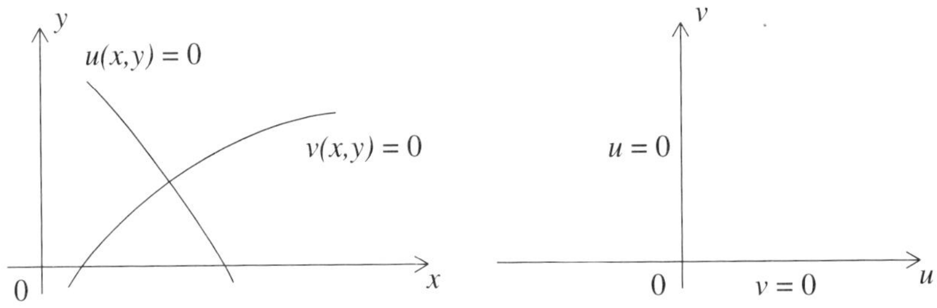

FIGURE 1.12. Image of curves $u(x, y) = 0$ and $v(x, y) = 0$ in the w-plane.

Letting $z = x + iy$ in part (c) of this example, we obtain

$$w = z^2 = x^2 - y^2 + 2xyi,$$

that is, the function $w = z^2$ is given by two real functions of two real variables

$$u(x, y) = x^2 - y^2, \qquad v(x, y) = 2xy.$$

In particular, this function maps the point $z_0 = 5 + i$ to the point

$$w_0 = 5^2 - 1^2 + i \times 2 \times 5 \times 1 = 24 + 10i.$$

In general, a function of a complex variable,

$$w = f(z) = u(x, y) + iv(x, y), \tag{1.3.1}$$

is equivalently defined by two real functions of two real variables,

$$u = u(x, y) = \Re f(z), \qquad v = v(x, y) = \Im f(z).$$

The curves $u(x, y) = \Re f(z) = 0$ and $v(x, y) = \Im f(z) = 0$ in the z-plane lie on the vertical and horizontal axes, respectively, of the complex w-plane (see Fig 1.12). The function (1.3.1) maps every point z of its domain of definition in the complex z-plane to some point w of the complex w-plane, that is, if $z_0 = x_0 + iy_0$ then

$$w_0 = f(z_0) = u(x_0, y_0) + iv(x_0, y_0).$$

1.3.2. Limit and continuity of a function of a complex variable. Firstly, we define the limit of a function, $f(z)$, by means of sequences of values of f.

DEFINITION 1.3.2. A number w_0 is called the *limit of a function* of a complex variable, $w = f(z)$, as $z \to z_0$, if for each sequence $\{z_n\}$ converging to z_0 as $n \to \infty$, the corresponding sequence, $\{f(z_n)\}$, converges to w_0 as $n \to \infty$.

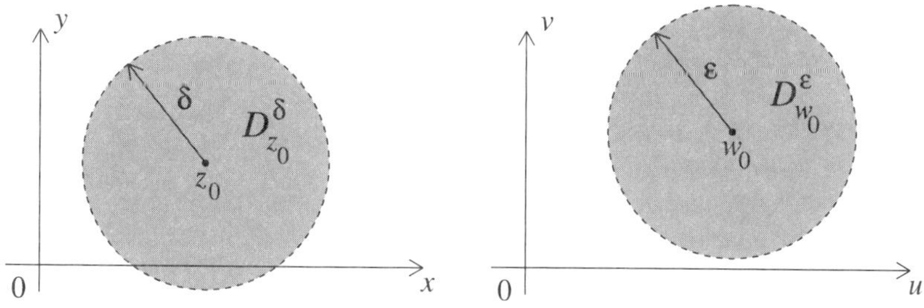

FIGURE 1.13. Interior of the disks $D_{z_0}^{\delta}$ and $D_{w_0}^{\varepsilon}$ in the z- and w-planes, respectively.

Secondly, we define the limit of a function using the Cauchy "ε-δ" terminology.

DEFINITION 1.3.3. A number w_0 is called the *limit of a function* $w = f(z)$ as $z \to z_0$ if, for every $\varepsilon > 0$, there exists $\delta_{z_0,\varepsilon} > 0$ such that, for all z satisfying the inequality

$$|z - z_0| < \delta_{z_0,\varepsilon}, \tag{1.3.2}$$

$f(z)$ satisfies the inequality

$$|f(z) - w_0| < \varepsilon. \tag{1.3.3}$$

In this case, we write

$$w_0 = \lim_{z \to z_0} f(z). \tag{1.3.4}$$

It can easily be shown that the previous two definitions are equivalent.

Geometrically, inequality (1.3.2) represents the interior of the disk $D_{z_0}^{\delta}$ in the z-plane while inequality (1.3.3) represents the interior of the disk $D_{w_0}^{\varepsilon}$ in the w-plane (see Fig 1.13). Hence, $\lim_{z \to z_0} f(z) = w_0$ if, for every $\varepsilon > 0$ there exists $\delta = \delta_{z_0,\varepsilon} > 0$ such that for all $z \in D_{z_0}^{\delta}$, $w = f(z) \in D_{w_0}^{\varepsilon}$.

Note 1.3.1. It follows from Definition 1.3.3 that the limit of $f(z)$ at $z = z_0$ does not depend upon the direction of the ray along which z approaches z_0. If z approaches z_0 along any ray, then as soon as it gets into the disk $D_{z_0}^{\delta}$, the corresponding values of w gets into the disk $D_{w_0}^{\varepsilon}$. This fact will often be used in this book.

The following theorem relates the convergence of a function of $z \in \mathbb{C}$ to the convergence of two functions of $(x, y) \in \mathbb{R}^2$.

THEOREM 1.3.1. *The limit of a complex function $f(z) = u(x, y) + iv(x, y)$ exists as $z \to z_0 = x_0 + iy_0$ and is equal to*

$$w_0 = u_0 + iv_0 = \lim_{z \to z_0} f(z), \tag{1.3.5}$$

if and only if the limits of its real and imaginary parts exist and are equal to

$$u_0 = \lim_{(x,y)\to(x_0,y_0)} u(x,y), \qquad v_0 = \lim_{(x,y)\to(x_0,y_0)} v(x,y). \tag{1.3.6}$$

PROOF. (1) Suppose that the limit in (1.3.5) exists, that is, inequality (1.3.3) is satisfied for all z satisfying (1.3.2). We rewrite (1.3.2) and (1.3.3) in the form

$$\sqrt{(x-x_0)^2 + (y-y_0)^2} < \delta \tag{1.3.7}$$

and

$$\sqrt{(u-u_0)^2 + (v-v_0)^2} < \varepsilon, \tag{1.3.8}$$

respectively. It follows from (1.3.8) that

$$|u-u_0| < \varepsilon, \qquad |v-v_0| < \varepsilon, \tag{1.3.9}$$

for all (x, y) satisfying (1.3.7). But (1.3.9) implies the existence of limits (1.3.6).

(2) Suppose that the limits in (1.3.6) exist, that is, for every $\varepsilon > 0$ there exists $\delta > 0$ such that for all $(x, y) \in D^{\delta}_{(x_0,y_0)}$ the following inequalities are fulfilled:

$$|u-u_0| < \frac{\varepsilon}{\sqrt{2}}, \qquad |v-v_0| < \frac{\varepsilon}{\sqrt{2}}. \tag{1.3.10}$$

Inequality (1.3.8) follows from (1.3.10) for all $(x, y) \in D^{\delta}_{z_0}$. Hence the limit in (1.3.5) exists. □

DEFINITION 1.3.4. A function $w = f(z)$ is said to be *continuous* at the point $z_0 = x_0 + iy_0$ if

$$f(z_0) = \lim_{z\to z_0} f(z). \tag{1.3.11}$$

Using the difference notation,

$$\Delta f(z)\big|_{z_0} = f(z) - f(z_0), \qquad \Delta z = z - z_0, \tag{1.3.12}$$

we rewrite (1.3.11) in the equivalent form

$$\lim_{\Delta z\to 0} \Delta f(z)\big|_{z_0} = 0. \tag{1.3.13}$$

The following theorem holds.

THEOREM 1.3.2. *A function $f(z) = u(x,y) + iv(x,y)$ is continuous at the point $z_0 = x_0 + iy_0$ if and only if its real and imaginary parts, $u(x,y)$ and $v(x,y)$, are continuous at the point (x_0, y_0).*

The proof of this theorem is similar to the proof of the previous one and is left as an exercise to the reader.

Exercises for Section 1.3

Describe the domain of definition of each of the given functions.

1. $f(z) = \dfrac{1}{z^2+4}$.
2. $f(z) = \dfrac{z+2}{z+\bar{z}}$.
3. $f(z) = \operatorname{Arg}\left(\dfrac{1}{z}\right)$, where $-\pi < \operatorname{Arg} z \leq \pi$.
4. $f(z) = \dfrac{1}{1-|z|^2}$.

Find the real and imaginary parts of the following functions.

5. $f(z) = 3z^2 - 2iz$.
6. $f(z) = z + \dfrac{1}{z}$.
7. $f(z) = z^3 + z + 2$.
8. $f(z) = \dfrac{1-z}{1+z}$.
9. $f(z) = \bar{z} - iz^2$.

10. Let $z = x + iy$. Express the right-hand side of $f(z) = x^2 - y^2 - 2y + i(2x - 2y)$ in terms of z and simplify.

Find the following limits.

11. $\lim\limits_{z\to 2i} \dfrac{iz^3 - 8}{z - 2i}$.
12. $\lim\limits_{z\to 3} \dfrac{z^2 + 4z - 21}{z - 3}$.
13. $\lim\limits_{z\to 2-3i} |z|$.
14. $\lim\limits_{z\to\infty} \dfrac{z^2 + 3z + 2}{4z^2 + 2z - 1}$.

Find the following limits, if they exist.

15. $\lim\limits_{z\to 0} \dfrac{|z|}{z}$.

16. $\displaystyle\lim_{z\to 0} \frac{|z|^2}{z}$.

17. $\displaystyle\lim_{z\to 0} \frac{z}{|z|}$.

18. $\displaystyle\lim_{z\to 0} \frac{z - \Re z}{\Im z}$.

19. Consider the rational function

$$f(z) = \frac{a_m z^m + \cdots + a_1 z + a_0}{b_n z^n + \cdots + b_1 z + b_0}, \qquad a_m \neq 0,\ b_n \neq 0,$$

and discuss the possible values of $\displaystyle\lim_{z\to\infty} f(z)$.

Find all points of discontinuity of the following functions.

20. $f(z) = \dfrac{z-2}{z^2 + 4z + 10}$.

21. $f(z) = \dfrac{1}{z(z^2+1)}$.

22. $f(z) = \dfrac{z^2+3}{z^3 - 27}$.

23. $f(z) = \dfrac{1}{z^4+1}$.

24. Let

$$f(z) = \begin{cases} \frac{1+z^2}{z-i}, & \text{if } z \neq i, \\ 4i, & \text{if } z = i. \end{cases}$$

(a) Prove that $\displaystyle\lim_{z\to i} f(z)$ exists and determine its value.

(b) Is $f(z)$ continuous at $z = i$? Explain.

(c) Is $f(z)$ continuous at $z \neq i$? Explain.

25. Where is the rational function $f(z)$ of Exercise 19 continuous?

Prove that the following functions are continuous.

26. $f(z) = \Re z$.

27. $f(z) = \Im z$.

28. $f(z) = |z|^2$.

29. $f(z) = z + |z|$.

30. The functions

$$\frac{\Re z}{|z|}, \qquad \frac{z}{|z|}, \qquad \frac{\Re(z^2)}{|z|^2}, \qquad \frac{z\Re z}{|z|}$$

are all defined for $z \neq 0$. Which of them can be defined at the point $z = 0$ in such a way that the extended functions are continuous at $z = 0$?

1.4. Analytic functions

1.4.1. Analytic or holomorphic functions. We give two equivalent definitions of differentiability of a function of a complex variable.

DEFINITION 1.4.1. If the limit

$$\lim_{\Delta z \to 0} \frac{f(z_0 + \Delta z) - f(z_0)}{\Delta z}$$

exists and is finite, then it is called the *derivative* of the function $f(z)$ at z_0 and is denoted by $f'(z_0)$. In this case, we write

$$f'(z_0) = \lim_{\Delta z \to 0} \frac{f(z_0 + \Delta z) - f(z_0)}{\Delta z} \tag{1.4.1}$$

and say that $f(z)$ is *differentiable* at z_0.

We recall that the limit in (1.4.1), if it exists, is independent of the direction along which the point $z = z_0 + \Delta z$ approaches z_0 (in particular, the point z can approach z_0 along any ray).

DEFINITION 1.4.2. A function $f : D \to \mathbb{C}$ is *differentiable* at $z = a$ if there exists a function $f_1 : D \to \mathbb{C}$ that is continuous at a and such that

$$f(z) = f(a) + (z - a)f_1(z), \quad \text{for all } z \in D.$$

If f_1 exists, it is determined by f,

$$f_1(z) = \frac{f(z) - f(a)}{z - a}, \qquad \text{for } z \in D \setminus \{a\}.$$

Setting $h = z - a$, the continuity of f_1 implies that

$$f_1(a) = \lim_{h \to 0} \frac{f(a + h) - f(a)}{h}.$$

The number $f_1(a) \in \mathbb{C}$ is called the *derivative of f at a* and we write

$$f_1(a) = f'(a) = \frac{df}{dz}(a).$$

Differentiable functions of a complex variable are important and carry special names.

DEFINITION 1.4.3. A function $f(z)$ that is differentiable at every point of a domain D is said to be *analytic* (or *holomorphic*) in D; f is said to be *analytic at* z_0 if it is analytic in a neighborhood of z_0.

1.4.2. The Cauchy–Riemann equations. Necessary and sufficient conditions for the differentiability of a function $f(z)$ at a point z_0 are given in the following Theorems 1.4.1 and 1.4.2.

We shall use indifferently the following notation to denote the partial derivatives of u and v:

$$u_x = \frac{\partial u}{\partial x}, \qquad u_y = \frac{\partial u}{\partial y}, \qquad v_x = \frac{\partial v}{\partial x}, \qquad v_y = \frac{\partial v}{\partial y}.$$

The following partial differential equations play a central role in the theory of analytic (holomorphic) functions.

DEFINITION 1.4.4 (CAUCHY–RIEMANN EQUATIONS). The partial differential equations

$$\frac{\partial u}{\partial x} = \frac{\partial v}{\partial y}, \qquad \frac{\partial u}{\partial y} = -\frac{\partial v}{\partial x}, \tag{1.4.2}$$

are called the *Cauchy–Riemann equations.*

THEOREM 1.4.1. *If a function $f(z) = u(x,y) + iv(x,y)$ is differentiable at a point $z_0 = x_0 + iy_0$, then the partial derivatives of $u(x,y)$ and $v(x,y)$ with respect to x and y exist at the point $M_0 = (x_0, y_0)$. Moreover, u and v satisfy the Cauchy–Riemann equations*

$$u_x(x_0, y_0) = v_y(x_0, y_0), \qquad u_y(x_0, y_0) = -v_x(x_0, y_0), \tag{1.4.3}$$

at M_0.

PROOF. It follows from the existence of limit (1.4.1) that this limit does not depend on the direction of the ray along which $\Delta z \to 0$. We now show that the Cauchy–Riemann equations hold.

Firstly, let $\Delta z = \Delta x$ in (1.4.1), that is $z \to z_0$ along a ray parallel to the x-axis (see Fig 1.14). Thus

$$\begin{aligned} f'(z_0) &= \lim_{\Delta x \to 0} \frac{1}{\Delta x}[u(x_0 + \Delta x, y_0) + iv(x_0 + \Delta x, y_0) \\ &\qquad\qquad - u(x_0, y_0) - iv(x_0, y_0)] \\ &= \lim_{\Delta x \to 0} \frac{1}{\Delta x}[u(x_0 + \Delta x, y_0) - u(x_0, y_0)] \\ &\qquad + i \lim_{\Delta x \to 0} \frac{1}{\Delta x}[v(x_0 + \Delta x, y_0) - v(x_0, y_0)]. \end{aligned} \tag{1.4.4}$$

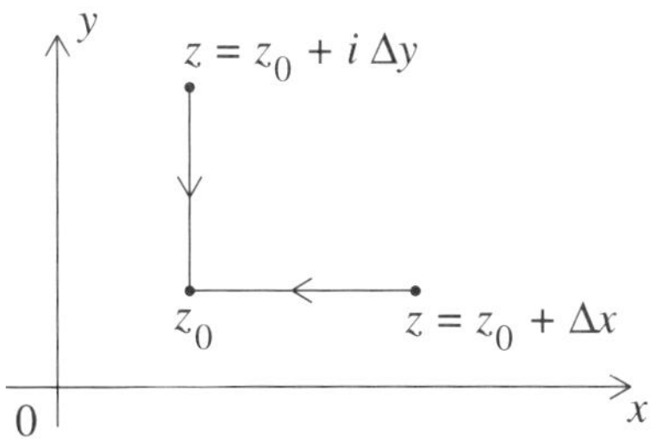

FIGURE 1.14. Approaching z_0 along the real and imaginary axes in the derivation of the Cauchy–Riemann equations (1.4.3).

Since, by assumption, $f'(z_0)$ exists and is finite, the three limits in (1.4.4) exist. This implies that u_x and v_x also exist at the point $M_0 = (x_0, y_0)$ so that (1.4.4) can be written in the form

$$f'(z_0) = \left.\frac{\partial u}{\partial x}\right|_{M_0} + i \left.\frac{\partial v}{\partial x}\right|_{M_0}. \tag{1.4.5}$$

Secondly, suppose that $\Delta z = i\Delta y$ in (1.4.1), that is, the point $z = z_0 + i\Delta y$ approaches z_0 along a ray parallel to the imaginary axis (see Fig 1.14). Then we have

$$\begin{aligned} f'(z_0) &= \lim_{\Delta y \to 0} \frac{1}{i\Delta y}[u(x_0, y_0 + \Delta y) + iv(x_0, y_0 + \Delta y) \\ &\qquad - u(x_0, y_0) - iv(x_0, y_0)] \\ &= \lim_{\Delta y \to 0} \frac{1}{\Delta y}[v(x_0, y_0 + \Delta y) - v(x_0, y_0)] \\ &\qquad - i \lim_{\Delta y \to 0} \frac{1}{\Delta y}[u(x_0, y_0 + \Delta y) - u(x_0, y_0)]. \end{aligned} \tag{1.4.6}$$

We see, as for (1.4.4), that the three limits in (1.4.6) exist, so that u_y and v_y exist at the point M_0. Hence (1.4.6) has the form

$$f'(z_0) = \left.\frac{\partial v}{\partial y}\right|_{M_0} - i \left.\frac{\partial u}{\partial y}\right|_{M_0}. \tag{1.4.7}$$

Finally, since the left-hand sides of (1.4.5) and (1.4.7) are equal, then their right-hand sides also are equal. Equating the real and imaginary parts of the right-hand sides of (1.4.5) and (1.4.7) we obtain the Cauchy–Riemann equations (1.4.3). □

We prove the following converse to Theorem 1.4.1.

THEOREM 1.4.2. *If the functions of two variables, $u(x, y)$ and $v(x, y)$, are differentiable at the point (x_0, y_0) and their partial derivatives are continuous and satisfy the Cauchy–Riemann equations* (1.4.3), *then the function*

$$f(z) = u(x, y) + iv(x, y)$$

is differentiable at the point $z_0 = x_0 + iy_0$.

PROOF. Since u and v have continuous first-order partial derivatives, as shown in advanced calculus, we can write

$$u(x+h, y+k) - u(x, y) = \frac{\partial u}{\partial x} h + \frac{\partial u}{\partial y} k + \varepsilon_1,$$
$$v(x+h, y+k) - v(x, y) = \frac{\partial v}{\partial x} h + \frac{\partial v}{\partial y} k + \varepsilon_2,$$

where the remainders ε_1 and ε_2 tend to zero more rapidly than $h+ik$, that is,

$$\varepsilon_1/(h+ik) \to 0, \qquad \varepsilon_2/(h+ik) \to 0, \qquad \text{as } h+ik \to 0.$$

With the notation $f(z) = u(x, y) + iv(x, y)$, by the Cauchy–Riemann equations, we obtain

$$f(z+h+ik) - f(z) = \left(\frac{\partial u}{\partial x} + i\frac{\partial v}{\partial x}\right)(h+ik) + \varepsilon_1 + i\varepsilon_2,$$

and hence,

$$\lim_{h+ik\to 0} \frac{f(z+h+ik) - f(z)}{h+ik} = \frac{\partial u}{\partial x} + i\frac{\partial v}{\partial x}.$$

It then follows that $f(z)$ is analytic. □

Because of Theorems 1.4.1 and 1.4.2, the Cauchy–Riemann equations (1.4.2) are also known as conditions of analyticity of a function.

Using the Cauchy–Riemann equations, one can express the derivative of an analytic function in the following equivalent forms:

$$f'(z) = \frac{\partial u}{\partial x} + i\frac{\partial v}{\partial x} = \frac{\partial v}{\partial y} - i\frac{\partial u}{\partial y} = \frac{\partial u}{\partial x} - i\frac{\partial u}{\partial y} = \frac{\partial v}{\partial y} + i\frac{\partial v}{\partial x}. \quad (1.4.8)$$

1.4.3. Basic properties of analytic functions. Using the expression (1.4.1) for the derivative, f', of f, one can transfer some properties of differentiable functions to analytic functions. Let D be a domain. Then we have the following properties of analytic functions.

(1) If $f(z)$ is analytic in D, then it is continuous in D since it follows from (1.4.1) that

$$\Delta f(z)\big|_{z_0} = f'(z_0)\Delta z + \alpha\Delta z,$$

where $\alpha \to 0$ as $\Delta z \to 0$; thus, $\Delta f(z)\big|_{z_0} \to 0$ as $\Delta z \to 0$.

(2) If $f_1(z)$ and $f_2(z)$ are analytic in D, then their sum, difference, product and quotient, $f_1 \pm f_2$, $f_1 f_2$ and f_1/f_2 (if $f_2 \neq 0$), are analytic in D. Moreover,

$$(f_1 \pm f_2)' = f_1' \pm f_2', \qquad (f_1 f_2)' = f_1' f_2 + f_1 f_2',$$

and

$$\left(\frac{f_1}{f_2}\right)' = \frac{f_1' f_2 - f_1 f_2'}{f_2^2}, \qquad \text{provided} \quad f_2 \neq 0. \tag{1.4.9}$$

(3) Let $w = f(z)$ be analytic in D and $f'(z) \neq 0$ in D. If $\zeta = g(w)$ is defined and analytic on the range, $G = \{w = f(z);\ z \in D\}$, of f in the w-plane, then the *composite function* $g[f(z)]$ is analytic in D, in the z-plane, and $\zeta'(z)$ is expressed by the chain rule,

$$\frac{d\zeta}{dz} = \frac{d\zeta}{dw}\,\frac{dw}{dz}. \tag{1.4.10}$$

(4) If $w = f(z)$ is analytic in D and $f'(z) \neq 0$ at the point z_0 and hence, by continuity, in some neighborhood U of z_0, then an *inverse function* $z = g(w)$ is defined in a neighborhood of the point $w_0 = f(z_0)$ of the range of f over U. Moreover g is an analytic function of the complex variable w and

$$g'(w_0) = \frac{1}{f'(z_0)}. \tag{1.4.11}$$

(5) If the real part, $u(x, y)$, of an analytic function, $f(z)$, is given in a simply connected domain D of the (x, y)-plane, then the imaginary part, $v(x, y)$, of $f(z)$ is determined by the Cauchy–Riemann equations (1.4.3) to within an arbitrary constant. In fact, we have

$$\begin{aligned} v(x, y) &= \int_{M_0}^{M} \left(\frac{\partial v}{\partial x}\,dx + \frac{\partial v}{\partial y}\,dy\right) + C \\ &= \int_{M_0}^{M} \left(-\frac{\partial u}{\partial y}\,dx + \frac{\partial u}{\partial x}\,dy\right) + C, \end{aligned} \tag{1.4.12}$$

where the points $M_0 = (x_0, y_0)$ and $M = (x, y)$ can be joined by any curve in D. It is more convenient to join M_0 and M by a polygonal line whose segments are parallel to the x- and y-axes.

An analytic function, f, can be conveniently expressed by means of its real part, $u(x, y)$,

$$f(z) = 2u\left(\frac{z + \bar{z}_0}{2}, \frac{z - \bar{z}_0}{2i}\right) - \overline{f(z_0)}, \tag{1.4.13}$$

or its imaginary part, $v(x, y)$,

$$f(z) = 2iv\left(\frac{z + \bar{z}_0}{2}, \frac{z - \bar{z}_0}{2i}\right) + \overline{f(z_0)}, \tag{1.4.14}$$

where the bar indicates complex conjugation.

(6) If

$$f(z) = u(x, y) + iv(x, y)$$

is analytic in a domain D, then the family of curves

$$u(x, y) = c, \qquad v(x, y) = d,$$

are *orthogonal*. In fact, by the Cauchy–Riemann equations (1.4.2), we have

$$\begin{aligned} \nabla u \cdot \nabla v &= \left(\frac{\partial u}{\partial x}, \frac{\partial u}{\partial y}\right) \cdot \left(\frac{\partial v}{\partial x}, \frac{\partial v}{\partial y}\right) \\ &= \frac{\partial u}{\partial x}\frac{\partial v}{\partial x} + \frac{\partial u}{\partial y}\frac{\partial v}{\partial y} \\ &= -\frac{\partial u}{\partial x}\frac{\partial u}{\partial y} + \frac{\partial u}{\partial y}\frac{\partial u}{\partial x} \\ &= 0, \end{aligned}$$

that is, the vectors ∇u and ∇v are orthogonal. But, since these vectors are orthogonal to the families of curves $u(x, y) = c$ and $v(x, y) = d$, respectively, these families also are orthogonal.

(7) Suppose that $f(z)$ is analytic in a domain D. We represent $f(z)$ in the form

$$\begin{aligned} f(z) &= u(x, y) + iv(x, y) = |f(z)|\, e^{i \arg f(z)} \\ &= r(x, y)\, e^{i\theta(x,y)} = r\cos\theta + ir\sin\theta, \end{aligned} \tag{1.4.15}$$

where $r(x, y) = |f(z)|$ and $\theta(x, y) = \arg f(z)$ are the modulus and argument of $f(z)$, respectively. We prove that $r(x, y)$ and $\theta(x, y)$ satisfy the following equations:

$$\frac{\partial r}{\partial x} = r\,\frac{\partial \theta}{\partial y}, \qquad \frac{\partial r}{\partial y} = -r\,\frac{\partial \theta}{\partial x}. \tag{1.4.16}$$

In fact, since $f(z)$ is analytic, then the functions

$$u(x,y) = r\cos\theta, \qquad v(x,y) = r\sin\theta, \tag{1.4.17}$$

satisfy the Cauchy–Riemann equations (1.4.2). Hence $u_x = v_y$ implies that

$$\frac{\partial r}{\partial x}\cos\theta - r\sin\theta\,\frac{\partial\theta}{\partial x} = \frac{\partial r}{\partial y}\sin\theta + r\cos\theta\,\frac{\partial\theta}{\partial y}. \tag{1.4.18}$$

Equating the coefficients of $\cos\theta$ and $\sin\theta$, respectively, in (1.4.18), we obtain (1.4.16). Similarly, $u_y = -v_x$ implies (1.4.16).

1.4.4. Complex and real differentiability. We rederive the Cauchy–Riemann equations by means of real differentiation. For this purpose, we identify the complex number $z = x + iy$ with a particular 2×2 real matrix as follows:

$$x + iy \leftrightarrow \begin{bmatrix} x & -y \\ y & x \end{bmatrix}.$$

Moreover, we have the correspondence

$$f : z \mapsto u + iv \qquad \leftrightarrow \qquad f : (x,y) \mapsto \big(u(x,y,), v(x,y)\big).$$

Let $a = \alpha + i\beta$. We have the $\mathbb{C}$-linear application

$$z \mapsto f'(a)z$$

and the $\mathbb{R}$-linear application

$$df(\alpha,\beta) = \begin{bmatrix} u_x(a) & u_y(a) \\ v_x(a) & v_y(a) \end{bmatrix}.$$

We identify $\mathbb{C}$ and $\mathbb{R}^2$. Then the complex number

$$f'(a) = u_x(a) + iv_x(a)$$

is the derivative of $f(z)$ at the point a if and only if

$$df(\alpha,\beta) = \begin{bmatrix} u_x(a) & u_y(a) \\ v_x(a) & v_y(a) \end{bmatrix} = \begin{bmatrix} u_x(a) & -v_x(a) \\ v_x(a) & u_x(a) \end{bmatrix}.$$

This establishes the Cauchy–Riemann equations.

Remark 1.4.1. Let

$$T : \mathbb{C} \to \mathbb{C}$$

be a $\mathbb{C}$-linear application. Identifying $\mathbb{C}$ and $\mathbb{R}^2$ via the correspondence

$$1 \leftrightarrow \begin{bmatrix} 1 \\ 0 \end{bmatrix}, \quad \text{and} \quad i \leftrightarrow \begin{bmatrix} 0 \\ 1 \end{bmatrix},$$

we have

$$T = \begin{bmatrix} \alpha & -\beta \\ \beta & \alpha \end{bmatrix}$$

if

$$Tz = T(x+iy) \leftrightarrow (\alpha + i\beta)(x+iy).$$

This holds since

$$(\alpha + i\beta)(x+iy) = (\alpha x - \beta y) + (\alpha y + \beta x)i \leftrightarrow \begin{bmatrix} \alpha & -\beta \\ \beta & \alpha \end{bmatrix} \begin{bmatrix} x \\ y \end{bmatrix}.$$

1.4.5. Harmonic functions. It will be shown later that the derivative of an analytic function $f(z) = u(x,y)+iv(x,y)$ is itself analytic. Thus u and v will have continuous partial derivatives of all orders, and, in particular, the mixed derivatives, say u_{xy} and u_{yx}, will be equal. It then follows from the Cauchy–Riemann equations (1.4.2) that

$$\Delta u = \frac{\partial^2 u}{\partial x^2} + \frac{\partial^2 u}{\partial y^2} = 0, \tag{1.4.19}$$

$$\Delta v = \frac{\partial^2 v}{\partial x^2} + \frac{\partial^2 v}{\partial y^2} = 0. \tag{1.4.20}$$

DEFINITION 1.4.5. A function $u(x,y)$ which satisfies the Laplace equation $\Delta u = 0$ is said to be *harmonic*.

Thus, we see that the real and imaginary parts of an analytic function are harmonic. If two harmonic functions u and v satisfy the Cauchy–Riemann equations (1.4.2), then v is *conjugate harmonic* to u, and u is conjugate harmonic to $-v$, since $i(u+iv) = -v + iu$.

Hence, we have proved the following theorem.

THEOREM 1.4.3. *The real and imaginary parts of an analytic function are harmonic functions.*

It follows from (1.4.12) that given a harmonic function u, its conjugate harmonic v is uniquely determined up to an arbitrary constant of integration. Similarly, if v is harmonic, its conjugate harmonic, $-u$, will be determined up to an arbitrary constant.

It was observed in the previous subsection that an analytic function f could be conveniently expressed by its real part u in (1.4.13) and by its imaginary part v in (1.4.14).

Exercises for Section 1.4

Find the first derivative of the following functions.

1. $f(z) = \dfrac{z+2}{3z+4}, \qquad z \neq -\dfrac{4}{3}.$

2. $f(z) = 3z^2 + 4z + 5.$

3. $f(z) = \dfrac{z^2+3}{(2z^3+7)^2}, \qquad 2z^3 + 7 \neq 0.$

4. $f(z) = \dfrac{(1-z^2)^3}{z^4}, \qquad z \neq 0.$

5. Determine whether the function $f(z) = \Re z$ has a derivative at every point.

6. Show that the functions $\bar{z}$, $\Im z$, $|z|$ and $\arg z$ are nowhere differentiable.

7. Prove that $f(z) = |z|^2$ is differentiable but not analytic at $z = 0$.

Determine the domain of analyticity of the following functions:

8. $f(z) = z^6 + 9z^3 + 1.$

9. $f(z) = \dfrac{1}{(z^6-1)^2}.$

10. $f(z) = z + \dfrac{1}{z(z^2-1)}.$

11. $f(z) = \dfrac{z}{(z^2+4)^2}.$

12. Derive the polar-coordinate form of the Cauchy–Riemann equations:

$$r u_r = v_\theta, \qquad r v_r = -u_\theta.$$

Are the following functions harmonic? If so, find a corresponding analytic function $f(z) = u(x,y) + iv(x,y)$.

13. $u(x,y) = \dfrac{x}{x^2+y^2}.$

14. $v(x,y) = 2xy.$

15. $v(x,y) = e^{2x} \sin 3y.$

16. $u(x,y) = \dfrac{1}{\sqrt{x^2+y^2}}.$

17. $u(x, y) = x^3 - 3xy^2$.

18. $v(x, y) = \cos x \sinh y$.

19. Show that

$$f(z) = \begin{cases} z^5/|z|^4, & \text{if } z \neq 0, \\ 0, & \text{if } z = 0, \end{cases}$$

satisfies the Cauchy–Riemann equations at $z = 0$ but is not differentiable there.

20. Show that $f(z) = x^3 + iy^3$ satisfies the Cauchy–Riemann equations at the point $z = 0$, but is not analytic there.

21. Let $f(z)$ be a polynomial in $z \in \mathbb{C}$. Prove that the function given by $g(z) = \overline{f(\bar{z})}$ is differentiable everywhere, but that $h(z) = \overline{f(z)}$ is differentiable at 0 if, and only if, $f'(0) = 0$.

22. Determine the analytic function $f(z) = u + iv$ for which $u + v$ is known. (Hint: Put $F = u + v$.)

23. If $f(z)$ is continuously differentiable in a domain D and $f' \equiv 0$ in D, prove that $f(z)$ is constant in D.

24. If $f(z) = u + iv$ is analytic, and u and v possess continuous second partial derivatives, show that

$$\left(\frac{\partial^2}{\partial x^2} + \frac{\partial^2}{\partial y^2}\right) |f|^2 = 4|f'|^2.$$

1.5. Elementary analytic functions

1.5.1. The exponential function $w = e^z$. To define the function e^z, we use the definition of the function e^x of the real variable x:

$$e^x = \lim_{n \to \infty} \left(1 + \frac{x}{n}\right)^n. \tag{1.5.1}$$

Replacing x by z in (1.5.1) (that is, continuing the right-hand side of (1.5.1) to the complex plane), we have

$$e^z = \lim_{n \to \infty} \left(1 + \frac{z}{n}\right)^n, \tag{1.5.2}$$

if the limit exists. Now, expressing powers of complex numbers by formula (1.1.28), we obtain

$$\left(1 + \frac{z}{n}\right)^n = \left|1 + \frac{z}{n}\right|^n (\cos n\theta + i \sin n\theta), \tag{1.5.3}$$

where

$$\left|1+\frac{z}{n}\right|^n = \left|1+\frac{x+iy}{n}\right|^n = \left[\left(1+\frac{x}{n}\right)^2 + \frac{y^2}{n^2}\right]^{n/2}, \tag{1.5.4}$$

and

$$n\theta = n\arctan\frac{y/n}{1+x/n}. \tag{1.5.5}$$

Let us find the limits of (1.5.4) and (1.5.5) as $n \to \infty$.

Firstly,

$$\begin{aligned}\lim_{n\to\infty}\left|\frac{1+z}{n}\right|^n &= \lim_{n\to\infty}\left(1+\frac{2x}{n}+\frac{x^2+y^2}{n^2}\right)^{n/2}\\ &= \lim_{n\to\infty}\left(1+\frac{2x}{n}\right)^{n/2}\\ &= e^x. \end{aligned}\tag{1.5.6}$$

To obtain (1.5.6) we have discarded the infinitely small value $(x^2+y^2)/n^2$ of higher order with respect to $2x/n$ and used (1.5.1).

Secondly,

$$\begin{aligned}\lim_{n\to\infty} n\arg\left(1+\frac{z}{n}\right) &= \lim_{n\to\infty} n\arctan\frac{y/n}{1+x/n}\\ &= \lim_{n\to\infty} n\frac{y/n}{1+x/n}\\ &= y. \end{aligned}\tag{1.5.7}$$

To obtain (1.5.7) we have used the approximation $\arctan x \sim x$ if $x \sim 0$. Substituting (1.5.6) and (1.5.7) into (1.5.3) we finally obtain

$$\lim_{n\to\infty}\left(1+\frac{z}{n}\right)^n = e^x(\cos y + i\sin y).$$

Thus we have the following definition.

DEFINITION 1.5.1. The *exponential function*, e^z, is defined by the expression

$$e^z = e^{x+iy} = e^x(\cos y + i\sin y). \tag{1.5.8}$$

The previously used *Euler's formula*,

$$e^{iy} = \cos y + i\sin y, \qquad y \in \mathbb{R}, \tag{1.5.9}$$

is derived by setting $x = 0$ in (1.5.8).

We show that e^z possesses the following four properties:

(a) For real $z = x$, definition (1.5.8) coincides with the usual definition of e^x.

(b) The function e^z is everywhere analytic in the z-plane.

(c) The usual formula of differentiation is still valid:

$$\frac{d}{dz}e^z = e^z. \tag{1.5.10}$$

(d) The law of exponents holds:

$$e^{z_1}e^{z_2} = e^{z_1+z_2}. \tag{1.5.11}$$

Property (a) follows from (1.5.8) with $y = 0$.

Property (b) follows from the fact that the functions $u(x, y) = \Re e^z = e^x \cos y$ and $v(x, y) = \Im e^z = e^x \sin y$ are everywhere continuously differentiable and satisfy everywhere the Cauchy–Riemann equations

$$\begin{aligned}\frac{\partial}{\partial x}(e^x \cos y) &= \frac{\partial}{\partial y}(e^x \sin y),\\ \frac{\partial}{\partial y}(e^x \cos y) &= -\frac{\partial}{\partial x}(e^x \sin y).\end{aligned}$$

Since e^z is analytic by (b), to prove (c) we use the independence of the derivative upon the direction of Δz in (1.4.1) and compute the derivative of e^z with $\Delta z = \Delta x$ (see (1.4.8)):

$$\frac{d}{dz}e^z = \frac{\partial}{\partial x}e^x(\cos y + i \sin y) = e^x(\cos y + i \sin y) = e^z.$$

We, of course, obtain the same result by calculating $(e^z)'$ with $\Delta z = i\Delta y$:

$$\frac{d}{dz}e^z = \frac{1}{i}\frac{\partial}{\partial y}e^x(\cos y + i \sin y) = \frac{1}{i}e^x(-\sin y + i \cos y) = e^z.$$

Finally, to prove property (d), we let $z_1 = x_1 + iy_1$, $z_2 = x_2 + iy_2$ and use formula (1.5.8) and the rule (1.1.21) for the multiplication of complex numbers; thus

$$\begin{aligned}e^{z_1}e^{z_2} &= e^{x_1}(\cos y_1 + i \sin y_1)e^{x_2}(\cos y_2 + i \sin y_2)\\ &= e^{x_1+x_2}[(\cos y_1 \cos y_2 - \sin y_1 \sin y_2) + i(\sin y_1 \cos y_2 + \cos y_1 \sin y_2)]\\ &= e^{x_1+x_2}[\cos(y_1 + y_2) + i \sin(y_1 + y_2)]\\ &= e^{z_1+z_2}.\end{aligned}$$

Properties (a)–(d) are valid for both real and complex arguments of e^z. But for a complex argument, the function e^z has pure imaginary period $2\pi i$ since, by Euler's formula (1.5.9), we have

$$e^{z+2k\pi i} = e^z e^{2k\pi i} = e^z(\cos 2k\pi + i\sin 2k\pi) = e^z$$

for any integer k.

The well known *De Moivre's formula*,

$$(\cos\theta + i\sin\theta)^n = \cos n\theta + i\sin n\theta, \tag{1.5.12}$$

follows from formula (1.1.28) with $|z| = r = 1$ or from Euler's formula (1.5.9) with $y = n\theta$.

1.5.2. Logarithm of z.

DEFINITION 1.5.2. Given a nonzero complex number z, a complex number w such that $e^w = z$ is called a *logarithm of* z, written

$$w = \log z. \tag{1.5.13}$$

Suppose that $w = u + iv$ and let $z = e^w = e^{u+iv}$. Then

$$e^u = |z| \quad\Longrightarrow\quad u = \ln|z|,$$

where $\ln|z|$ is the *natural logarithm*, to the base e, of a real number, and

$$v = \arg z = \operatorname{Arg} z + 2k\pi.$$

Thus the expression

$$\log z = u + iv = \ln|z| + i(\operatorname{Arg} z + 2k\pi), \qquad k = 0, \pm1, \pm2, \dots, \tag{1.5.14}$$

has infinitely many values at each point z, that is, one for each value of k. For a fixed value of k, the right-hand side of (1.5.14) defines a branch of the logarithm and it is a function of z.

DEFINITION 1.5.3. The function

$$w = \ln|z| + i\operatorname{Arg} z$$

is called the *principal value* (or principal branch) of $\log z$ and is denoted by $\operatorname{Log} z$,

$$\operatorname{Log} z = \ln|z| + i\operatorname{Arg} z, \tag{1.5.15}$$

where $-\pi < \operatorname{Arg} z \le \pi$ by (1.1.7) or $0 \le \operatorname{Arg} z < 2\pi$ by (1.1.8).

Using (1.5.15), we rewrite (1.5.14) in the form

$$\log z = \operatorname{Log} z + 2k\pi i, \tag{1.5.16}$$

which leads to the following definition.

DEFINITION 1.5.4. The functions

$$w_k = \operatorname{Log} z + 2k\pi i, \qquad k = 1, 2, \dots,$$

are the branches of $\log z$.

We give two simple examples.

EXAMPLE 1.5.1. *Find all the values of* $\log 3$.

SOLUTION. As $\operatorname{Arg} 3 = 0$, then according to (1.5.14) we have

$$\log 3 = \ln 3 + 2k\pi i, \qquad k = 0, \pm 1, \pm 2, \dots \quad \square$$

EXAMPLE 1.5.2. *Evaluate* $\operatorname{Log}\left(\sqrt{3} + i\right)$, *where* $-\pi < \operatorname{Arg} z \le \pi$.

SOLUTION. Let us transform $\sqrt{3} + i$ to the exponential form:

$$\left|\sqrt{3+i}\right| = 2, \quad \tan \operatorname{Arg}\left(\sqrt{3} + i\right) = \frac{1}{\sqrt{3}}, \quad \operatorname{Arg}\left(\sqrt{3} + i\right) = \frac{\pi}{6}.$$

Then, according to (1.5.15),

$$\operatorname{Log}\left(\sqrt{3} + i\right) = \operatorname{Log}\left(2e^{\pi i/6}\right) = \ln 2 + \frac{\pi}{6}\, i. \quad \square$$

1.5.3. The trigonometric and hyperbolic functions. The trigonometric and hyperbolic functions can be expressed by means of the exponential function. Using Euler's formula (1.5.9) for real x,

$$e^{ix} = \cos x + i \sin x, \qquad e^{-ix} = \cos x - i \sin x, \tag{1.5.17}$$

we have

$$\cos x = \frac{e^{ix} + e^{-ix}}{2}, \qquad \sin x = \frac{e^{ix} - e^{-ix}}{2i}. \tag{1.5.18}$$

Now we use the analytic continuation of the right-hand sides of (1.5.18) from the real axis to the complex plane to define $\cos z$ and $\sin z$ as functions of the complex variable z,

$$\cos z = \frac{e^{iz} + e^{-iz}}{2}, \qquad \sin z = \frac{e^{iz} - e^{-iz}}{2i}, \tag{1.5.19}$$

which coincide with the functions $\cos x$ and $\sin x$ for real $z = x$. It can be shown that this continuation is unique. We leave as an exercise for the reader to show from (1.5.19) that $\cos z$ and $\sin z$

(a) are analytic everywhere,

(b) satisfy the usual rules of differentiation

$$\frac{d}{dz} \cos z = -\sin z, \qquad \frac{d}{dz} \sin z = \cos z,$$

(c) are 2π-periodic,

(d) satisfy the usual trigonometric identities,

$$\sin^2 z + \cos^2 z = 1, \quad \cos 2z = \cos^2 z - \sin^2 z, \quad \sin 2z = 2\sin z\cos z,$$

etc.

The other trigonometric and hyperbolic functions are similarly defined as follows:

$$\tan z = -i\,\frac{e^{iz} - e^{-iz}}{e^{iz} + e^{-iz}}, \tag{1.5.20}$$

$$\cot z = i\,\frac{e^{iz} + e^{-iz}}{e^{iz} - e^{-iz}}, \tag{1.5.21}$$

$$\cosh z = \frac{e^{z} + e^{-z}}{2}, \tag{1.5.22}$$

$$\sinh z = \frac{e^{z} - e^{-z}}{2}, \tag{1.5.23}$$

$$\tanh z = \frac{e^{z} - e^{-z}}{e^{z} + e^{-z}}, \tag{1.5.24}$$

$$\coth z = \frac{e^{z} + e^{-z}}{e^{z} - e^{-z}}. \tag{1.5.25}$$

Comparing (1.5.19), (1.5.22) and (1.5.23) we obtain

$$\cos z = \cosh iz, \qquad \sinh iz = i\sin z. \tag{1.5.26}$$

Changing z to iz in (1.5.26) we get

$$\cos iz = \cosh z, \qquad \sin iz = i\sinh z. \tag{1.5.27}$$

Note that the inequalities

$$|\sin z| > 1, \qquad |\cos z| > 1 \tag{1.5.28}$$

can hold in the complex plane. For example, if $z = iy$ for $y \in \mathbb{R}$, then

$$|\cos iy| = |\cosh y| > \frac{1}{2}\,e^{y} > 1$$

for $y > \ln 2$.

1.5.4. The inverse trigonometric functions. The inverse trigonometric functions can be expressed in terms of appropriate branches of the logarithm. We recall that the term function implies a given domain of definition D such that the map $f : z \mapsto f(z)$ is single valued in D.

DEFINITION 1.5.5. We say that $w = \arcsin z$ if $\sin w = z$.
We have

$$\begin{aligned}\sin w = z \quad &\Longrightarrow \quad \frac{e^{iw} - e^{-iw}}{2i} = z \\ &\Longrightarrow \quad e^{iw} - \frac{1}{e^{iw}} - 2iz = 0 \\ &\Longrightarrow \quad \left(e^{iw}\right)^2 - 2iz\, e^{iw} - 1 = 0.\end{aligned} \tag{1.5.29}$$

It follows from this quadratic equation that

$$e^{iw} = iz + \sqrt{1 - z^2} \tag{1.5.30}$$

(we omit the $\pm$ sign before the square root in (1.5.30) because $\sqrt{1 - z^2}$ is understood to have two branches, each of which is a function). We find from (1.5.30) that arcsine is given by the formula

$$w = \arcsin z = \frac{1}{i} \log \left(iz + \sqrt{1 - z^2}\right). \tag{1.5.31}$$

DEFINITION 1.5.6. The function

$$\operatorname{Arcsin} z = \frac{1}{i} \operatorname{Log} \left(iz + \sqrt{1 - z^2}\right), \tag{1.5.32}$$

where the principal value of the square root is chosen, is called the *principal value of* $\arcsin z$.

If we let $z = x$ in (1.5.32) and assume that $|x| < 1$, then the following formula, which is known from real analysis, follows from (1.5.31):

$$\arcsin x = (-1)^n \operatorname{Arcsin} x + n\pi. \tag{1.5.33}$$

If, for example, $z = 1/2$, then we obtain from (1.5.31) that

$$\arcsin \frac{1}{2} = \frac{1}{i} \log \left(\frac{1}{2}\, i \pm \frac{\sqrt{3}}{2}\right). \tag{1.5.34}$$

We consider the plus and minus signs in (1.5.34) separately. Since

$$\frac{1}{2}\, i + \frac{\sqrt{3}}{2} = \sqrt{\frac{1}{4} + \frac{3}{4}}\, e^{\pi i/6} = e^{\pi i/6},$$

then

$$\arcsin \frac{1}{2} = \frac{1}{i} \log e^{\pi i/6} = \frac{1}{i} \left(\frac{\pi}{6} i + 2k\pi i\right) = \frac{\pi}{6} + 2k\pi.$$

Now, since

$$\frac{1}{2}i - \frac{\sqrt{3}}{2} = e^{5\pi i/6},$$

then

$$\arcsin\frac{1}{2} = \frac{1}{i}\log e^{5\pi i/6} = \frac{1}{i}\left(\frac{5\pi}{6}i + 2k\pi i\right) = -\frac{\pi}{6} + (2k+1)\pi.$$

Hence

$$\arcsin\frac{1}{2} = \begin{cases} \frac{\pi}{6} + 2k\pi, \\ -\frac{\pi}{6} + (2k+1)\pi, \end{cases}$$

or

$$\arcsin\frac{1}{2} = (-1)^n\frac{\pi}{6} + n\pi,$$

which corresponds to formula (1.5.33).

DEFINITION 1.5.7. We say that $w = \arccos z$ if $\cos w = z$.

It is a simple exercise to prove that arccosine is given by the formula

$$\arccos z = \frac{1}{i}\log\left(z + \sqrt{z^2 - 1}\right), \tag{1.5.35}$$

and that its *principal value*,

$$\operatorname{Arccos} z = \frac{1}{i}\operatorname{Log}\left(z + \sqrt{z^2 - 1}\right), \tag{1.5.36}$$

is determined by the principal value of the logarithm.

Similarly, one can express arctangent and arccotangent through the logarithm:

$$\arctan z = \frac{1}{2i}\log\frac{1+iz}{1-iz}, \qquad z \neq \pm i, \tag{1.5.37}$$

$$\operatorname{arccot} z = \frac{1}{2i}\log\frac{iz-1}{iz+1}, \qquad z \neq \pm i. \tag{1.5.38}$$

It is left as an exercise to show that the principal values of $\arctan z$ and $\operatorname{arccot} z$ are determined by the principal values of the logarithm on the right-hand sides of formulae (1.5.37) and (1.5.38).

From the previous considerations, we see that the power, exponential and logarithmic functions can be considered as *basic elementary functions* because all the trigonometric functions can be expressed in terms of the exponential function while the inverse trigonometric functions can be expressed through the logarithmic function. As in real analysis one can introduce the concept of elementary functions.

DEFINITION 1.5.8. A function of the complex variable z is called an *elementary function* if it is obtained from the basic elementary functions, namely, z^n, e^z and $\log z$, by a finite number of the four arithmetic operations and a finite number of compositions of elementary functions.

For example, the function $w = \sin\left[\cos\left(e^{1+z^2}\right)\right]$ is an elementary function.

The following theorem holds.

THEOREM 1.5.1. *Elementary functions of a complex variable are analytic in their domains of definition.*

Exercises for Section 1.5

Represent the following numbers in the form $x + iy$.

1. $e^{3+\pi i/2}$.
2. e^{i}.
3. $e^{e^{i}}$.
4. $e^{1+(\pi i/4)}/e^{\pi i/3}$.

5. Prove that there cannot be any finite values of z such that $e^z = 0$.

6. Show that $|1 + e^z| \leq 1 + e^x$.

7. Prove that the function

$$f(z) = \begin{cases} e^{-1/z^4}, & \text{if } z \neq 0, \\ 0, & \text{if } z = 0, \end{cases}$$

satisfies the Cauchy–Riemann equations at every point of the plane without being analytic in the whole plane.

8. Describe the limiting behavior of e^z as $z \to \infty$ along the ray $\arg z = \alpha$.

If $-\pi < \operatorname{Arg} z \leq \pi$, evaluate the following expressions.

9. $\operatorname{Log}(3i)$.
10. $\operatorname{Log}(-2i)$.
11. $\log(1 + i)$.
12. $\log(z^5)$, where $z = 3\,e^{\pi i/6}$.

Find all the roots of the following equations.

13. $e^z = -4$.

14. $e^z = 2i$.

15. $e^{iz} = -\sqrt{3} + i$.

16. $\log z = (\pi/2)i$.

Derive formulae for the real and imaginary parts of the following functions and check that they satisfy the Cauchy–Riemann equations.

17. $\sin z$.

18. $\cos z$.

19. $\cosh z$.

20. $\sinh z$.

Prove the following identities.

21. $\cos 2z = \cos^2 z - \sin^2 z$.

22. $1 + \tan^2 z = \dfrac{1}{\cos^2 z}$.

23. $\sin(z_1 + z_2) = \sin z_1 \cos z_2 + \cos z_1 \sin z_2$.

24. $\cos(z_1 - z_2) = \cos z_1 \cos z_2 + \sin z_1 \sin z_2$.

25. Show that neither $\sin \bar{z}$ nor $\cos \bar{z}$ is an analytic function of z anywhere.

26. Prove that $|\sin z| \geq |\sin x|$ and $|\cos z| \geq |\cos x|$.

Where are the following functions analytic?

27. $\dfrac{e^z}{z^2(z+1)}$.

28. $\cos \dfrac{1}{z}$.

29. $\dfrac{e^z - 1}{e^z + 1}$.

30. $\dfrac{e^z}{\cos z}$.

Find all possible solutions of the following equations.

31. $\cos z = i$.

32. $\sin z = 32$.

33. $\cosh z = 1/4$.

34. $\sinh z = 2i$.

35. Find all the zeros of the functions $\cosh z$ and $\sinh z$.

36. Is $\sin |z|^2$ anywhere differentiable? Is it anywhere analytic?

37. Prove that all the roots of the equations $\sin z = a$ and $\cos z = a$ are real, if $-1 \leq a \leq 1$.

38. Prove that if $z \in \mathbb{C}$ and $|\sin z| \leq 1$, then $z \in \mathbb{R}$.

39. Find the principal value of i^i, $(1-i)^{2i}$. (Hint. By definition, $z^a = e^{a \log z}$, if $z \in \mathbb{C}$ and $a \in \mathbb{C}$.)

40. Find the real and imaginary parts of z^z where $z = x + iy$.

If $-\pi < \operatorname{Arg} z \leq \pi$, represent the following functions in the form $z = x + iy$.

41. $\operatorname{Arcsin} i$.

42. $\operatorname{Arccos} \pi i$.

If $-\pi < \operatorname{Arg} z \leq \pi$, find all the values of

43. $\arcsin 2$.

44. $\arccos 100$.

CHAPTER 2

Elementary Conformal Mappings

2.1. Geometric meaning of $f'(z)$

2.1.1. Geometric meaning of the argument of $f'(z)$. Consider the analytic function $w = f(z)$ which maps a point $z_0 = x_0 + iy_0$ of the z-plane into a point $w_0 = f(z_0) = u_0 + iv_0$ of the w-plane with real u-axis and imaginary v-axis (see Fig 2.1).

In this chapter, unless otherwise stated, *positive angles* are measured counterclockwise from the real positive semi-axes, and *branch cuts* are taken along the same semi-axes.

Let γ_1 be a differentiable curve passing through the point z_0. The function $w = f(z)$ maps γ_1 into a curve Γ_1 in the w-plane. We assume that $f'(z_0) \neq 0$ and find the modulus and argument of $f'(z_0)$. For given values z_0 and $w_0 = f(z_0)$ that are kept fixed, let

$$\Delta z = z - z_0 \quad \text{and} \quad \Delta w = w - w_0 = f(z) - f(z_0)$$

denote the increments in z and $f(z)$, respectively. Then by definition,

$$\begin{aligned} f'(z_0) &= \lim_{\Delta z \to 0} \frac{\Delta w}{\Delta z} \\ &= \lim_{\Delta z \to 0} \left\{ \frac{|\Delta w|}{|\Delta z|}\, e^{i \arg(\Delta w/\Delta z)} \right\}, \end{aligned} \tag{2.1.1}$$

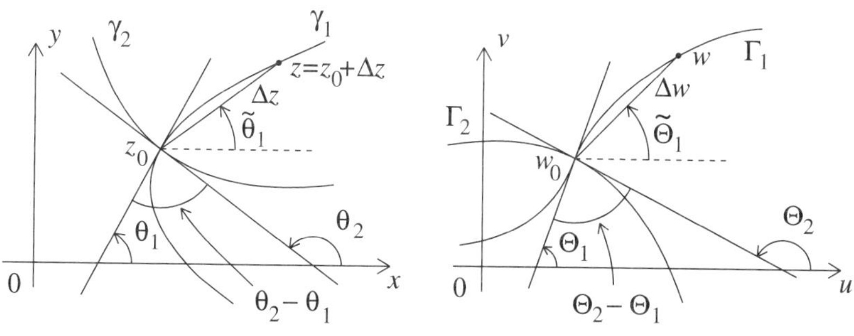

FIGURE 2.1. Geometric meaning of $|f'(z)|$ and $\arg f'(z)$

where we have used the exponential form of the complex number $\frac{\Delta w}{\Delta z}$. Thus

$$f'(z_0) = \lim_{\Delta z \to 0} \left\{ \frac{|\Delta w|}{|\Delta z|} \right\} \left\{ \exp\left(i \lim_{\Delta z \to 0} [\arg \Delta w - \arg \Delta z] \right) \right\}, \quad (2.1.2)$$

since the argument of a fraction is equal to the difference of the arguments of the numerator and denominator. It follows from (2.1.2) that

$$\arg f'(z_0) = \lim_{\Delta z \to 0} \arg \Delta w - \lim_{\Delta z \to 0} \arg \Delta z. \quad (2.1.3)$$

Consider the point $z = z_0 + \Delta z$ on the curve γ_1 and its image $w = w_0 + \Delta w$ on the curve Γ_1. The vector $\Delta z = z - z_0$ joining the points z_0 and z on γ_1 goes over the vector $\Delta w = w - w_0$ joining the points $w_0 = f(z_0)$ and $w = f(z)$ on Γ_1 (see Fig 2.1). Here we have used the geometric representation of the difference of two complex numbers. Let

$$\widetilde{\theta}_1 = \arg \Delta z$$

be the angle between the vector Δz and the x-axis and

$$\widetilde{\Theta}_1 = \arg \Delta w$$

be the angle between the vector Δw and the u-axis. If $\Delta z \to 0$ while $z \in \gamma_1$, then the direction of the vector Δz tends to the direction of the tangent to γ_1 at the point z_0, that is,

$$\lim_{\Delta z \to 0} \arg \Delta z = \theta_1, \quad (2.1.4)$$

where θ_1 is the angle between the tangent to γ_1 at the point z_0 and the x-axis.

Similarly, as $\Delta z \to 0$, $\Delta w \to 0$ while $w \in \Gamma_1$, so that the direction of the vector Δw tends to the direction of the tangent to Γ_1, that is,

$$\lim_{\Delta w \to 0} \arg \Delta w = \Theta_1, \quad (2.1.5)$$

where Θ_1 is the angle between the tangent to Γ_1 at w_0 and the u-axis. Substituting (2.1.4) and (2.1.5) into (2.1.3) we obtain

$$\arg f'(z_0) = \Theta_1 - \theta_1, \quad (2.1.6)$$

that is, geometrically, the argument of the derivative is the difference between the angles Θ_1 and θ_1.

Let us draw another curve γ_2 through the point z_0. This curve is mapped by the function $f(z)$ into the curve Γ_2 through the point w_0 in the w-plane. Repeating the previous argument, we get

$$\arg f'(z_0) = \Theta_2 - \theta_2, \quad (2.1.7)$$

where Θ_2 and θ_2 are the angles formed by the tangents to Γ_2 and γ_2 and the u- and x-axes, respectively (see Fig 2.1).

Since the left-hand sides of (2.1.6) and (2.1.7) are equal (the derivative $f'(z_0)$ does not depend on how Δz approaches zero), the right-hand sides also are equal, namely, $\Theta_1 - \theta_1 = \Theta_2 - \theta_2$, which we rewrite in the form

$$\Theta_2 - \Theta_1 = \theta_2 - \theta_1. \tag{2.1.8}$$

But $\Theta_2 - \Theta_1$ is the angle between the tangents to Γ_2 and Γ_1 while $\theta_2 - \theta_1$ is the angle between the tangents to γ_2 and γ_1. Therefore, the angle between two curves that intersect at a point z_0 remains constant under the mapping by an analytic function $f(z)$, provided $f'(z_0) \neq 0$.

Note that angles between curves are preserved not only in absolute value but also in direction. In fact, by (2.1.8),

$$\theta_2 - \theta_1 > 0 \implies \Theta_2 - \Theta_1 > 0$$

and

$$\theta_2 - \theta_1 < 0 \implies \Theta_2 - \Theta_1 < 0.$$

This property is called the *angle-preserving property.*

2.1.2. Geometric meaning of $|f'(z)|$**.** Taking the modulus in (2.1.1) we have

$$|f'(z_0)| = \lim_{\Delta z \to 0} \frac{|\Delta w|}{|\Delta z|}, \tag{2.1.9}$$

where we have omitted the symbol $|_{z_0}$ on the right-hand side. We suppose that $|f'(z_0)| = k > 0$. We know that a function f, which is continuous at a point z_0, is equal to its limit at z_0 plus a function g which goes to zero as $z \to z_0$. Then taking (2.1.9) into account, we get

$$\frac{|\Delta w|}{|\Delta z|} = k + g(z) \to k, \qquad \text{as } \Delta z \to 0. \tag{2.1.10}$$

Thus, to within higher order infinitesimal terms with respect to $|\Delta z|$, we have

$$|\Delta w|\big|_{z_0} = k|\Delta z|\big|_{z_0}, \qquad k = |f'(z_0)| = \text{constant } > 0. \tag{2.1.11}$$

Therefore, the length of each sufficiently small vector originating from the point z_0 is dilated by the factor $k = |f'(z_0)|$ under the mapping by an analytic function $w = f(z)$. This property is known as the *property of constant dilation.*

It follows from (2.1.11) that any circle with sufficiently small radius δ centered at z_0 is mapped into a circle of radius $k\delta$ centered at w_0; each sufficiently small triangle with a vertex at z_0 is mapped into a similar curvilinear triangle with a vertex at w_0 with similarity coefficient k.

Note 2.1.1. Besides the condition $f'(z_0) \neq 0$, in order to satisfy the angle-preserving property and the property of constant dilation, one has to require that the function $f(z)$ should be univalent, that is, injective.

A function is *univalent* or *injective* if different points of the z-plane are mapped into different points of the w-plane, that is, for every pair of points z_1, z_2 in a domain D, we have the implication

$$z_1 \neq z_2 \implies f(z_1) \neq f(z_2).$$

The concept of univalent function and the determination of *domains of univalence* will be illustrated later by means of examples of concrete mappings.

DEFINITION 2.1.1. A mapping of a neighborhood of a point z_0 onto a neighborhood of a point w_0 that satisfies the angle-preserving property and the constant dilation property is called a *conformal mapping*.

The previous arguments lead to the following necessary and sufficient conditions for a function $f(z)$ to produce a *conformal mapping* of a domain D:

(a) univalence condition,
(b) analyticity of f,
(c) for all $z \in D$, $f'(z) \neq 0$.

It can be shown that the univalence of f in D implies that $f'(z) \neq 0$ everywhere in D, so that condition (c) can be omitted. The converse, in general, is not true, that is, it does not follow that f is univalent in D if $f'(z) \neq 0$ in D. For example, the mapping $w = z^4$ is not univalent on the half-annulus,

$$1 < |z| < 2, \qquad 0 < \operatorname{Arg} z < \pi,$$

because the annulus is mapped onto the domain

$$1 < |w| < 16, \qquad 0 < \arg w < 4\pi,$$

that is, on two copies of the annulus $1 < |w| < 16$, $0 < \arg w < 2\pi$, but $w' = 4w^3 \neq 0$ in the annulus.

2.2. Basic problems and principles of conformal mappings

2.2.1. Forward and inverse problems. We mention two basic problems related to conformal mappings.

Forward problem. Given a domain D in the z-plane and a function $w = f(z)$ analytic and univalent in D, find a domain $\widetilde{D}$ in the w-plane such that $\widetilde{D}$ will be the image of D under the mapping $w = f(z)$.

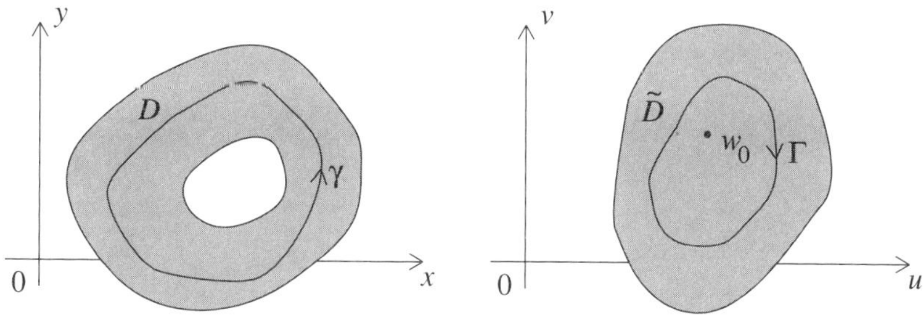

FIGURE 2.2. The impossibility of mapping a doubly connected domain D onto a simply connected domain $\widetilde{D}$.

This problem always has a solution, but it is not so important for the applications. A more important problem is the following inverse problem.

Inverse problem. Let domains D and $\widetilde{D}$ be given, in the z- and w-planes, respectively. Find an analytic function $w = f(z)$ which maps D onto $\widetilde{D}$.

This second problem is very important in the applications, but it does not always have a solution. For example, it is not possible to map a *multiply connected domain* onto a *simply connected domain* (see Fig 2.2). Indeed, a closed contour γ in D is mapped into a closed contour Γ in $\widetilde{D}$. It follows from the shrinking of the contour Γ to a point $w_0 \in \widetilde{D}$ that the contour γ (by the continuity of the mapping) should shrink to a point $z_0 \in D$, but this is impossible.

It can be shown that is not possible to map the whole complex z-plane onto a bounded domain $\widetilde{D}$ in the complex w-plane. Moreover, the following theorem holds.

THEOREM 2.2.1 (RIEMANN MAPPING THEOREM). *Given any simply connected domains D and $\widetilde{D}$ (with boundaries consisting of more than one point), any points $z_0 \in D$ and $w_0 \in \widetilde{D}$ and any real number α_0, there exists a unique conformal mapping*

$$w = f(z) \tag{2.2.1}$$

of D onto $\widetilde{D}$ such that

$$f(z_0) = w_0, \qquad \operatorname{Arg} f'(z_0) = \alpha_0. \tag{2.2.2}$$

We shall not prove this theorem. The uniqueness condition (2.2.2) of the mapping function (2.2.1) can be changed to the following: three given points z_1, z_2 and z_3 of D have to map into three given points w_1, w_2 and w_3 of $\widetilde{D}$.

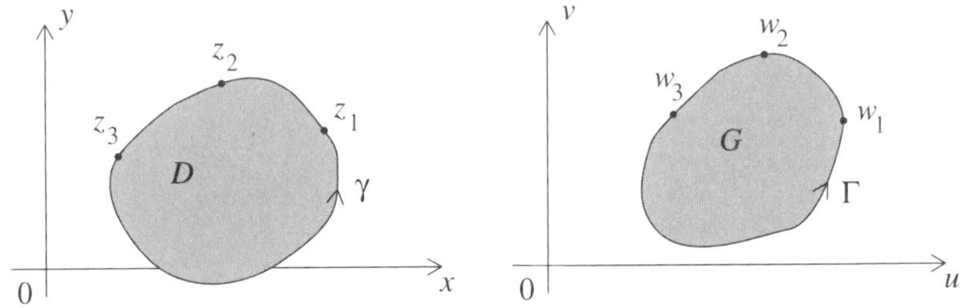

FIGURE 2.3. Mapping of domain D onto domain G if the closed contours γ and Γ have the same orientation.

2.2.2. Boundary-to-boundary and symmetry principles. We mention two basic principles of conformal mappings.

PRINCIPLE 2.2.1. *Boundaries are mapped onto boundaries.*

Consider a simply connected domain D in the z-plane bounded by a closed curve γ. Suppose that $w = f(z)$ is a nonconstant function analytic on the region $D \cup \gamma$ which contains the region of univalence of $f(z)$. Let $f(z)$ map γ into a closed contour Γ in the w-plane. Then there are two cases to be considered.

Case 1. If three distinct points, z_1, z_2, z_3, of γ are mapped into three distinct points, w_1, w_2, w_3, of Γ with the same orientation as the points z_1, z_2, z_3, then the domain D is mapped onto the domain G lying inside Γ (see Fig 2.3).

Case 2. If three distinct points, z_1, z_2, z_3, of γ are mapped into three distinct points, w_1, w_2, w_3, of Γ with orientation opposite to the orientation of the points z_1, z_2, z_3, then the domain D is mapped by the function $w = f(z)$ onto the domain G lying outside Γ (see Fig 2.4).

This principle simplifies considerably the solution of the forward problem of a conformal mapping. In order to map a given simply connected domain D by a given analytic function $w = f(z)$, it is sufficient to map the boundary γ of D onto the closed contour Γ. In this case the image of D will lie either inside Γ or outside Γ.

PRINCIPLE 2.2.2. *The symmetry principle.*

Suppose that an analytic function $w = f(z)$ maps a straight segment γ (or an arc γ of a circle) onto a straight segment Γ (or an arc Γ of a circle). Let z_1 and z_2 be two points of the z-plane that are symmetric with respect to γ (symmetry with respect to an arc of a circle will be defined in Subsection 2.3.2). Then z_1 and z_2 are mapped into points w_1 and w_2

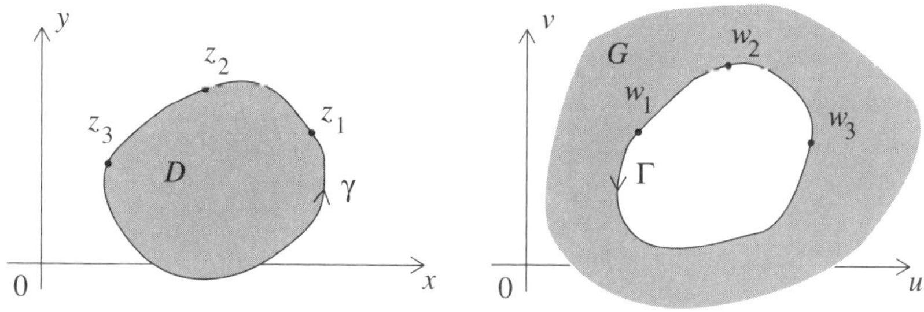

FIGURE 2.4. Mapping of domain D onto domain G if the closed contours γ and Γ have opposite orientations.

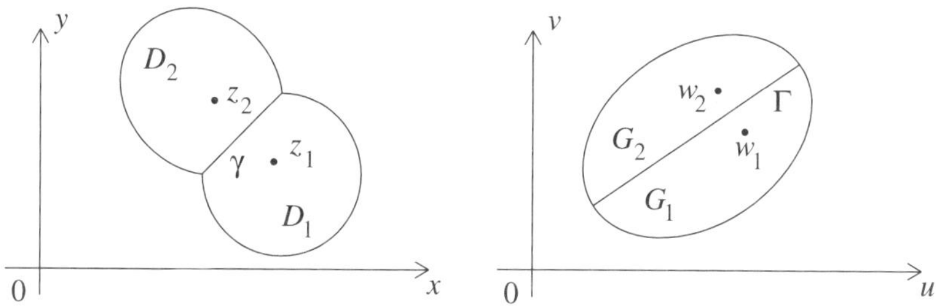

FIGURE 2.5. The symmetry principle.

which are symmetric with respect to Γ (see Fig 2.5); any two sets D_1 and D_2 that are symmetric with respect to γ are mapped onto sets G_1 and G_2 that are symmetric with respect to Γ.

Exercises for Sections 2.1 and 2.2

Represent the following curves in the z-plane in the form $z = z(t)$ and compute the corresponding tangent vectors.

1. $y = x^2, \qquad 1 \le x \le 3.$

2. $y = x^3, \qquad -2 \le x \le -1.$

3. $x^2 + y^2 = 4, \qquad 0 \le x \le 1,\ \sqrt{3} \le y \le 2.$

4. $\dfrac{x^2}{4} + \dfrac{y^2}{9} = 1, \qquad 0 \le x \le \sqrt{2},\ -3 \le y \le -\dfrac{3}{\sqrt{2}}.$

5. $y = 1/x^2, \qquad 1 \le x \le 4.$

6. $y = 4 - x^2, \qquad -1 \le x \le 0.$

Find the angle through which a curve drawn from the point z_0 is rotated under the mapping $w = f(z)$, and find the corresponding scale factor of the transformation.

7. $z_0 = -1, \quad w = z^2.$
8. $z_0 = 1 + 2i, \quad w = z^2 + 2z.$
9. $z_0 = -1 + i, \quad w = 1/z.$
10. $z_0 = -i, \quad w = z^3.$

Determine all the points in the z-plane for which the following mappings are not conformal.

11. $\sin z.$
12. $\sinh z.$
13. $z^2 + 4z + 3.$
14. $z^3 + 3z^2 - 9z.$
15. $\dfrac{z^2+4}{2z+3}, \quad z \neq -\dfrac{3}{2}.$
16. $\dfrac{2z^2+6}{z-2}, \quad z \neq 2.$
17. $e^{z^2+2z}.$
18. $e^{z^4-32z}.$
19. Consider the two curves

$$\gamma_1 : z = t + i, \quad 0 \leq t \leq 1; \qquad \gamma_2 : z = \tau + i\tau, \quad 0 \leq \tau \leq 1.$$

   (a) Find the point of intersection, P, of the curves and the angle, α, between the curves at P.
   (b) Find the image of each curve under the mapping $w = z^2$ and determine the angle between the images of the curves in the w-plane. Is this angle equal to α? Explain your answer.

20. Consider the two curves

$$\gamma_1 : z = t, \quad 0 \leq t \leq 1; \qquad \gamma_2 : z = \tau + i\tau, \quad 0 \leq \tau \leq 1.$$

   (a) Find the point of intersection, P, of the two curves and the angle, α, between the curves at P.
   (b) Find the image of each curve under the mapping $w = \bar{z}$ and determine the angle between the images of the curves in the w-plane. Is this angle equal to α? Explain your answer.

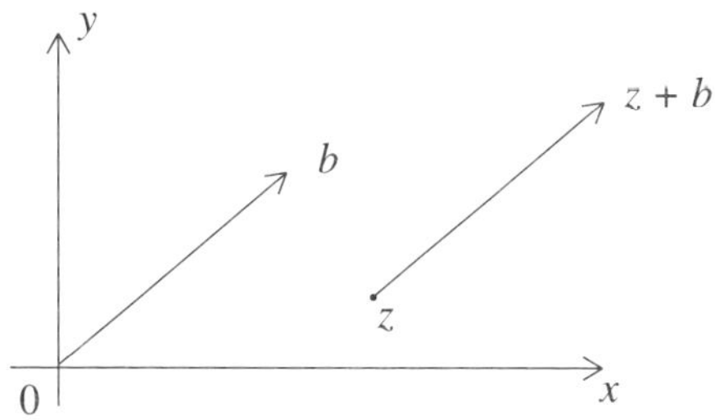

FIGURE 2.6. Transformation by a parallel translation.

2.3. Linear mapping and inversion

2.3.1. Mapping by a linear function $w = az + b$. Let $a \neq 0$ and b be two complex constants and consider the linear function

$$w = f(z) = az + b. \tag{2.3.1}$$

Since $w' = a \neq 0$ and $z_1 \neq z_2$ implies that $f(z_1) \neq f(z_2)$, then the region of univalence of f is the extended complex plane and the mapping is conformal in the extended plane.

We consider three particular cases where, for convenience, the z- and the w-planes are identified in Fig 2.6 and Fig 2.7.

Case 1. $a = 1$, $b = \beta_1 + i\beta_2$. By the geometric meaning of the sum of two complex numbers, the transformation

$$w = z + \beta_1 + i\beta_2 \tag{2.3.2}$$

is a *parallel translation*. It sends the point z to the point w along the vector b (see Fig 2.6).

Case 2. $a > 0$, $b = 0$. Writing

$$z = |z| \exp(i \operatorname{Arg} z) \quad \text{and} \quad w = a\,|z| \exp(i \operatorname{Arg} z),$$

we have

$$|w| = a|z|, \qquad \operatorname{Arg} w = \operatorname{Arg} z,$$

that is, the points w and z lie on the same ray emanating from the origin, but the length, $|w|$, of w is a times the length, $|z|$, of z (see Fig 2.7(a)). The mapping

$$w = az, \qquad a > 0, \tag{2.3.3}$$

is a *similarity transformation* with factor a. It maps a figure in the z-plane into a similar figure in the w-plane. In particular, the circle $|z - z_0| = \rho$

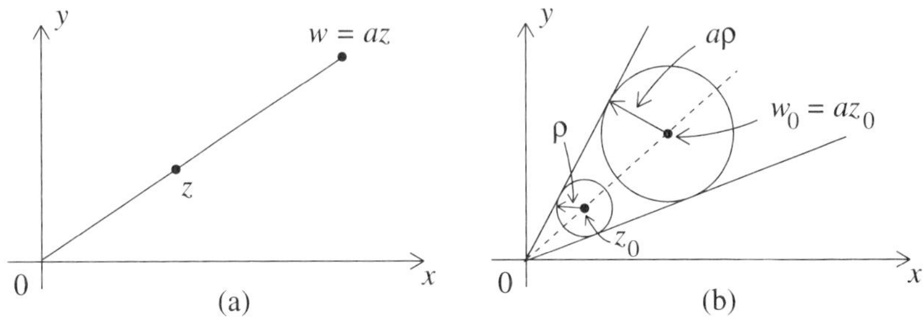

FIGURE 2.7. The similarity transformation $w = az$, $a > 0$: (a) similarity transformation with coefficient $a > 1$, (b) mapping of a circle by a similarity transformation.

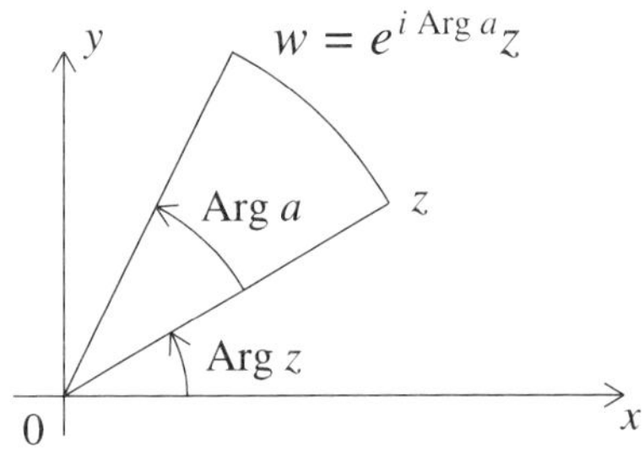

FIGURE 2.8. Rotation of z through Arg a.

centered at z_0 with radius ρ is transformed into the circle $|w - w_0| = a\rho$ centered at $w_0 = az_0$ with radius $a\rho$ (see Fig 2.7(b)).

Case 3. $|a| = 1$, $b = 0$. Writing $z = |z| \exp(i \operatorname{Arg} z)$, we have the transformation

$$w = |z|\, e^{i(\operatorname{Arg} z + \operatorname{Arg} a)}, \tag{2.3.4}$$

so that

$$|w| = |z|, \qquad \arg w = \operatorname{Arg} z + \operatorname{Arg} a. \tag{2.3.5}$$

The transformation

$$w = az, \qquad a = e^{i\alpha}, \tag{2.3.6}$$

is a *rotation*. It rotates every point z through the angle $\alpha = \operatorname{Arg} a$ around the origin (see Fig 2.8).

The general case $w = az + b$ can be obtained by successive applications of the transformations of the previous cases 1, 2 and 3:

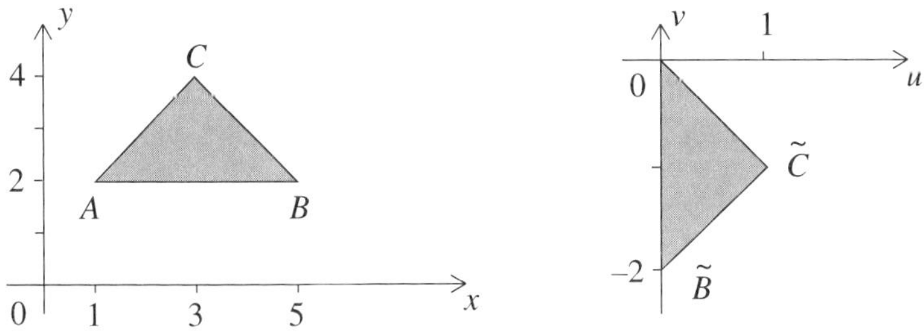

FIGURE 2.9. Mapping of a triangle into a similar triangle.

(a) a similarity: $w_1 = |a|z$,

(b) a rotation: $w_2 = w_1\, e^{i \operatorname{Arg} a} \quad (= |a|\, e^{i \operatorname{Arg} a} z = az)$,

(c) a translation: $w = w_2 + b \quad (= az + b)$.

EXAMPLE 2.3.1. *Find the function that maps the triangle ABC in the z-plane into a similar triangle $O\widetilde{B}\widetilde{C}$ in the w-plane (see Fig 2.9), if*

$$A = (1, 2),\ B = (5, 2),\ C = (3, 4),$$

and

$$O = (0, 0),\ \widetilde{B} = (0, -2),\ \widetilde{C} = (1, -1).$$

SOLUTION. Since the triangles ABC and $O\widetilde{B}\widetilde{C}$ are similar, then the mapping is given by a linear function. We perform the mapping in three stages:

(a) a parallel translation by the vector $-(1 + 2i)$, so that the vertex A is mapped into the origin of the w_1-plane (see Fig 2.10(a)),

(b) a rotation through the angle $-\pi/2$ (see the w_2-plane in Fig 2.10(b)),

(c) a contraction with coefficient $1/2$ (see the w-plane in Fig 2.10(c)).

Hence these three steps can be described as follows:

(1) $w_1 = z - (1 + 2i)$,

(2) $w_2 = e^{-i\pi/2} w_1 = -i[z - (1 + 2i)]$,

(3) $w = w_2/2 = -i[z - (1 + 2i)]/2$.

Thus the mapping is given by the linear function

$$w = -\frac{1}{2} i[z - (1 + 2i)]. \quad \square$$

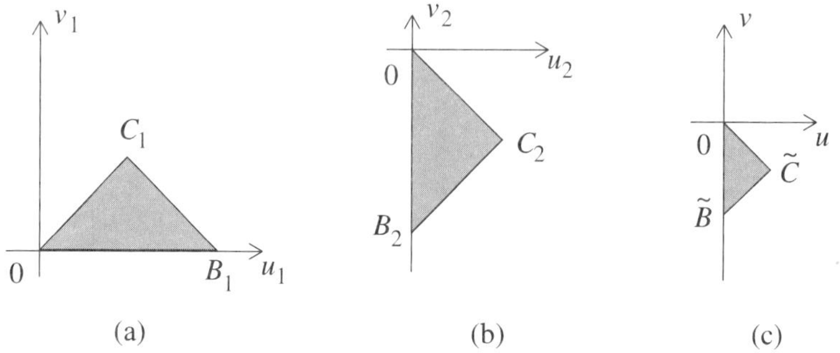

FIGURE 2.10. Mappings $z \to w_1 \to w_2 \to w$.

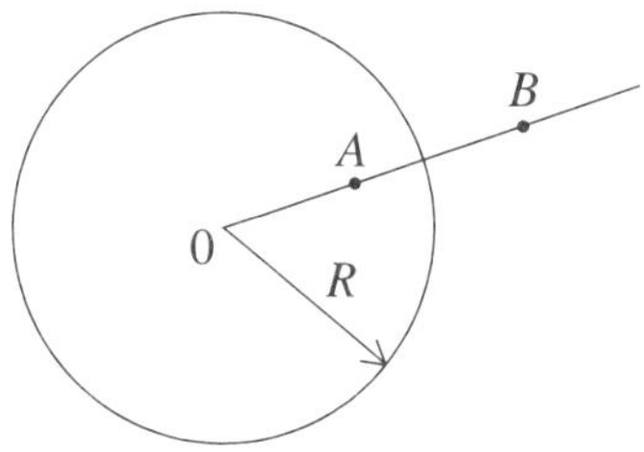

FIGURE 2.11. Symmetry with respect to a circle.

2.3.2. Mapping by the function $w = 1/z$. The transformation

$$w = \frac{1}{z} \tag{2.3.7}$$

is called an *inversion.*

One sees that the inversion $w = 1/z$ maps the circle $z = Re^{i\theta}$ into the circle $w = (1/R)e^{-i\theta}$.

We shall need the following definition of symmetric points with respect to a circle.

DEFINITION 2.3.1. Two points A and B are said to be *symmetric with respect to a circle* of center 0 and radius R (see Fig 2.11) if

(a) they lie on the same ray emanating from the origin, and
(b) the product of their distances from the center of the circle is equal to the square of the radius of the circle:

$$|\overrightarrow{OA}| \cdot |\overrightarrow{OB}| = R^2. \tag{2.3.8}$$

It follows from (2.3.8) that if the point A approaches the circle, that is, $|\overrightarrow{OA}| \to R$, then the point B also approaches the circle ($|\overrightarrow{OB}| \to R$), and if the point A approaches the center of the circle ($|\overrightarrow{OA}| \to 0$), then the

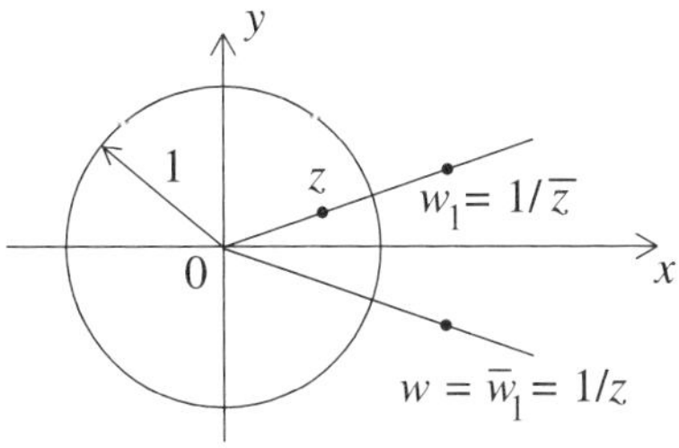

FIGURE 2.12. Reflection of the point z with respect to the unit circle $|z| = 1$.

point B moves away to infinity. This means that the points 0 and ∞ are symmetric with respect to the circle.

We prove that the inversion $w = 1/z$ is the successive application of two reflections:

(1) A reflection of the point z with respect to the circle $|z| = 1$ (see Fig 2.12),

$$w_1 = \frac{1}{\bar{z}}. \tag{2.3.9}$$

Indeed, if $z = |z| \exp(i \operatorname{Arg} z)$, then

$$\bar{z} = |z| \exp(-i \operatorname{Arg} z), \qquad \frac{1}{\bar{z}} = \frac{1}{|z|} e^{i \operatorname{Arg} z},$$

that is,

$$\operatorname{Arg} \frac{1}{\bar{z}} = \operatorname{Arg} z, \qquad |z| \frac{1}{|\bar{z}|} = 1;$$

thus, the points z and $1/\bar{z}$ are symmetric with respect to the circle $|z| = 1$.

(2) A reflection of the point $1/\bar{z}$ with respect to the x-axis:

$$w = \overline{\left(\frac{1}{\bar{z}}\right)} = \frac{1}{z}. \tag{2.3.10}$$

Here, we have used the relation

$$\overline{\left(\frac{z_1}{z_2}\right)} = \frac{\bar{z}_1}{\bar{z}_2},$$

whose proof is left to the reader.

The inversion (2.3.10) maps the interior of the upper half-disk,

$$|z| < 1, \qquad \Im z > 0,$$

of the z-plane into the exterior of the lower half-disk,

$$|w| > 1, \qquad \Im w < 0,$$

in the lower half of the w-plane, and conversely. Similarly, the interior of the lower half-disk is mapped into the exterior of the upper half-disk in the upper half of the w-plane, and conversely. Since

$$\left(\frac{1}{z}\right)' = -\frac{1}{z^2} \neq 0, \qquad \text{if } z \neq 0,$$

and, since, for any two distinct points $z_1 \neq z_2$, $1/z_1 \neq 1/z_2$, then the region of univalence is the whole complex plane, and the mapping is everywhere conformal, except at the point $z = 0$ which is mapped into the point $w = \infty$. If we assume that two curves, γ_1 and γ_2, intersecting at $z = 0$ at an angle α, are mapped into two curves that intersect in the same angle at $w = \infty$, then the mapping will be conformal also at $z = 0$, that is, in the extended complex plane.

Exercises for Section 2.3

Describe the geometrical meaning, in terms of translations, dilations and rotations, of the following mappings.

1. $w = z - i$.
2. $w = z + 8$.
3. $w = -iz$.
4. $w = 2z + 1$.
5. $w = e^{i\pi/3}z$.
6. $w = 3 + 4i + (1 + i)z$.

Find a linear transformation $w = az + b$ which has fixed point z_0 (that is, $w(z_0) = z_0$) and maps the point z_1 into the point w_1.

7. $z_0 = -1, \quad z_1 = 1 + i, \quad w_1 = 3 - i$.
8. $z_0 = i, \quad z_1 = 3 + 2i, \quad w_1 = 4 - 3i$.
9. $z_0 = 1 + i, \quad z_1 = 2 - i, \quad w_1 = 6 + i$.
10. $z_0 = 1 - 2i, \quad z_1 = 3 + 4i, \quad w_1 = -1 + i$.

Find the image of the following regions D under the given mapping $w = f(z)$.

11. $D = \{z;\ \Re z \geq 0\}$, $\quad w = iz$.
12. $D = \{(x, y) \in \mathbb{R}^2;\ -\infty < x < +\infty,\ 0 < y < 1\}$, $\quad w = iz + 2$.
13. $D = \{(x, y) \in \mathbb{R}^2;\ 0 < x < 1,\ -\infty < y < +\infty\}$, $\quad w = (1 + i)z$.
14. $D = \{z;\ \Im z \geq 0\}$, $\quad w = (1 - i)z + 1 + i$.
15. $D = \{(x, y) \in \mathbb{R}^2;\ 1 \leq x \leq 2,\ 0 \leq y \leq 1\}$, $\quad w = 3z + 1$.
16. $D = \{(x, y) \in \mathbb{R}^2;\ x^2 + y^2 = 1\}$, $\quad w = 2z + i$.
17. $D = \{z;\ |z| < 3,\ 0 \leq \operatorname{Arg} z \leq \pi/4\}$, $\quad w = \dfrac{1 - i}{\sqrt{2}} z$.
18. $D = \{z;\ |z| < 1,\ \Re z > 0\}$, $\quad w = 2iz + i$.

Find a linear transformation $w = az + b$ which maps the strip contained between the given straight lines, L_1 and L_2, onto the strip $0 < \Re w < 1$ with the given normalization.

19. $L_1 :\ x = 1$, $\quad L_2 :\ x = 2$, $\quad w(1) = 0$.
20. $L_1 :\ y = x$, $\quad L_2 :\ y = x - 2$, $\quad w(0) = 0$.

Find the images of the following curves under the inversion $w = 1/z$.

21. $|z + 1| = 1$.
22. $|z - 1| = 2$.
23. $x^2 + y^2 = 4x$.
24. $x^2 + y^2 = 6y$.
25. $y = x + 2$.
26. $y = 3x$.

Find the images of the following regions under the inversion $w = 1/z$.

27. $D = \{(x, y) \in \mathbb{R}^2;\ 0 < x < 2,\ -\infty < y < +\infty\}$.
28. $D = \{(x, y) \in \mathbb{R}^2;\ -\infty < x < +\infty,\ 0 < y < 1\}$.
29. $D = \{(x, y) \in \mathbb{R}^2;\ x > 0,\ y < 1\}$.
30. $D = \{(x, y) \in \mathbb{R}^2;\ x > 1,\ y < 0\}$.

31. Prove that the reflection $z \mapsto \bar{z}$ is not a linear transformation.

32. Prove that the most general linear transformation which leaves the origin fixed and preserves all distances is a rotation.

33. Show that any linear transformation which transforms the real axis into itself can be written with real coefficients.

2.4. Linear fractional transformations

2.4.1. Definition and properties.

DEFINITION 2.4.1. A *linear fractional transformation* is a transformation of the form

$$w = \frac{az+b}{cz+d}, \qquad ad \neq bc, \tag{2.4.1}$$

where a, b, c, d are complex constants.

We note that if $ad = bc$, then $w = \text{constant}$.

A linear fractional transformation (2.4.1) is also called a *bilinear* or *Möbius* transformation.

The derivative of (2.4.1),

$$w' = \frac{a(cz+d) - c(az+b)}{(cz+d)^2} = \frac{ad-bc}{(cz+d)^2} \neq 0,$$

exists everywhere, except at the point $z = -d/c$, which is mapped to the point $w = \infty$.

Linear fractional transformations are univalent and conformal in the extended complex plane. In fact, since the inverse of (2.4.1) is again a linear fractional transformation,

$$z = -\frac{dw-b}{cw-a}, \tag{2.4.2}$$

it has the same properties. Hence, (2.4.1) is a univalent mapping of the extended z-plane onto the extended w-plane.

It can be proved (see [**33**], p. 128) that the linear fractional transformations are the only analytic functions with this property (of course, parallel translations and the inversion $w = 1/z$ have the same properties since they are particular cases of linear fractional transformations).

We show that the function (2.4.1) can be obtained by the successive applications of two linear transformations and one inversion $w = 1/z$. To this end we rewrite (2.4.1) as follows:

$$w = \frac{(cz+d)a/c + b - ad/c}{cz+d} = \frac{a}{c} + \frac{b - ad/c}{cz+d}. \tag{2.4.3}$$

This function can be obtained by successive applications of the following three mappings:

(a) $w_1 = cz + d$,

(b) $w_2 = \dfrac{1}{w_1}$,

(c) $w = \dfrac{a}{c} + \left(b - \dfrac{ad}{c}\right) w_2$.

Hence if a property is satisfied by a linear function or the inversion $w = 1/z$, then it will also be true for the function (2.4.1).

The following theorem holds.

THEOREM 2.4.1. *Every circle is mapped into a circle by a linear fractional transformation, provided a line is considered as a circle with radius* $R = \infty$.

PROOF. We prove the theorem only for the inversion mapping

$$w = \frac{1}{z}, \tag{2.4.4}$$

since obviously any linear mapping preserves circles (see Section 2.3 and Fig 2.7(b)).

Let $z_0 = x_0 + iy_0$. The Cartesian equation of the circle $|z - z_0| = R$ in the z-plane is

$$A(x^2 + y^2) + Bx + Cy + D = 0, \tag{2.4.5}$$

where A, B, C and D are real constants chosen such that (2.4.5) is the equation of the circle $(x - x_0)^2 + (y - y_0)^2 = R^2$ in $\mathbb{R}^2$.

Let $w = u + iv$ and $z = x + iy$. Then the equation $w = 1/z$ can be written in the form

$$x + iy = \frac{1}{u + iv}.$$

Separating the real and imaginary parts, we obtain

$$x = \frac{u}{u^2 + v^2}, \qquad y = -\frac{v}{u^2 + v^2}. \tag{2.4.6}$$

Substituting (2.4.6) into (2.4.5), we get

$$A\frac{1}{u^2 + v^2} + B\frac{u}{u^2 + v^2} - C\frac{v}{u^2 + v^2} + D = 0,$$

or

$$D(u^2 + v^2) + Bu - Cv + A = 0. \tag{2.4.7}$$

This equation is the equation of a circle in the w-plane.

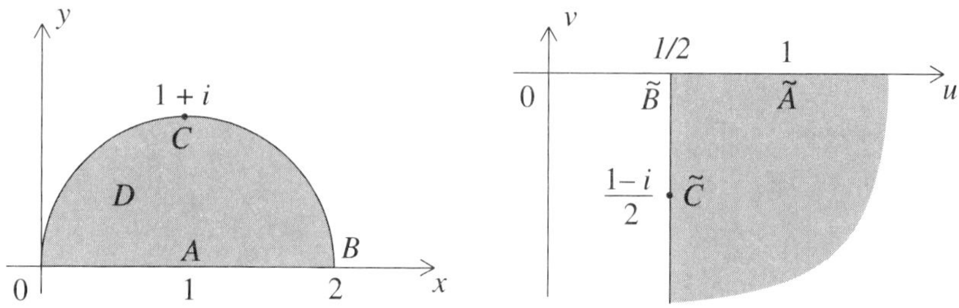

FIGURE 2.13. Mapping of the upper half-disk, D, by the inversion $w = 1/z$.

There are four cases:

(a) If $A \neq 0, D \neq 0$, circles are mapped into circles.
(b) If $A \neq 0, D = 0$, circles are mapped into straight lines.
(c) If $A = 0, D \neq 0$, straight lines are mapped into circles.
(d) If $A = 0, D = 0$, straight lines are mapped into straight lines.

This completes the proof. □

2.4.2. Examples. We first present two examples of solutions of the forward problem, that is, given a domain D in the z-plane and a function f analytic on D, find the image, $\widetilde{D}$, of D under the mapping $w = f(z)$ in the w-plane.

EXAMPLE 2.4.1. *Find the image of the region* $D = \{|z-1| \leq 1, \Im z \geq 0\}$ *by the inversion* $w = 1/z$.

SOLUTION. Since D is bounded by a semi-circle and a straight line, and since circles are mapped into circles by linear fractional transformations, then the boundary of the image will also be bounded by arcs of circles or by straight segments. Since the point $z = 0$ is mapped into the point $z = \infty$, then both arcs of circles pass through the point $w = \infty$, that is, the images of the lines OAB and OCB are straight lines (see Fig 2.13).

In order to obtain the direction of the lines it is sufficient to find the images of the points A, B and C. We have

$$A = (1, 0) \mapsto \widetilde{A} = (1, 0), \qquad B = (2, 0) \mapsto \widetilde{B} = \left(\frac{1}{2}, 0\right),$$

and

$$C = (1, 1) \mapsto \widetilde{C} = \left(\frac{1}{2}, -\frac{1}{2}\right),$$

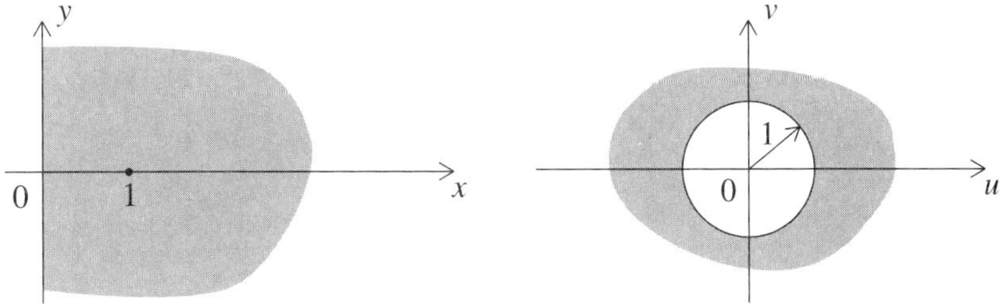

FIGURE 2.14. Mapping of the upper half-plane $\Re z \geq 0$ by the linear fractional transformation $w = (2 + z)/(2 - z)$.

where the the expression $z \mapsto w$ means the point z *is mapped to* the point w. Hence the image of the straight segment BAO is the semi-infinite ray defined by $\widetilde{B}\widetilde{A}$ in Fig 2.13 and the image of the semi-circle BCO is the semi-infinite ray defined by $\widetilde{B}\widetilde{C}$. If we traverse the contour ABC so that interior points of D lie on the left-hand side, then the image, $\widetilde{D}$, of D, as we traverse the contour $\widetilde{A}\widetilde{B}\widetilde{C}$, also lies on the left-hand side. Hence $\widetilde{D}$ lies inside the shaded right-angled wedge shown in Fig 2.13. □

EXAMPLE 2.4.2. *Find the image, $\widetilde{D}$, of the right half-plane,*

$$D = \{\Re z \geq 0\},$$

(see Fig 2.14*) by the linear fractional transformation*

$$w = \frac{2 + z}{2 - z}. \tag{2.4.8}$$

SOLUTION. We first find the image of the straight boundary $x = 0$. Since $z = iy$ is not a zero of the denominator of (2.4.8), we know, by Theorem 2.4.1, that the image of this boundary is a circle of finite radius.

Solving (2.4.8) for z, we have

$$z = 2\,\frac{w - 1}{w + 1}. \tag{2.4.9}$$

Substituting this expression for z in the equation $\Re z = 0$, we obtain

$$\Re\left(2\frac{w - 1}{w + 1}\right) = 0, \tag{2.4.10}$$

and letting $w = u + iv$ in (2.4.10), we have

$$\Re\left[\frac{(u - 1 + iv)(u + 1 - iv)}{(u + 1)^2 + v^2}\right] = 0,$$

that is,

$$u^2 + v^2 = 1, \tag{2.4.11}$$

which is the equation of a circle of radius 1 centered at $w = 0$.

The following question arises: Does the image $\widetilde{D}$ of the right half-plane in Fig 2.14 lie inside or outside the given circle? To answer this question it suffices to know where the image of any point in D lies. For instance, we may consider the point $z = 1$. In this case, we have

$$z = 1 \implies \frac{2+1}{2-1} = 3 = w.$$

Since the point $w = 3$ lies outside the circle $|w| \leq 1$, then $\widetilde{D}$ lies outside this circle. □

Note that by the method used in the previous solution one can find the image, expressed in Cartesian coordinates, of any line or circle under a linear fractional transformation.

EXAMPLE 2.4.3. *Find the image of the circle* $|z - 1| = 1$ *under the mapping* (2.4.8).

SOLUTION. Since (2.4.8) sends $z = 0$ to $w = 1$ and $z = 2$ to $w = \infty$, we know by Theorem 2.4.1 that the circle goes into a straight line. To find the direction of this line, we verify that the point $z = 1 + i$ on the circle goes to the point

$$w = \frac{2+1+i}{2-1-i} = \frac{3+i}{1-i}\,\frac{1+i}{1+i} = 1 + 2i.$$

Hence the image of the circle is the vertical straight line $\Re w = 1$.

We obtain the same result by applying the method of the previous example. It follows from (2.4.9) that the image of the circle is the curve

$$\left| 2\,\frac{w-1}{w+1} - 1 \right| = 1. \tag{2.4.12}$$

Separating the real and imaginary parts of the expression

$$\begin{aligned} 2\,\frac{w-1}{w+1} - 1 &= \frac{2(w-1) - w - 1}{w+1} \\ &= \frac{w-3}{w+1} \\ &= \frac{(u-3+iv)(u+1-iv)}{(u+1)^2+v^2} \\ &= \frac{(u-3)(u+1)+v^2}{(u+1)^2+v^2} + i\frac{(u+1)v-(u-3)v}{(u+1)^2+v^2} \end{aligned}$$

$$= \frac{(u-3)(u+1)+v^2}{(u+1)^2+v^2} + i\frac{4v}{(u+1)^2+v^2}, \tag{2.4.13}$$

inserting these into (2.4.12) and chasing the denominators, we obtain

$$\left[(u-3)(u+1)+v^2\right]^2 + 16v^2 = \left[(u+1)^2+v^2\right]^2. \tag{2.4.14}$$

Expanding both sides of (2.4.14), we get

$$(u+1-4)^2(u+1)^2+2v^2(u-3)(u+1)+v^4+16v^2 = (u+1)^4+2v^2(u+1)^2+v^4,$$

which, upon simplification, becomes

$$[(u+1)^2+v^2](u-1)=0,$$

so that, finally,

$$u = 1. \tag{2.4.15}$$

Hence the image of the circle $|z-1|=1$ is the straight line $\Re w = 1$. □

Images of straight lines and circles by elementary linear fractional transformations (2.4.1) can be found, for instance, in [**15**], pp. 345–352, [**33**] and [**44**], pp. 132–133. We present these formulae for reference purposes.

(a) A straight line

$$\Re(\lambda z) = \alpha \tag{2.4.16}$$

that does not pass through the point $z = -d/c$ (that is, if $\Re(\lambda d/c) \neq -\alpha$) is mapped into the circle $|w - w_0| = R$, where

$$w_0 = \frac{2a\alpha\bar{c} + ad\bar{\lambda} + b\lambda\bar{c}}{2\alpha|c|^2 + 2\Re\left(cd\bar{\lambda}\right)}, \qquad R = \left|\frac{a}{c} - w_0\right|. \tag{2.4.17}$$

In order to use formula (2.4.17) one has to determine the parameters λ and α by means of the Cartesian equation of a line,

$$Ax + By + C = 0. \tag{2.4.18}$$

Letting $\lambda = \lambda_1 + i\lambda_2$ and $z = x + iy$ in (2.4.16), we have

$$\Re[(\lambda_1 + i\lambda_2)(x+iy)] = \alpha,$$

and simplifying the last relation, we get

$$\lambda_1 x - \lambda_2 y = \alpha. \tag{2.4.19}$$

It follows from (2.4.18) and (2.4.19) that

$$\lambda_1 = A, \qquad \lambda_2 = -B, \qquad \alpha = -C,$$

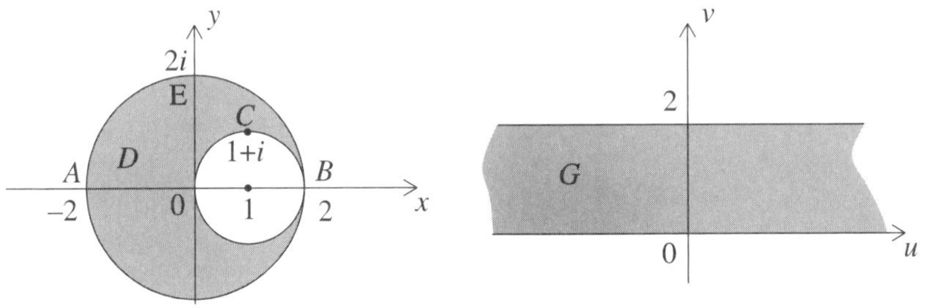

FIGURE 2.15. Shaded regions D and G of Example 2.4.4.

that is,

$$\lambda = A - Bi, \qquad \alpha = -C.$$

(b) The straight line

$$\Re(\lambda z) = -\Re\left(\frac{\lambda d}{c}\right)$$

passing through the point $z = -d/c$ is mapped into the straight line

$$\Re\left(\frac{ad - bc}{c^2}\,\lambda\bar{w}\right) = \Re\left(\frac{ad - bc}{c^2}\,\lambda\,\frac{\bar{a}}{\bar{c}}\right). \tag{2.4.20}$$

(c) The circle $|z - z_0| = r$ that does not pass through the point $z = -d/c$, for $r \neq |z_0 + d/c|$, is mapped into the circle $|w - w_0| = R$, where

$$w_0 = \frac{(az_0 + b)(\bar{c}\bar{z}_0 + \bar{d}) - \bar{c}r^2}{|cz_0 + d|^2 - |c|^2r^2}, \qquad R = \frac{r|ad - bc|}{\big||cz_0 + d|^2 - |c|^2r^2\big|}. \tag{2.4.21}$$

(d) A circle $|z - z_0| = |z_0 + d/c|$ is mapped into the straight line

$$\Re\left(\frac{ad - bc}{c(cz_0 + d)}\bar{w}\right) = \frac{|ad - bc|^2 + 2\Re[c(az_0 + b)(\bar{a}\bar{d} - \bar{b}\bar{c})]}{2|c(cz_0 + d)|^2}. \tag{2.4.22}$$

We present an example of a solution to an inverse problem, that is, given domains D and $\widetilde{D}$ in the z- and w-planes, respectively, find the mapping $w = f(z)$ which sends D univalently onto $\widetilde{D}$.

EXAMPLE 2.4.4. *Given the following region D between two tangent circles and the strip G (see Fig 2.15):*

$$D = \{|z| \leq 2, |z - 1| \geq 1\}, \qquad G = \{0 \leq \Im z \leq 2\},$$

find the linear fractional transformation which maps D onto G.

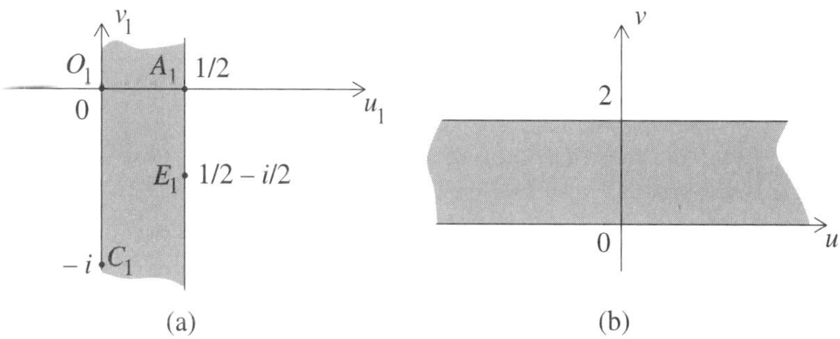

FIGURE 2.16. Mappings $z \to w_1 \to w$ of Example 2.4.4.

SOLUTION. First of all, it is necessary that the point, $B = (2, 0)$, common to both circles be mapped to the point $w = \infty$, for, it is only in this case that the images of both circles will be straight lines. Thus the general form of the mapping is

$$w_1 = \frac{az + b}{z - 2},$$

where a and b are arbitrary constants. For example, we can choose $a = 1$ and $b = 0$ so that

$$w_1 = \frac{z}{z - 2}. \tag{2.4.23}$$

As it is already clear that both circles will map into straight lines (moreover, the images of the circles will be parallel lines because they intersect at the point $w = \infty$) then it is sufficient to find the images of two points on each circle.

We choose the two points, $A = (-2, 0)$ and $E = (0, 2)$, on the large circle. Their images by transformation (2.4.23) are

$$A = (-2, 0) \mapsto A_1 = (1/2, 0), \qquad E = (0, 2) \mapsto E_1 = (1/2, -1/2).$$

Similarly, the images of the two points, $O = (0, 0)$ and $C = (1, 1)$, on the small circle are

$$O = (0, 0) \mapsto O_1 = (0, 0), \qquad C = (1, 1) \mapsto C_1 = (0, -1).$$

Since $z = -1 \mapsto w_1 = 1/3$, the region D is mapped in the w_1-plane onto the strip that is bounded by the straight lines passing through the points A_1, E_1 and O_1, C_1 (see Fig 2.16(a)).

In order to map the strip in Fig 2.16(a) onto the strip in Fig 2.16(b), it is sufficient to perform a rotation through an angle $\pi/2$ (that is, to multiply

w_1 by $\exp(i\pi/2) = i$) and a similarity with dilation factor 4:

$$w = 4e^{i\pi/2} w_1 = 4i\,\frac{z}{z-2}. \quad \square \tag{2.4.24}$$

Note 2.4.1. The mapping (2.4.24) is not unique; one can add an arbitrary real constant c to the right-hand side of (2.4.24). In this case the strip will be shifted parallel to itself in the w-plane along the vector $(c, 0)$.

Exercises for Section 2.4

Find the image of the following domain D under the mapping $w = \dfrac{az+b}{cz+d}$.

1. $D = \{z;\ |z| < 1\}, \qquad w = \dfrac{z+i}{z-i}$.
2. $D = \{z;\ |z-1| > 1\}, \qquad w = \dfrac{z}{z-2}$.
3. $D = \{(x,y) \in \mathbb{R}^2;\ x > 0,\ y < 0\}, \qquad w = \dfrac{z+1}{z-1}$.
4. $D = \{z;\ |z| < 1,\ 0 < \operatorname{Arg} z < \pi/4\}, \qquad w = \dfrac{2z+1}{z+i}$.
5. $D = \{z;\ 1 < |z| < 2\}, \qquad w = \dfrac{z}{z+1}$.
6. $D = \{z;\ 0 < \Im z < 1\}, \qquad w = \dfrac{z}{z-i}$.

Find the linear fractional transformation which transforms the points z_1, z_2, z_3 into the points w_1, w_2, w_3, respectively.

7. $z_1 = 1,\ z_2 = 0,\ z_3 = i, \qquad w_1 = -1,\ w_2 = \infty,\ w_3 = 1$.
8. $z_1 = i,\ z_2 = 1-i,\ z_3 = 1, \qquad w_1 = 0,\ w_2 = -1,\ w_3 = \infty$.
9. $z_1 = 1+i,\ z_2 = 1-i,\ z_3 = -1, \qquad w_1 = 0,\ w_2 = 1,\ w_3 = i$.
10. $z_1 = -i,\ z_2 = i,\ z_3 = 0, \qquad w_1 = 1,\ w_2 = i,\ w_3 = 1-i$.

A point z_0 is called a *fixed point* of the transformation $w = f(z)$ if $f(z_0) = z_0$. Find the fixed points of the following transformations.

11. $w = \dfrac{z}{z+2}$.
12. $w = \dfrac{z-i}{z+i}$.
13. $w = 1/z$.

14. $w = \dfrac{2z+1}{z-2}$.

15. $w = az + b, \qquad a \neq 0.$

16. $w = \dfrac{az+b}{cz+d}, \qquad ad - bc \neq 0.$

Find the linear fractional transformation which maps the region D of the z-plane onto the region G of the w-plane with the given normalization.

17. $D = \{z;\ \Im z \geq 0\}, \qquad G = \{w;\ |w| \leq 1\},$
with $w(0) = 1, \qquad w(1) = i, \qquad w(-1) = -i.$

18. $D = \{z;\ |z| \leq 1\}, \qquad G = \{w;\ |w-1| \leq 1\},$
with $w(1) = 0, \qquad w(i) = 2, \qquad w(-i) = 1 + i.$

Find a linear fractional transformation which maps the region D of the z-plane onto the region G of the w-plane.

19. $D = \{z;\ |z| < 2\}, \qquad G = \{w;\ \Im w > 0\}.$

20. $D = \{z;\ |z-1| < 1\}, \qquad G = \{w;\ \Re w > 0\}.$

21. $D = \{z;\ |z+1| < 2\}, \qquad G = \{w;\ \Re w < 0\}.$

22. $D = \{z;\ |z-i| < 1\}, \qquad G = \{w;\ \Im w > 0\}.$

23. Find the general form of a linear fractional transformation which maps the upper half-plane onto itself.

24. Find the general form of a linear fractional transformation which maps the upper half-plane onto the lower half-plane.

2.5. Symmetry and linear fractional transformations

2.5.1. Mapping of the upper half-plane onto the unit disk. We want to map the upper half-plane, $\Im z \geq 0$, onto the unit disk, $|w| \leq 1$, so that a given point, z_0, in the half-plane is mapped to the center of the disk (see Fig 2.17).

Since the point $z = z_0$ is mapped to the point $w = 0$, then by the symmetry principle, the point $\bar{z}_0$, symmetric to the point z_0 with respect to the x-axis, has to be mapped to the point $w = \infty$, symmetric to the point $w = 0$ with respect to any circle centered at $w = 0$. Hence the desired mapping has the form

$$w = k\frac{z - z_0}{z - \bar{z}_0}, \qquad k = \text{constant}. \tag{2.5.1}$$

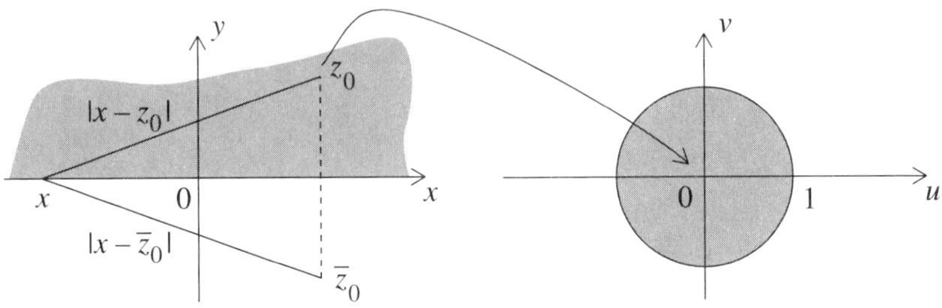

FIGURE 2.17. Mapping of the upper half-plane onto the unit disk, such that z_0 maps to 0.

Let us choose k so that the disk is the unit disk, that is, we assume that $|w| = 1$ if $z = x$ is real:

$$|w|\Big|_{z=x} = |k|\frac{|x - z_0|}{|x - \bar{z}_0|} = 1. \tag{2.5.2}$$

It follows from Fig 2.17 that $|x - z_0| = |x - \bar{z}_0|$, so that from (2.5.2) we obtain

$$|k| = 1, \quad \text{or} \quad k = e^{i\alpha},$$

where α is any real constant.

Hence our problem is solved by the linear fractional transformation

$$w = e^{i\alpha}\,\frac{z - z_0}{z - \bar{z}_0} \tag{2.5.3}$$

(the factor $e^{i\alpha}$ rotates the disk $|w| \le 1$ through the angle α about the point $w = 0$).

2.5.2. Mapping of the unit disk onto the unit disk. We want to map the disk $|z| \le 1$ onto the disk $|w| \le 1$ so that a given point, z_0, of the first disk is mapped to the center of the second disk (see Fig 2.18).

We use the fact that z_0 and $1/\bar{z}_0$ are symmetric with respect to the circle $|z| = 1$ (see Definition 2.3.1). As z_0 is mapped to $w = 0$, then $1/\bar{z}_0$ has to map to $w = \infty$, which is symmetric to $w = 0$ with respect to any circle centered at $w = 0$. Hence the desired mapping has the form

$$w = k\frac{z - z_0}{z - \frac{1}{\bar{z}_0}} = \tilde{k}\,\frac{z - z_0}{1 - z\bar{z}_0}, \qquad \tilde{k} = -\bar{z}_0 k, \tag{2.5.4}$$

where $\tilde{k}$ is an arbitrary constant to be chosen so that the disk in the w-plane is the unit disk.

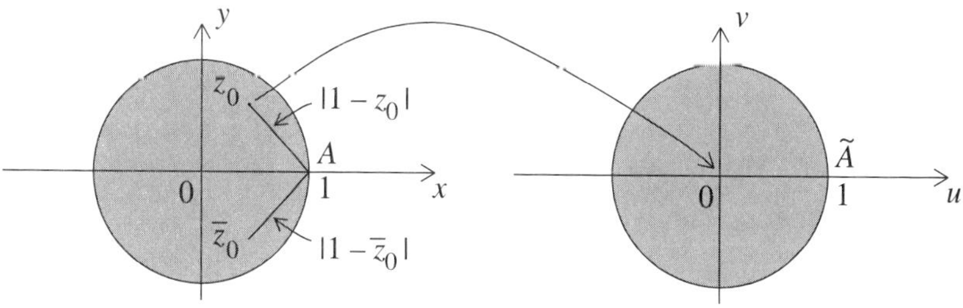

FIGURE 2.18. Mapping of the unit disk onto the unit disk, such that z_0 maps to 0.

For this purpose we require that $z = 1 \mapsto 1 = w$ so that

$$|\tilde{k}|\frac{|1-z_0|}{|1-\bar{z}_0|} = 1.$$

Since $|1-z_0| = |1-\bar{z}_0|$ (see Fig 2.18), then the last relation implies that

$$|\tilde{k}| = 1, \qquad \tilde{k} = e^{i\alpha}.$$

Hence the desired mapping has the form

$$w = e^{i\alpha}\frac{z-z_0}{1-z\bar{z}_0}. \tag{2.5.5}$$

2.5.3. Mapping of an eccentric annulus onto a concentric one. We discuss this mapping by means of a concrete example. The general case is considered in [**33**], pp. 148–149, where one circle lies inside or outside the other circle.

EXAMPLE 2.5.1. *Map the eccentric annulus*

$$|z-3| > 9, \qquad |z-8| < 16$$

onto the concentric annulus

$$1 < |w| < 3/2.$$

We shall see later that the external radius 3/2 is unique in this example if the internal radius is equal to 1.

SOLUTION. Let us find two points $P_1 = (x_1, 0)$ and $P_2 = (x_2, 0)$ on the real axis that are symmetric with respect to both circles simultaneously. In that case the coordinates x_1 and x_2 must satisfy the system of equations

$$\begin{aligned}(x_2-3)(x_1-3) &= 9^2,\\ (x_2-8)(x_1-8) &= 16^2.\end{aligned} \tag{2.5.6}$$

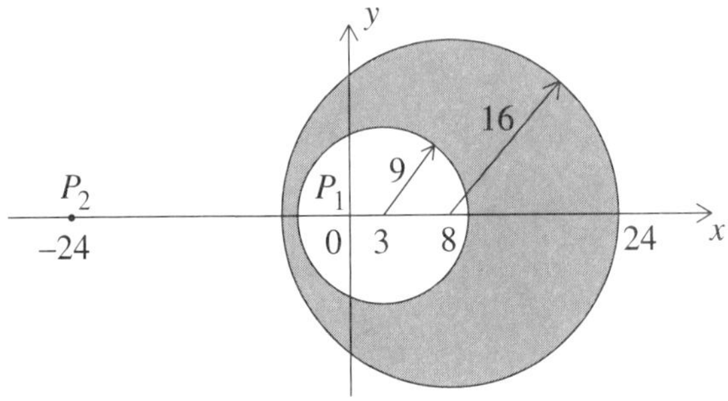

FIGURE 2.19. The eccentric annulus of Example 2.5.1.

Here $|x_1 - 3|$ and $|x_2 - 3|$ are the distances from the points P_1 and P_2 to the center of the inner circle while $|x_1 - 8|$ and $|x_2 - 8|$ are the distances from the same points to the center of the outer circle, respectively.

It follows from (2.5.6) that

$$\begin{aligned} x_1 x_2 - 3(x_1 + x_2) + 9 &= 81, \\ x_1 x_2 - 8(x_1 + x_2) + 64 &= 256. \end{aligned} \tag{2.5.7}$$

Using the substitutions $x_1 x_2 = \xi$ and $x_1 + x_2 = \eta$, from (2.5.7) we get

$$\begin{aligned} \xi - 3\eta &= 72, \\ \xi - 8\eta &= 192. \end{aligned}$$

The solution to this system is $\xi = 0, \eta = -24$, so that

$$\begin{aligned} x_1 x_2 &= 0, \\ x_1 + x_2 &= -24. \end{aligned}$$

Hence, $x_1 = 0$ and $x_2 = -24$ (or $x_2 = 0$ and $x_1 = -24$). If we map the point $x_1 = 0$ to the point $w = 0$ and the point $x_2 = -24$ to the point $w = \infty$, then by the symmetry of the points P_1 and P_2 with respect to both circles (see Fig 2.19) each of these circles is mapped into a circle centered at $w = 0$ in the w-plane. Hence we let

$$w = k \frac{z - 0}{z + 24}. \tag{2.5.8}$$

The parameter k in (2.5.8) is determined (up to an exponential factor $e^{i\alpha}$) if, for example, one requires that the inner circle in Fig 2.19 be mapped into the circle $|w| = 1$:

$$|w|\big|_{z=12} = 1.$$

In this case, $|k| \times 12/36 = 1$, $k = 3\,e^{i\alpha}$, so that (2.5.8) can be written in the form

$$w = e^{i\alpha}\frac{3z}{z+24}. \tag{2.5.9}$$

Therefore the function (2.5.9) maps the eccentric annulus in Fig 2.19 onto the concentric annulus $1 \le |w| \le 3/2$. □

Exercises for Section 2.5

Find the point symmetric to the point z_0 with respect to the given curve.

1. $z_0 = 2 - i$, $\Re z = 0$.
2. $z_0 = 4 + 3i$, $\Im z = 0$.
3. $z_0 = 1 + i$, $|z + i| = 1$.
4. $z_0 = 2 - 2i$, $|z - 1| = 2$.

Find the linear fractional transformation which maps the region D of the z-plane onto the region G of the w-plane with the given normalization.

5. $D = \{z;\ |z| \le 1\}$, $G = \{w;\ \Im w \ge 0\}$, with $w(0) = i$, $\arg w'(0) = \pi/4$.
6. $D = \{z;\ \Re z > 0\}$, $G = \{w;\ \Im w > 0\}$, with $w(1) = i$, $\arg w'(1) = \pi/3$.

Map the upper half-plane, $\Im z > 0$, onto the unit disk, $|w| < 1$, in such a way that:

7. $w(-1 + i) = 0$, $\operatorname{Arg} w'(-1 + i) = \pi/4$.
8. $w(i) = 0$, $\operatorname{Arg} w'(i) = -\pi/2$.
9. $w(3i) = 0$, $\operatorname{Arg} w'(3i) = 0$.
10. $w(1 + i) = 0$, $\operatorname{Arg} w'(1 + i) = \pi/3$.

11. Map the eccentric ring bounded by the circles $|z| = 4$, $|z + 1| = 1$ onto the ring $1 < |w| < R$. Find R.

12. Map the eccentric ring bounded by the circles $|z| = 1$, $|z - 1| = 5/2$ onto the ring $1 < |w| < R$. Find R.

2.6. Mapping by z^n and $w = z^{1/n}$

Since the linear fractional transformations (2.4.1) are the only analytic functions which are univalent from the extended complex z-plane onto the

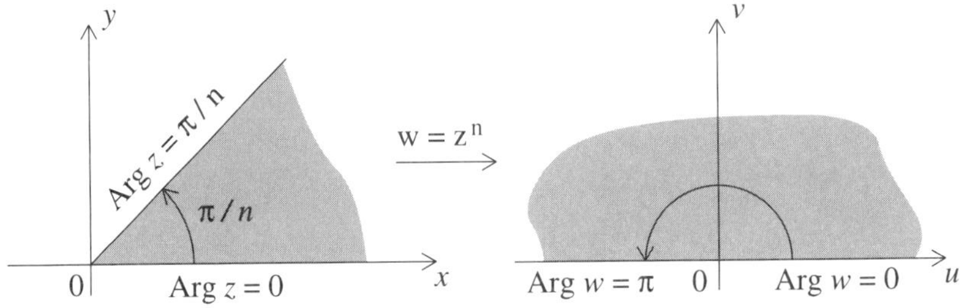

FIGURE 2.20. Mapping of the wedge of angle π/n, $0 \le \operatorname{Arg} z \le \pi/n$, by the function $w = z^n$.

extended complex w-plane, then the other functions considered in the sequel do not possess this property. These functions are univalent maps either of a finite part of the z-plane to the whole complex w-plane with a cut or, conversely, of the whole complex z-plane with a cut on a finite part of the w-plane.

In this section, we consider the mappings by the functions z^n and $z^{1/n}$, with particular attention to the univalence of the domains (because any domain of the z-plane which extends beyond the region of univalence cannot be mapped conformally).

2.6.1. The power function. Consider the *power function*

$$w = z^n, \qquad n \in \mathbb{N}. \tag{2.6.1}$$

Since $z = |z|\, e^{i \arg z}$, then $w = |z|^n\, e^{in \arg z}$. Hence

$$|w| = |z|^n, \qquad \arg w = n \arg z. \tag{2.6.2}$$

It follows from (2.6.2) that each ray $\theta_0 = \arg z$ is mapped into the ray $\arg w = n\theta_0$. The ray $\theta = 0$ is mapped into the ray $\operatorname{Arg} w = 0$, but the ray $\operatorname{Arg} z = \pi/n$ is mapped into the ray $\operatorname{Arg} w = \pi$, that is, the wedge $0 \le \operatorname{Arg} z \le \pi/n$ is mapped onto the upper half-plane (see Fig 2.20).

Similarly, the function

$$w = z^\alpha, \tag{2.6.3}$$

to be understood as

$$e^{\alpha \log z},$$

where α is any real number, maps the wedge $0 \le \arg z \le \pi/\alpha$ onto the upper half-plane.

In general, the wedge $0 \le \operatorname{Arg} z \le \theta_0$ (if $n\theta_0 < 2\pi$) is mapped into the wedge $0 \le \operatorname{Arg} w \le n\theta_0 < 2\pi$.

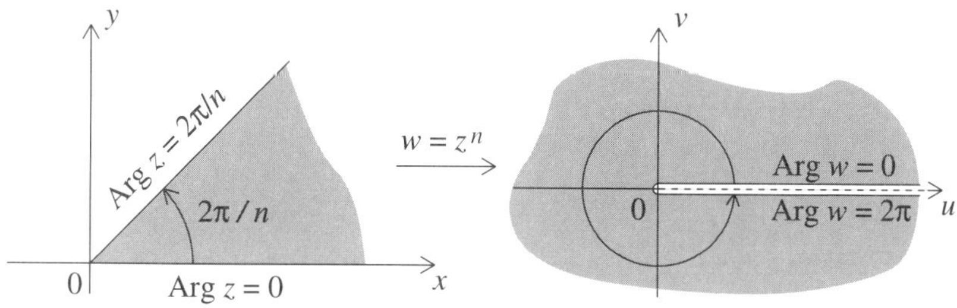

FIGURE 2.21. Mapping of the wedge of angle $2\pi/n$, $0 \le \operatorname{Arg} z \le 2\pi/n$, by the function $w = z^n$.

The wedge $0 \le \theta \le \operatorname{Arg} z \le 2\pi/n$ is mapped onto the whole w-plane less the real positive axis. The rays $\theta = 0$ and $\theta = 2\pi/n$ are mapped onto this axis. To have a univalent mapping, it is necessary to cut the plane along, say, the real axis $\Re w = 0$ (that is, to consider this line as a double line) and assume that the ray $\theta = 0$ is mapped onto the upper part of the cut and the ray $\theta = 2\pi/n$ is mapped onto the lower part of the cut (see Fig 2.21).

Hence the largest angle, in absolute value, measured from the positive x-axis and such that $w_1 \neq w_2$ if $z_1 \neq z_2$ is the angle $0 \le \operatorname{Arg} z \le 2\pi/n$. This angle defines the region of univalence of the function $w = z^n$.

We can ask the following question "What is the image of other parts of the z-plane?" Let us consider, for example, the wedge $2\pi/n \le \arg z \le 4\pi/n$. As the ray $\theta = 2\pi/n$ is mapped onto the ray $\arg w = 2\pi$ and the ray $\theta = 4\pi/n$ is mapped onto the ray $\arg w = 4\pi$, the wedge $2\pi/n \le \arg z \le 4\pi/n$ is also mapped onto the whole complex w-plane with a cut, but, in this case, the ray $\theta = 2\pi/n$ is mapped onto the upper part of the cut while the ray $\theta = 4\pi/n$ is mapped onto the lower part of the cut. In general, for the function $w = z^n$, the whole z-plane is divided into n regions of univalence of the form

$$\frac{2k\pi}{n} \le \arg z \le \frac{2(k+1)\pi}{n}, \qquad k = 0, 1, 2, \dots, n-1.$$

We see that each of these regions is mapped univalently onto the whole plane, except for one cut. Such regions have a special name (see [**2**], pp. 98–99).

DEFINITION 2.6.1. A region which is mapped univalently onto the whole plane, except for one or more cuts, by a function $f(z)$ is called a *fundamental region* of f.

2.6.2. The nth root of z. Since the power function $z = w^n$ has n fundamental regions in the w-plane, its inverse $w = z^{1/n}$ has n branches,

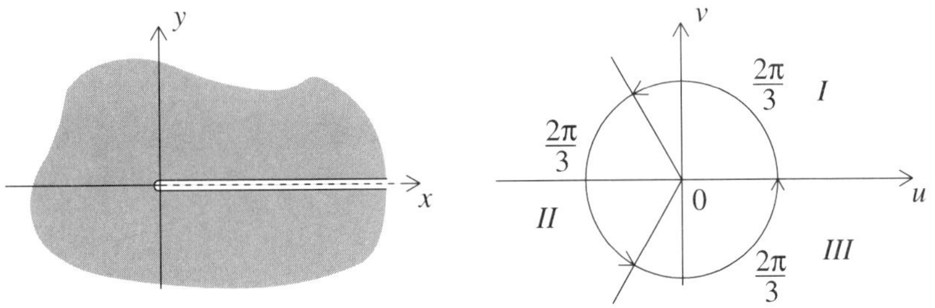

FIGURE 2.22. Mapping of the complex z-plane with a cut along the positive real axis by the branches w_0, w_1 and w_2 of $w = z^{1/3}$ to the regions I, II and III, respectively.

each of which is a function. More precisely, each branch of $z^{1/n}$,

$$w_k = z^{1/n} = |z|^{1/n}\, e^{i(\operatorname{Arg} z + 2k\pi)/n}, \qquad 0 \le \operatorname{Arg} z \le 2\pi, \quad k = 0, 1, \dots, n-1, \tag{2.6.4}$$

maps the wedge $0 \le \operatorname{Arg} z \le \theta_0$ $(\theta_0 \le 2\pi)$ onto the wedge $0 \le \operatorname{Arg} w \le \theta_0/n$ which is n times smaller, and the whole complex z-plane with a cut along the positive real axis (this is the region of univalence of the branch w_0) is mapped onto the wedge $0 \le \operatorname{Arg} w \le 2\pi/n$.

EXAMPLE 2.6.1. *Consider in detail the three branches of* $z^{1/3}$, *namely,*

$$w_k = z^{1/3} = |z|^{1/3}\, e^{i(\operatorname{Arg} z + 2k\pi)/3}, \qquad k = 0, 1, 2. \tag{2.6.5}$$

SOLUTION. The principal branch

$$w_0 = |z|^{1/3}\, e^{i(\operatorname{Arg} z)/3} \tag{2.6.6}$$

maps the whole complex plane with a cut along the positive x-axis, that is, the domain D, onto region I shown in Fig 2.22.

Since the branch w_1 is related to w_0 by (2.6.5), that is,

$$w_1 = w_0\, e^{2\pi i/3}, \tag{2.6.7}$$

it maps D onto region II obtained by rotating region I through an angle $2\pi/3$. Similarly, the branch

$$w_2 = w_1 e^{2\pi i/3} \tag{2.6.8}$$

maps D onto region III obtained by rotating region II through an angle $2\pi/3$.

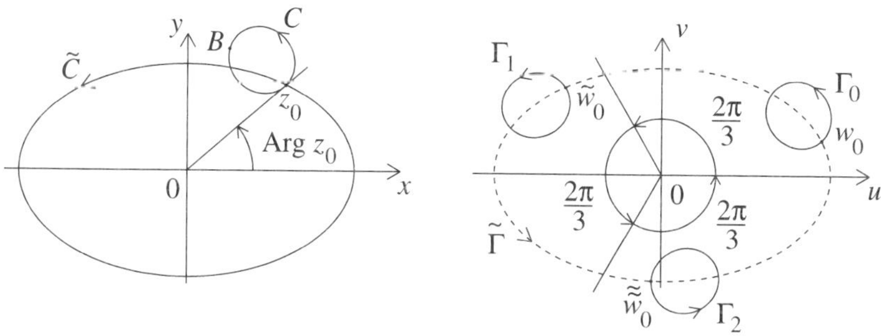

FIGURE 2.23. Mapping of the closed contours C and $\widetilde{C}$ by the different branches of $w = z^{1/3}$.

Let us study the image of a closed contour C in the z-plane by each branch w_k of $w = z^{1/3}$. There are two cases:

(a) The point $z = 0$ is not in the finite region enclosed by the contour C (see Fig 2.23). If we start from the point z_0 and go along the contour C, $\arg z$ increases until we reach the point B, then decreases to its initial value $\arg z_0$ as we return to the point z_0. Therefore the branch w_0 (see (2.6.6)) maps the closed contour C into a closed contour Γ_0 lying in region I in the w-plane and the point w_0 corresponds to the point z_0. Since the branch w_1 is related to the branch w_0 by relation (2.6.7), the contour C is mapped by w_1 onto the contour Γ_1 obtained by rotating Γ_0 through an angle $2\pi/3$. Similarly, the branch w_2 maps the contour C onto the contour Γ_2 obtained by rotating Γ_1 through an angle $2\pi/3$.

(b) The point $z = 0$ lies in the finite region enclosed by the contour $\widetilde{C}$ (see Fig 2.23). If we go along the closed contour $\widetilde{C}$, then $\arg z_0$ increases to 2π. Therefore the initial value of the branch w_0 at z_0 differs from the value of the branch w_0 at z_0 after $\widetilde{C}$ is traversed once, denoted by $z_0 + 1\widetilde{C}$:

$$\begin{aligned} w\big|_{z_0+1\widetilde{C}} &= |z_0|^{1/3} e^{i(\operatorname{Arg} z_0 + 2\pi)/3} \\ &= w\big|_{z_0+0\widetilde{C}}\, e^{i2\pi/3} \\ &\neq w\big|_{z_0+0\widetilde{C}}. \end{aligned}$$

Hence the closed contour $\widetilde{C}$ is mapped into an arc $w_0\widetilde{w}_0$ of the closed contour $\widetilde{\Gamma}$. If $\widetilde{C}$ is traversed a second time, denoted by $z_0 + 2\widetilde{C}$, then $\arg z_0$ increases by 4π and therefore

$$\begin{aligned} w_0\big|_{z_0+2\widetilde{C}} &= |z_0|^{1/3} e^{i(\operatorname{Arg} z_0 + 4\pi)/3} \\ &= w_0\big|_{z_0+1\widetilde{C}}\, e^{i2\pi/3}, \end{aligned}$$

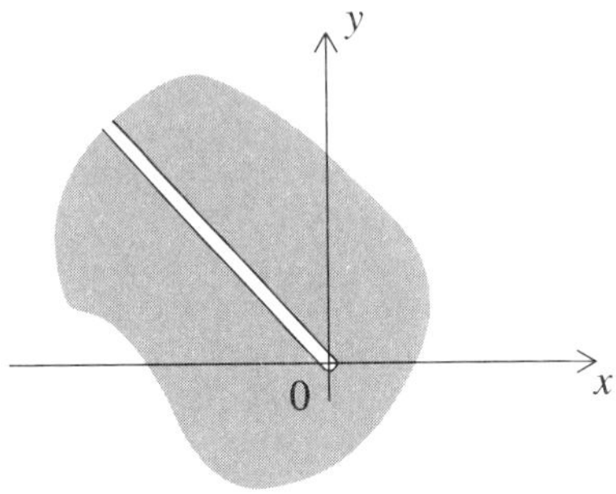

FIGURE 2.24. The region of univalence for the branches of $w = z^{1/n}$.

that is, an arc $\widetilde{w}_0\widetilde{\widetilde{w}}_0$ of the contour $\widetilde{\Gamma}$ corresponds to the second time along $\widetilde{C}$. If $\widetilde{C}$ is traversed a third time, then $\arg z_0$ increases by 6π and

$$\begin{aligned} w_0\big|_{z_0+3\widetilde{C}} &= |z_0|^{1/3}\, e^{i(\operatorname{Arg} z_0+6\pi)/3} \\ &= w_0\big|_{z_0+0\widetilde{C}}. \end{aligned}$$

This means that traversing the closed contour $\widetilde{C}$ three times corresponds to traversing the closed contour $\widetilde{\Gamma}$ only once in the w-plane. Similarly, traversing the contour $\widetilde{C}$ n times for the function $w = z^{1/n}$ corresponds to traversing the contour $\widetilde{\Gamma}$ once. □

2.6.3. Algebraic branch points.

DEFINITION 2.6.2. A point z_0 is called a *branch point* of the function $w = f(z)$ if the argument of w changes as z goes around z_0 along any sufficiently small closed contour.

DEFINITION 2.6.3. If the increment of the argument of a function $f(z)$ is equal to zero when a branch point is encircled n times, $n < \infty$, then the branch point is called an *algebraic branch point* of order n.

We see that the point $z = 0$ is an algebraic branch point of order n for each branch of $w = z^{1/n}$. So is the point $z = \infty$. Indeed, by the inversion $z = 1/z_1$, $z^{1/n} = 1/z_1^{1/n}$. Since $z_1 = 0$ is a branch point, then $z = \infty$ is also a branch point. If we join the points $z = 0$ and $z = \infty$ by a cut, we obtain a region of univalence for each branch of $w = z^{1/n}$. The cut can be any arc from 0 to ∞ (see Fig 2.24). As soon as the cut is fixed, each branch of $w = z^{1/n}$ is uniquely determined.

Consider, for example, the branch w_0 of $w = z^{1/3}$,

$$w_0(z) = |z|^{1/3}\, e^{i(\operatorname{Arg} z/3)},$$

and let $z = -i$. If we cut the complex z-plane along the negative real axis,

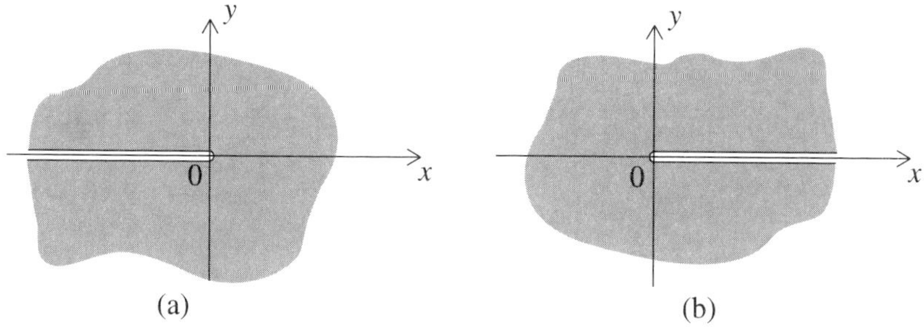

FIGURE 2.25. Variants of a cut: (a) $-i = e^{-\pi i/2}$, (b) $-i = e^{3\pi i/2}$.

as shown in Fig 2.25(a), then $-i = e^{-\pi i/2}$ (because $-\pi < \operatorname{Arg} z \leq \pi$) and $w_0(-i) = e^{-\pi i/6}$.

If we cut the complex z-plane along the positive real axis, as shown in Fig 2.25(b), then

$$0 \leq \operatorname{Arg} z < 2\pi, \quad -i = e^{3\pi i/2}, \quad w_0(-i) = e^{\pi i/2} = i \neq e^{-\pi i/6}.$$

2.6.4. Examples of Riemann surfaces. We introduce the concept of the *Riemann surface* of a function $f(z)$. Consider, for example, the branch w_0 of $w = z^{1/3}$:

$$w_0(z) = |z|^{1/3} e^{i(\operatorname{Arg} z)/3}.$$

We superpose three copies of the z-plane above each other and cut them along the positive real axis (see Fig 2.26). We then glue together the lower part of the first sheet with the upper part of the second sheet, the lower part of the second sheet with the upper part of the third sheet and the lower part of the third sheet with the upper part of the first sheet (the last glue is abstract). Such a surface is called a Riemann surface.

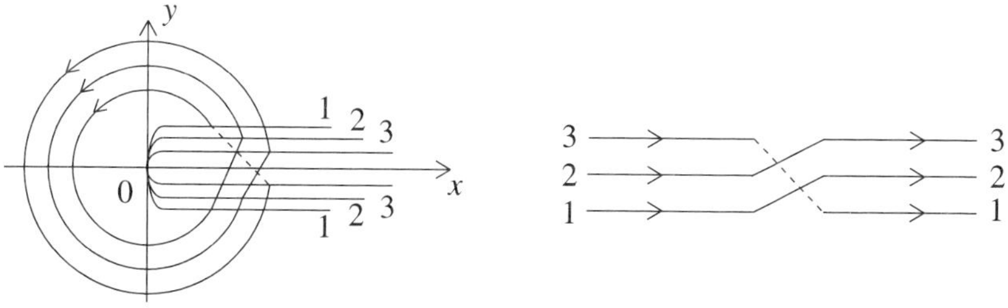

FIGURE 2.26. The Riemann surface of the function $z^{1/3}$.

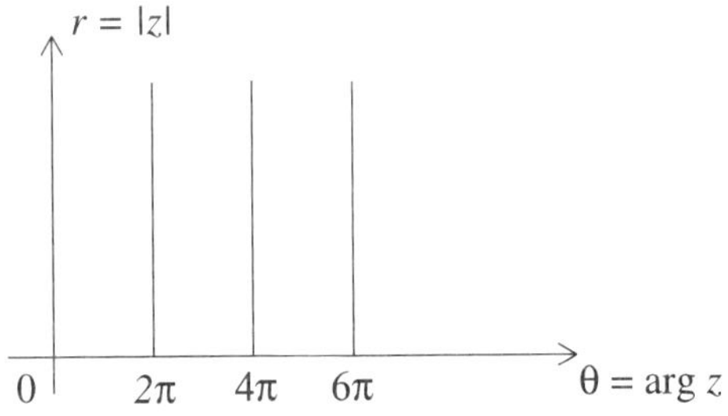

FIGURE 2.27. A non-abstract Riemann surface for the function $w = z^{1/3}$.

On its Riemann surface, $w = z^{1/3}$ is single-valued and analytic and hence a function because under the transition from the first sheet to the second sheet the branch w_0 continuously passes to the branch w_1. Similarly, under the transition from the second sheet to the third sheet, the branch w_1 continuously passes to the branch w_2 and under the transition from the third sheet to the first sheet the branch w_2 continuously passes to the branch w_0. The total angle along the closed contour on this surface is equal to 6π.

A variant of the Riemann surface for $w = z^{1/3}$ without any abstract glue along the cut between the third and the first sheets is shown in Fig 2.27. We take $\theta = \arg z$ and $r = |z|$ as the horizontal and vertical axes, respectively. If we glue such a plane along the lines $\theta = 0$ and $\theta = 6\pi$, we obtain a cylindrical surface which represents a nonabstract variant of the Riemann surface.

2.6.5. Mappings by composition of linear fractional functions and power functions. Composing mappings by linear fractional functions and power functions, one can map a region bounded by arcs of circles onto the upper half-plane, and a wedge $0 \leq \arg z \leq \pi/\alpha$ onto a disk.

EXAMPLE 2.6.2. *Map the wedge* $0 \leq \operatorname{Arg} z \leq \pi/6$ *onto the unit disk* $|w| \leq 1$ *such that the point* $z_1 = e^{i\pi/12}$ *is mapped to the point* $w = 0$ *and the point* $z_2 = 0$ *is mapped to the point* $w = 1$ *(see Fig 2.28).*

SOLUTION. First, we map the wedge $0 \leq \operatorname{Arg} z \leq \pi/6$ onto the upper half-plane (see Fig 2.29):

$$w^{(1)} = z^6.$$

In this case the point $z_1 = e^{\pi i/12}$ is mapped to the point

$$w_1^{(1)} = \left(e^{\pi i/12}\right)^6 = e^{\pi i/2} = i,$$

and the point $z_2 = 0$ is mapped to the point $w_2^{(1)} = 0$.

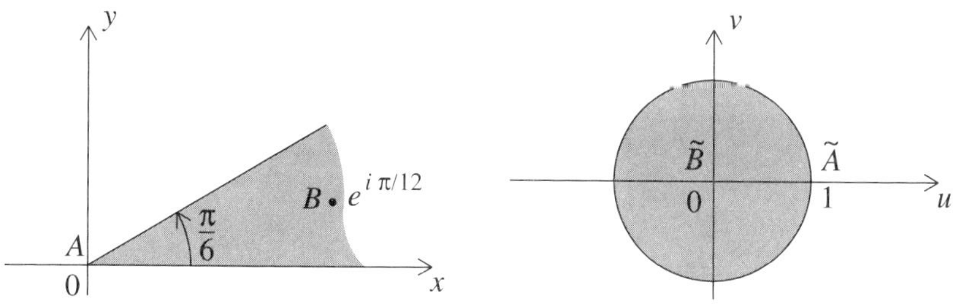

FIGURE 2.28. Mapping of the wedge $0 \leq \text{Arg}\, z \leq \pi/6$ onto the unit disk $|w| \leq 1$ in Example 2.6.2.

Then one has to map the upper half-plane $\Im w^{(1)} \geq 0$ onto the unit disk $|w| \leq 1$ so that the point $w_1^{(1)} = i$ is mapped to the point $w = 0$ and the point $w_2^{(1)} = 0$ is mapped to the point $w = 1$. Letting $z_0 = i$ in (2.5.3), we obtain

$$w = e^{i\alpha} \frac{w^{(1)} - i}{w^{(1)} + i} = e^{i\alpha} \frac{z^6 - i}{z^6 + i}. \tag{2.6.9}$$

Since the point $z = 0$ is mapped to the point $w = 1$, it follows from (2.6.9) that $1 = e^{i\alpha}(-1)$, so that

$$e^{i\alpha} = -1.$$

Finally, (2.6.9) has the form

$$w = \frac{i - z^6}{i + z^6}. \quad \square \tag{2.6.10}$$

EXAMPLE 2.6.3. *Map the region bounded by the arcs of two circles intersecting at the points z_1 and z_2 at an angle α onto the upper half-plane $\Im w \geq 0$ (see Fig 2.30).*

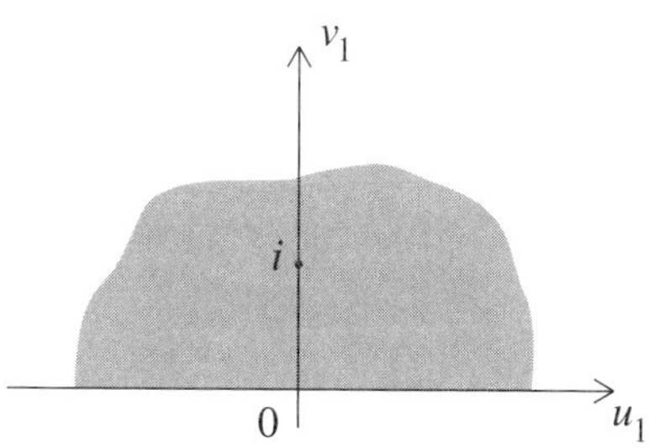

FIGURE 2.29. Intermediate mapping of the wedge $0 \leq \text{Arg}\, z \leq \pi/6$ onto the upper half-plane $\Im w^{(1)} \geq 0$ in Example 2.6.2.

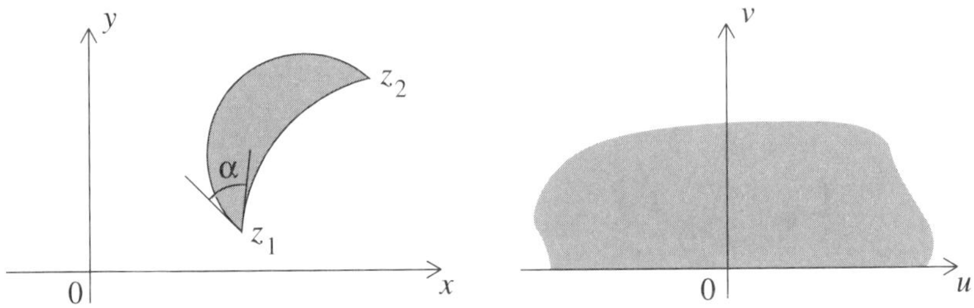

FIGURE 2.30. The initial and final regions under the mapping in Example 2.6.3.

SOLUTION. Since linear fractional transformations map circles into circles, and circles through infinity are straight lines, we map z_1 to $w_1 = 0$ and z_2 to $w = \infty$ by the linear fractional transformation,

$$w_1 = \frac{z - z_1}{z - z_2}, \tag{2.6.11}$$

so that the arcs of the circles will be mapped into straight lines that intersect in the angle α (see Fig 2.31).

The angle θ_0 in Fig 2.31(a) depends on the points z_1 and z_2. We rotate the domain in Fig 2.31(a) through the angle $-\theta_0$ (see Fig 2.31(b)),

$$w_2 = e^{-i\theta_0} w_1 = e^{-i\theta_0} \frac{z - z_1}{z - z_2}. \tag{2.6.12}$$

Finally, we use the property of generalized power functions (formula (2.6.3)) to map the region in Fig 2.31(b) onto the upper half-plane,

$$w = w_2^{\pi/\alpha} = \left(e^{-i\theta_0} \frac{z - z_1}{z - z_2} \right)^{\pi/\alpha}. \quad \square \tag{2.6.13}$$

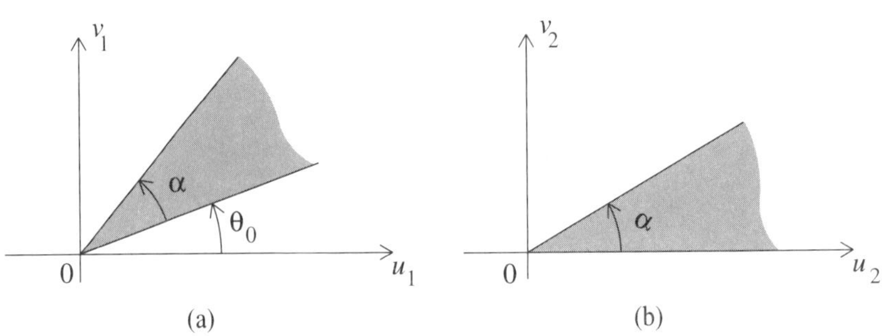

FIGURE 2.31. Images of mappings (2.6.11) and (2.6.12).

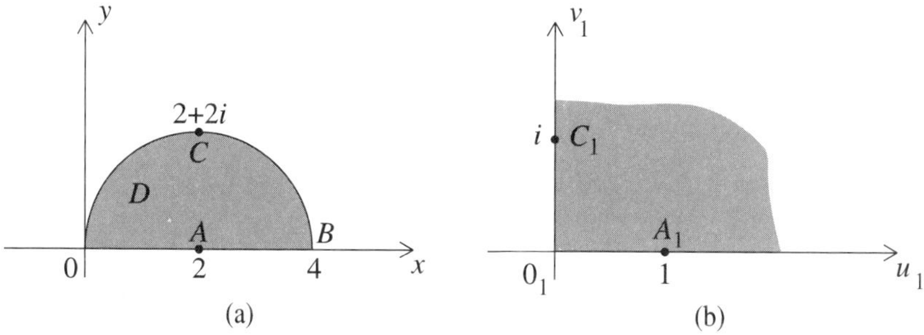

FIGURE 2.32. The initial and intermediate regions in Example 2.6.4 for the transformation (2.6.14).

EXAMPLE 2.6.4. *Map the region bounded by the upper half-disk,*

$$|z - 2| \leq 2, \qquad \Im z \geq 0,$$

onto the upper half-plane (see Fig 2.32*(a)).*

SOLUTION. This example is a particular case of Example 2.6.3. First, we map the points $z = 0$ and $z = 4$ into the points $w_1 = 0$ and $w_1 = \infty$, respectively, by the linear fractional transformation

$$w_1 = \frac{z}{4 - z}. \tag{2.6.14}$$

In order to find the image of the region D by transformation (2.6.14) we obtain the images of the points $z = 0$, $z = 2$ and $z = 2 + 2i$ (it is already clear that the image of D is a right-angled wedge, so that we have to know where the wedge is located). Using (2.6.14) we obtain

$$O = (0, 0) \mapsto O_1 = (0, 0), \qquad A = (2, 0) \mapsto A_1 = (1, 0),$$

and

$$C = (2, 2) \mapsto C_1 = (0, 1).$$

Therefore the edges of the right angle pass through the points $O_1 = (0, 0)$, $A_1 = (1, 0)$ and $B_1 = (0, 1)$, that is, the image of D under transformation (2.6.14) is the first quadrant (see Fig 2.32(b)).

In order to map this quadrant onto the upper half-plane it is sufficient to square (2.6.14):

$$w = w_1^2 = \left(\frac{z}{4 - z}\right)^2. \quad \square \tag{2.6.15}$$

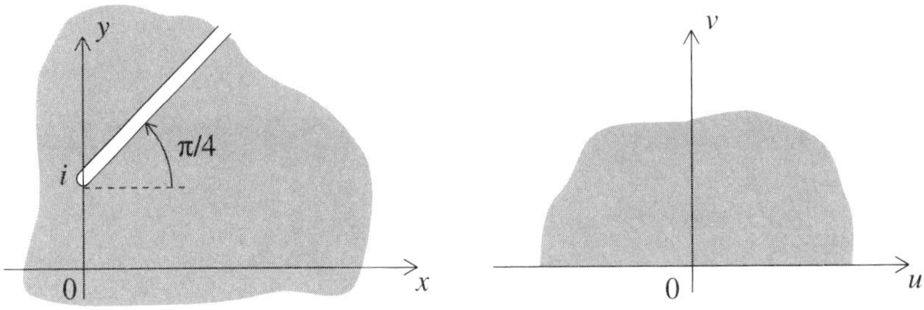

FIGURE 2.33. The initial and final regions in Example 2.6.5.

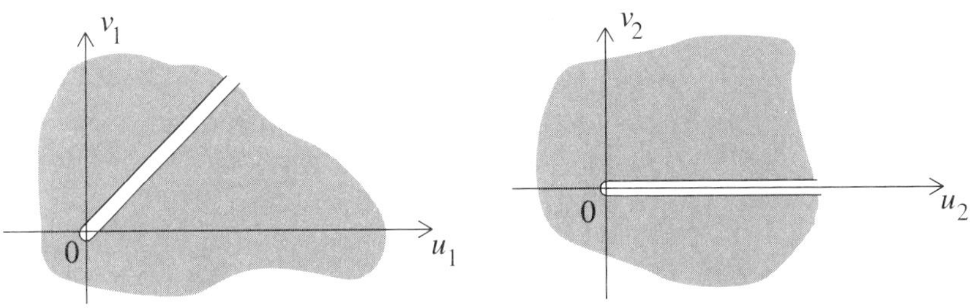

FIGURE 2.34. The second mapping in Example 2.6.5.

EXAMPLE 2.6.5. *Map the z-plane, with a cut from the point $z = i$ to the point $z = \infty$ making an angle $\pi/4$ with the positive x-axis, onto the upper half-plane (see Fig 2.33).*

SOLUTION. The problem can be solved in three stages:

$$\begin{aligned}
&\text{(a)} \qquad w_1 = z - i,\\
&\text{(b)} \qquad w_2 = e^{-i\pi/4} w_1 = e^{-i\pi/4}(z - i),\\
&\text{(c)} \qquad w = \sqrt{w_2} = \sqrt{e^{-i\pi/4}(z - i)}.
\end{aligned}$$

The second mapping is shown in Fig 2.34. □

EXAMPLE 2.6.6. *Map the upper half-plane, $\Im z \geq 0$, with a cut from the point $z = i$ to the point $z = 0$, onto the upper half-plane, $\Im w \geq 0$, without a cut (see Fig 2.35).*

SOLUTION. If we let

$$w_1 = z^2,$$

then the boundary of the region D in Fig 2.35 is mapped into a cut going from the point $w_1 = -1$ to the point $w_1 = 0$ in the positive direction of the u_1-axis (see Fig 2.36(a)). Indeed, the point $z = i$ is mapped to the point $w_1 = i^2 = -1$.

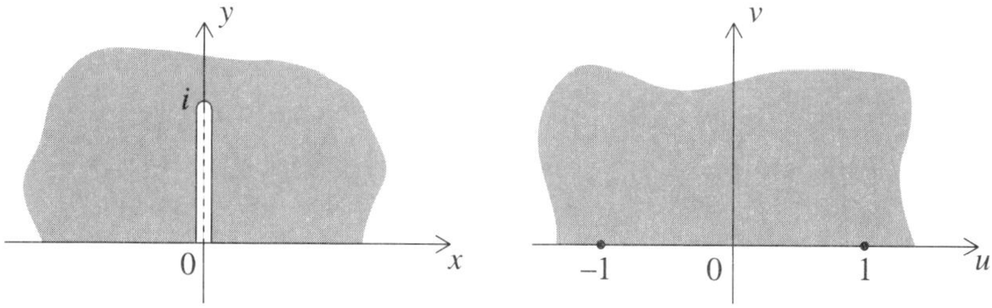

FIGURE 2.35. The initial and final regions in Example 2.6.6.

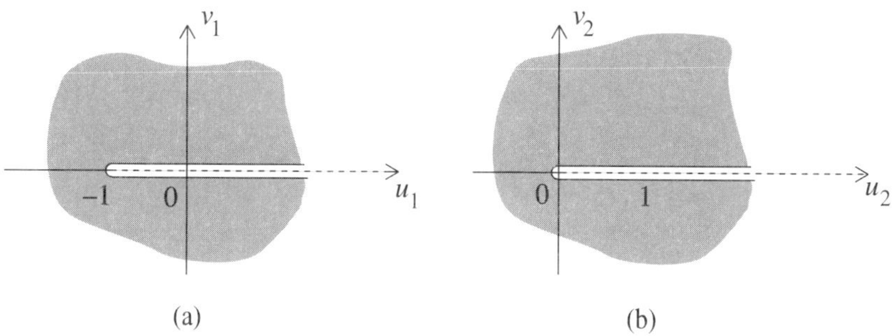

FIGURE 2.36. The sequence of mappings in Example 2.6.6.

The negative real axis, $-\infty < x < 0$, is mapped into the lower part of a cut going from $w_1 = 0$ to $w_1 = \infty$, and the positive real axis, $0 < x < +\infty$, is mapped into the upper part of the cut from $w_1 = 0$ to $w_1 = +\infty$.

Hence the cut from i to 0 in Fig 2.35 is mapped into the part of a cut in Fig 2.36(a) that is located from $w_1 = -1$ to $w_1 = 0$. The remaining mappings are elementary:

$$w_2 = w_1 + 1,$$
$$w = \sqrt{w_1 + 1} = \sqrt{z^2 + 1}. \quad \square$$

Exercises for Section 2.6

Find the image $\widetilde{D}$ of the following region D under the mapping $w = f(z)$.

1. $D = \{z;\ 0 \leq \operatorname{Arg} z \leq \pi/3\}, \qquad w = z^2.$
2. $D = \{z;\ -\pi/8 \leq \operatorname{Arg} z \leq \pi/8\}, \qquad w = z^4.$
3. $D = \{z;\ 1 < |z| < 2,\ 0 < \operatorname{Arg} z < \pi/4\}, \qquad w = z^3.$
4. $D = \{z;\ 2 < |z| < 4,\ -\pi/4 < \operatorname{Arg} z < \pi/2\}, \qquad w = z^2.$

Map the following domain D of the z-plane onto the domain G of the w-plane.

5. $D = \{z;\ 0 < \operatorname{Arg} z < \pi/3\}, \qquad G = \{w;\ \Re w > 0\}$.
6. $D = \{z;\ -\pi/4 < \operatorname{Arg} z < \pi/4\}, \qquad G = \{w;\ \Im w > 0\}$.
7. $D = \{z;\ 0 < \operatorname{Arg} z < \pi\alpha,\ 0 < \alpha \leq 2\}, \qquad G = \{w;\ \Im w > 0\}$.
8. $D = \{z;\ -\pi/3 < \operatorname{Arg} z < \pi/2\}, \qquad G = \{w;\ \Re w > 0\}$.

Map the circular lunes (two-angles) onto the upper half-plane.

9. $|z| < 1, \qquad |z + i| < 1$.
10. $|z| > 1, \qquad |z + i| < 1$.

Map the given domains onto the upper half-plane.

11. The plane with a cut along the segment $[1, 2]$.
12. The plane with a cut along the segment $[i, 3i]$.
13. The plane with a cut along the segment $[-1+i, -2+2i]$ which lies on the ray $y = -x$.
14. The plane with a cut along the ray $y = \sqrt{3}\,x$ in the first quadrant with initial point $z = 1 + i\sqrt{3}$.
15. The half-plane $\Im z > 0$ with a cut along the segment $[-2i, 0]$.
16. The half-plane $\Im z > 0$ with a cut along the ray from 1 to ∞.

2.7. Exponential and logarithmic mappings

2.7.1. The exponential function $w = e^z$. Let us find the *region of univalence*, R, of the mapping $w = e^z$, that is, the region that satisfies the following property:

$$z_1 \neq z_2 \quad \Longrightarrow \quad e^{z_1} \neq e^{z_2}, \tag{2.7.1}$$

for any pair of points z_1 and z_2 in R.

Since $e^{2k\pi i} = 1$, (2.7.1) implies that the following inequality holds:

$$z_1 \neq z_2 + 2k\pi i. \tag{2.7.2}$$

For example, the interior points of the strip

$$0 \leq \Im z < 2\pi \tag{2.7.3}$$

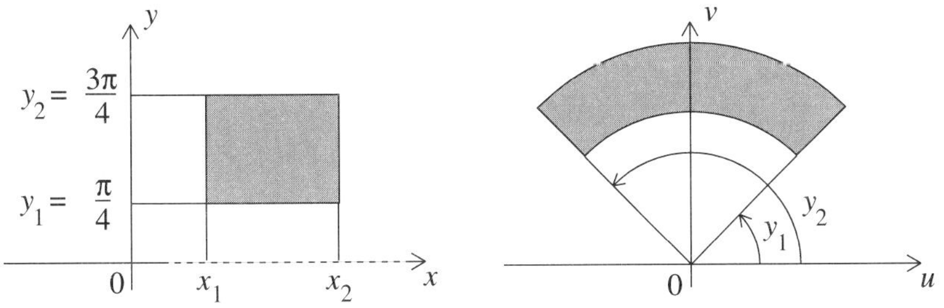

FIGURE 2.37. Mapping of a rectangle by the function $w = e^z$, where $y_1 = \pi/4$ and $y_2 = 3\pi/4$.

satisfy inequality (2.7.2). The region (2.7.3) is one of the regions of univalence for the function $w = e^z$. The whole complex plane z can be covered by similar regions of univalence,

$$2k\pi \leq \Im z < 2(k+1)\pi, \qquad k = 0, \pm 1, \pm 2, \dots . \tag{2.7.4}$$

If $z = x + iy$ and $0 \leq y < 2\pi$, then

$$w = e^{x+iy} = e^x e^{iy},$$

so that

$$|w| = e^x, \qquad \arg w = y. \tag{2.7.5}$$

It follows from (2.7.5) that the straight segment, $x = x_0$, $0 \leq y < 2\pi$, is mapped into the circle

$$|w| = e^{x_0}, \tag{2.7.6}$$

and the straight line $y = y_0$ is mapped into the ray

$$\arg w = y_0. \tag{2.7.7}$$

It thus follows from (2.7.6) and (2.7.7) that a rectangle bounded by the lines $x = x_1$, $x = x_2$, $y = y_1$, $y = y_2$ and located, for example, in the strip $0 \leq \Im z < 2\pi$ of the z-plane is mapped onto a curvilinear rectangle in the w-plane bounded by the rays $\arg w = y_1$, $\arg w = y_2$ and by arcs of the circles $|w| = e^{x_1}$ and $|w| = e^{x_2}$ (see Fig 2.37).

Returning to (2.7.7), we see that the function $w = e^z$ maps the strip $0 \leq \Im z \leq \alpha < 2\pi$ onto the wedge $0 \leq \arg w \leq \alpha$ (see Fig 2.38). The lower part of the strip, $y = 0$, $-\infty < x < +\infty$, is mapped into the ray $w = e^x$, $-\infty < x < +\infty$, so that $u = e^x$, $-\infty < x < +\infty$, $v = 0$ (the positive u-axis), but the upper part of the strip, $y = \alpha$, $-\infty < x < +\infty$, is mapped into the ray $w = e^x e^{i\alpha}$, $-\infty < x < +\infty$, that is, into the ray $\arg w = \alpha$.

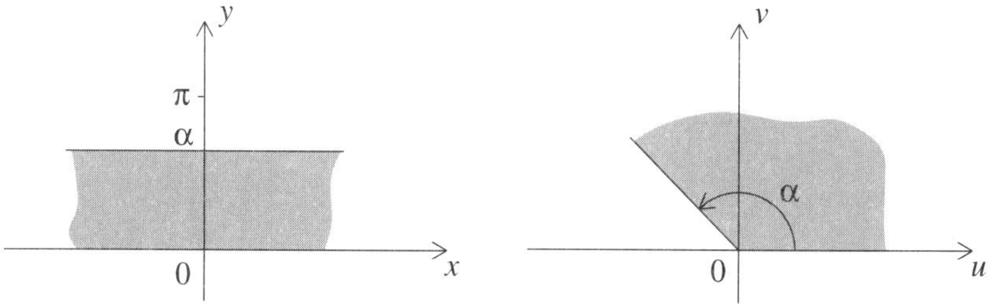

FIGURE 2.38. Mapping of a strip by the function $w = e^z$.

In particular, the strip $0 \le \Im z \le \pi$ is mapped into the upper half-plane $\Im w \ge 0$, and the strip $0 \le \Im z \le 2\pi$ is mapped into the whole w-plane with a cut along the positive real axis (see Fig 2.39). It is seen that the strip $0 \le \Im z < 2\pi$ is a fundamental region of the function $w = e^z$.

The following question arises: what are the images of other fundamental regions, $2\pi n \le y < 2(n+1)\pi$, of $w = e^z$ (see (2.7.4))? The answer is the following: they will map on the same region in the w-plane in Fig 2.39 as the strip $0 \le \Im z < 2\pi$.

For example, the strip $2\pi \le \Im z < 4\pi$ is mapped onto the whole w-plane with a cut along the positive real axis (compare with Fig 2.39), so that the side $y = 2\pi$ is mapped onto the upper part of the cut and the side $y = 4\pi$ is mapped onto the lower part of the cut in the w-plane.

2.7.2. The logarithm of z. We recall that the logarithm of z is given by the formula

$$\begin{aligned} w &= \log z = \operatorname{Log} z + 2k\pi i \\ &= \ln|z| + i \operatorname{Arg} z + 2k\pi i. \end{aligned} \tag{2.7.8}$$

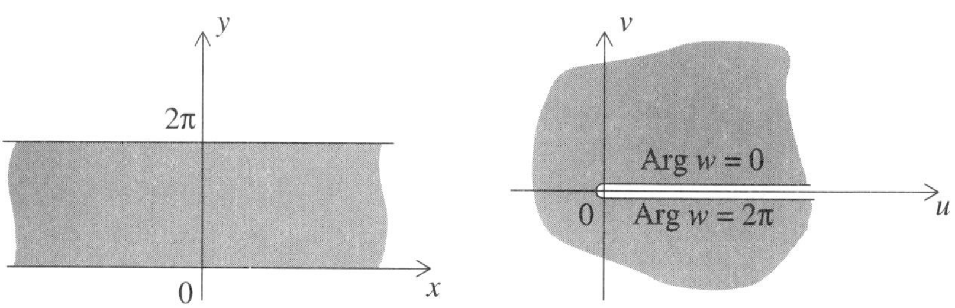

FIGURE 2.39. Mapping of the strip $0 \le y \le 2\pi$, of width 2π, by the function $w = e^z$.

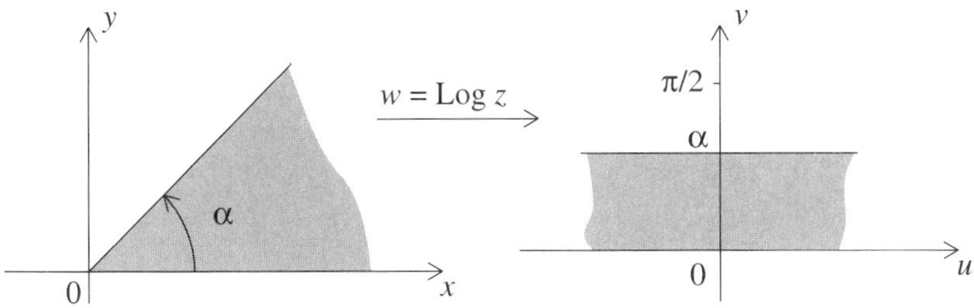

FIGURE 2.40. Mapping of a wedge onto a strip by the function $w = \text{Log } z$.

Each branch of (2.7.8), for instance, the principal value $w = \text{Log}\, z$, is the inverse of the exponential function $z = e^w$, that is, $w = \text{Log}\, z$ maps the wedge $0 \le \text{Arg}\, z \le \alpha < 2\pi$ onto the horizontal strip $0 \le \Im w \le \alpha$ (see Fig 2.40).

In particular, the wedge $0 \le \text{Arg}\, z \le \pi$ (that is, the upper half-plane $\Im z \ge 0$) is mapped onto the strip $0 \le \Im w \le \pi$, and the whole complex z-plane with a cut along the positive real axis is mapped onto the strip $0 \le \Im w < 2\pi$ (see Fig 2.41).

The image of a closed contour C encircling the point $z = 0$, when traversed once, is the straight line segment $w_0 \widetilde{w}_0$ in the w-plane (see Fig 2.41). In this case,

$$\text{Log}\, z_{z_0+1C} = \ln|z_0| + i \,\text{Arg}\, z_0 + 2\pi i,$$

that is, the branch $w_0 = \text{Log}\, z$ goes into the branch $w_1 = \text{Log}\, z + 2\pi i$ when C is traversed once. Hence $z = 0$ is a branch point of the function $w = \text{Log}\, z$. The point $\widetilde{w}_0$ will go into the point $\widetilde{\widetilde{w}}_0$ after a second time around the contour C, that is, the branch w_1 will go into the branch w_2, and so on.

We see that the Riemann surface of the function $w = \log z$ contains

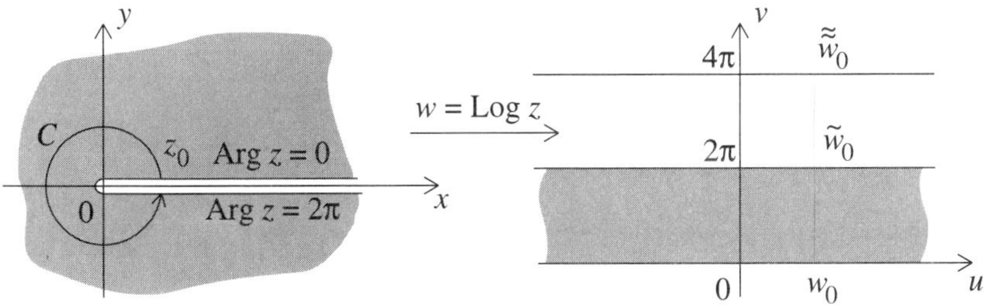

FIGURE 2.41. Mapping of the complex plane with a cut by the function $w = \text{Log } z$.

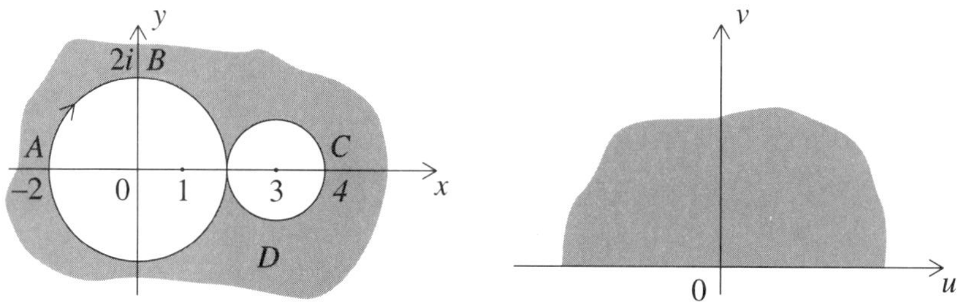

FIGURE 2.42. The initial and final regions in Example 2.7.1.

infinitely many sheets of the z-plane cut along the positive x-axis and these are glued in the same manner as for the function $w = z^{1/n}$ (the lower part of the cut of the first sheet is glued to the upper part of the second sheet, the lower part of the second sheet is glued to the upper part of the third sheet, etc.).

No matter how many times we go around the contour C we cannot obtain a closed contour in the w-plane. Such branch point is called a *logarithmic branch point.*

2.7.3. Examples of composite mappings. Combining linear fractional, logarithmic and exponential functions, one can map the region outside two tangent circles or outside two intersecting circles onto the upper half-plane or onto a strip, and a strip with a cut can be mapped onto the upper half-plane.

We consider several examples.

EXAMPLE 2.7.1. *Map the region*

$$D = \{|z| \geq 2\} \cap \{|z - 3| \geq 1\},$$

(consisting of the complement of the union of two tangent open disks) onto the upper half-plane $\Im w \geq 0$ *(see Fig* 2.42*).*

SOLUTION. To send the two tangent circles into parallel straight lines, it suffices to map the point $z = 2$ to the point $w_1 = \infty$ by the linear fractional transformation

$$w_1 = \frac{z+2}{z-2}. \tag{2.7.9}$$

Thus, the region D is mapped onto the strip, S, whose position is determined by the images of three boundary points, that is,

$$A = (-2, 0) \mapsto A_1 = (0, 0), \qquad B = (0, 2) \mapsto B_1 = (0, -1),$$

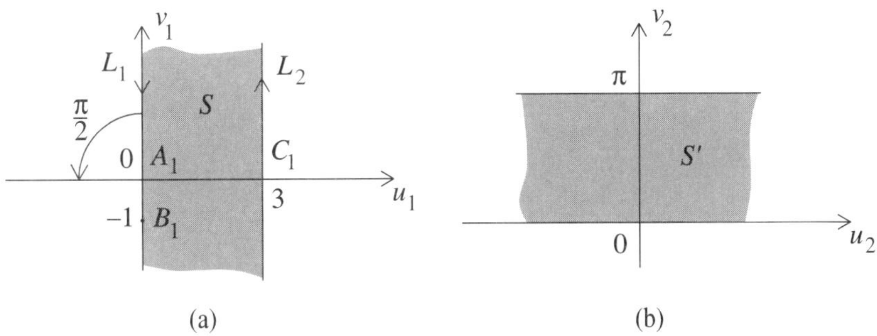

FIGURE 2.43. Intermediate regions in Example 2.7.1.

and

$$C = (4, 0) \mapsto C_1 = (3, 0).$$

Hence the strip S is bounded by the two parallel lines L_1, passing through the points A_1 and B_1, and L_2, passing through the point C_1 (see Fig 2.43(a)). Finally, considering the orientation of the boundary of D and of its image, we see that D is mapped inside the strip S of width 3 bounded by the lines L_1 and L_2.

Next, the linear transformation

$$w_2 = e^{\pi i/2} \frac{\pi}{3} w_1 = i \frac{\pi}{3} \frac{z+2}{z-2} \tag{2.7.10}$$

maps the vertical strip S of Fig 2.43(a) onto the vertical strip S' defined by the inequations

$$0 \le \Im w_2 \le \pi$$

(see Fig 2.43(b)). Finally, using the mapping properties of the exponential function, we map the strip S' onto the upper half-plane $\Im w \ge 0$ (see Fig 2.42),

$$w = \exp\left(i \frac{\pi}{3} \frac{z+2}{z-2}\right). \quad \square \tag{2.7.11}$$

EXAMPLE 2.7.2. *Map the strip $0 \le \Re z \le 1$ with a cut joining the points $z_1 = 1/2 + ih$ and $z_2 = 1/2 + i\infty$, onto the upper half-plane $\Im w \ge 0$ (see Fig 2.44).*

SOLUTION. First, we map the strip shown in Fig 2.44 onto the strip $0 \le \Im w_1 \le 2\pi$ by the linear transformation

$$w_1 = e^{i\pi/2} 2\pi z = 2\pi i z. \tag{2.7.12}$$

Now, we determine the new position of the cut. The initial point of the cut, $z_1 = 1/2 + ih$, is mapped to the point $w_1 = 2\pi i(1/2 + ih) = -2\pi h + \pi i$, so

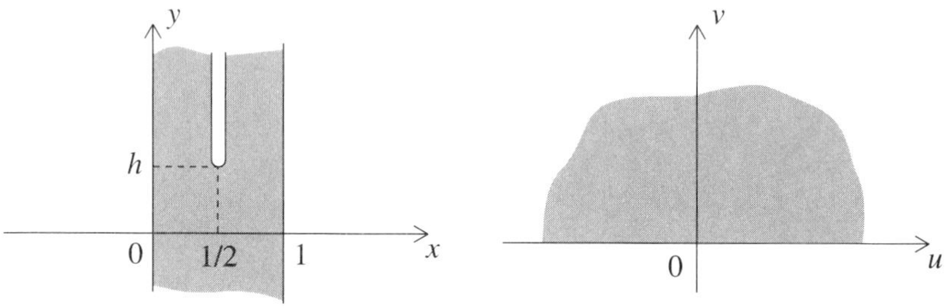

FIGURE 2.44. The initial and final regions in Example 2.7.2.

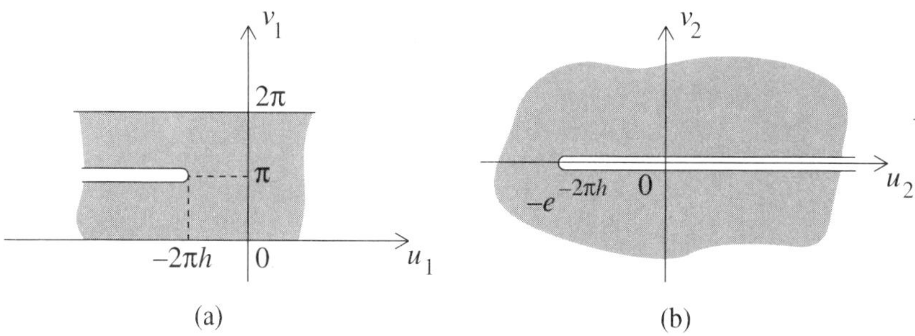

FIGURE 2.45. Intermediate regions in Example 2.7.2.

that the cut joins the points $-\infty + \pi i$ and $-2\pi h + \pi i$ in the w_1-plane (see Fig 2.45(a)).

We map the region shown in Fig 2.45(a) onto the upper half-plane $\Im w_2 \geq 0$ with a cut along the positive real axis by

$$w_2 = e^{w_1} = e^{2\pi i z}. \tag{2.7.13}$$

The initial point, $w_1 = -2\pi h + \pi i$, of the cut in Fig 2.45(a) is mapped to the point

$$w_2 = e^{-2\pi h + \pi i} = -e^{-2\pi h},$$

and the point $w_1 = -\infty + \pi i$ is mapped to the point $w_2 = e^{-\infty + \pi i} = 0$. Hence, the semifinite cut in Fig 2.45(a) is mapped to the finite cut joining the points $-e^{-2\pi h}$ and 0 in Fig 2.45(b). The boundaries $\Im w_1 = 0$ and $\Im w_1 = 2\pi$ of the region shown in Fig 2.45(a) are mapped to the upper and lower parts of the cut joining the points $w_2 = 0$ and $w_2 = \infty$, respectively, shown in Fig 2.45(b). The mapping of the region shown in Fig 2.45(b) onto the upper half-plane is elementary:

$$w_3 = w_2 + e^{-2\pi h} = e^{2\pi i z} + e^{-2\pi h}.$$

Hence

$$w = \sqrt{w_3} = \sqrt{e^{2\pi iz} + e^{-2\pi h}}. \quad \square \tag{2.7.14}$$

Exercises for Section 2.7

Find the images of the following domains under the mapping $w = e^z$.

1. $D = \{(x, y) \in \mathbb{R}^2;\ 0 < x < 1,\ 0 < y < \pi\}$.
2. $D = \{(x, y) \in \mathbb{R}^2;\ -\infty < x < +\infty,\ 0 < y < \pi/2\}$.
3. $D = \{(x, y) \in \mathbb{R}^2;\ -\infty < x < 0,\ 0 < y < \pi/4\}$.
4. $D = \{(x, y) \in \mathbb{R}^2;\ 0 < x < +\infty,\ 0 < y < \pi/3\}$.
5. $D = \{(x, y) \in \mathbb{R}^2;\ 0 < x < +\infty,\ 0 < y < \pi\}$.
6. $D = \{(x, y) \in \mathbb{R}^2;\ -\infty < x < 0,\ 0 < y < 2\pi\}$.

Find the images of the following regions under the given mapping.

7. $D = \{(x, y) \in \mathbb{R}^2;\ 0 < x < \pi,\ 0 < y < +\infty\}$, $\quad w = e^{iz}$.
8. $D = \{(x, y) \in \mathbb{R}^2;\ -\infty < x < 0,\ 0 < y < \pi/2\}$, $\quad w = e^{-z} + 2$.
9. $D = \{(x, y) \in \mathbb{R}^2;\ 0 < x < +\infty,\ 0 < y < \pi/6\}$, $\quad w = e^{3z}$.
10. $D = \{(x, y) \in \mathbb{R}^2;\ 0 < x < \pi/2,\ -\infty < y < 0\}$, $\quad w = e^{2iz}$.

Map the region D of the z-plane onto the region G of the w-plane.

11. $D = \{(x, y) \in \mathbb{R}^2;\ 0 < x < +\infty,\ 0 < y < \pi/2\}$,
    $G = \{w;\ |w| < 1,\ \Im w > 0\}$.
12. $D = \{(x, y) \in \mathbb{R}^2;\ -\infty < x < +\infty,\ x < y < x + 1\}$,
    $G = \{w;\ \Im w > 0\}$.
13. $D = \{(x, y) \in \mathbb{R}^2;\ 0 < x < \pi/3,\ -\infty < y < +\infty\}$,
    $G = \{w;\ \Im w > 0,\ \Re w > 0\}$.
14. $D = \{(x, y) \in \mathbb{R}^2;\ -\infty < x < +\infty,\ 0 < y < +\infty\}$,
    $G = \{w;\ \Re w > 0,\ 0 < \Im w < \pi/2\}$.

Find the images of the following regions under the mapping $w = \operatorname{Log} z$.

15. $D = \{z;\ 0 < \operatorname{Arg} z < \pi/2\}$.
16. $D = \{z;\ |z| < e,\ 0 < \operatorname{Arg} z < \pi/4\}$.
17. $D = \{z;\ 1 < |z| < 2,\ 0 < \operatorname{Arg} z < \pi\}$.
18. $D = \{z;\ 2 < |z| < 4,$ with the cut along the segment $[2, 4]\}$.

Find the images of the following regions under the given mapping. (Hint: Consider each problem as a composite mapping.)

19. $D = \{z;\ 1 < |z| < 2,\ 0 < \operatorname{Arg} z < \pi/2\}, \qquad w = \operatorname{Log} z + 2 + i.$

20. $D = \{z;\ |z| < 1,\ 0 < \operatorname{Arg} z < \pi/4\}, \qquad w = \operatorname{Log}(z^2).$

21. $D = \{z;\ \Re z > 0,\ \Im z > 0\}, \qquad w = \operatorname{Log}(-iz).$

22. $D = \{z;\ \Re z > 0,\ \Im z > 0\}, \qquad w = \operatorname{Log}\left(\dfrac{z-i}{z+i}\right).$

Map the region D of the z-plane onto the region G of the w-plane.

23. $D = \{z;\ |z| < e,\ 0 < \operatorname{Arg} z < 3\pi/4\}$,
$G = \{w;\ \Re w < 1,\ 0 < \Im w < 3\pi/4\}$.

24. $D = \{z;\ \Re z > 0,\ \Im z > 0\}$,
$G = \{w;\ -\infty < \Re w < +\infty,\ 1 < \Im w < \pi/2 + 1\}$.

25. $D = \{z;\ \Re w + \Im w > -1\}$,
$G = \{w;\ 0 < \Re w < 1\}$.

26. $D = \{z;\ |z| < 2,\ 0 < \operatorname{Arg} z < \pi/4\}$,
$G = \{w;\ 0 < \Re w < +\infty,\ 0 < \Im w < 1\}$.

2.8. Mapping by Joukowsky's function

2.8.1. Joukowsky's function. Joukowsky's function has the form

$$w = \frac{a}{2}\left(z + \frac{1}{z}\right), \qquad a = \text{constant}. \tag{2.8.1}$$

Since

$$w' = \frac{a}{2}\left(1 - \frac{1}{z^2}\right) = 0$$

only if $z = \pm 1$, then the mapping (2.8.1) is conformal in any region not containing the points $z = \pm 1$.

Letting $z = re^{i\theta}$ and separating the real and imaginary parts of (2.8.1), we have

$$\begin{aligned} w &= \frac{a}{2}\left(re^{i\theta} + \frac{1}{r}e^{-i\theta}\right) \\ &= \frac{a}{2}\left[\left(r + \frac{1}{r}\right)\cos\theta + i\left(r - \frac{1}{r}\right)\sin\theta\right] \\ &=: u + iv. \end{aligned}$$

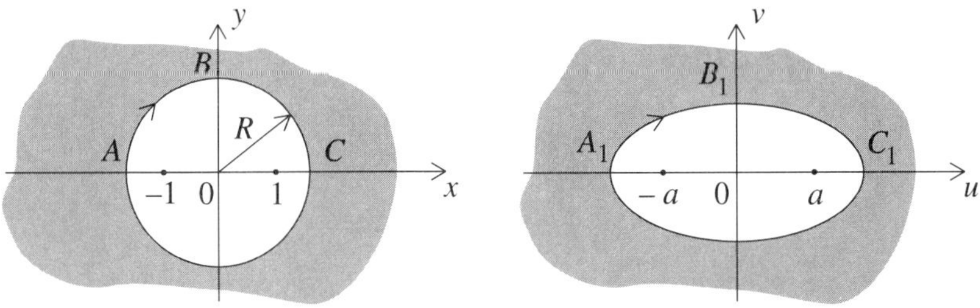

FIGURE 2.46. Mapping of the region $|z| \geq R > 1$ by Joukowsky's function: the upper (lower) half-plane outside the disk is mapped onto the upper (lower) half-plane outside the ellipse.

Thus

$$u = \frac{a}{2}\left(r + \frac{1}{r}\right)\cos\theta, \qquad v = \frac{a}{2}\left(r - \frac{1}{r}\right)\sin\theta. \tag{2.8.2}$$

Let us find the image of the circle $|z| = R$. Letting $r = R$ in (2.8.2), we get

$$u = \frac{a}{2}\left(R + \frac{1}{R}\right)\cos\theta, \qquad v = \frac{a}{2}\left(R - \frac{1}{R}\right)\sin\theta, \tag{2.8.3}$$

and eliminating θ we obtain

$$\frac{u^2}{\frac{a^2}{4}\left(R + \frac{1}{R}\right)^2} + \frac{v^2}{\frac{a^2}{4}\left(R - \frac{1}{R}\right)^2} = 1, \tag{2.8.4}$$

that is, the circle $|z| = R$ is mapped onto the ellipse with semi-axes

$$\tilde{a} = \frac{a}{2}\left(R + \frac{1}{R}\right), \qquad \tilde{b} = \frac{a}{2}\left|R - \frac{1}{R}\right|. \tag{2.8.5}$$

The coordinates of the foci of the ellipse are

$$c = \pm\sqrt{\tilde{a}^2 - \tilde{b}^2} = \pm a, \tag{2.8.6}$$

that is, the ellipses (2.8.4) are confocal with foci at the points $\pm a$.

We consider the two cases: $R > 1$ and $R < 1$.

(a) The case $R > 1$. In this case, the points $z = \pm 1$ are located inside the disk $|z| \leq R$ and therefore the mapping is conformal in the region $|z| \geq R$.

Let us find the image of the region $|z| \geq R$ if $R > 1$ (see Fig 2.46). We

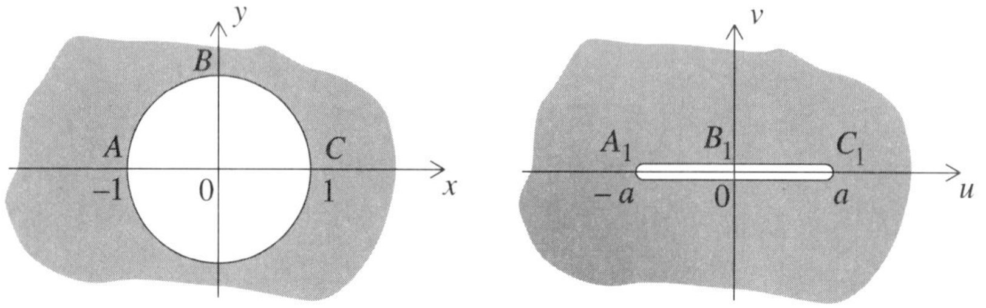

FIGURE 2.47. Mapping of the exterior of the unit disk onto the complex plane with a cut by Joukowsky's function.

first find the images of the points A, B and C on the boundary of the disk by using formulae (2.8.3):

$$\begin{aligned} A=Re^{i\pi} &\mapsto A_1 = \left(-\frac{a}{2}\left(R+\frac{1}{R}\right), 0\right), \\ B=Re^{i\pi/2} &\mapsto B_1 = \left(0, \frac{a}{2}\left(R-\frac{1}{R}\right)\right), \\ C=Re^{i0} &\mapsto C_1 = \left(\frac{a}{2}\left(R+\frac{1}{R}\right), 0\right). \end{aligned} \tag{2.8.7}$$

Since $R-(1/R)>0$ if $R>1$, then the point B_1 is located in the upper part of the ellipse. Therefore, going once along the circle is the same as going once along the ellipse, as shown in Fig 2.46. Hence the exterior of the disk is mapped onto the exterior of the ellipse.

If $R \to 1$, it follows from (2.8.5) that

$$\tilde{a} \to a, \qquad \tilde{b} \to 0, \tag{2.8.8}$$

so that the ellipse degenerates into a cut joining the foci $w=-a$ and $w=a$ (see Fig 2.47). Hence the region $|z| \geq 1$ is a fundamental region of Joukowsky's function.

(b) The case $R<1$. In this case, the points $z=\pm 1$ are located outside the disk $|z| \leq R$, and therefore the mapping is conformal in the region $|z| \leq R$.

To find the image of the region $|z| \leq R < 1$ (see Fig 2.48) we use the images of the points A, B and C (formulae (2.8.7)). The difference with the case (a) is that $R-(1/R)<0$ if $R<1$, and therefore the point B_1 is located in the lower part of the ellipse, that is, the directions along the circle and the ellipse in Fig 2.48 are opposite to each other. Hence the interior of the disk $|z| \leq R$ is mapped onto the exterior of the ellipse, where

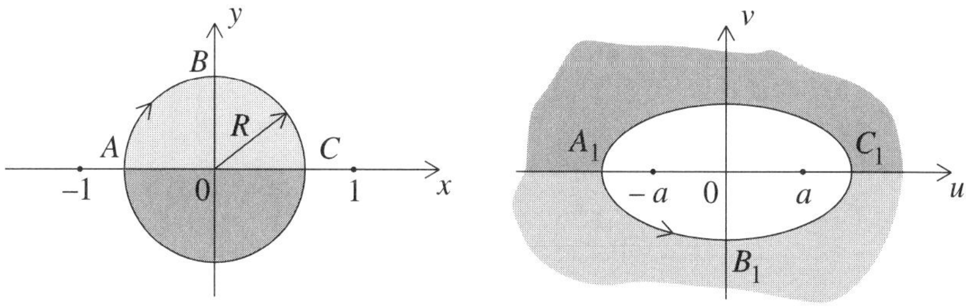

FIGURE 2.48. Mapping of the interior of the disk $|z| \le R < 1$ onto the exterior of the ellipse by Joukowsky's function.

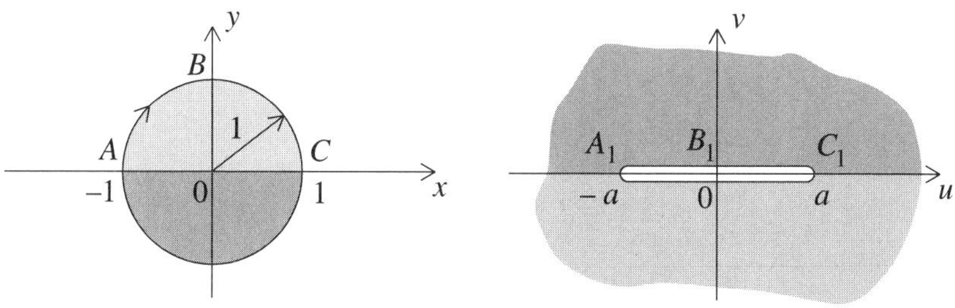

FIGURE 2.49. Mapping of the closed disk $|z| \le 1$ by Joukowsky's function.

the lower half-disk is mapped onto the upper part of the half-plane outside the ellipse, but the upper half-disk is mapped onto the lower part of the half-plane outside the ellipse (see Fig 2.48).

If $R \to 1$, the ellipse, as in the case (a), degenerates into a cut joining the points $-a$ and a (see Fig 2.47), but the upper semicircle ABC, in this case, is mapped onto the lower part of the cut while the lower semicircle is mapped onto the upper part of the cut (see Fig 2.49).

2.8.2. Examples of Joukowsky's mapping. Joukowsky's mapping will be illustrated by means of examples.

EXAMPLE 2.8.1. *Map the open disk* $|z| < 1$ *with two cuts along the segments* $[1/2, 1]$ *and* $[-1, -1/2]$ *of the real axis as shown in Fig* 2.50, *onto the upper half-plane* $\Im w > 0$.

SOLUTION. Since Joukowsky's function

$$w_1 = \frac{1}{2}\left(z + \frac{1}{z}\right) \tag{2.8.9}$$

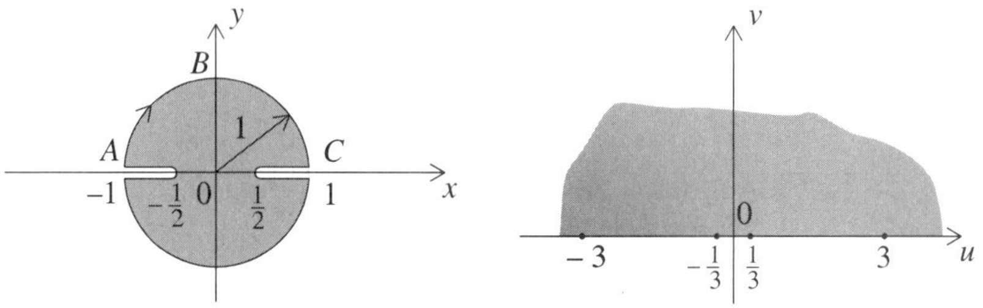

FIGURE 2.50. The initial and final regions under the mapping in Example 2.8.1.

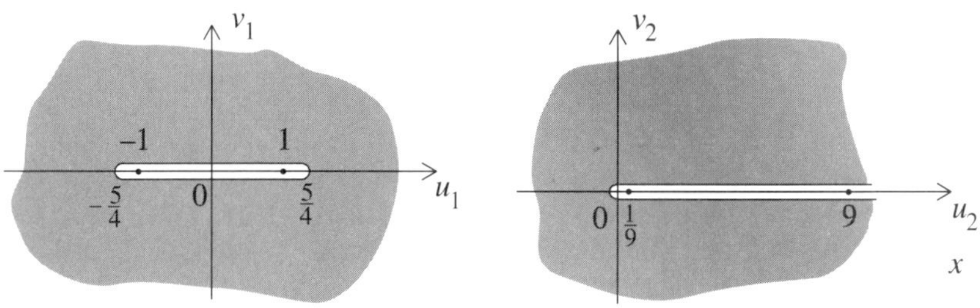

(a) (b)

FIGURE 2.51. Intermediate regions under the mapping in Example 2.8.1.

maps the disk $|z| \leq 1$ onto the w_1-complex plane with a cut from $w_1 = -1$ to $w_1 = 1$, then the endpoints of the cuts are sent to the points

$$w_1\Big|_{z=-1} = -1, \quad w_1\Big|_{z=-1/2} = -\frac{5}{4}, \quad w_1\Big|_{z=1/2} = \frac{5}{4}, \quad w_1\Big|_{z=1} = 1.$$

Hence the cuts in the disk are mapped into the cuts in the w_1-plane joining the points $-5/4$ and -1 and the points 1 and $5/4$. These cuts are continuations to the left and to the right of the cut joining the points -1 and 1 (see Fig 2.51(a)).

Next, we map the point $-5/4$ to 0 and the point $5/4$ to ∞ by the linear fractional transformation

$$w_2 = \frac{w_1 + \frac{5}{4}}{\frac{5}{4} - w_1}. \tag{2.8.10}$$

In order to determine the direction of the cut we compute the images of the points $w_1 = \mp 1$:

$$w_2\Big|_{w_1=-1} = \frac{1}{9}, \qquad w_2\Big|_{w_1=1} = 9.$$

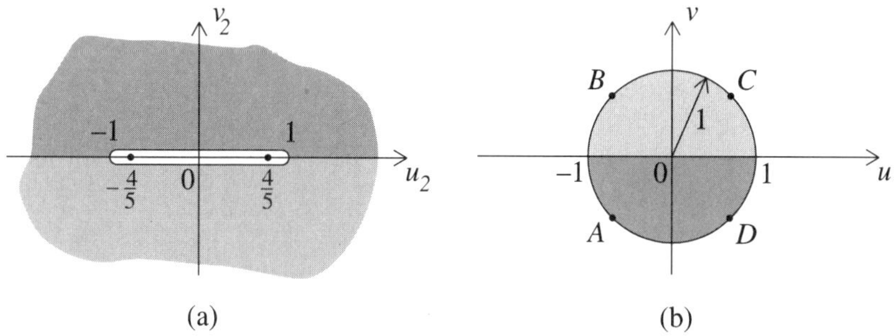

FIGURE 2.52. Intermediate and final regions under the mapping in Example 2.8.2.

Hence the cut goes to the right along the positive real axis (see Fig 2.51(b)). We map the region in Fig 2.51(b) onto the upper half-plane $\Im w \geq 0$:

$$w = \sqrt{w_2} = \sqrt{\frac{w_1 + \frac{5}{4}}{\frac{5}{4} - w_1}}$$

$$= \sqrt{\frac{\frac{1}{2}(z + \frac{1}{z}) + \frac{5}{4}}{\frac{5}{4} - \frac{1}{2}(z + \frac{1}{z})}}. \tag{2.8.11}$$

The desired mapping is given by (2.8.11). The lower semicircle is mapped onto the segment $(1/3, 3)$ of the u-axis (see Fig 2.50). The upper semicircle is mapped onto the segment $(-3, -1/3)$ of the u-axis. The right cut is mapped onto the segment $(3, +\infty)$. Finally, the left cut is mapped onto the segment $(-\infty, -3)$. □

EXAMPLE 2.8.2. *Map the disk with the two cuts shown in Fig* 2.50 *of Example* 2.8.1 *onto the disk* $|w| \leq 1$ *without cuts, that is "straighten the cuts."*

SOLUTION. As in Example 2.8.1, we use Joukowsky's function (2.8.9) and get the region shown in Fig 2.51(a). Next, we map the cut in Fig 2.51(a) onto the cut joining the points -1 and 1 by the linear transformation (see Fig 2.52(a)):

$$w_2 = \frac{4}{5} w_1.$$

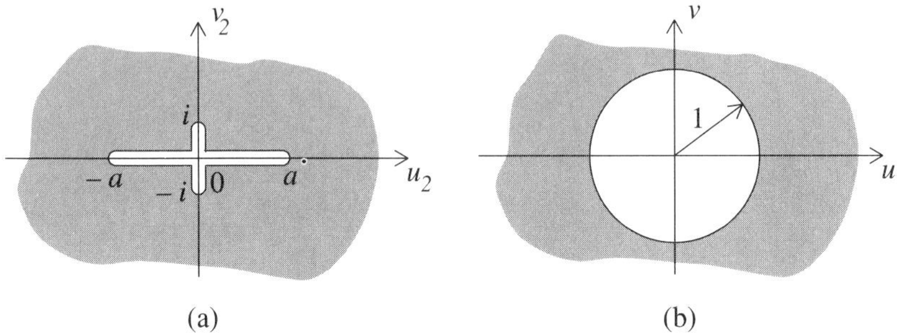

FIGURE 2.53. The initial and final regions in Example 2.8.3.

Finally, we use the fact that the function inverse to Joukowsky's function (2.8.9),

$$\begin{aligned} w &= w_2 + \sqrt{w_2^2 - 1} \\ &= \frac{2}{5}\left(z + \frac{1}{z}\right) + \sqrt{\frac{16}{25} \times \frac{1}{4}\left(z + \frac{1}{z}\right)^2 - 1}, \end{aligned} \tag{2.8.12}$$

maps the w_2-plane with a cut from $w_2 = -1$ to $w_2 = 1$ onto the region $|w| \le 1$ (see Fig 2.49, where the roles of the z- and the w-planes have to be interchanged). We find the images of the different parts of the cut in Fig 2.52(a):

$$w\big|_{w_2=-4/5} = -\frac{4}{5} \pm \frac{3}{5}i, \qquad w\big|_{w_2=4/5} = \frac{4}{5} \pm \frac{3}{5}i.$$

Hence the left cut in Fig 2.50 is mapped onto the arc AB in Fig 2.52(b), where $A = (-4/5, -3/5)$ and $B = (-4/5, 3/5)$. The right cut in Fig 2.50 is mapped onto the arc CD, where $C = (4/5, 3/5)$ and $D = (4/5, -3/5)$. The upper and lower semicircles in Fig 2.50 are mapped onto the arcs AD and BC, respectively. The desired mapping is given by (2.8.12). □

EXAMPLE 2.8.3. *Map the exterior of the cross shown in Fig* 2.53 *onto the exterior of the unit disk.*

SOLUTION. Since the function

$$w_1 = \sqrt{z^2 + 1} \tag{2.8.13}$$

maps the upper half-plane with a cut joining the points $z = 0$ and $z = i$ onto the upper half-plane $\Im w \ge 0$ without the cut (see Example 2.6.6 and Fig 2.35), then points $z = -a$ and $z = a$ are mapped by (2.8.13) into

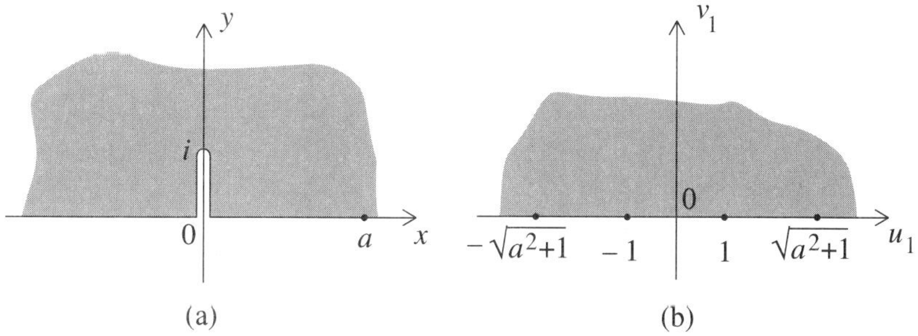

FIGURE 2.54. Initial and intermediate regions under the mapping in Example 2.8.3.

the points $-\sqrt{a^2+1}$ and $\sqrt{a^2+1}$, respectively (see Fig 2.54). In fact, if $z = -a = e^{i\pi}a$, then

$$\begin{aligned} z^2 + 1 &= a^2\, e^{2\pi i} + e^{2\pi i} \\ &= e^{2\pi i}(a^2 + 1) \end{aligned}$$

and

$$\begin{aligned} \sqrt{z^2+1}\,\Big|_{z=-a} &= e^{\pi i}\sqrt{a^2+1} \\ &= -\sqrt{a^2+1}, \end{aligned}$$

where we have taken the branch of $\sqrt{a^2+1}$ for which $\sqrt{1} = 1$.

By the symmetry principle, the lower half-plane, $\Im z \leq 0$, with a cut from the point $z = 0$ to the point $z = -i$ is mapped onto the region $\Im w_1 \leq 0$ by the function (2.8.13). Hence the function (2.8.13) maps the cross in Fig 2.53(a) onto the cut shown in Fig 2.55(a).

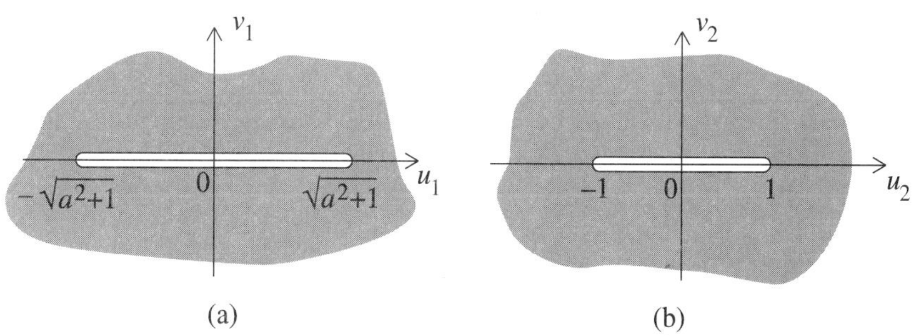

FIGURE 2.55. Intermediate regions under the mapping in Example 2.8.3.

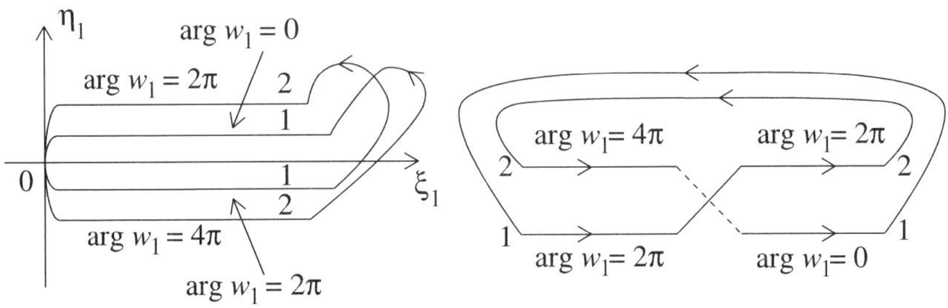

FIGURE 2.56. Two-sheeted Riemann surface of the mapping $\zeta_1 = \xi_1 + i\eta_1 = z^2$.

The remaining mappings are elementary:

$$w_2 = \frac{w_1}{\sqrt{a^2+1}} \qquad \text{(see Fig 2.55(b))}$$

and

$$\begin{aligned} w &= w_2 + \sqrt{w_2^2 - 1} \\ &= \frac{\sqrt{z^2+1}}{\sqrt{a^2+1}} + \sqrt{\frac{z^2+1}{a^2+1} - 1} \\ &= \frac{1}{\sqrt{a^2+1}} \left(\sqrt{z^2+1} + \sqrt{z^2 - a^2} \right), \end{aligned} \tag{2.8.14}$$

which is the desired mapping. □

Note 2.8.1. One can raise the question: "Why does (2.8.12) map a given domain onto the interior of the unit disk $|w| \le 1$ in Example 2.8.2 and onto the exterior of the same disk in Example 2.8.3?" The answer is that (2.8.12) defines a function with two branches, one branch mapping onto the domain $|w| < 1$ and the other onto the domain $|w| > 1$.

Note 2.8.2. In the previous Example 2.8.3, in considering the mapping $w_1 = \sqrt{z^2+1}$ as a sequence of the three intermediate mappings,

$$z \mapsto \zeta_1 = z^2 \mapsto \zeta_2 = \zeta_1 + 1 \mapsto \zeta_3 = \sqrt{\zeta_2},$$

one needs to consider the Riemann surface of the mapping $\zeta = z^2$ whose fundamental regions are the upper and lower half-planes,

$$\Im z > 0 \quad \text{and} \quad \Im z < 0.$$

In the first step, $\zeta_1 = z^2$ maps the whole z-plane with a cut in the form of a cross, shown in Fig 2.53(a). This map is possible only if the ζ_1-plane consists of a two-sheeted Riemann surface (see Fig 2.56). The upper half-plane $\Im z > 0$ with the cut from $z = 0$ to $z = i$ is mapped on the first sheet, where $0 < \arg \zeta_1 < 2\pi$. The lower half-plane $\Im z < 0$ with the cut from $z = 0$ to $z = -i$ is mapped on the second sheet, where $2\pi < \arg \zeta_1 < 4\pi$.

The second mapping, $\zeta_2 = \zeta_1 + 1$, shifts both sheets of the Riemann surface shown in Fig 2.56 to the right by 1.

The third mapping, $\zeta_3 = \sqrt{\zeta_2}$, sends the first and second sheets of the Riemann surface of Fig 2.55(a) onto the upper and lower half-planes, respectively, so that the regions in Fig 2.56 are mapped onto the whole complex plane in Fig 2.55(a) with a cut along the real axis from $-\infty$ to $+\infty$. In particular, the exterior of the circle in Fig 2.53(b) is mapped onto the region in Fig 2.55(a) with a cut from $-\sqrt{a^2+1}$ to $\sqrt{a^2+1}$ along the real axis.

2.9. Mapping by trigonometric functions

Each trigonometric function can be represented as a composition of the exponential and Joukowsky's functions. For example,

$$\begin{aligned} w = \sin z &= \frac{1}{2i}\left(e^{iz} - e^{-iz}\right) \\ &= \frac{1}{2}\left(w_1 + \frac{1}{w_1}\right), \end{aligned}$$

where $w_1 = e^{iz}/i$. Therefore we consider only the mapping by the function

$$\begin{aligned} w = \cos z &= \cos(x + iy) \\ &= \cos x \cosh y - i \sin x \sinh y. \end{aligned} \tag{2.9.1}$$

It follows from (2.9.1) that

$$\begin{aligned} u &= \cos x \cosh y \\ v &= -\sin x \sinh y. \end{aligned} \tag{2.9.2}$$

Let us find the image of the upper semi-strip bounded by the sides A, B, and C shown in Fig 2.57 under the mapping (2.9.2):

$$\begin{aligned} A &= \{x = 0,\ 0 \le y < +\infty\} \Rightarrow A' = \{u = \cosh y,\ v = 0,\ 0 \le y < +\infty\}, \\ B &= \{y = 0,\ 0 \le x \le \pi\} \quad\ \Rightarrow B' = \{u = \cos x, v = 0,\ 0 \le x \le \pi\}, \\ C &= \{x = \pi,\ 0 \le y < +\infty\} \Rightarrow C' = \{u = -\cosh y,\ v = 0,\ 0 \le y < +\infty\}. \end{aligned}$$

Since the upper semi-strip lies on our left as we traverse the sides A, B, C, then, if we go along A', B' and C', the region $\Im w \le 0$ also lies on the left. This means that the upper semi-strip is mapped onto the lower half-plane $\Im w \le 0$.

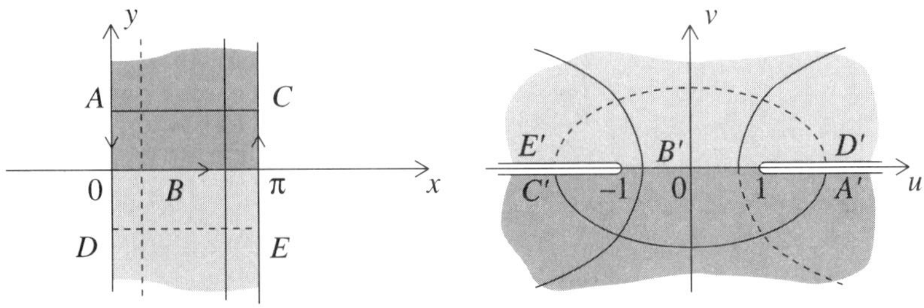

FIGURE 2.57. Mapping of the strip by the function $w = \cos z$.

Let us find the image of the lower semi-strip D, B, E, shown in Fig 2.57, under the same mapping:

$$\begin{aligned}
D &= \{x = 0,\ -\infty < y < 0\} \Rightarrow D' = \{u = \cosh y,\ v = 0,\ -\infty \le y < 0\},\\
B &= \{y = 0,\ 0 \le x \le \pi\} \quad\ \Rightarrow B' = \{u = \cos x, v = 0,\ 0 \le x \le \pi\},\\
E &= \{x = \pi,\ -\infty < y < 0\} \Rightarrow E' = \{u = -\cosh y,\ v = 0,\ -\infty < y < 0\}.
\end{aligned}$$

Since the half-lines D' and E' are mapped onto the same segments $(1, +\infty)$ and $(-\infty, -1)$ of the u-axis in the w-plane as the lines A' and C', then one has to cut the u-axis along these segments and consider that the sides A' and C' are attached to the lower parts of these cuts while the sides D' and E' are attached to the upper parts of the cuts. If we compare the directions of the sides D, B, E and D', B', E', it will be clear that the lower semi-strip is mapped onto the upper half-plane. Hence the strip

$$S = \{0 \le x \le \pi,\ -\infty < y < +\infty\}$$

is mapped onto the whole complex w-plane with two cuts. Therefore S is a fundamental region of the function $w = \cos z$. The other fundamental regions are the strips

$$k\pi \le x \le (k+1)\pi, \qquad -\infty < y < +\infty, \qquad k = 0, \pm 1, \pm 2, \dots .$$

We show that each straight line, $x = x_0$, in the strip S is mapped into one of the branches of a hyperbola. Letting $x = x_0$ in (2.9.2) we obtain the parametric equations of a parabola,

$$\begin{aligned}
u &= \quad \cosh y \cos x_0, \qquad\quad 0 < x_0 < \pi,\\
v &= -\sinh y \sin x_0, \qquad -\infty < y < +\infty,
\end{aligned} \tag{2.9.3}$$

which, rewritten in Cartesian coordinates, becomes

$$\frac{u^2}{\cos^2 x_0} - \frac{v^2}{\sin^2 x_0} = 1. \tag{2.9.4}$$

If $0 < x_0 < \pi/2$, it follows from (2.9.3) that $u > 0$, that is, (2.9.3) describes the right branch of the hyperbola, while in the case $\pi/2 < x_0 < \pi$, we have $u < 0$, so that (2.9.3) describes the left branch of the hyperbola. The points $u = \pm 1$ are the foci of the hyperbolae (2.9.4).

We show that each straight segment $y = y_0$ in the strip S is mapped into the lower or upper part of the ellipse that is confocal with the hyperbola (2.9.4). Letting $y = y_0$ in (2.9.2), we obtain the parametric equations of an ellipse,

$$\begin{aligned} u &= \cosh y_0 \cos x, & 0 < x < \pi, \\ v &= -\sinh y_0 \sin x, & -\infty < y_0 < +\infty, \end{aligned} \tag{2.9.5}$$

which, rewritten in Cartesian coordinates, becomes

$$\frac{u^2}{\cosh^2 y_0} + \frac{v^2}{\sinh y_0} = 1. \tag{2.9.6}$$

If $y_0 > 0$, it follows from (2.9.5) that $v < 0$, that is, (2.9.5) describes the lower part of the ellipse, but if $y_0 < 0$, then $v > 0$, so that (2.9.5) describes the upper part of the ellipse. The points $u = \pm 1$ are the foci of the ellipse (2.9.6), that is, this ellipse is confocal with the hyperbola (2.9.4).

Exercises for Sections 2.8 and 2.9

Find the image of the following domain D under the mapping $w = \frac{1}{2}\left(z + \frac{1}{z}\right)$.

1. $D = \{z;\ |z| < 1\}$.
2. $D = \{z;\ |z| > 1\}$.
3. $D = \{z;\ \Im z > 0\}$.
4. $D = \{z;\ |z| < 1,\ \Im z > 0\}$.
5. $D = \{z;\ R < |z| < 1,\ \Im z > 0\}$.
6. $D = \{z;\ 1 < |z| < R,\ \Im z > 0\}$.

Find the image of the given region under the given mapping.

7. $D = \{z;\ 0 < \Re z < \pi/2,\ \Im z > 0\}$, $\quad w = \sin z$.
8. $D = \{z;\ 0 < \Re z < \pi/2,\ \Im z > 0\}$, $\quad w = \cos z$.

9. $D = \{z;\ 0 < \Re z < \pi,\ -\infty < \Im z < +\infty\}$, $\quad w = \cos z$.

10. $D = \{z;\ 0 < \Re z < \pi,\ -\infty < \Im z < +\infty\}$, $\quad w = \sin z$.

11. $D = \{z;\ 0 < \Re z < \pi/2,\ 0 < \Im z < \pi/2\}$, $\quad w = \cos z$.

12. $D = \{z;\ 0 < \Re z < \pi/2,\ 0 < \Im z < \pi/2\}$, $\quad w = \sin z$.

Map the region D of the z-plane onto the region G of the w-plane.

13. $D = \{z;\ 0 < \Re z < \pi,\ \Im z < 0\}$, $\quad G = \{w;\ \Im w > 0\}$.

14. $D = \{z;\ \Re z > 0,\ 0 < \Im z < \pi/2\}$, $\quad G = \{w;\ \Im w > 0\}$.

15. $D = \{z;\ \Re z < 2,\ 0 < \Im z < 1\}$, $\quad G = \{w;\ \Im w > 0\}$.

16. $D = \{z;\ |z-1| > 1,\ |z-3| > 1,\ \Im z > 0\}$, $\quad G = \{w;\ \Im w > 0\}$.

CHAPTER 3

Complex Integration and Cauchy's Theorem

3.1. Paths in the complex plane

The integration of a function of a complex variable is done along a path in the complex plane as shown in Fig 3.1. For this purpose, we define piecewise differentiable paths and related terminology.

DEFINITION 3.1.1. Let $I = [\alpha, \beta]$, where $\alpha < \beta$, be a closed interval in $\mathbb{R}$. A *path* C is given by a continuous mapping,

$$\gamma : I \to \mathbb{C},$$

which is piecewise continuously differentiable; that is, $\gamma'(t)$ is piecewise continuous and

$$\gamma(t) = \alpha + \int_{\alpha}^{t} \gamma'(s)\, ds.$$

As t varies from α to β, the point $\gamma(t)$ describes a *curve* or *contour* or *trajectory* $\gamma(I)$ in $\mathbb{C}$. At the points $\gamma(t)$ where $\gamma'(t)$ is continuous and nonzero, the trajectory has a tangent in the direction $\gamma'(t) \in \mathbb{C}$. The points t where $\gamma'(t)$ is discontinuous but has both left and right nonzero limits are called *angular points* (see Fig 3.2).

Remark 3.1.1. A path C can have multiple points, that is, $\gamma(t_1) = \gamma(t_2)$ for some $t_1 \neq t_2$ (see Fig 3.2). Every point of a path can be a multiple point, as will be illustrated in the following example.

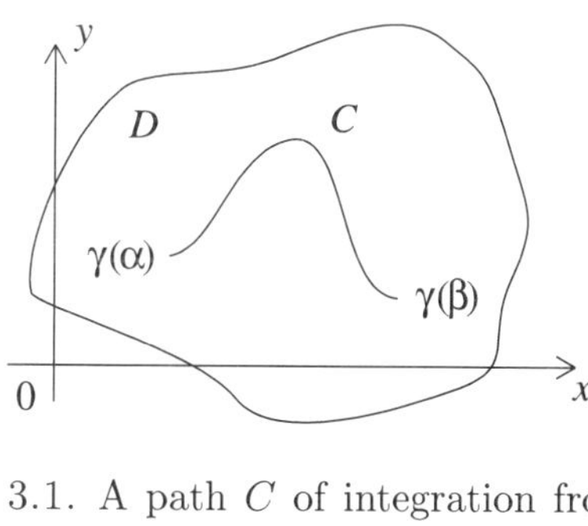

FIGURE 3.1. A path C of integration from $\gamma(\alpha)$ to $\gamma(\beta)$ in the complex z-plane.

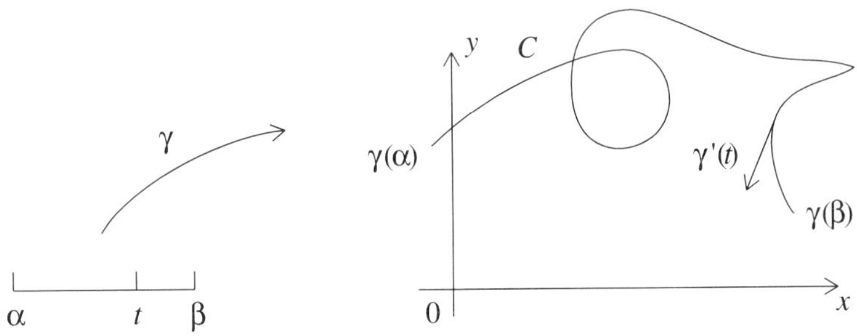

FIGURE 3.2. A path in the complex plane.

EXAMPLE 3.1.1. *Consider the path C given by the continuous mapping $\gamma : [0,1] \to \mathbb{C}$ defined by*

$$t \mapsto e^{2\pi i \nu t}, \qquad \nu \in \mathbb{R}, \qquad \nu \neq 0.$$

One sees that $\gamma(I)$ is a subset of the unit circle. However, if $\nu = n$ is a positive integer, the unit circle is traversed n times.

EXAMPLE 3.1.2. *Consider the path given by the mapping $\gamma : [0,2] \to \mathbb{C}$ defined by*

$$\gamma(t) = \begin{cases} c(1-t) + dt, & 0 \leq t \leq 1, \\ d(2-t) - c(1-t), & 1 \leq t \leq 2. \end{cases}$$

One sees that the path C is the segment with endpoints c and d, traversed from c to d and from d to c. One also sees that $t = 1$ is a point of discontinuity of γ'.

It is important to distinguish between a path C given by γ and the corresponding curve $\gamma(I)$ which is the pointset covered by C. In fact, a path is a *parametrized curve* and the parametrization is as important as the curve itself.

We have the following terminology.

DEFINITION 3.1.2.

(a) A path C given by γ is contained in an open set D if $\gamma(I) \subset D$.
(b) If $I = [\alpha, \beta]$, then $\gamma(\alpha)$ and $\gamma(\beta)$ are the *initial* and *terminal* points of C.
(c) C is a *closed path* if $\gamma(\alpha) = \gamma(\beta)$.
(d) If the paths C_1 and C_2 are given by $\gamma_1 : [\alpha, \beta] \to \mathbb{C}$ and $\gamma_2 : [\xi, \eta] \to \mathbb{C}$, respectively, such that $\gamma_1(\beta) = \gamma_2(\xi)$, then the path $C = C_1 + C_2$ is the *juxtaposition* of C_1 and C_2 defined as follows:

$$\gamma : [\alpha, \eta + \beta - \xi] \to \mathbb{C},$$

where

$$\gamma(t) = \begin{cases} \gamma_1(t), & \text{for } \alpha \le t \le \beta, \\ \gamma_2(t - \beta + \xi), & \text{for } \beta \le t \le \eta + \beta - \xi. \end{cases}$$

(e) Given a path C, the *opposite path*, denoted by $-C$, is given by

$$-\gamma(t) = \gamma(\alpha + \beta - t),$$

which traverses C in the opposite direction.

We remark that any path C given by $\gamma : [\alpha, \beta] \to \mathbb{C}$ is the juxtaposition of its restrictions C_1 and C_2; that is, for $a \le \xi \le \eta$,

$$\gamma_1 : [\alpha, \xi] \to \mathbb{C}, \qquad \gamma_2 : [\xi, \eta] \to \mathbb{C}.$$

Every closed path is the juxtaposition of two paths in an infinite number of ways.

DEFINITION 3.1.3. A *simple closed curve* is a curve whose only double points are its initial and terminal points.

We state without proof the following theorem.

THEOREM 3.1.1 (JORDAN CURVE THEOREM). *Let C be a simple closed curve in $\mathbb{C}$. Then $\mathbb{C} \backslash C$ has exactly two connected components, one bounded and the other unbounded.*

3.2. Complex line integrals

3.2.1. Definition of the complex line integral. Let C be a path given by $\gamma : [\alpha, \beta] \to \mathbb{C}$ and f a complex-valued function which is continuous on the curve $\gamma(I)$. Then the composite function $I : [\alpha, \beta] \to \mathbb{C}$ defined by

$$t \mapsto f\big(\gamma(t)\big)\gamma'(t)$$

is piecewise continuous on $[\alpha, \beta]$. Therefore its integral is defined.

DEFINITION 3.2.1. The complex number

$$\int_C f(z)\,dz = \int_\alpha^\beta f\big(\gamma(t)\big)\gamma'(t)\,dt \tag{3.2.1}$$

is called the *integral of f along the path C.*

Path or line integrals depend not only on the curve $\gamma(I)$ but also on the parametrization of C.

EXAMPLE 3.2.1. *Let C_k be the paths given by the continuous mappings $\gamma_k : [0, 1] \to \mathbb{C}$ defined by*

$$t \mapsto e^{2\pi i k t}, \qquad k \in \mathbb{Z}.$$

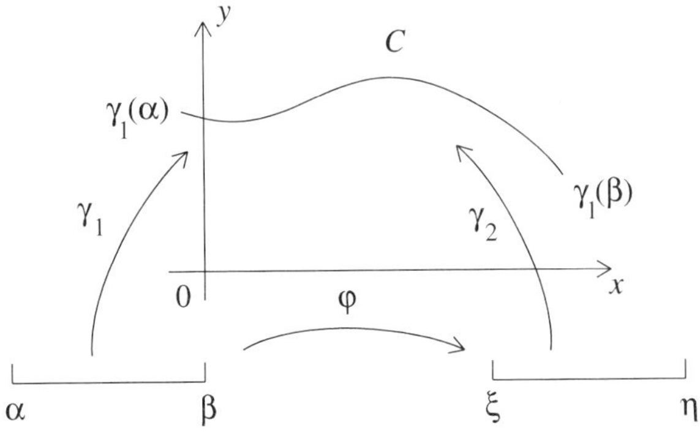

FIGURE 3.3. Equivalent paths in the complex plane.

Given the function $f(z) = 1/z$, *we have the path integral*

$$\int_{C_k} f(z)\,dz = \int_0^1 \frac{1}{e^{2\pi ikt}}\left(2\pi ik\, e^{2\pi ikt}\right)\,dt = 2\pi ik \int_0^1 dt = 2\pi ik.$$

It is seen that the value of the integral depends upon the number of times the path traverses the unit circle.

Two paths, C_1 and C_2 given by $\gamma_1 : [\alpha, \beta] \to \mathbb{C}$ and $\gamma_2 : [c, d] \to \mathbb{C}$, respectively, are said to be *equivalent* if there exists a strictly increasing bijection $\varphi : [\alpha, \beta] \to [\widetilde{\alpha}, \widetilde{\beta}]$ (see Fig 3.3) which is continuous and piecewise continuously differentiable together with its inverse φ^{-1}, such that

$$\gamma_1(t) = \gamma_2 \circ \varphi(t) =: \gamma_2(\varphi(t)), \qquad t \in [\alpha, \beta].$$

The following useful invariance theorem holds for integrals along equivalent paths.

THEOREM 3.2.1. *Let* C_1 *and* C_2 *be two equivalent paths. Then*

$$\int_{C_1} f(z)\,dz = \int_{C_2} f(z)\,dz. \tag{3.2.2}$$

PROOF. Letting φ denote the bijection between the two paths and applying the definitions, we have

$$\begin{aligned}
\int_{C_1} f(z)\,dz &= \int_\alpha^\beta f\big(\gamma_1(t)\big)\gamma_1'(t)\,dt \\
&= \int_\alpha^\beta f\big(\gamma_2 \circ \varphi(t)\big)\gamma_2'\big(\varphi(t)\big)\varphi'(t)\,dt \\
&= \int_c^d f\big(\gamma_2(u)\big)\,du \qquad (u = \varphi(t))
\end{aligned}$$

$$= \int_{C_2} f(z)\,dz. \quad \square$$

3.2.2. Properties of the line integral. We establish a few basic properties of complex line integrals, where the path C is given by the mapping $\gamma : [\alpha, \beta] \to \mathbb{C}$ such that $t \mapsto \gamma(t)$.

(a) Direction dependence.

$$\int_{-C} f(z)\,dz = -\int_C f(z)\,dz, \tag{3.2.3}$$

where, by part (e) of Definition 3.1.2, the opposite path $-C$ is given by $\widetilde{\gamma} = -\gamma : [\alpha, \beta] \to \mathbb{C}$ defined by $t \mapsto \gamma(\alpha + \beta - t)$. This property is derived as follows:

$$\begin{aligned}
\int_{-C} f(z)\,dz &= \int_\alpha^\beta f\left(\widetilde{\gamma}(t)\right) \widetilde{\gamma}'(t)\,dty \\
&= \int_\alpha^\beta f\left(\gamma\alpha + \beta - t)\right)\left[-\gamma'(\alpha + \beta - t)\,dt \right.\\
&= \int_b^a f\left(\gamma(u)\right)\gamma'(u)\,du \qquad (\text{putting } u = \alpha + \beta - t) \\
&= -\int_\alpha^\beta f\left(\gamma(u)\right)\gamma'(u)\,du \\
&= -\int_C f(z)\,dz.
\end{aligned}$$

(b) Additivity. If the path C is the juxtaposition of the paths C_1 and C_2, as shown in Fig 3.4, then

$$\int_C f(z)\,dz = \int_{C_1} f(z)\,dz + \int_{C_2} f(z)\,dz. \tag{3.2.4}$$

(c) Linearity.

$$\int_C [af(z) + bg(z)]\,dz = a\int_C f(z)\,dz + b\int_C g(z)\,dz. \tag{3.2.5}$$

(d) Integral of a constant. If $f(z) = 1$, then

$$\int_C dz = \int_\alpha^\beta \gamma'(t)\,dt = \gamma(\beta) - \gamma(\alpha). \tag{3.2.6}$$

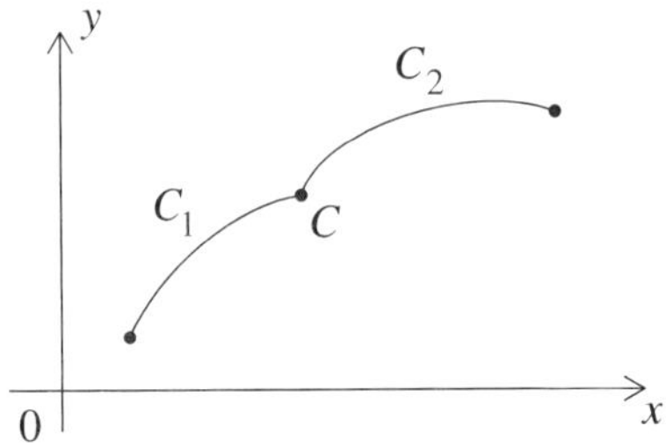

FIGURE 3.4. Additivity of the line integral.

(e) Upper bound for the modulus of an integral. If

$$|f(z)| \leq M, \quad \text{for all } z \in \gamma(I), \tag{3.2.7}$$

then

$$\left| \int_C f(z)\, dz \right| \leq M \int_\alpha^\beta |\gamma'(t)|\, dt = ML, \tag{3.2.8}$$

where L is the length of C.

To prove (e), we use the inequality

$$\left| \int_\alpha^\beta w(t)\, dt \right| \leq \int_\alpha^\beta |w(t)|\, dt,$$

where it is assumed that the function $w : [\alpha, \beta] \to \mathbb{C}$ is piecewise continuous. If

$$\int_\alpha^\beta w(t)\, dt = 0,$$

the inequality is obvious. Otherwise,

$$\int_\alpha^\beta w(t)\, dt = r_0\, e^{i\varphi_0},$$

whence

$$\begin{aligned} r_0 &= \int_\alpha^\beta e^{-i\varphi_0} w(t)\, dt \\ &= \Re\left(\int_\alpha^\beta e^{-i\varphi_0} w(t)\, dt \right) \\ &= \int_\alpha^\beta \Re\left(e^{-i\varphi_0} w(t)\right)\, dt \\ &\leq \int_\alpha^\beta \left|e^{-i\varphi_0} w(t)\right|\, dt \end{aligned}$$

$$= \int_{\alpha}^{\beta} |w(t)|\, dt.$$

Remark 3.2.1. Other types of complex line integrals have been defined. In the language of differential geometry, one can define integrals of 0- and 1-forms along a curve C on a two-dimensional manifold.

Let C be a smooth curve given by the equation

$$z = \gamma(t), \qquad \alpha \le t \le \beta.$$

If $f(z)$ is a 0-form, that is, a smooth function defined on C, the line integral of f along C is

$$\int_C f\, ds = \int_C f(z)\, |dz| = \int_{\alpha}^{\beta} f(z(t))|\gamma'(t)|\, dt. \tag{3.2.9}$$

This integral is independent of the parametrization of the curve C. In the Russian mathematical literature, a line integral of a function is called an integral of the first kind.

A line integral of the second kind along C is the integral of a 1-form, that is, the integral of a vector field along the curve. Integral of 1-forms can be expressed by the formula

$$\int_C a \cdot ds, \tag{3.2.10}$$

where a is a vector and ds is a differential element tangent to the curve. Our definition of integral (3.2.1) is an integral of the second kind, as can be seen from the following formulation (3.2.12). The sign of an integral of the second kind depends upon the direction along which the curve is traversed, as shown in property (c).

Note 3.2.1. The upper bound (3.2.8) can be sharpened by using the definition (3.2.9). In this case, we obtain the estimate

$$\left| \int_C f(z)\, dz \right| \le \int_C |f(z)|\, |dz| = \int_C |f(z)|\, ds, \tag{3.2.11}$$

where ds is the differential of an arc on C and the integral on the right-hand side of (3.2.11) is the line integral of a real positive function along C.

3.2.3. Integration methods. Separating the real and the imaginary parts of a line integral, we obtain two real integrals,

$$\begin{aligned} \int_C f(z)\, dz = \int_C [u(x,y)\, dx - v(x,y)\, dy] \\ + i \int_C [v(x,y)\, dx + u(x,y)\, dy]. \end{aligned} \tag{3.2.12}$$

Therefore the complex line integral of the function $f(z) = u(x,y) + iv(x,y)$ of a complex variable exists simultaneously with the real line integrals of the real functions $u(x,y)$ and $v(x,y)$. These line integrals exist, for example, if the curve C is piecewise smooth and the functions u and v are piecewise continuous on C.

Note that the line integrals in (3.2.12) can be reduced to definite integrals. We consider two such cases.

(1) If the path C is given by the parametric equations

$$x = x(t), \qquad y = y(t), \qquad \alpha \le t \le \beta,$$

so that $a = x(\alpha) + iy(\alpha)$ and $b = x(\beta) + iy(\beta)$, and $f(z)$ is piecewise smooth on C, then

$$\begin{aligned}\int_C f(z)\,dz &= \int_\alpha^\beta \left\{u\big(x(t),y(t)\big)x'(t) - v\big(x(t),y(t)\big)y'(t)\right\}dt \\ &\quad + i\int_\alpha^\beta \left\{v\big(x(t),y(t)\big)x'(t) + u\big(x(t),y(t)\big)y'(t)\right\}dt. \qquad (3.2.13)\end{aligned}$$

(2) If the path C is given by the equation $y = y(x)$ on $\alpha \le x \le \beta$ and $f(z)$ is piecewise smooth on C, then

$$\begin{aligned}\int_C f(z)\,dz &= \int_\alpha^\beta \left\{u\big(x,y(x)\big) - v\big(x,y(x)\big)y'(x)\right\}dx \\ &\quad + i\int_\alpha^\beta \left\{v\big(x,y(x)\big) + u\big(x,y(x)\big)y'(x)\right\}dx. \qquad (3.2.14)\end{aligned}$$

EXAMPLE 3.2.2. *For each of the following curves C with initial and terminal points $(0,0)$ and $(1,1)$, respectively, as shown in Fig 3.5, compute the line integral*

$$\begin{aligned}I_1 &= \int_C \bar{z}\,dz = \int_C (x - iy)(dx + i dy) \\ &= \int_C (x\,dx + y\,dy) + i\int_C (-y\,dx + x\,dy), \qquad (3.2.15)\end{aligned}$$

where

(a) C is a segment of the straight line $y = x$,
(b) C is a part of the parabola $y = x^2$,
(c) C is the polygonal line OAB.

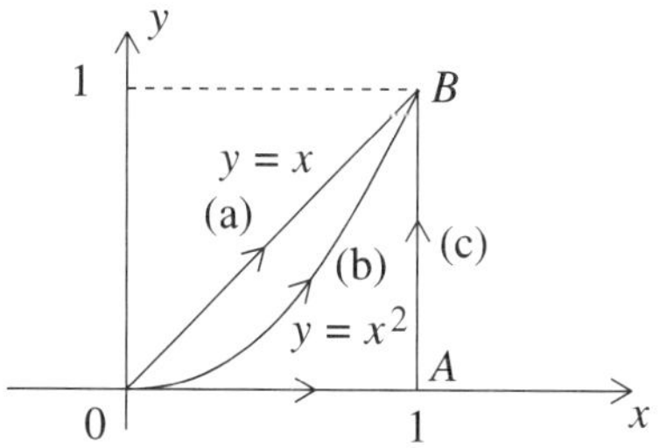

FIGURE 3.5. Paths of integration (a), (b) and (c) in Examples 3.2.2 and 3.2.3

SOLUTION. (a) If $y = x$ on $0 \le x \le 1$, then $dy = dx$ and formula (3.2.15) gives

$$I_1 = \int_0^1 (x\,dx + x\,dx) + i\int_0^1 (-x\,dx + x\,dx) = 2\frac{x^2}{2}\bigg|_0^1 = 1.$$

(b) If $y = x^2$ on $0 \le x \le 1$, then $dy = 2x\,dx$ and we have

$$\begin{aligned} I_1 &= \int_0^1 (x\,dx + x^2 \times 2x\,dx) + i\int_0^1 (-x^2\,dx + x \times 2x\,dx) \\ &= \left(\frac{x^2}{2} + \frac{2x^4}{4}\right)\bigg|_0^1 + i\left(\frac{-x^3}{3} + \frac{2x^3}{3}\right)\bigg|_0^1 = 1 + \frac{1}{3}\,i. \end{aligned}$$

(c) Integrating along the polygonal line OAB, we obtain

$$I_1 = \int_{OAB} \bar{z}\,dz = \int_{OA} \bar{z}\,dz + \int_{AB} \bar{z}\,dz.$$

On the line segment OA, we have $y = 0$, $dy = 0$, $0 \le x \le 1$, so that

$$\int_{OA} \bar{z}\,dz = \int_0^1 x\,dx + i\int_0^1 (-0\,dx + x \times 0) = \frac{1}{2}.$$

On the line segment AB, we have $x = 1$, $dx = 0$, $0 \le y \le 1$, hence

$$\int_{AB} \bar{z}\,dz = \int_0^1 (1 \times 0 + y\,dy) + i\int_0^1 [(-y) \times 0 + 1 \times dy] = \frac{1}{2} + i.$$

Therefore

$$I_1 = \left(\int_{OA} + \int_{AB}\right)\bar{z}\,dz = 1 + i. \quad \square$$

As can be seen from this example, the value of the integral depends on the path joining the points (0,0) and (1,1).

EXAMPLE 3.2.3. *For the three paths, (a), (b) and (c), joining the points* $(0,0)$ *and* $(1,1)$, *given in the previous example and shown in Fig 3.5, compute the integral*

$$\begin{aligned} I_2 &= \int_C z^2\,dz = \int_C (x^2 - y^2 + 2ixy)(dx + i dy) \\ &= \int_C [(x^2 - y^2)\,dx - 2xy\,dy] + i\int_C [2xy\,dx + (x^2 - y^2)\,dy]. \end{aligned} \tag{3.2.16}$$

SOLUTION. (a) If $y = x$ on $0 \le x \le 1$, then formula (3.2.16) gives

$$\begin{aligned} I_2 &= \int_0^1 [(x^2 - x^2)\,dx - 2x^2\,dx] + i\int_0^1 [2x^2\,dx + 0 \times dx] \\ &= -2\frac{x^3}{3}\bigg|_0^1 + i\,\frac{2x^3}{3}\bigg|_0^1 = -\frac{2}{3} + \frac{2}{3}\,i. \end{aligned}$$

(b) If $y = x^2$ on $0 \le x \le 1$, then formula (3.2.16) gives

$$\begin{aligned} I_2 &= \int_0^1 [(x^2 - x^4)\,dx - 2x \times x^2 \times 2x\,dx] \\ &\quad + i\int_0^1 [2x \times x^2\,dx + (x^2 - x^4)2x\,dx] \\ &= \left(\frac{x^3}{3} - x^5\right)\bigg|_0^1 + i\left(x^4 - \frac{x^6}{3}\right)\bigg|_0^1 = -\frac{2}{3} + \frac{2}{3}\,i. \end{aligned}$$

(c) Integrating along the polygonal line segment OAB, we obtain

$$I_2 = \int_{OAB} z^2\,dz = \int_{OA} z^2\,dz + \int_{AB} z^2\,dz.$$

On the line segment OA, we have from (3.2.16) that

$$\int_{OA} z^2\,dz = \int_0^1 x^2\,dx = \frac{1}{3}.$$

On the line AB, $x = 1$, $dx = 0$, $0 \le y \le 1$, so that

$$\begin{aligned} \int_{AB} z^2\,dz &= \int_0^1 (-2 \times 1 \times y)\,dy + i\int_0^1 (1 - y^2)\,dy \\ &= -y^2\bigg|_0^1 + \left(y - \frac{y^3}{3}\right)\bigg|_0^1 \\ &= -1 + \frac{2}{3}\,i. \end{aligned}$$

Hence

$$I_2 = \left(\int_{OA} + \int_{AB}\right) z^2\,dz = -\frac{2}{3} + \frac{2}{3}i. \quad \sqcup$$

In this example the value of the integral I_2 is independent of the path of integration.

The following question arises: why does I_1 depend on the path of integration joining the end points of the curve C in Example 3.2.2, while, in Example 3.2.3, I_2 does not depend on this path? This question will be answered by Cauchy's Theorem, considered in the next section.

3.2.4. Complex line integral of non-parametric curve. In this subsection we give a definition of the integral of a function $f(z)$ along a curve C in non-parametric form (see, for example, [**20**], p. 15 and [**44**], p. 92). This definition may be useful for numerical integration.

A curve is said to be *rectifiable*, or of *bounded variation*, if it is of finite length.

Suppose that C is a rectifiable curve in the complex plane, with initial and terminal points a and b, respectively, and let $w = f(z)$ be a continuous function defined on C (see Figure 3.6). We subdivide C into n arcs, γ_k,

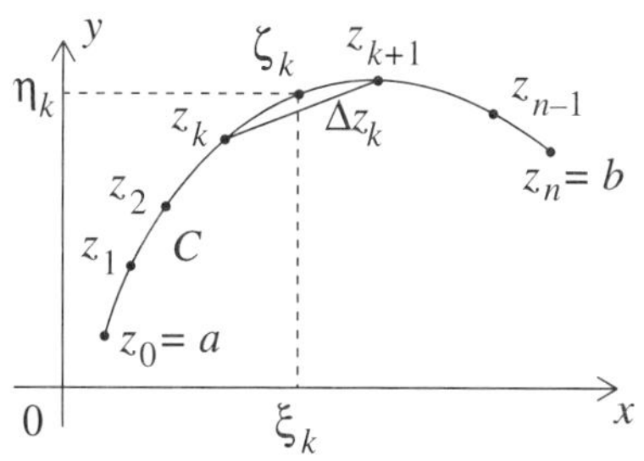

FIGURE 3.6. Partition of the curve C.

$k = 0, 1, \dots, n-1$, by means of $n-1$ succesive points, $z_1, z_2, \dots, z_{n-1}$, chosen arbitrarily, and set $z_0 = a$ and $z_n = b$. On each arc γ_k joining z_k to z_{k+1} we choose an arbitrary point $\zeta_k = (\xi_k, \eta_k) \in \gamma_k$ and form the *integral sum*

$$S_n = \sum_{k=0}^{n-1} f(\zeta_k)\Delta z_k, \tag{3.2.17}$$

where $\Delta_k = z_{k+1} - z_k$.

DEFINITION 3.2.2. Given a rectifiable curve C in $\mathbb{C}$ and a continuous function $f(z)$ defined on C, if the integral sum (3.2.17) converges to a finite limit as $\max|\Delta z_k| \to 0$ independently of any particular subdivision of C

and of the choice of the points ζ_k, then this limit is called the complex line integral of f on C and is denoted by

$$\int_C f(z)\,dz = \lim_{\max|\Delta z_k|\to 0} \sum_{k=0}^{n-1} f(\zeta_k)\Delta z_k. \tag{3.2.18}$$

We note that the *length*, L, of a rectifiable curve C is given by the integral

$$L = \int_C |dz| = \lim_{\max|\Delta z_k|\to 0} \sum_{k=0}^{n-1} |\Delta z_k|.$$

THEOREM 3.2.2 (EXISTENCE OF LINE INTEGRALS). *If the curve C is piecewise smooth and the function $f(z)$ is piecewise continuous on C with a finite number of finite jumps, then the line integral* (3.2.18) *exits.*

The properties of line integrals for parametric curves listed in Subsection 3.2.2 also holds for the line integral (3.2.18). Moreover, the integral (3.2.18) of a complex-valued function $f(z)$ can be expressed in terms of integrals of real functions of two real variables in the form (3.2.12).

Exercises for Section 3.2

Integrate $\displaystyle\int_C \bar{z}\,dz$ along:

1. The line segment joining the point $z = 1+i$ to the point $z = 3+2i$.
2. The semicircle $|z| = 2$, $0 \le \operatorname{Arg} z \le \pi$, with initial point $z = 2$.
3. The parabola $y = x^2$ joining the point $z = 0$ to the point $z = 1+i$.
4. The polygonal line through the points $z = 0$, $z = 2$ and $z = 2+2i$, with initial point $z = 0$.
5. The circle $|z-1| = 1$ taken counterclockwise.

Integrate $\displaystyle\int_C |z|^2\,dz$ along each of the following curves joining the point $z = 0$ to the point $z = 2+2i$.

6. $y = x$.
7. $y = x^2/2$.
8. $x = y^2/2$.
9. The polygonal line through the points $z = 0$, $z = 2i$ and $z = 2+2i$, with initial point $z = 0$.

10. The polygonal line through the points $z = 0$, $z = 2$ and $z = 2 + 2i$, with initial point $z = 0$.

Evaluate $\int_C f(z)\,dz$ for each given pair f and C.

11. $f(z) = \dfrac{z^2 - 1}{z^2}$, $\quad C:\ z = 1 + (1+i)t, \quad 0 \le t \le 1.$

12. $f(z) = z^2$, $\quad C:\ z = e^{it}, \quad 0 \le t \le \pi.$

13. $f(z) = |z|^4$, $\quad C:\ |z| = 4, \quad 0 \le \arg z \le 2\pi.$

14. $f(z) = z^2 \Re z$, $\quad C:\ z = 1 + (2+i)t, \quad 0 \le t \le 1.$

15. $f(z) = (\Im z)^2$, $\quad C:\ z = e^{it}, \quad -\pi/2 \le t \le \pi/2.$

16. $f(z) = \bar{z}$, $\quad C:\ z = 2\,e^{it}, \quad 0 \le t \le \pi.$

17. $f(z) = \operatorname{Arg} z$, $\quad C:\ |z| = R, \quad 0 \le \arg z \le \pi.$

18. $f(z) = \bar{z}|z|$, $\quad C:\ |z - i| = 1$, taken counterclockwise.

19. $f(z) = \dfrac{\operatorname{Log}^2 z}{z}$, $\quad C:$ the line segment joining the point $z = 1$ to the point $z = 2 + i$.

20. $f(z) = z\cos z$, $\quad C:$ the arc $z = it$ with $0 \le t \le \pi$.

Use the ML-inequality to obtain an upper bound for the following integrals, where M is an upper bound for the modulus of the integrand and L is the length of the curve of integration.

21. $\displaystyle\int_C \frac{1}{z - 1 + i}\,dz$, $\quad$ where $C:\ |z - 1 + i| = 2.$

22. $\displaystyle\int_C [(2+i)z^2 + 3iz]\,dz$, $\quad$ where $C:\ |z| = 1.$

23. $\displaystyle\int_C \frac{1}{z^2(z^2+4)}\,dz$, $\quad$ where $C:\ |z| = 2, \quad 0 \le \operatorname{Arg} z \le \pi/4.$

24. $\displaystyle\int_C \frac{e^z - 1}{z}\,dz$, $\quad$ where $C:\ |z| = 1, \quad 0 \le \operatorname{Arg} z \le \pi.$

25. Find an upper bound for the integral $\int_{C_R} u(z)/z^2\,dz$, where C_R is the circle $|z| = R$, and $u(z)$ is a continuous function which is bounded for all z. Find $\lim_{R\to\infty} \int_{C_R} u(z)/z^2\,dz$.

26. Find an upper bound for the integral $\displaystyle\int_{C_R} \frac{1}{z^2} \operatorname{Log} z \, dz$, where C_R is the semicircle $|z| = R$, $0 \leq \operatorname{Arg} z \leq \pi$. Find $\displaystyle\lim_{R\to\infty} \int_{C_R} \frac{1}{z^2} \operatorname{Log} z \, dz$.

3.3. Cauchy's Theorem

Cauchy's Theorem is one of the fundamental theorems in complex analysis.

3.3.1. Cauchy's Theorem for simply connected domains. Since the sign of a line integral depends on the direction of integration along the closed path C, the *positive direction* along C will be the direction for which the interior region, R, lies on our left as we traverse the curve. The other direction is the *negative direction.* If a closed path C is simple, then the positive and negative directions of C, corresponding to the bounded domain enclosed by C, may be said to be *counterclockwise* and *clockwise*, respectively.

On occasions, positively and negatively oriented closed paths will be denoted by C^+ and C^-, respectively. Thus, when necessary, integration in the positive and negative directions will be denoted by

$$\oint_{C^+} f(z)\,dz \quad \text{and} \quad \oint_{C^-} f(z)\,dz,$$

respectively (see Fig 3.7).

We shall use the following auxiliary theorem from the theory of real line integrals.

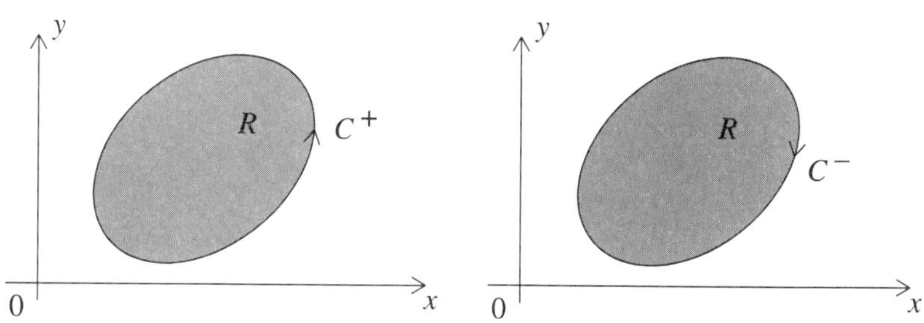

FIGURE 3.7. (a) Positive and (b) negative directions of integration along C.

THEOREM 3.3.1 (GREEN'S FORMULA). *Given that the real-valued functions $P(x,y)$ and $Q(x,y)$ and their partial derivatives Q_x and P_y are continuous in a closed simply connected region, D, bounded by a closed path C, then*

$$\oint_C P(x,y)\,dx + Q(x,y)\,dy = \iint_D \left(\frac{\partial Q}{\partial x} - \frac{\partial P}{\partial y}\right) dx\,dy. \qquad (3.3.1)$$

Formula (3.3.1) is known as Green's Theorem (see [**35**], p. 407).

We now state and prove the main theorem of this chapter under the condition that the derivative of an analytic function is continuous. However, this continuity assumption on $f'(z)$ will be removed by Goursat's Theorem 3.5.1 in Section 3.5.

THEOREM 3.3.2. (CAUCHY'S THEOREM FOR SIMPLY CONNECTED DOMAINS). *If $f(z)$ is analytic in a simply connected domain D and $f'(z)$ is continuous in D, then*

$$\oint_C f(z)\,dz = 0, \qquad (3.3.2)$$

where C is any closed path lying entirely in D.

PROOF. Using (3.2.12), we express the left-hand side of (3.3.2) as the sum of two real integrals:

$$\int_C f(z)\,dz = \int_C [u(x,y)\,dx - v(x,y)\,dy] + i\int_C [v(x,y)\,dx + u(x,y)\,dy]. \qquad (3.3.3)$$

Since $f(z)$ is analytic in D, the Cauchy–Riemann equations,

$$\frac{\partial u}{\partial x} = \frac{\partial v}{\partial y}, \qquad \frac{\partial u}{\partial y} = -\frac{\partial v}{\partial x}, \qquad (3.3.4)$$

hold everywhere in D. Moreover, by the continuity of $f'(z)$, the functions u_x, u_y, v_x and v_y are continuous in the closed region, R, bounded by the path C. Hence one can apply Green's formula (3.3.1) and the Cauchy–Riemann equations (3.3.4) to the two integrals on the right-hand side of (3.3.3). Therefore

$$\begin{aligned}\oint_C u\,dx - v\,dy &= \iint_R \left(-\frac{\partial v}{\partial x} - \frac{\partial u}{\partial y}\right) dx\,dy \\ &= \iint_R \left(\frac{\partial u}{\partial y} - \frac{\partial u}{\partial y}\right) dx\,dy = 0, \end{aligned} \qquad (3.3.5)$$

and

$$\begin{aligned}\oint_C v\,dx + u\,dy &= \iint_R \left(\frac{\partial u}{\partial x} - \frac{\partial v}{\partial y}\right) dx\,dy \\ &= \iint_R \left(\frac{\partial v}{\partial y} - \frac{\partial v}{\partial y}\right) dx\,dy = 0. \qquad (3.3.6)\end{aligned}$$

It then follows that

$$\oint_C f(z)\,dz = 0. \quad \square$$

Note 3.3.1. If a line integral is equal to zero along every closed path lying in a simply connected domain D, then the value of the integral does not depend on the path joining any two points in D and lying entirely in D (see [**32**], p. 510).

Equivalently, the following corollary can be derived from Cauchy's Theorem.

COROLLARY 3.3.1. *If $f(z)$ is analytic in a simply connected domain D then, for any two points z_0 and z lying in D, the integral*

$$F(z) = \int_{z_0}^{z} f(\zeta)\,d\zeta \qquad (3.3.7)$$

does not depend on the path, in D, joining z_0 and z and is a function of the upper limit z.

In particular, this corollary explains why the integral

$$\int_C \bar{z}\,dz,$$

in Example 3.2.2, depends on the path of integration since $f(z) = \bar{z}$ is not analytic (the Cauchy–Riemann equations are not satisfied). On the other hand, the integral

$$\int_C z^2\,dz,$$

in Example 3.2.3, is independent of the path of integration and depends only on the endpoints 0 and $1+i$ of C since $f(z) = z^2$ is analytic in the whole complex plane, and, in this case, the integral can be simply evaluated as follows:

$$\int_C z^2\,dz = \int_0^{1+i} z^2\,dz = \left.\frac{z^3}{3}\right|_0^{1+i} = \frac{1}{3}(1+i)^3 = -\frac{2}{3} + \frac{2}{3}\,i.$$

THEOREM 3.3.3. *If $f(z)$ is defined and continuous in a simply connected domain D and the integral of $f(z)$ along any path, lying entirely in D, is equal to zero, then the function*

$$F(z) = \int_{z_0}^{z} f(\zeta)\, d\zeta, \qquad z_0, z \in D, \tag{3.3.8}$$

called an indefinite integral, primitive *or* antiderivative *of $f(z)$, is analytic in D and $F'(z) = f(z)$.*

PROOF. Consider the difference quotient

$$\begin{aligned} \frac{F(z+\triangle z) - F(z)}{\triangle z} &= \frac{1}{\triangle z}\left[\int_{z_0}^{z+\triangle z} - \int_{z_0}^{z}\right] f(\zeta)\, d\zeta \\ &= \frac{1}{\triangle z}\left[\int_{z_0}^{z} + \int_{z}^{z+\triangle z} - \int_{z_0}^{z}\right] f(\zeta)\, d\zeta \\ &= \frac{1}{\triangle z}\int_{z}^{z+\triangle z} f(\zeta)\, d\zeta. \end{aligned} \tag{3.3.9}$$

To derive (3.3.9) we have used the additivity property of the integral and have assumed that both integrals from z_0 to z have been computed along the same arbitrary path. This path can be arbitrary since the integral of $f(z)$ along any path in D is equal to zero.

By formula (3.2.6), we obtain

$$\int_{z}^{z+\triangle z} d\zeta = z + \triangle z - z = \triangle z \tag{3.3.10}$$

for any path lying entirely in D and joining z and $z + \triangle z$. Then

$$\int_{z}^{z+\triangle z} f(z)\, d\zeta = f(z)\triangle z. \tag{3.3.11}$$

Using (3.3.9) and (3.3.11) and assuming that z and $z + \triangle z$ are joined by a straight line segment, we obtain the estimate

$$\begin{aligned} \left|\frac{F(z+\triangle z) - F(z)}{\triangle z} - f(z)\right| &= \frac{1}{|\triangle z|}\left|\int_{z}^{z+\triangle z} [f(\zeta) - f(z)]\, d\zeta\right| \\ &\le \max_{\zeta \in [z, z+\triangle z]} |f(\zeta) - f(z)|. \end{aligned}$$

Since $f(\zeta)$ is continuous at the point z, then for every $\varepsilon > 0$ there exists $\delta > 0$ such that, if $|\triangle z| < \delta$, then

$$\max_{\zeta \in [z, z+\triangle z]} |f(\zeta) - f(z)| < \varepsilon,$$

and hence

$$\left|\frac{F(z+\triangle z)-F(z)}{\triangle z}-f(z)\right|<\varepsilon.$$

This last inequality means that the derivative

$$F'(z)=\lim_{\triangle z\to 0}\frac{F(z+\triangle z)-F(z)}{\triangle z}=f(z)$$

exists and is equal to $f(z)$. □

The analog of Newton–Leibniz' formula,

$$\int_{z_1}^{z_2} f(z)\,dz = F(z_2)-F(z_1), \tag{3.3.12}$$

can be derived in a standard way.

Since all elementary functions of a complex variable are analytic in their domains of definition, then (3.3.12) is valid for all elementary functions over a simply connected domain.

EXAMPLE 3.3.1. *Evaluate the integral*

$$I_1=\int_1^i \frac{\operatorname{Log} z}{z}\,dz.$$

SOLUTION. By (3.3.12),

$$\begin{aligned}\int_1^i \frac{\operatorname{Log} z}{z}\,dz &= \int_1^i \operatorname{Log} z\, d(\operatorname{Log} z)\\ &= \left.\frac{\operatorname{Log}^2 z}{2}\right|_1^i = \frac{1}{2}\left(\operatorname{Log} e^{i\pi/2}\right)^2\\ &= \frac{1}{2}\left(i\,\frac{\pi}{2}\right)^2 = -\frac{\pi^2}{8}. \quad \square\end{aligned}$$

EXAMPLE 3.3.2. *Compute the integral*

$$I_2=\int_{|z|=1}\sqrt{z}\,dz, \tag{3.3.13}$$

where one selects the branch of $\sqrt{z}$ for which $\sqrt{1}=1$.

SOLUTION. To choose a branch of the function $\sqrt{z}$ we need to cut the complex plane from $z=0$ to $z=\infty$. Let us choose a cut along the negative real semi-axis (see Fig 3.8). Then $-\pi<\operatorname{Arg} z\le\pi$, and $\sqrt{e^{i\theta}}=e^{i\theta/2}$.

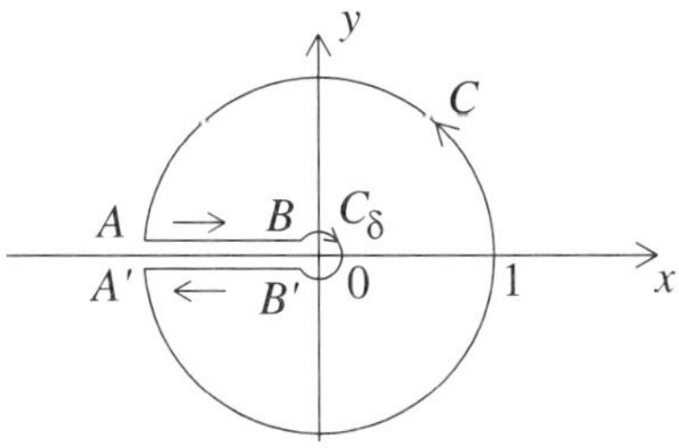

FIGURE 3.8. Closed path of integration in the complex plane with a cut along the negative real semi-axis for Example 3.3.2.

Since the path is $|z| = 1$, that is, $z = e^{i\theta}$ with $-\pi < \theta \le \pi$, then $z = e^{i\theta}$ and $dz = ie^{i\theta}\,d\theta$ in (3.3.13). Thus

$$\begin{aligned} I_2 &= \int_{-\pi}^{\pi} e^{i\theta} e^{i\theta/2} i\,d\theta = i\int_{-\pi}^{\pi} e^{3\theta i/2}\,d\theta \\ &= \frac{2}{3}\,e^{i3\theta/2}\Big|_{-\pi}^{\pi} = \frac{2}{3}\left(e^{3\pi i/2} - e^{-3\pi i/2}\right) \\ &= -\frac{4}{3}\,i \neq 0. \end{aligned} \tag{3.3.14}$$

We note that the integral is not equal to zero since the path $|z| = 1$ is not closed (see Fig 3.8). In order to close the path, one has to integrate along: (a) the upper cut from A to B, (b) the circle C_δ, of small radius δ, taken in the clockwise direction, and (c) the lower cuts from B' to A':

$$\int_{ABC_\delta B'A'} \sqrt{z}\,dz = \left(\int_{AB} + \int_{C_\delta} + \int_{B'A'}\right)\sqrt{z}\,dz.$$

On the segment AB, we have

$$z = re^{i\pi}, \qquad dz = e^{i\pi}\,dr, \qquad \sqrt{z} = e^{i\pi/2}\sqrt{r},$$

$$\int_{AB} \sqrt{z}\,dz = \int_1^{\delta} e^{i\pi} e^{i\pi/2}\sqrt{r}\,dr = e^{3\pi i/2}\,\frac{2}{3}\,r^{3/2}\Big|_1^{\delta} \to \frac{2}{3}\,i, \quad \text{as } \delta \to 0.$$

On the circle C_δ, we have

$$z = \delta\, e^{i\varphi}, \qquad dz = \delta i\, e^{i\varphi}\,d\varphi, \qquad \sqrt{z} = \sqrt{\delta}\, e^{i\varphi/2},$$

$$\int_{C_\delta} \sqrt{z}\,dz = \int_{\pi}^{-\pi} \sqrt{\delta}\, e^{i\varphi/2} \delta i\, e^{i\varphi}\,d\varphi \to 0, \quad \text{as } \delta \to 0.$$

On the segment $B'A'$, we have

$$z = re^{-i\pi}, \qquad dz = e^{-i\pi}\,dr, \qquad \sqrt{z} = e^{-i\pi/2}\sqrt{r},$$

$$\int_{B'A'} \sqrt{z}\,dz = \int_\delta^1 e^{-i\pi}e^{-i\pi/2}\sqrt{r}\,dr = e^{-3\pi i/2}\frac{2}{3}\,r^{3/2}\bigg|_\delta^1 \to \frac{2}{3}\,i, \quad \text{as } \delta \to 0.$$

Then

$$\int_{ABC_\delta B'A'} \sqrt{z}\,dz = \frac{2}{3}\,i + \frac{2}{3}i = \frac{4}{3}\,i. \tag{3.3.15}$$

Adding (3.3.15) and (3.3.14), we obtain $4i/3 + (-4i/3) = 0$, as it should be by Cauchy's Theorem. □

EXAMPLE 3.3.3. *Compute the integral*

$$I_3 = \oint_{|z|=2} z^2\,dz.$$

SOLUTION. Since the equation of the path $|z| = 2$ is $z = 2e^{i\theta}$ with $0 \le \theta \le 2\pi$, then $dz = 2e^{i\theta}\,i\,d\theta$ and

$$\begin{aligned} I_3 &= \int_0^{2\pi} 2^2e^{2i\theta}2e^{i\theta}i\,d\theta \\ &= 8i\int_0^{2\pi} e^{3i\theta}\,d\theta = \frac{8i}{3i}\,e^{3i\theta}\bigg|_0^{2\pi} \\ &= \frac{8}{3}(e^{6\pi i} - 1) = 0, \end{aligned}$$

as it should be by Cauchy's Theorem. □

EXAMPLE 3.3.4. *Compute the integral*

$$I_4 = \oint_{|z|=2} \frac{dz}{z}.$$

SOLUTION. As in the previous example, $z = 2\,e^{i\theta}$ and $dz = 2\,e^{i\theta}i\,d\theta$. Hence we have

$$\begin{aligned} I_4 &= \int_0^{2\pi} \frac{2ie^{i\theta}}{2e^{i\theta}}\,d\theta \\ &= i\int_0^{2\pi} d\theta = 2\pi i \ne 0. \end{aligned}$$

The nonzero value comes from the fact that $z = 0$ is a singular point of the integrand $f(z) = 1/z$ inside the path $|z| = 2$ and therefore the conditions of Cauchy's Theorem are not satisfied. □

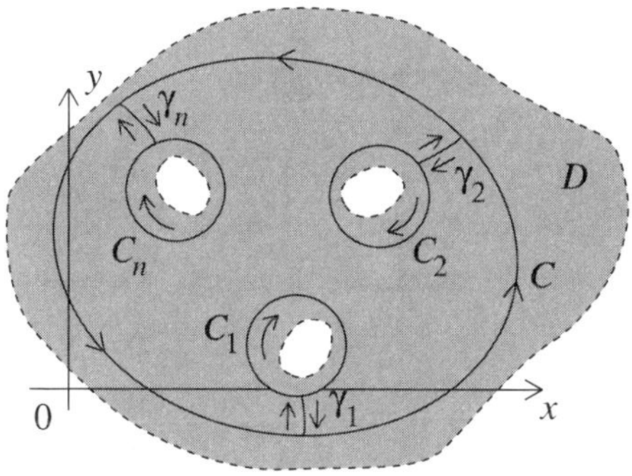

FIGURE 3.9. A multiply connected domain.

In the next section, Theorem 3.3.3 and Cauchy's integral formula will be used to prove a converse to Cauchy's Theorem called Morera's Theorem.

3.3.2. Cauchy's Theorem for multiply connected domains. Suppose that $f(z)$ is analytic in a multiply connected domain containing an external closed path, C, and internal closed paths, $C_1, C_2, \ldots, C_n$ (see Fig 3.9).

If the path C and the paths $C_1, C_2, \ldots, C_n$ are joined by the n arcs $\gamma_1, \gamma_2, \ldots, \gamma_n$, respectively, then D contains a simply connected region R bounded by the paths $C, C_1, C_2, \ldots, C_n$ and the arcs $\gamma_1, \gamma_2, \ldots, \gamma_n$. We recall that a region is said to be simply connected if any closed curve lying entirely in D can be shrunk to a point in D, that is, the region has no holes.

Using Cauchy's Theorem for simply connected domains, we have

$$\oint_C f(z)\,dz + \sum_{k=1}^{n} \oint_{C_k} f(z)\,dz + \sum_{k=1}^{n} \left(\int_{\gamma_k} + \int_{-\gamma_k} \right) f(z)\,dz = 0. \tag{3.3.16}$$

The two integrals along the arcs γ_k and $-\gamma_k$ add up to zero since γ_k is traversed twice, but in opposite directions. Therefore from (3.3.16) we obtain

$$\oint_C f(z)\,dz + \sum_{k=1}^{n} \oint_{C_k} f(z)\,dz = 0, \tag{3.3.17}$$

where C and all the C_k are traversed either in the positive or in the negative direction. More specifically, formula (3.3.17) can be written in the form

$$\oint_{C^+} f(z)\,dz = \sum_{k=1}^{n} \oint_{C_k^+} f(z)\,dz. \tag{3.3.18}$$

Cauchy's Theorem for multiply connected domains follows from (3.3.18).

THEOREM 3.3.4. (CAUCHY'S THEOREM FOR MULTIPLY CONNECTED DOMAINS). *If $f(z)$ is analytic in a domain D containing the simple closed path C and the simple closed paths $C_1, C_2, \dots, C_n$ all interior to C, then the integral along C is equal to the sum of the integrals along all the C_k, provided all the paths are traversed either counterclockwise or clockwise.*

Note 3.3.2. One can obtain (3.3.18) without joining C by arcs with internal closed paths by using Green's formula for multiply connected domains (see [**13**], p. 172, [**35**], p. 408):

$$\left[\oint_{C^+} - \sum_{k=1}^{n} \oint_{C_k^+}\right] [P(x,y)\,dx + Q(x,y)\,dy] = \iint_D \left(\frac{\partial Q}{\partial x} - \frac{\partial P}{\partial y}\right) dx\,dy. \quad (3.3.19)$$

EXAMPLE 3.3.5. *Show that Cauchy's Theorem holds for the function $f(z) = 1/z$ and the closed paths $|z| = 2$ and $|z| = 1$, that is, prove that*

$$\oint_{|z|=2} \frac{dz}{z} = \oint_{|z|=1} \frac{dz}{z} = 2\pi i. \quad (3.3.20)$$

SOLUTION. The integral along $|z| = 2$ has already been computed in Example 3.3.4 of the previous subsection. In computing the integral along the path $|z| = 1$ given by $z = e^{i\theta}$, $0 \le \theta \le 2\pi$, we have $dz = ie^{i\theta}\,d\theta$ and hence

$$\oint_{|z|=1} \frac{dz}{z} = \int_0^{2\pi} \frac{e^{i\theta} i\,d\theta}{e^{i\theta}} = i\int_0^{2\pi} d\theta = 2\pi i.$$

Thus, formula (3.3.20) is valid in this particular case. □

It is left as an exercise to show that

$$\oint_C \frac{dz}{z} = 2\pi i,$$

if the path C is given by the following contours taken in the positive direction:

(a) $|z| = R$,

(b) a square centered at $z = 0$ with sides of length 2 parallel to the coordinate axes.

Exercises for Section 3.3

Use Cauchy's Theorem to show that the following integrals are zero.

1. $\oint_C e^{z^2}\,dz$, where C is the unit circle.

2. $\oint_C \frac{\sin(z/3)}{1-\cos z}\,dz$, where C is the square with vertices at $z_1 = 1$, $z_2 = 2$, $z_3 = 2+i$, $z_4 = 1+i$.

3. $\oint_C \frac{\tan z}{z-1}\,dz$, where C is the circle $|z-2| = 0.1$.

4. $\oint_C \frac{\cosh z}{z^2+1}\,dz$, where C is the circle $|z| = 1/2$.

Without computing integrals, find which of the following integrals are equal to zero. In each case, the path of integration, C, is the unit circle in the positive direction.

5. $\oint_C \cos^2 z\,dz$.

6. $\oint_C \frac{e^z}{z^3+8}\,dz$.

7. $\oint_C \frac{z^2+4z+1}{z^3+0.125}\,dz$.

8. $\oint_C \frac{\cos z}{(z^2+0.25)^2}\,dz$.

3.4. Cauchy's integral formula and applications

3.4.1. Derivation of Cauchy's integral formula. In the derivation of Cauchy's integral formula we shall use the following theorem from real analysis (see, for example, [**29**], Vol. 2, p. 269).

THEOREM 3.4.1. *If $f(x,y)$ is continuous in a rectangle, $a \le x \le b$, $c \le y \le d$, then the function*

$$F(y) = \int_\alpha^\beta f(x,y)\,dx \tag{3.4.1}$$

is continuous on the segment $c \le y \le d$; moreover,

$$\lim_{y\to y_0} F(y) = \int_\alpha^\beta f(x,y_0)\,dx. \tag{3.4.2}$$

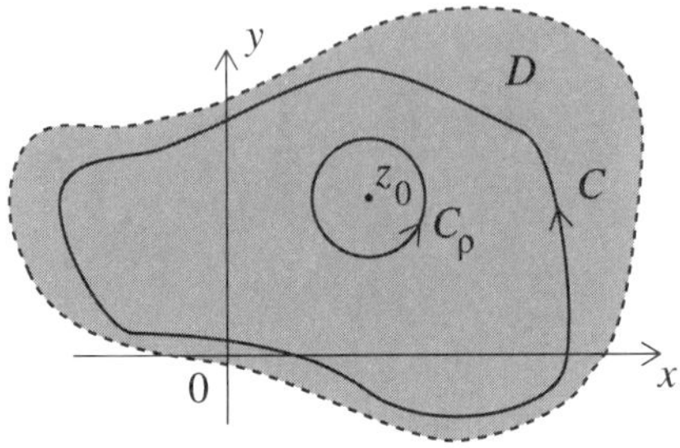

FIGURE 3.10. Simply connected domain for Cauchy's integral formula.

THEOREM 3.4.2 (CAUCHY'S INTEGRAL FORMULA). *Let $f(z)$ be analytic in a simply connected domain D containing the closed path C taken in the positive direction, and let z_0 be any point interior to C (see Fig 3.10). Then*

$$f(z_0) = \frac{1}{2\pi i} \oint_C \frac{f(z)}{z - z_0} \, dz. \tag{3.4.3}$$

This formula is known as Cauchy's integral formula.

PROOF. Let C_ρ be a circle of radius ρ centered at z_0 where ρ is taken so small that C_ρ is interior to C. Then $f(z)/(z - z_0)$ is analytic in the doubly connected domain containing C and C_ρ. By Cauchy's Theorem for multiply connected domains, we have

$$\oint_C \frac{f(z)}{z - z_0} \, dz = \oint_{C_\rho} \frac{f(z)}{z - z_0} \, dz, \tag{3.4.4}$$

where C_ρ is taken counterclockwise.

Since the path C_ρ is given by $z - z_0 = \rho e^{i\theta}$ with $0 \le \theta \le 2\pi$, then $dz = \rho i e^{i\theta} \, d\theta$ and from (3.4.4) we obtain

$$\oint_C \frac{f(z)}{z - z_0} \, dz = i \int_0^{2\pi} f\left(z_0 + \rho e^{i\theta}\right) \, d\theta. \tag{3.4.5}$$

We now take the limit in (3.4.5) as $\rho \to 0$. Since $f(z) = u(x, y) + iv(x, y)$ is analytic in D, it is continuous in D. The last statement is equivalent to the continuity of $u(x, y)$ and $v(x, y)$ in D (see Theorem 1.3.2).

Therefore we can use Theorem 3.4.1 to go to the limit as $\rho \to 0$ in the integral on the right-hand side of (3.4.5):

$$\lim_{\rho \to 0} i \int_0^{2\pi} f(z_0 + \rho e^{i\theta}) \, d\theta = i \int_0^{2\pi} f(z_0) \, d\theta = 2\pi i f(z_0).$$

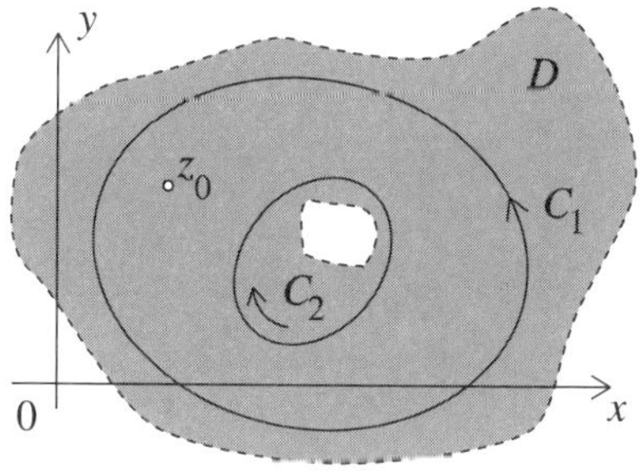

FIGURE 3.11. Doubly connected domain for Cauchy's integral formula.

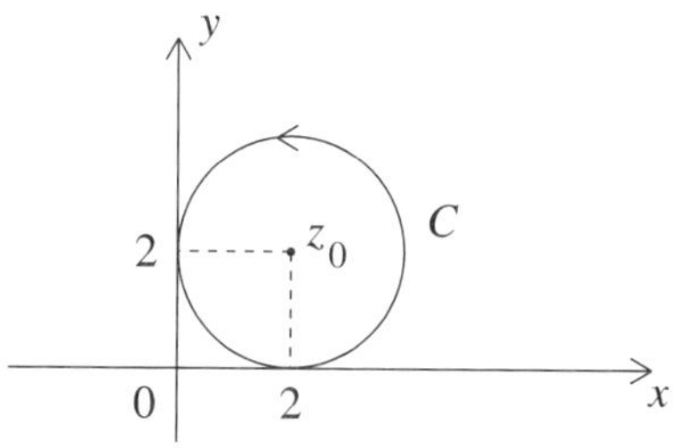

FIGURE 3.12. Path of integration for Example 3.4.1.

Since the integral on the left-hand side of (3.4.5) does not depend on ρ, then, in the limit as $\rho \to 0$, the formula

$$\oint_C \frac{f(z)}{z - z_0}\, dz = 2\pi i f(z_0)$$

follows from (3.4.5). □

Multiply connected domains can be handled by Cauchy's integral formula as in Subsection 3.3.2. For instance, let $f(z)$ be analytic in a doubly connected domain, D, containing the outer and inner closed paths, C_1 and C_2, respectively, shown in Fig 3.11. If the point z_0 lies in the region bounded by C_1 and C_2, then

$$f(z_0) = \frac{1}{2\pi i} \oint_{C_1} \frac{f(z)}{z - z_0}\, dz + \frac{1}{2\pi i} \oint_{C_2} \frac{f(z)}{z - z_0}\, dz,$$

where C_1 and C_2 are both taken in the positive direction.

EXAMPLE 3.4.1. *Consider the function $f(z) = z^2$ analytic in the complex plane and the point $z_0 = 2 + 2i$. Let C be the circle of radius 2 centered at z_0 (see Fig 3.12). By Cauchy's integral formula (3.4.3), one already* *knows that*

$$I = \oint_C \frac{z^2}{z-(2+2i)}\, dz = 2\pi i(2+2i)^2, \tag{3.4.6}$$

where the contour C shown in the figure is given by the equation

$$z = (2+2i) + 2\, e^{i\theta}, \qquad 0 \le \theta \le 2\pi.$$

Obtain (3.4.6) *by computing the integral directly.*

SOLUTION. Since $dz = 2i\, e^{i\theta}\, d\theta$, (3.4.6) becomes

$$\begin{aligned} I &= \int_0^{2\pi} \frac{[(2+2i)+2\, e^{i\theta}]^2}{2\, e^{i\theta}}\, 2i\, e^{i\theta}\, d\theta \\ &= i \int_0^{2\pi} \left[(2+2i)^2 + 2(2+2i)2\, e^{i\theta} + 4\, e^{2i\theta}\right] d\theta \\ &= i \left[(2+2i)^2\theta + 4(2+2i)\frac{1}{i}\, e^{i\theta} + 4 \times \frac{1}{2i}\, e^{2i\theta}\right]\Bigg|_{\theta=0}^{\theta=2\pi} \\ &= 2\pi i(2+2i)^2. \quad \square \end{aligned}$$

EXAMPLE 3.4.2. *Show by direct integration that*

$$\oint_C \frac{z^2}{z-(1+i)}\, dz = 2\pi i(1+i)^2,$$

if the path C is a square centered at $z_0 = 1+i$, with sides of length 2 (see Fig 3.13).

In the following subsections, several important results for analytic functions will be derived by means of Cauchy's integral formula.

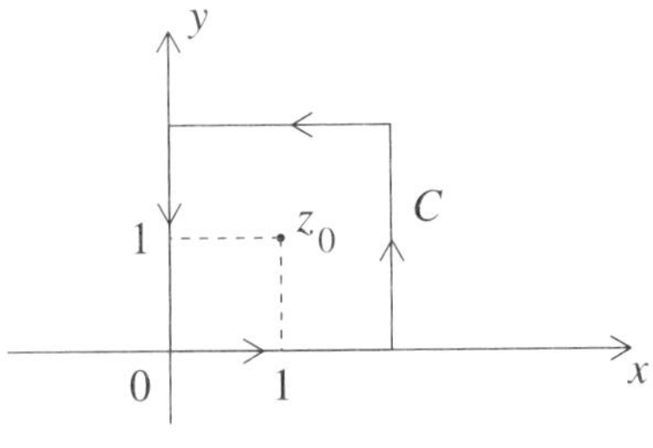

FIGURE 3.13. Square of sides 2 centered at $1+i$ for Example 3.4.2.

3.4.2. Infinite differentiability of analytic functions. As a first consequence of Cauchy's integral formula, we prove the infinite differentiability of analytic functions.

THEOREM 3.4.3. *An analytic function is infinitely often differentiable.*

PROOF. Let us replace z with ζ and z_0 with z in (3.4.3):

$$f(z) = \frac{1}{2\pi i} \oint_C \frac{f(\zeta)}{\zeta - z} \, d\zeta. \tag{3.4.7}$$

Let D be a simply connected domain containing the simple closed path C. We shall prove that, if $f(z)$ is analytic in D, then the integral (3.4.7) can be differentiated an arbitrary number of times with respect to z and

$$f^{(n)}(z) = \frac{n!}{2\pi i} \oint_C \frac{f(\zeta)}{(\zeta - z)^{n+1}} \, d\zeta. \tag{3.4.8}$$

In fact, for any complex h such that $z + h \in D$, we obtain from (3.4.7) that

$$\begin{aligned} \frac{f(z+h) - f(z)}{h} &= \frac{1}{h}\frac{1}{2\pi i} \oint_C f(\zeta) \left[\frac{1}{\zeta - z - h} - \frac{1}{\zeta - z} \right] d\zeta \\ &= \frac{1}{2\pi i} \oint_C \frac{f(\zeta)}{(\zeta - z - h)(\zeta - z)} \, d\zeta, \end{aligned}$$

so that

$$\frac{f(z+h) - f(z)}{h} = \frac{1}{2\pi i} \oint_C \frac{f(\zeta)}{(\zeta - z - h)(\zeta - z)} \, d\zeta. \tag{3.4.9}$$

Since $f(\zeta)$ is analytic on C, it is continuous there. Furthermore, if

$$|h| < \frac{1}{2} |\zeta - z|, \tag{3.4.10}$$

then the function

$$\frac{f(\zeta)}{(\zeta - z - h)(\zeta - z)}$$

is continuous on C with respect to the variables ζ and h for fixed z. Therefore (see Theorem 3.4.3) we can take the limit under the integral sign as $h \to 0$ in (3.4.9). Moreover, the integral

$$\oint_C \frac{f(\zeta)}{(\zeta - z)^2} \, d\zeta$$

exists since $f(\zeta)/(\zeta - z)^2$ is analytic on C if z is an internal point of D. Therefore the limit of the left-hand side of (3.4.9) exists as $h \to 0$. Hence, taking the limit in (3.4.9) as $h \to 0$, we have

$$f'(z) = \frac{1}{2\pi i} \oint_C \frac{f(\zeta)}{(\zeta - z)^2} \, d\zeta. \tag{3.4.11}$$

Similar arguments applied to (3.4.11) give

$$\frac{f'(z+h) - f'(z)}{h} = \frac{1}{h} \frac{1}{2\pi i} \oint_C f(\zeta) \left[\frac{1}{(\zeta - z - h)^2} - \frac{1}{(\zeta - z)^2} \right] d\zeta.$$

Simplifying the expression inside the square brackets in the previous formula, we obtain

$$\frac{f'(z+h) - f'(z)}{h} = \frac{1}{2\pi i} \oint_C \frac{2(\zeta - z - h/2) f(\zeta)}{(\zeta - z)^2 (\zeta - z - h)^2} \, d\zeta. \tag{3.4.12}$$

The integrand in (3.4.12) is continuous with respect to the variables ζ and h on a neighborhood of C if z is fixed and $|h| < |\zeta - z|/2$. Therefore, we can take the limit in (3.4.12) as $h \to 0$; moreover, the integral

$$\oint_C \frac{f(\zeta)}{(\zeta - z)^3} \, d\zeta$$

exists. Hence, as $h \to 0$, from (3.4.12) we obtain

$$f''(z) = \frac{2!}{2\pi i} \oint_C \frac{f(\zeta)}{(\zeta - z)^3} \, d\zeta. \tag{3.4.13}$$

This argument can be repeated as often as we please, if we use the fact that

$$a^n - (a - b)^n = nba^{n-1} - \frac{n(n-1)}{2!} b^2 a^{n-2} + \cdots + (-1)^{n+1} b^n, \tag{3.4.14}$$

where $a = \zeta - z$ and $b = h$.

Assuming that the formula

$$f^{(n-1)}(z) = \frac{(n-1)!}{2\pi i} \oint_C \frac{f(\zeta)}{(\zeta - z)^n} \, d\zeta \tag{3.4.15}$$

holds for a given n, by induction we obtain the same formula for $n + 1$. From (3.4.15), we have

$$\frac{f^{(n-1)}(z+h) - f^{(n-1)}(z)}{h} = \frac{(n-1)!}{2\pi i h} \oint_C f(\zeta) \left[\frac{1}{(\zeta - z - h)^n} - \frac{1}{(\zeta - z)^n} \right] d\zeta.$$

Hence, simplifying the expression inside the square brackets and using (3.4.14), we have

$$\frac{f^{(n-1)}(z+h)-f^{(n-1)}(z)}{h}=\frac{(n-1)!}{2\pi i h}\oint_C f(\zeta)$$
$$\times\frac{nh(\zeta-z)^{n-1}-\frac{n(n-1)}{2!}h^2(\zeta-z)^{n-2}+\cdots+(-1)^{(n+1)}h^n}{(\zeta-z-h)^n(\zeta-z)^n}\,d\zeta. \quad (3.4.16)$$

The integrand in (3.4.16) is continuous with respect to variables ζ and h on a neighborhood of C if z is fixed and $|h|<|\zeta-z|/2$. Hence, as $h\to 0$, from (3.4.16) we obtain

$$f^{(n)}(z)=\frac{n!}{2\pi i}\oint_C\frac{f(\zeta)}{(\zeta-z)^{n+1}}\,d\zeta. \quad \square$$

It follows from the previous theorem that if $f(z)$ is analytic in D (that is, if the first derivative of $f(z)$ exists in each point of D) then $f(z)$ has derivatives of all orders in D.

This is not true for functions of a real variable. For example, the function $f(x)=(x-1)^{7/3}$ is defined and continuous for all $x\in(-\infty,\infty)$. Moreover, the first and second derivatives exist at $x=1$:

$$f'(x)=\frac{7}{3}(x-1)^{4/3}, \qquad f'(1)=0,$$

$$f''(x)=\frac{7\times 4}{3^2}(x-1)^{1/3}, \qquad f''(1)=0.$$

But it is obvious that $f'''(1)$ does not exist.

For the function of a complex variable

$$f(z)=(z-1)^{7/3},$$

the point $z=1$ is a branch point and single analytic branches of $f(z)$ exist in each domain with a cut joining the points $z=1$ and $z=\infty$.

3.4.3. A converse to Cauchy's Theorem: Morera's Theorem. As a second consequence of Cauchy's integral formula, we prove Morera's Theorem, which is a converse to Cauchy's Theorem for simply connected domains.

THEOREM 3.4.4 (MORERA'S THEOREM). *Let $f(z)$ be a continuous function in a simply connected domain D and suppose that the integral of $f(z)$ along any closed path lying entirely in D is equal to zero. Then $f(z)$ is analytic in D.*

PROOF. By Theorem 3.3.3, the function

$$F(z) = \int_{z_0}^{z} f(\zeta)\, d\zeta, \tag{3.4.17}$$

where z_0, $z \in D$ and the integral (3.4.17) is computed along any path lying entirely in D, is analytic in D and $F'(z) = f(z)$. Then by Theorem 3.4.3, $F''(z) = f'(z)$ in D. Thus, $f(z)$ is analytin in D. □

3.4.4. Liouville's Theorem. As a third consequence of Cauchy's integral formula we prove Liouville's Theorem for bounded entire functions. An everywhere analytic function without singularities in the complex plane is said to be an *entire function.*

THEOREM 3.4.5 (LIOUVILLE'S THEOREM). *If the entire function $f(z)$ is uniformly bounded in the whole complex plane, then $f(z) =$ constant.*

PROOF. We use formula (3.4.8) with $n = 1$:

$$f'(z) = \frac{1}{2\pi i} \oint_C \frac{f(\zeta)}{(\zeta - z)^2}\, d\zeta,$$

and let C be a circle of radius R centered at z, that is, $\zeta = z + R\, e^{i\theta}$ with $0 \le \theta \le 2\pi$, and $d\zeta = Ri\, e^{i\theta}\, d\theta$. Then

$$f'(z) = \frac{1}{2\pi R} \int_0^{2\pi} f\left(z + R\, e^{i\theta}\right) e^{-i\theta}\, d\theta, \tag{3.4.18}$$

which, upon taking absolute values, becomes

$$\begin{aligned} |f'(z)| &\le \frac{1}{2\pi R} \int_0^{2\pi} \left| f\left(z + R\, e^{i\theta}\right) \right| d\theta \\ &< \frac{M}{2\pi R} \int_0^{2\pi} d\theta = \frac{M}{R}, \end{aligned} \tag{3.4.19}$$

since $|f(z)| < M$ for every z in $\mathbb{C}$. Letting $R \to \infty$ in (3.4.19) we have $|f'(z)| = 0$. Since z is arbitrary, then $|f'(z)| = 0$ for all z in $\mathbb{C}$. We conclude that $f(z) =$ constant. □

3.4.5. Mean-value theorems for analytic and harmonic functions. As a fourth consequence of Cauchy's integral formula, we prove the mean-value theorem for analytic and harmonic functions.

THEOREM 3.4.6 (MEAN-VALUE THEOREM). *Suppose that $f(z)$ is analytic in a domain containing a closed disk $D : |z - z_0| \le R$. Then the value,*

$f(z_0)$, of f at the center of the disk is equal to the arithmetic mean of its values on the boundary, $C : |z - z_0| = R$, of the disk:

$$f(z_0) = \frac{1}{2\pi R} \oint_C f\left(z_0 + R\,e^{i\theta}\right) dl, \tag{3.4.20}$$

where $dl = R\,d\theta$ is the differential of arc length along C and $2\pi R$ is the length of C.

PROOF. Substituting the equation $z - z_0 = R\,e^{i\theta}$ of C in Cauchy's integral formula,

$$f(z_0) = \frac{1}{2\pi i} \oint_C \frac{f(z)}{z - z_0}\, dz,$$

we have

$$\begin{aligned} f(z_0) &= \frac{1}{2\pi i} \oint_C \frac{f\left(z_0 + R\,e^{i\theta}\right)}{R\,e^{i\theta}} R\,e^{i\theta} i\, d\theta \\ &= \frac{1}{2\pi} \int_0^{2\pi} f\left(z_0 + R\,e^{i\theta}\right) d\theta \\ &= \frac{1}{2\pi R} \oint_C f\left(z_0 + R\,e^{i\theta}\right) dl, \end{aligned}$$

which is (3.4.20). □

The mean-value theorem for harmonic functions follows from Theorem 3.4.6.

THEOREM 3.4.7. (MEAN-VALUE THEOREM FOR HARMONIC FUNCTIONS). *Suppose $u(x, y)$ is a harmonic function of the real variables x and y in a closed disk of radius R and center (x_0, y_0) bounded by the circle*

$$C : (x - x_0)^2 + (y - y_0)^2 = R^2.$$

Then the value, $u(x_0, y_0)$, of u at the center of the circle is equal to the arithmetic mean of its values on the circle:

$$u(x_0, y_0) = \frac{1}{2\pi R} \oint_C u(\xi, \eta)\, dl, \tag{3.4.21}$$

where $dl = R\,d\theta$ is the differential of arc length of the circle.

PROOF. Let us write the equation of the circle in the form

$$\xi = x_0 + R\cos\theta, \qquad \eta = y_0 + R\sin\theta, \qquad 0 \le \theta \le 2\pi.$$

Since $u(x, y)$ is the real (or imaginary) part of an analytic function $f(z)$, taking the real part of $f\left(z_0 + R\,e^{i\theta}\right)$, we have

$$\Re f(x_0 + iy_0 + R\cos\theta + iR\sin\theta) = u(x_0 + R\cos\theta, y_0 + R\sin\theta).$$

Hence, taking the real part of both sides of (3.4.20), we obtain (3.4.21). □

Remark 3.4.1. Instead of formula (3.4.21), some authors (see, for example, [**42**], p. 68, formula (6)) use the form

$$u(z_0) = \frac{1}{2\pi}\int_0^{2\pi} u\left(z_0 + R\,e^{i\theta}\right)\,d\theta.$$

This form of the mean-value theorem for harmonic functions may be misleading since $u(x, y)$ is a function of the real variables x and y.

3.4.6. The maximum modulus theorem for analytic functions.

A fifth and last consequence of Cauchy's integral formula is the maximum modulus principle for analytic functions.

THEOREM 3.4.8 (MAXIMUM MODULUS PRINCIPLE). *If $f(z)$ is analytic and nonconstant in a domain D, then its absolute value, $|f(z)|$, has no maximum in D.*

PROOF. From Cauchy's integral formula for a circle of radius R inside D,

$$f(z_0) = \frac{1}{2\pi}\int_0^{2\pi} f\left(z_0 + R\,e^{i\theta}\right)\,d\theta, \tag{3.4.22}$$

we derive the inequality

$$|f(z_0)| \leq \frac{1}{2\pi}\int_0^{2\pi} \left|f\left(z_0 + R\,e^{i\theta}\right)\right|\,d\theta. \tag{3.4.23}$$

If $|f(z_0)|$ were a maximum, then we would have

$$|f(z_0 + R\,e^{i\theta})| \leq |f(z_0)|.$$

If strict inequality held for a single value of θ, by continuity it would hold on a whole arc. But then, the mean value of $|f(z_0+R\,e^{i\theta})|$ would be strictly less than $|f(z_0)|$, and (3.4.23) would lead to the contradiction $|f(z_0)| < |f(z_0)|$. Thus $|f(z_0)|$ must be constantly equal to $|f(z_0)|$ on all sufficiently small circles $|z - z_0| = R$ and, hence, in a neighborhood of z_0. It follows that $f(z)$ must reduce to a constant. □

Similarly, one can derive from Theorem 3.4.7 that a nonconstant harmonic function, $u(x, y)$, in a domain D does not take its maximum or its minimum inside D. This is called the *maximum principle for harmonic functions*.

3.4.7. Schwarz' Lemma. It follows from the maximum modulus principle for analytic functions that, if $f(z)$ is analytic in the open disk $|z| < R$ and continuous on the closed disk $|z| \le R$ and $|f(z)| \le M$ on $|z| = R$, then $|f(z)| \le M$ in the whole disk. The equality can hold only if $f(z)$ is a constant of modulus M. If, however, it is known that $f(z)$ takes some value of modulus smaller than M, it may be possible to have a better estimate, as shown in the following result, known as Schwarz' Lemma.

THEOREM 3.4.9 (SCHWARZ' LEMMA). *Let $f(z)$ be analytic for $|z| < 1$. If f satisfies the conditions*

$$|f(z)| \le 1, \qquad f(0) = 0,$$

then $|f(z)| \le |z|$ and $|f'(0)| \le 1$. On the other hand, if

$$|f(z)| = |z| \quad \text{for some } z \ne 0, \qquad \text{or} \quad f'(0) = 1,$$

then $f(z) = cz$ with a constant c of modulus 1.

PROOF. We apply the maximum modulus principle to the function

$$g(z) = \begin{cases} f(z)/z, & \text{if } z \ne 0, \\ f'(0), & \text{if } z = 0, \end{cases}$$

which is analytic in the open disk $|z| < 1$ and continuous on the closed disk $|z| \le 1$. On the circle $|z| = r < 1$, $|g(z)| \le 1/r$, and hence $|g(z)| \le 1/r$ for $|z| \le r$. Letting r tend to 1, we find that $|g(z)| \le 1$ for all z; this inequality is the assertion of the theorem. If equality holds at a single point, then $|g(z)|$ attains it maximum at an interior point and hence $g(z)$ reduces to a constant. □

Exercises for Section 3.4

Evaluate the following integrals where the path C is taken counterclockwise.

1. $\displaystyle\oint_C \frac{e^z}{(z+3)(z-1)}\,dz$, where C is the circle $|z| = 2$.

2. $\displaystyle\oint_C \frac{1+z^2}{(z^3+27)(z-i)}\,dz$, where C is the square with vertices

$$z_1 = 0, \quad z_2 = 2, \quad z_3 = 2+2i, \quad z_4 = 2i.$$

3. $\displaystyle\oint_C \frac{z^2 \cos z}{z^2-4}\,dz$, where C is the circle $|z-1| = 3/2$.

4. $\displaystyle\oint_C \frac{e^z(z+4)}{z^2+9}\,dz$, where C is the circle $|z-2| = 2$.

Evaluate the integral $\displaystyle\int_C \frac{dz}{z^2+4}$ along the following circles taken counterclockwise.

5. $|z-4|=1$.
6. $|z-1|=3/2$.
7. $|z+2|=1$.

Evaluate the integral $\displaystyle\int_C \frac{z\sin z}{z^3+8}\,dz$ along the following circles taken counterclockwise.

8. $|z|=1$.
9. $|z+2|=1$.
10. $|z-1-2i|=2$.

11. Let $f(z)$ be an analytic function in the region $|z-z_0|\le R$. Show that

$$f(z_0)=\frac{1}{2\pi}\int_0^{2\pi} f\left(z_0+Re^{i\theta}\right)\,d\theta.$$

12. Let $f(z)$ and $g(z)$ be analytic in a simply connected domain D. Prove that

$$\int_\alpha^\beta f(z)g(z)\,dz=[f(z)h(z)]\Big|_\alpha^\beta-\int_\alpha^\beta f'(z)h(z)\,dz,$$

where $h(z)$ is an indefinite integral of $g(z)$ in D and the path of integration lies in D.

13. Use formula (3.4.8) with the circle $C=\{z;\ |z-z_0|=r\}$ taken in the positive direction to establish *Cauchy's estimate*:

$$|f^{(n)}(z_0)|\le\frac{n!}{r^n}\max_{|z-z_0|=r}|f(z)|,\qquad n=0,1,2,\ldots,\tag{3.4.24}$$

whenever $f(z)$ is analytic on a domain containing the disk bounded by C.

14. Use Cauchy's estimate (3.4.24) of the previous exercise with $n=1$ to prove Liouville's Theorem 3.4.5 by showing that the derivative of a bounded entire function is identically zero.

15. Suppose that $f(z)$ is an entire function and $\Re f(z)\le c$ for all z. Show that $f(z)$ is a constant.
(Hint: Consider the function $e^{f(z)}$.)

16. Suppose that $f(z)$ is an entire function and $\Im f(z)\le c$ for all z. Show that $f(z)$ is a constant.

17. Let $f(z)$ be entire and $|f(r\,e^{i\theta})| < Mr$, where M is a constant. Prove that $f(z)$ is a polynomial of degree at most 1. Can this result be generalized to polynomials of higher degrees?

18. Consider the function $f(z) = (z+1)^2$ over the closed triangular region R with vertices at the points $z = 0$, $z = 2$ and $z = i$. Find points in R where $|f(z)|$ has its maximum and minimum values, thus illustrating the maximum modulus theorem (Theorem 3.4.8).

19. Consider the function $f(z) = e^z$ and the rectangular region R defined by $0 \le x \le 1$, $0 \le y \le \pi$. Illustrate the maximum principle for harmonic functions by finding the points in R where $u(x,y) = \Re f(z)$ reaches its maximum and minimum values.

20. The so-called *fundamental theorem of algebra* asserts that every polynomial,

$$p(z) = a_n z^n + \cdots + a_1 z + a_0,$$

of degree $n > 0$ has at least one zero. Use Liouville's Theorem to prove the fundamental theorem of algebra.
(Hint: Consider the function $1/p(z)$.)

21. Let the function $f(z)$ be analytic in a domain D containing the closed disk $|z| \le r$. If $|f(z)|$ is constant on $|z| = r$ and $f(z) \ne 0$ for $|z| < r$, show that $f(z)$ is constant.

22. If $f(z)$ is analytic for $|z| < 1$ and $|f(z)| \le 1/(1 - |z|)$, find the best estimate of $|f^{(n)}(0)|$ that Cauchy's estimate (3.4.24) will yield.

23. Show that the successive derivatives of an analytic function at a point can never satisfy the inequality $|f^{(n)}(z)| > n!n^n$. Formulate a sharper theorem of the same kind.

24. Prove that there is no function analytic in $|z| \le 1$ such that

$$|f(z)| \le 1 \quad \text{on } |z| = 1, \qquad f\left(\frac{1}{2}\right) = 0, \qquad f\left(-\frac{1}{2}\right) = \frac{19}{20}.$$

25. The function of a complex variable defined by $f(z) = \cos z$ is analytic everywhere and satisfies the inequality $|\cos x| \le 1$ for all real x. However, it is not a constant. Is there a contradiction with Liouville's Theorem?

3.5. Goursat's Theorem

In this last section, Morera's Theorem will be used to remove the continuity assumption on the derivative, $f'(z)$, of an analytic function, $f(z)$, used in the proof of Cauchy's Theorem and in the subsequent results in this chapter. The removal of this restriction is the contents of Goursat's Theorem.

THEOREM 3.5.1 (GOURSAT'S THEOREM). *Let G be an open set and let $f(z)$ be differentiable on G. Then $f'(z)$ is continuous on G.*

PROOF. We need only show that $f'(z)$ is continuous on each open disk contained in G, so that we may assume that G is itself an open disk. The continuity of $f'(z)$ will follow from Morera's Theorem 3.4.4, that is, we must show that

$$\int_S f(z)\,dz = 0,$$

for each triangular path S in G.

Let S be the triangular path A, B, C, A and let T be the closed set formed by S and its interior (see Fig 3.14).

Note that $S = \partial T$ is the boundary of T. Now using the midpoints of the sides of T, form four triangles T_1, T_2, T_3 and T_4 inside T. By giving the boundaries appropriate directions, we have that each $S_j = \partial T_j$ is a triangular path and

$$\oint_S f(z)\,dz = \sum_{j=1}^{4} \oint_{S_j} f(z)\,dz. \tag{3.5.1}$$

Among these four paths, there is one, called $S^{(1)}$, such that

$$\left|\oint_{S^{(1)}} f(z)\,dz\right| \geq \left|\oint_{S_j} f(z)\,dz\right|, \qquad j = 1, 2, 3, 4.$$

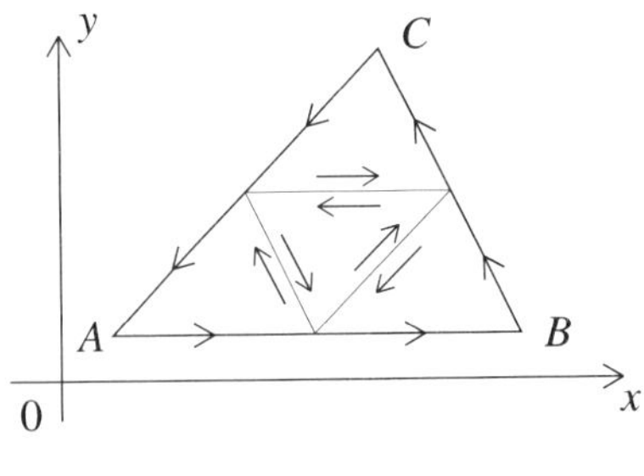

FIGURE 3.14. Triangular region T for the proof of Goursat's Theorem.

Let $L(S)$ and $D(T)$ denote the length of S and the diameter of T, respectively. Then, we have

$$L(S_j) = \frac{1}{2}\,L(S), \qquad D(T_j) = \frac{1}{2}\,D(T).$$

Finally, by (3.5.1), we have

$$\left|\oint_S f(z)\,dz\right| \le 4\left|\oint_{S^{(1)}} f(z)\,dz\right|.$$

Now performing the same process on $S^{(1)}$, we obtain a triangle $S^{(2)}$ with the analogous properties. By induction, we get a sequence $\{S^{(n)}\}$ of closed triangular paths and closed sets $\{T^{(n)}\}$, each consisting of the region enclosed by $S^{(n)}$ and its boundary. Thus we have

$$T^{(1)} \supset T^{(2)} \supset \dots, \tag{3.5.2}$$

$$\left|\oint_{S^{(n)}} f(z)\,dz\right| \le 4\left|\oint_{S^{(n+1)}} f(z)\,dz\right|, \tag{3.5.3}$$

$$L\left(S^{(n+1)}\right) = \frac{1}{2}\,L\left(S^{(n)}\right), \tag{3.5.4}$$

$$D\left(T^{(n+1)}\right) = \frac{1}{2}\,D\left(T^{(n)}\right). \tag{3.5.5}$$

These relations imply:

$$\left|\oint_S f(z)\,dz\right| \le 4^n\left|\oint_{S^{(n)}} f(z)\,dz\right|, \tag{3.5.6}$$

$$L\left(S^{(n)}\right) = \left(\frac{1}{2}\right)^n L(S), \tag{3.5.7}$$

$$D\left(T^{(n)}\right) = \left(\frac{1}{2}\right)^n D(T). \tag{3.5.8}$$

Since the $T^{(n)}$ are closed, then their intersection is non empty and consists of a single point z_0,

$$z_0 = \bigcap_{n=1}^{\infty} T^{(n)}.$$

Let $\varepsilon > 0$. Since $f(z)$ has a derivative at z_0, we can find a $\delta > 0$ such that $D_{z_0}^{\delta} \subset G$ and

$$\left|\frac{f(z) - f(z_0)}{z - z_0} - f'(z_0)\right| < \varepsilon,$$

whenever $0 < |z - z_0| < \delta$, that is,

$$|f(z) - f(z_0) - f'(z_0)(z - z_0)| < \varepsilon|z - z_0|, \tag{3.5.9}$$

whenever $0 < |z - z_0| < \delta$.

Choose n such that

$$D\left(T^{(n)}\right) = \left(\frac{1}{2}\right)^n D(T) < \delta.$$

Since $z \in T^{(n)}$, then $T^{(n)} \subset D_{z_0}^{\delta}$. Now, Cauchy's Theorem implies that

$$\oint_{S^{(n)}} dz = \oint_{S^{(n)}} z\,dz = 0.$$

Hence,

$$\begin{aligned}\left|\oint_{S^{(n)}} f(z)\,dz\right| &= \left|\oint_{S^{(n)}} [f(z) - f(z_0) - f'(z_0)(z - z_0)]\,dz\right| \\ &\leq \varepsilon \oint_{S^{(n)}} |z - z_0|\,|dz| \\ &\leq \varepsilon D\left(T^{(n)}\right) L\left(S^{(n)}\right) \\ &= \left(\frac{1}{4}\right)^n \varepsilon D(T) L(S).\end{aligned}$$

But by (3.5.6), we have

$$\left|\oint_S f(z)\,dz\right| \leq 4^n \left(\frac{1}{4}\right)^n \varepsilon D(T) L(S) = \varepsilon D(T) L(S).$$

Since ε was arbitrary, and $D(T)$ and $L(S)$ are fixed, then

$$\oint_S f(z)\,dz = 0.$$

The result follows by Morera's Theorem. $\square$

CHAPTER 4

Taylor and Laurent Series

4.1. Infinite series

Infinite series are the starting point of the Weierstrassian theory of analytic functions.

4.1.1. Series of complex numbers.

DEFINITION 4.1.1. If $\{z_n\}$ is a sequence of complex numbers,

$$z_n = x_n + iy_n, \qquad n = 1, 2, \dots, \tag{4.1.1}$$

the infinite sum

$$\sum_{n=1}^{\infty} z_n = \sum_{n=1}^{\infty} x_n + i \sum_{n=1}^{\infty} y_n = z_1 + z_2 + \dots + z_n + \dots \tag{4.1.2}$$

is called a *series of complex numbers* and

$$S_n = \sum_{k=1}^{n} z_k = \sum_{k=1}^{n} x_k + i \sum_{k=1}^{n} y_k, \qquad n = 1, 2, \dots, \tag{4.1.3}$$

denotes the *nth partial sum* of the series.

The next definition gives a useful meaning to a series of complex numbers.

DEFINITION 4.1.2. Let $\{S_n\}$ be the sequence of partial sums of the series (4.1.2). If the limit

$$S = S_x + iS_y = \lim_{n\to\infty} S_n \tag{4.1.4}$$

exists and is finite, then the series is said to be *convergent* and its sum is equal to S; otherwise it is said to be *divergent*.

From Theorem 1.2.1 of Chapter 1, the limit of sequence (4.1.1) exists if and only if the limits of the sequences $\{x_n\}$ and $\{y_n\}$ exist. Therefore the limit (4.1.4) exists if and only if the two limits

$$S_x = \lim_{n\to\infty} \sum_{k=1}^{n} x_k, \qquad S_y = \lim_{n\to\infty} \sum_{k=1}^{n} y_k, \tag{4.1.5}$$

exist and are finite. This justifies the notation $S_x = \Re S$ and $S_y = \Im S$ implicitely used in (4.1.4).

Thus, the convergence of a series of complex numbers can be reduced to the convergence of two series of real numbers. Therefore, we shall use known convergence tests for series of positive numbers, such as the comparison test, the ratio test and the root test (see, for example, [**50**], pp. 20–23).

THEOREM 4.1.1 (NECESSARY CONDITION FOR CONVERGENCE). *Let $\{a_n\}$ be a sequence of positive numbers. If the series $\sum_{n=1}^{\infty} a_n$ converges, then $\lim_{n\to\infty} a_n = 0$.*

PROOF. If the limit $S = \lim_{n\to\infty} S_n$ of the partial sums, $S_n = \sum_{k=1}^{n} a_k$, exists, then $S = \lim_{n\to\infty} S_{n-1}$. Hence

$$\lim_{n\to\infty} a_n = \lim_{n\to\infty} (S_n - S_{n-1}) = S - S = 0. \quad \square$$

THEOREM 4.1.2 (COMPARISON TEST). *Let $\{a_n\}$ and $\{b_n\}$ be two sequences of positive numbers, such that $a_n < b_n$ for all $n \in N$, and consider the two series*

$$\sum_{n=1}^{\infty} a_n, \qquad \sum_{n=1}^{\infty} b_n. \tag{4.1.6}$$

If the second series converges, so does the first. If the first series diverges, so does the second.

THEOREM 4.1.3 (RATIO TEST). *Let $\sum_{n=1}^{\infty} a_n$ be a series of positive numbers and suppose the limit L,*

$$L = \lim_{n\to\infty} \frac{a_{n+1}}{a_n},$$

is finite. Then:

(a) *If $L < 1$, the series converges.*
(b) *If $L > 1$, the series diverges.*
(c) *If $L = 1$, the question of convergence is open (the series may either diverge or converge).*

THEOREM 4.1.4 (ROOT TEST). *Let $\sum_{n=1}^{\infty} a_n$ be a series of positive numbers and suppose the limit L,*

$$L = \lim_{n\to\infty} a_n^{1/n},$$

is finite. Then:

(a) *If $L < 1$, the series converges.*
(b) *If $L > 1$, the series diverges.*
(c) *If $L = 1$, the question of convergence is open (the series may either diverge or converge).*

Note 4.1.1. The ratio and root tests can be formulated in a more general form by using the notion of *limit superior*:

$$\limsup \frac{a_{n+1}}{a_n}, \qquad \limsup a_n^{1/n},$$

which is the largest point of accumulation in case more than one such points exist. This formulation is useful when the sequences a_{n+1}/a_n and/or $a_n^{1/n}$ have no limit.

EXAMPLE 4.1.1. *Show that the series*

$$\sum_{n=1}^{\infty} a_n, \qquad \text{where} \quad a_n = \frac{1}{2^n}[1 + (-1)^n] + \frac{1}{3^n}[1 - (-1)^n],$$

converges.

SOLUTION. We have

$$a_{2n} = \frac{2}{2^{2n}}, \qquad a_{2n+1} = \frac{2}{3^{2n+1}}.$$

Therefore, the limit of a_{n+1}/a_n as $n \to \infty$ does not exist since the two subsequences

$$\frac{a_{2n+2}}{a_{2n}} = \frac{a_{2n+2}}{a_{2n+1}} \frac{a_{2n+1}}{a_{2n}}, \qquad \frac{a_{2n+3}}{a_{2n+1}} = \frac{a_{2n+3}}{a_{2n+2}} \frac{a_{2n+2}}{a_{2n+1}}$$

have different limits:

$$\lim_{n\to\infty} \frac{a_{2(n+1)}}{a_{2n}} = \lim_{n\to\infty} \frac{2 \times 2^{-2(n+1)}}{2 \times 2^{-2n}} = \frac{1}{4},$$

$$\lim_{n\to\infty} \frac{a_{2(n+1)+1}}{a_{2n+1}} = \lim_{n\to\infty} \frac{2 \times 3^{-(2n+3)}}{2 \times 3^{-(2n+1)}} = \frac{1}{9}.$$

In this case,

$$\limsup_{n\to\infty} \frac{a_{n+1}}{a_n} = \frac{1}{4} < 1,$$

so that the series $\sum_{n=1}^{\infty} a_n$ converges. □

Note 4.1.2. One can ask the following question: Can a series converge according to the ratio test but diverge according to the root test? The answer is in the negative by the following theorem (see [**29**], Vol. 1, p. 437).

THEOREM 4.1.5. *Consider a sequence $\{a_n\}$ of positive numbers. If the limit*

$$L = \lim_{n\to\infty} \frac{a_{n+1}}{a_n}$$

exists and is finite, then the limit

$$M = \lim_{n\to\infty} a_n^{1/n}$$

exists and is finite, and $M = L$.

DEFINITION 4.1.3. If the series of absolute terms $\sum_{n=1}^{\infty} |z_n|$ is convergent, then the series $\sum_{n=1}^{\infty} z_n$ is said to be *absolutely convergent*. On the other hand, a convergent series which is not absolutely convergent is said to be conditionally convergent.

In the particular case of a sequence of real numbers we have the following theorem (see [**29**], Vol. 1, p. 418).

THEOREM 4.1.6. *If the series $\sum_{n=1}^{\infty} a_n$ of real numbers is absolutely convergent, then it is convergent.*

We use this theorem to prove the next one for sequences of complex numbers.

THEOREM 4.1.7. *If the series $\sum_{n=1}^{\infty} z_n$ of complex numbers is absolutely convergent, then it is convergent.*

PROOF. We suppose that the series of absolute terms

$$S = \sum_{n=1}^{\infty} |z_n| = \sum_{n=1}^{\infty} \sqrt{x_n^2 + y_n^2}$$

is convergent. Hence, it follows from the inequalities

$$|x_n| \leq \sqrt{x_n^2 + y_n^2}, \qquad |y_n| \leq \sqrt{x_n^2 + y_n^2}$$

and the comparison test for series of positive numbers (Theorem 4.1.2) that the series

$$\sum_{n=1}^{\infty} |x_n|, \qquad \sum_{n=1}^{\infty} |y_n|$$

converge. Therefore, by Theorem 4.1.6, the two series

$$S_x = \sum_{n=1}^{\infty} x_n, \qquad S_y = \sum_{n=1}^{\infty} y_n$$

also converge, that is, the series

$$\sum_{n=1}^{\infty} z_n = \sum_{n=1}^{\infty} (x_n + iy_n) = S_x + iS_y$$

is convergent. □

For instance, the series

$$S = \sum_{n=1}^{\infty} \frac{5n + 7i}{n^3}$$

is absolutely convergent since

$$|S| \leq \sum_{n=1}^{\infty} \left| \frac{5n + 7i}{n^3} \right| \leq \sum_{n=1}^{\infty} \frac{5}{n^2} + \sum_{n=1}^{\infty} \frac{7}{n^3} < \infty.$$

On the other hand, the series

$$\sum_{n=1}^{\infty} (-1)^n \frac{2 + 3i}{n}$$

is only conditionally convergent since the series of absolute terms

$$\sum_{n=1}^{\infty} \left| \frac{2 + 3i}{n} \right|$$

diverges.

As in the real case, the converse of Theorem 4.1.7 is not true in general. For instance, the series

$$\sum_{n=1}^{\infty} (-1)^n \frac{2 + 3i}{n}$$

is convergent (its real and imaginary parts are conditionally convergent series), but the series

$$\sum_{n=1}^{\infty} \frac{|2 + 3i|}{n} = \sqrt{13} \sum_{n=1}^{\infty} \frac{1}{n}$$

is divergent.

4.1.2. Series of functions. Let $w_n(z)$, $n = 1, 2, 3, \dots$, be a sequence of complex-valued functions.

DEFINITION 4.1.4. The series

$$\sum_{n=1}^{\infty} w_n(z) = w_1(z) + w_2(z) + \cdots + w_n(z) + \dots \tag{4.1.7}$$

is called a *series of functions.*

By giving different values to the complex variable z in (4.1.7) we obtain different series of complex numbers which may either converge or diverge.

DEFINITION 4.1.5. The set of all values of z for which the series (4.1.7) is convergent is called the *domain of convergence* of (4.1.7).

The sum of (4.1.7) is a function, $S(z)$, of the complex variable z in the domain of convergence.

If D is the domain of convergence of the series (4.1.7), then for every $\varepsilon > 0$ and for every $z \in D$ there exists a number $N = N_{\varepsilon,z}$ such that for all $n > N_{\varepsilon,z}$ the following inequality is satisfied:

$$|S(z) - S_n(z)| < \varepsilon. \tag{4.1.8}$$

It is important to remark that the number N depends on both ε and z.

The concept of uniformly convergent series (see Definition 4.1.6), plays a central role in real and complex analysis, since a uniformly convergent series can be integrated termwise and a uniformly convergent series of continuous functions converges to a continuous function. Thus, tests of uniform convergence are important.

DEFINITION 4.1.6. If the series (4.1.7) converges to $S(z)$ in D and for every $\varepsilon > 0$ there exists $N = N_\varepsilon$ independent of $z \in D$ such that for all $n > N_\varepsilon$ inequality (4.1.8) is satisfied, then the series (4.1.7) is said to *converge uniformly* to $S(z)$ in D.

In this book it will suffice to use Weierstrass' M-test of uniform convergence, for which we need the following definition.

DEFINITION 4.1.7. The series (4.1.7) is said to be *majorizable* in D if there exists a convergent series $\sum_{n=1}^{\infty} a_n$ of nonnegative real numbers such that

$$|w_n(z)| \leq a_n, \qquad n = 1, 2, \dots, \tag{4.1.9}$$

for all z in D.

Since the formulation and proof of the following theorems in complex analysis are almost the same as in the real case, we state the theorems without proofs. The interested reader can consult, for example, [**45**].

THEOREM 4.1.8 (WEIERSTRASS' M-TEST). *If the series* (4.1.7) *is majorizable in D then it is uniformly convergent in D.*

We present another theorem for uniformly convergent series of continuous functions.

THEOREM 4.1.9. *If the series (4.1.7) of continuous functions, $w_n(z)$, converges uniformly to $S(z)$ in D, then:*
(a) *$S(z)$ is continuous in D,*
(b) *for any contour C in D, we have*

$$\int_C S(z)\,dz = \sum_{n=1}^{\infty} \int_C w_n(z)\,dz.$$

PROOF. See, for example, [**45**], pp. 61–62. □
The following theorem, which we prove completely, has no analog in the real case.

THEOREM 4.1.10 (WEIERSTRASS' THEOREM). *Consider a sequence $\{w_n(z)\}$ of analytic functions in a (simply or multiply connected) domain D and suppose that the series*

$$\sum_{n=1}^{\infty} w_n(z) \tag{4.1.10}$$

converges uniformly to $S(z)$ in any closed subregion D' of D. Then
(a) *the function $S(z)$ is analytic in D,*
(b) *$S^{(k)}(z) = \sum_{n=1}^{\infty} w_n^{(k)}(z)$ for any positive integer k and for all $z \in D$,*
(c) *the series $\sum_{n=1}^{\infty} w_n^{(k)}(z)$ is uniformly convergent in D' for any $k \in \mathbb{N}$.*

PROOF. For the first and second parts, let z_0 be an arbitrary interior point in a simply connected subregion D' of D, and let C be an arbitrary closed path in D' encircling z_0, as shown in Fig 4.1(a).

(a) The function $S(z)$ is continuous in D according to Theorem 4.1.9. By Theorem 4.1.9,

$$\oint_C S(z)\,dz = \sum_{n=1}^{\infty} \oint_C w_n(z)\,dz = 0$$

since the functions $w_n(z)$ are analytic in D. Since the conditions of Morera's Theorem 3.4.4 are satisfied, $S(z)$ is analytic in D.

(b) Since z_0 is located inside C and C is a closed set, then

$$\min_{z\in C} |z - z_0| = d > 0. \tag{4.1.11}$$

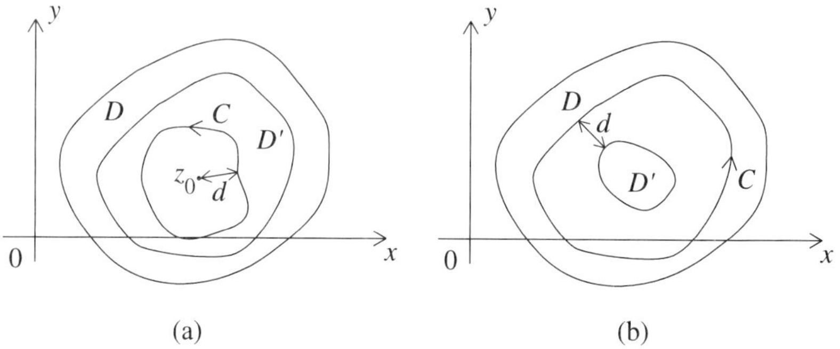

FIGURE 4.1. Subregions D' (a) containing, and (b) enclosed by, respectively, the curve C in the region D.

Consider the series

$$\frac{S(z)}{(z-z_0)^{k+1}} = \sum_{n=1}^{\infty} \frac{w_n(z)}{(z-z_0)^{k+1}}. \tag{4.1.12}$$

Since the series (4.1.10) is uniformly convergent on C and in some neighborhood of C not containing the point z_0, then series (4.1.12) has the same property. Indeed, if S_n denotes the nth partial sum of the uniformly convergent series (4.1.10), then for all $\varepsilon > 0$ there exists $N = N_\varepsilon$ such that for all $n > N_\varepsilon$ the inequality

$$|S(z) - S_n(z)| < \varepsilon d^{k+1} \tag{4.1.13}$$

holds. Then by (4.1.11) and (4.1.13), for all $n > N_\varepsilon$ we have

$$\begin{aligned} \left| \frac{S(z)}{(z-z_0)^{k+1}} - \frac{S_n(z)}{(z-z_0)^{k+1}} \right| &= \frac{1}{|z-z_0|^{k+1}} |S(z) - S_n(z)| \\ &< \frac{1}{d^{k+1}} \varepsilon d^{k+1} = \varepsilon. \end{aligned}$$

This inequality implies that series (4.1.12) is uniformly convergent on C and in some neighborhood of C and, therefore, it can be integrated termwise along C:

$$\frac{k!}{2\pi i} \oint_C \frac{S(z)}{(z-z_0)^{k+1}}\, dz = \sum_{n=1}^{\infty} \frac{k!}{2\pi i} \oint_C \frac{w_n(z)}{(z-z_0)^{k+1}}\, dz. \tag{4.1.14}$$

Since $S(z)$ and $w_n(z)$ are analytic in D', and C lies inside D', then by (3.4.8) for the kth derivative of an analytic function, (4.1.14) gives

$$S^{(k)}(z_0) = \sum_{n=1}^{\infty} w_n^{(k)}(z_0).$$

Since z_0 is an arbitrary point in D', the second statement of the theorem is proved.

(c) Let C be an arbitrary closed path lying entirely in a simply connected subregion of D and D' an arbitrary simply connected closed subregion surrounded by C at distance at least $d > 0$ (see Fig 4.1(b)),

$$d = \min_{\substack{z \in D' \\ \zeta \in C}} |z - \zeta|. \tag{4.1.15}$$

Since $S(z)$ is analytic in D by part (a), then the remainder

$$r_n(z) = \sum_{k=n+1}^{\infty} w_k(z) = S(z) - \sum_{k=1}^{n} w_k(z),$$

which is the sum of a finite number of analytic functions, is analytic in D. Therefore, for all $z \in D'$, we have

$$r_n^{(k)}(z) = \frac{k!}{2\pi i} \oint_C \frac{r_n(\zeta)}{(\zeta - z)^{k+1}}\, d\zeta. \tag{4.1.16}$$

Moreover, by part (b),

$$r_n^{(k)}(z) = \sum_{l=n+1}^{\infty} w_l^{(k)}(z)$$

is the remainder of the series $\sum_{n=1}^{\infty} w_n^{(k)}(z)$. Since the series $\sum_{n=1}^{\infty} w_n(z)$ is uniformly convergent, then for every $\varepsilon > 0$ there exists an integer N_ε such that for all $\zeta \in C$ and all $n > N_\varepsilon$, the inequality

$$|r_n(\zeta)| < \varepsilon \frac{2\pi d^{k+1}}{k!L}, \tag{4.1.17}$$

is satisfied, where L is the length of C. It then follows from (4.1.16), (4.1.15) and (4.1.17) that

$$\begin{aligned} \left| r_n^{(k)}(z) \right| &= \left| \frac{k!}{2\pi i} \oint_C \frac{r_n(\zeta)}{(\zeta - z)^{k+1}}\, d\zeta \right| \\ &\leq \frac{k!}{2\pi} \oint_C \left| \frac{r_n(\zeta)}{(\zeta - z)^{k+1}} \right| |d\zeta| \end{aligned}$$

$$\leq \frac{k!}{2\pi}\,\varepsilon\frac{2\pi d^{k+1}}{k!L}\,\frac{1}{d^{k+1}}\oint_C |d\zeta| = \varepsilon.$$

This last estimate implies that the series $\sum_{n=1}^{\infty} w_n^{(k)}(z)$ is uniformly convergent in any closed subregion D' of D. □

Note 4.1.3. Theorem 4.1.10 is valid only for closed subregions D' of D even if the original series is uniformly convergent in the closure of D.

For example, the series $\sum_{n=1}^{\infty} z^n/n^2$ is uniformly convergent in the region $|z| \leq 1$ since it is majorizable there by the convergent series $\sum_{n=1}^{\infty} 1/n^2$, but the series $\sum_{n=1}^{\infty} z^{n-1}/n$, obtained by termwise differentiation of the original series, is convergent only in the domain $|z| < 1$; in fact, it diverges at $z = 1$.

Since the statement of part (c) of the theorem is about the uniform convergence of a *termwise differentiated series* in a closed subregion D' of the given domain D, D' cannot, in general, be extended.

Note 4.1.4. In the real case, the second statement of the theorem is not true in general. In fact, one cannot differentiate a uniformly convergent series of continuous functions termwise an arbitrary number of times.

EXAMPLE 4.1.2. *Show that the series*

$$S(x) = \sum_{n=1}^{\infty} \frac{\sin nx}{n^3}, \qquad x \in \mathbb{R}, \tag{4.1.18}$$

is uniformly convergent, but cannot be differentiated termwise an arbitrary number of times.

SOLUTION. The given series is uniformly convergent for all real x since it is periodic and majorized by the convergent series $\sum_{n=1}^{\infty} 1/n^3$.

Similarly, the termwise differentiated series $\sum_{n=1}^{\infty}(\cos nx)/n^2$ is also uniformly convergent. Hence, by periodicity,

$$S'(x) = \sum_{n=1}^{\infty} \frac{\cos nx}{n^2}, \qquad x \in \mathbb{R}. \tag{4.1.19}$$

However, the series obtained by termwise derivation of (4.1.19),

$$-\sum_{n=1}^{\infty} \frac{\sin nx}{n}, \tag{4.1.20}$$

is only conditionally convergent, by Theorem 4.2.3 and Example 4.2.1, while the series obtained by termwise derivation of (4.1.20),

$$-\sum_{n=1}^{\infty} \cos nx,$$

is divergent. □

At first glance, this fact seems to contradict Weierstrass' Theorem, since the function $\sin nz$ is differentiable an arbitrary number of times in the whole complex plane and the series

$$\sum_{n=1}^{\infty} \frac{\sin nz}{n^3} \tag{4.1.21}$$

should be convergent for some values of z. We show that no such contradiction exists. The reason is that, if $z = x + iy$ and $y \neq 0$, then

$$\begin{aligned} |\sin n(x+iy)| &= \sqrt{\sin^2 nx \cosh^2 ny + \cos^2 nx \sinh^2 ny} \\ &> |\sinh^2 ny| \to \infty, \qquad \text{as } n \to \infty, \end{aligned}$$

for any $y \neq 0$, no matter how small. Therefore the series (4.1.21) is divergent in the whole complex plane except on the real axis $z = x$. Since the line $z = x$ is not a domain, Weierstrass' Theorem on the possibility of differentiating the series (4.1.21) termwise cannot be applied.

In the case of series of functions of a real variable, termwise differentiation is more restrictive, as stated in the following theorem.

THEOREM 4.1.11. *The series*

$$\sum_{n=1}^{\infty} u_n(x), \qquad a \leq x \leq b, \tag{4.1.22}$$

of functions of the real variable x is termwise differentiable if

(a) *it converges uniformly to a differentiable function $f(x)$ on $[a, b]$, and*

(b) *the differentiated series is uniformly convergent on $[a, b]$.*

In that case,

$$f'(x) = \sum_{n=1}^{\infty} u_n'(x).$$

Note 4.1.5. The converse of Weierstrass' Theorem is not true in general. That is, if a series can be differentiated termwise any number of times in a domain D, it does not follow, in general, that the series is uniformly convergent in D.

EXAMPLE 4.1.3. *Show that the series*

$$S(z) = \frac{1}{1 - e^{-z}} = \sum_{n=0}^{\infty} e^{-nz} \tag{4.1.23}$$

converges pointwise in the half-plane $0 < \Re z < \infty$, but not uniformly.

SOLUTION. For any fixed $z = x + iy$, with $x > 0$,

$$|S(z) - S_n(z)| = \left| \sum_{k=n+1}^{\infty} e^{-kz} \right| = \left| \frac{e^{-(n+1)z}}{1 - e^{-z}} \right| \to 0$$

as $n \to \infty$. Hence $S(z)$ converges pointwise in $\Re z > 0$.

On the other hand, if $\Re z > 0$ and $y \neq 0$, say $y = \pi/2$, the inequality

$$\left| \frac{e^{-(N+1)z}}{1 - e^{-z}} \right| < e^{-(N+1)x} < \varepsilon$$

implies

$$(N + 1)x > -\log \varepsilon = \log \frac{1}{\varepsilon},$$

that is,

$$N + 1 > \frac{1}{x} \log \frac{1}{\varepsilon}.$$

Hence, in the last inequality, N depends on x since $N + 1 \to \infty$ as $x \to 0+$. Therefore the series (4.1.23) is not uniformly convergent in the right-hand half-plane $0 < \Re z < +\infty$, but it can be differentiated termwise any number of times there. □

It is left as an exercise to show that the series (4.1.23) is uniformly convergent in the closed region $D = \{0 < \sigma_0 \leq \Re z < +\infty\}$. (Hint: Use the fact that the series is majorizable in D.)

Exercises for Section 4.1

Show whether the following sequences are convergent or divergent, as $n \to \infty$. In the case of convergence, find the limit.

1. $z_n = \dfrac{n - 1}{(1 + i)n + 5}$.

2. $z_n = \dfrac{n^2 - 4in + 2}{3n^2 + 4in - 2}$.

3. $z_n = \dfrac{\text{Log}(in)}{n}$.

4. $z_n = \dfrac{\tan(in)}{n}$.

5. $z_n = \dfrac{\sin[\pi/(in)]}{\sin[\pi/(2in)]}$.

6. $z_n = \text{Log}\left(1 + \dfrac{1}{in}\right)$.

7. $z_n = \dfrac{(2i)^n}{(2i)^n + 3^n}$.

8. $z_n = \left(\dfrac{1+i}{\sqrt{3}-i}\right)^n$.

Show whether the following series are convergent or divergent.

9. $\displaystyle\sum_{n=1}^{\infty} \frac{i^n}{n(n+1)}$.

10. $\displaystyle\sum_{n=1}^{\infty} \frac{n^2 \sin(in)}{2^n (n+1)}$.

11. $\displaystyle\sum_{n=1}^{\infty} \frac{e^{in}}{\sqrt{n(n+1)}}$.

12. $\displaystyle\sum_{n=1}^{\infty} \frac{\operatorname{Log} n}{e^{i\pi/n}}$.

13. $\displaystyle\sum_{n=1}^{\infty} \frac{(in)^n}{n!}$.

14. $\displaystyle\sum_{n=1}^{\infty} \frac{(1+i)^n}{(2-2i)^n}$.

15. $\displaystyle\sum_{n=1}^{\infty} \frac{\sin(in^3)}{n^3+1}$.

16. $\displaystyle\sum_{n=2}^{\infty} \frac{\sin(i/n)}{n \operatorname{Log} n}$.

Let $A = \sum_{n=1}^{\infty} z_n$ and $B = \sum_{n=1}^{\infty} \zeta_n$ be two convergent series. Show that the following relations hold.

17. $A + B = \displaystyle\sum_{n=1}^{\infty} (z_n + \zeta_n)$.

18. $cA - dB = \displaystyle\sum_{n=1}^{\infty} (cz_n - d\zeta_n)$, where c and d are constants.

Suppose that $A = \lim_{n \to \infty} z_n$.

19. Show that $|A| = \lim_{n \to \infty} |z_n|$.

20. Does $|A| = \lim_{n \to \infty} |z_n|$ imply that $A = \lim_{n \to \infty} z_n$?

21. Does $A = \lim_{n \to \infty} z_n$ imply that $\operatorname{Arg} A = \lim_{n \to \infty} \operatorname{Arg} z_n$, where $A \neq 0$?

Find the set on which each of the following series converges.

22. $\displaystyle\sum_{n=1}^{\infty} \frac{\cos nz}{n^3}$.

23. $\displaystyle\sum_{n=1}^{\infty} \frac{\sin nz}{n}$. (Hint: Use Corollary 4.2.3.)

24. $\displaystyle\sum_{n=1}^{\infty} \frac{z^n}{1 + z^{2n}}$.

25. $\displaystyle\sum_{n=1}^{\infty} \frac{4^n}{1 + z^n}$.

Find the regions of uniform convergence of the following series.

26. $\displaystyle\sum_{n=1}^{\infty} \frac{z^n}{n!}$.

27. $\displaystyle\sum_{n=1}^{\infty} \frac{n(n+2)}{(z+1)^n}$.

28. $\displaystyle\sum_{n=1}^{\infty} \sin\left(\frac{\pi}{n}\right) z^n$.

29. $\displaystyle\sum_{n=1}^{\infty} \frac{\sin(n|z|)}{n^2}$.

4.2. Integer power series

Convergent series in powers of z are the starting point of the local theory of analytic functions.

4.2.1. Definition and convergence theorem.

DEFINITION 4.2.1. A series of the form

$$\sum_{n=0}^{\infty} a_n z^n = a_0 + a_1 z + a_2 z^2 + \cdots + a_n z^n + \ldots, \qquad a_n \in \mathbb{C}, \quad (4.2.1)$$

is called an (integer) *power series.* The complex numbers a_n are called the coefficients of the series.

As in the real case, the following theorem plays an important role in the investigation of the convergence of complex power series.

THEOREM 4.2.1 (ABEL'S THEOREM). *(1) If the power series* (4.2.1) *is convergent at the point* z_1*, then it is absolutely convergent in any open disk* $|z| < |z_1|$*. Moreover, in each closed disk* $|z| \leq q|z_1|$*,* $0 < q < 1$*, the series converges uniformly and absolutely.*

(2) If the series (4.2.1) *is divergent at a point* z_2*, then it is divergent in the region* $|z| > |z_2|$*.*

PROOF. **Part (1).** Since the series (4.2.1) is convergent at the point z_1, it follows from the necessary condition of convergence (see Theorem 4.1.1) that

$$\lim_{n\to\infty} a_n z_1^n = 0. \qquad (4.2.2)$$

Therefore

$$\lim_{n\to\infty} |a_n z_1^n| = 0. \qquad (4.2.3)$$

It follows from (4.2.3) that there exists a positive number, M, such that for all $n = 1, 2, \ldots,$ the inequality

$$|a_n z_1^n| < M \qquad (4.2.4)$$

is satisfied. Let z be an arbitrary interior point of the disk $D_0^{|z_1|}$, that is, $|z| \leq q|z_1|$ for $0 < q < 1$. Then, using (4.2.4), we have

$$|a_n z^n| = |a_n|\,|z_1|^n \left|\frac{z}{z_1}\right|^n < M \left|\frac{z}{z_1}\right|^n = Mq^n.$$

Therefore, the series $\sum_{n=1}^{\infty} a_n z^n$ is clearly majorized by the convergent series $M\sum_{n=1}^{\infty} q^n$. Hence, by Weierstrass' Theorem 4.1.8, the former series, $\sum_{n=1}^{\infty} a_n z^n$, is absolutely and uniformly convergent in the disk $|z| < q|z_1|$.

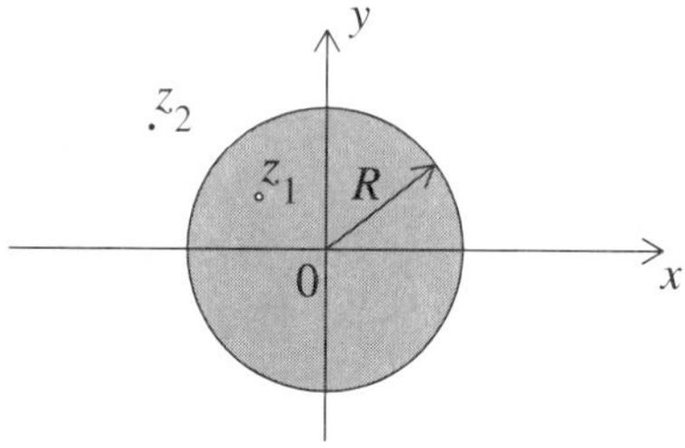

FIGURE 4.2. Shaded disk of convergence and unshaded region of divergence of a power series about $z_0 = 0$.

Since the number q can be as close to 1 as we please, we can conclude that series (4.2.1) is convergent for all z in the disk $|z| < |z_1|$.

Part (2). Suppose that the series (4.2.1) is divergent at $z = z_2$. Then it is divergent for all z such that $|z| > |z_2|$, for, if the series is convergent for such z, it is convergent also at the point z_2 by the already proved first part of the theorem. This contradicts the assumption of the present part. □

Note 4.2.1. One can apply Abel's Theorem to the series

$$\sum_{n=0}^{\infty} a_n (z - z_0)^n, \tag{4.2.5}$$

simply by changing z to $z - z_0$ in the proof of the theorem.

From the convergence of series (4.2.1) at the point z_1 it follows that the power series (4.2.5) is convergent at all the interior points of the disk $D_{z_0}^{|z_1|}$ centered at z_0 with radius $|z_1|$, that is, in the region $|z - z_0| < |z_1|$. On the other hand, if the series (4.2.1) is divergent at the point z_2, then (4.2.5) is divergent outside the disk $D_{z_0}^{|z_2|}$, that is, in the region $|z - z_0| > |z_2|$.

4.2.2. Radius of convergence of a power series. We give a few consequences of Abel's Theorem.

In view of the definition of radius of convergence (see Definition 4.2.2), we reformulate part of Abel's Theorem in the following corollary.

COROLLARY 4.2.1. *Suppose that the power series* (4.2.1) *about* $z_0 = 0$ *is convergent at the point* z_1 *and divergent at the point* z_2. *Then it is convergent in the disk* $|z| < |z_1|$ *and divergent outside the disk of radius* $|z_2|$ *(see Fig* 4.2*).*

Therefore, there exists a real number R such that $|z_1| \le R \le |z_2|$ with the following property: for all z such that $|z| < R$, the series (4.2.1) is convergent and for all z such that $|z| > R$, it is divergent.

DEFINITION 4.2.2. The number $R \geq 0$ having the property that the power series (4.2.1) is convergent in the region $|z| < R$ and divergent in the region $|z| > R$ is called the *radius of convergence* of the power series (4.2.1).

The ratio or root tests can be used to determine the radius of convergence, that is Theorems 4.1.3 or 4.1.4, as in the real case.

COROLLARY 4.2.2. *The radius of convergence, R, of the power series (4.2.1) is given by the following limits, if they exist, or the limits superior:*

$$R = \lim_{n\to\infty} \left|\frac{a_n}{a_{n+1}}\right|, \qquad R = \limsup_{n\to\infty} \left|\frac{a_n}{a_{n+1}}\right| \tag{4.2.6}$$

or

$$R = \lim_{n\to\infty} \frac{1}{|a_n|^{1/n}}, \qquad R = \limsup_{n\to\infty} \frac{1}{|a_n|^{1/n}}. \tag{4.2.7}$$

PROOF. We derive only the first formula in (4.2.6), for which we assume that the limit exists. Consider the series of absolute values of the terms in (4.2.1) for a fixed value of z:

$$\sum_{n=0}^{\infty} |a_n z^n|. \tag{4.2.8}$$

Since (4.2.8) is a series with positive numbers, then, for every fixed z, we can use the ratio test (Theorem 4.1.3) to investigate the region of convergence of the series.

We assume that the limit

$$L = \lim_{n\to\infty} \left|\frac{a_{n+1} z^{n+1}}{a_n z^n}\right| = |z| \lim_{n\to\infty} \left|\frac{a_{n+1}}{a_n}\right| \tag{4.2.9}$$

exists. In order to have convergence of (4.2.8) it suffices to satisfy the inequality $L < 1$,

$$|z| \lim_{n\to\infty} \left|\frac{a_{n+1}}{a_n}\right| < 1, \quad \text{that is,} \quad |z| < \lim_{n\to\infty} \left|\frac{a_n}{a_{n+1}}\right|. \tag{4.2.10}$$

Hence the series (4.2.1) is absolutely convergent in the open disk

$$|z| < R = \lim_{n\to\infty} \left|\frac{a_n}{a_{n+1}}\right|.$$

We now prove that series (4.2.1) is divergent in the region $|z| > R$. Indeed, since the inequality $|z| > R$ corresponds to the inequality $L > 1$, it follows from (4.2.10) that there exists a number N such that, for all $n > N$, the following inequality is satisfied:

$$\left|\frac{a_{n+1}z^{n+1}}{a_n z^n}\right| > 1 \quad \text{that is,} \quad |a_{n+1}z^{n+1}| > |a_n z^n|.$$

The last inequality implies that

$$\lim_{n\to\infty} |a_n z^n| \neq 0, \quad \text{that is,} \quad \lim_{n\to\infty} a_n z^n \neq 0,$$

so that the necessary condition of convergence of Theorem 4.1.1 is not satisfied and the series (4.2.1) diverges. This proves the first formula in (4.2.6). The first formula in (4.2.7) can be derived analogously. □

DEFINITION 4.2.3. The disk $|z| < R$, where R is the radius of convergence, is called the *disk of convergence* of the series (4.2.1).

The series (4.2.1) can either converge or diverge at points on the boundary of the disk.

For example, the series

$$\sum_{n=1}^{\infty} \frac{z^n}{n}$$

has radius of convergence $R = 1$ as can be seen from (4.2.6). On the circle $|z| = 1$ where $z = e^{i\theta}$, the series is divergent only at the point $z = 1$, that is, if $\theta = 0$ or $\theta = 2\pi$, but at all other points, $z = \cos\theta + i\sin\theta$, of the circle, the series

$$\sum_{n=1}^{\infty} \frac{\cos n\theta + i\sin n\theta}{n} = \sum_{n=1}^{\infty} \frac{\cos n\theta}{n} + i\sum_{n=1}^{\infty} \frac{\sin n\theta}{n} \tag{4.2.11}$$

is conditionally convergent.

Tests sharper than the ratio or the root tests can be used to prove the last statement. Such tests, like the Dirichlet–Abel Test (see [**29**], Vol. 1, p. 429), can be used to determine the conditional convergence of alternating series, that is, series with terms which change signs.

THEOREM 4.2.2 (DIRICHLET–ABEL TEST). *Let* $\{a_n\}$ *and* $\{b_n\}$ *be two sequences of complex numbers such that*

(a) $\lim_{n\to\infty} a_n = 0$,

(b) $\sum_{n=0}^{\infty} |a_{n+1} - a_n|$ *converges, and*

(c) *the partial sums of the series* $\sum b_n$ *are bounded, that is,*

$$|b_1 + b_2 + \cdots + b_n| = |S_n| \leq M, \qquad n = 1, 2, \ldots .$$

Then the series

$$\sum_{n=1}^{\infty} a_n b_n \tag{4.2.12}$$

converges.

PROOF. The proof follows by summation by parts:

$$\begin{aligned}\sum_{k=1}^{n} a_k b_k &= a_1 b_1 + a_2 b_2 + \cdots + a_n b_n \\ &= a_1 S_1 + a_2(S_2 - S_1) + \cdots + a_n(S_n - S_{n-1}) \\ &= (a_1 - a_2)S_1 + (a_2 - a_3)S_2 + \cdots + (a_{n-1} - a_n)S_{n-1} + a_n S_n \\ &= a_n S_n - \sum_{k=1}^{n-1}(a_{k+1} - a_k)S_k.\end{aligned}$$

Because of (a) and (c),

$$\lim_{n\to\infty} a_n S_n = 0.$$

Since

$$|(a_{k+1} - a_k)S_k| \le M|a_{k+1} - a_k|$$

and (b) holds, the series

$$\sum_{k=1}^{\infty}(a_{k+1} - a_k)S_k$$

is convergent. These together prove that $\lim_{n\to\infty} a_k b_k$ exists and $\sum_{n=1}^{\infty} a_k b_k$ converges. □

This theorem takes a simpler form if the sequence $\{a_n\}$ is monotone.

COROLLARY 4.2.3. *If the sequence $\{a_n\}$ is monotonic decreasing with $\lim_{n\to\infty} a_n = 0$, and the partial sums $\sum_{k=1}^{n} b_k$ are bounded, then $\sum_{n=1}^{\infty} a_n b_n$ converges.*

PROOF. Since

$$\begin{aligned}\sum_{k=1}^{n} |a_{k+1} - a_k| &= (a_1 - a_2) + (a_2 - a_3) + \cdots + (a_n - a_{n+1}) \\ &= a_1 - a_{n+1}\end{aligned}$$

and

$$\lim_{n\to\infty} \sum_{k=1}^{n} |a_{k+1} - a_k| = \lim_{n\to\infty}(a_1 - a_{n+1}) = a_1,$$

then hypothesis (b) of the theorem is satisfied. □

A special choice of the sequence $\{b_n\}$ yields the usual alternating series test.

COROLLARY 4.2.4. *If the sequence* $\{a_n\}$ *is monotonic decreasing with* $\lim_{n\to\infty} a_n = 0$, *then* $\sum_{n=1}^{\infty}(-1)^{n+1}a_n$ *converges.*

PROOF. With $b_n = (-1)^{n+1}$ the partial sums of $\sum_{n=1}^{\infty} b_n$ are always either 1 or 0. $\square$

Let us apply Corollary4.2.3 to the first series on the right-hand side of (4.2.11).

EXAMPLE 4.2.1. *Show that the series*

$$\sum_{n=1}^{\infty}\frac{\cos n\theta}{n}, \qquad 0<\theta<2\pi, \tag{4.2.13}$$

is conditionally convergent.

SOLUTION. Set $a_n = 1/n$ and $b_n = \cos n\theta$. Clearly, $a_n \geq a_{n+1} \to 0$ as $n \to \infty$. If $0 < \theta_0 \leq \theta \leq 2\pi - \theta_0$, the partial sum

$$S_n = \sum_{k=1}^{n}\cos k\theta = \frac{\sin([n+1]\theta/2) - \sin(\theta/2)}{2\sin(\theta/2)} \tag{4.2.14}$$

is bounded by

$$|S_n| < \frac{2}{2|\sin(\theta_0/2)|},$$

for each $n > 0$. Thus, the conditions of Corollary 4.2.3 are satisfied and the given series (4.2.13) is convergent. $\square$

Similarly, one can prove that the series

$$\sum_{n=1}^{\infty}\frac{\sin n\theta}{n}, \qquad 0<\theta<2\pi,$$

is conditionally convergent.

EXAMPLE 4.2.2. *Show that the two series*

$$\sum_{n=1}^{\infty}\frac{|\cos n\theta|}{n}, \qquad \sum_{n=1}^{\infty}\frac{|\sin n\theta|}{n}$$

are divergent.

SOLUTION. If the first series would be convergent then, by the obvious inequality

$$|\cos n\theta| \geq \cos^2 n\theta,$$

the series

$$2\sum_{n=1}^{\infty}\frac{\cos^2 n\theta}{n}=\sum_{n=1}^{\infty}\frac{1+\cos 2n\theta}{n}=\sum_{n=1}^{\infty}\frac{1}{n}+\sum_{n=1}^{\infty}\frac{\cos 2n\theta}{n} \tag{4.2.15}$$

would also be convergent. But this is false, since the last series converges for $\theta \neq k\pi$ but the second-last series diverges. Therefore, the two series

$$\sum_{n=1}^{\infty}\frac{\cos^2(n\theta)}{n}, \qquad \sum_{n=1}^{\infty}\frac{|\cos(n\theta)|}{n}$$

diverge. Thus the series

$$\sum_{n=1}^{\infty}\frac{\cos(n\theta)}{n}, \qquad \theta \neq k\pi,$$

is only conditionally convergent. The divergence of the second series follows in the same way. □

COROLLARY 4.2.5. *The sum, $S(z)$, of the power series (4.2.1) is analytic inside every disk $|z| \leq R_1 < R$ that lies entirely in the disk of convergence $|z| < R$.*

PROOF. Since the terms $a_n z^n$ of the power series are analytic in the whole complex plane and the series (4.2.1) is uniformly convergent in the region $|z| \leq R_1$, then, by the first part of Weierstrass' Theorem 4.1.10, $S(z)$ is analytic if $|z| \leq R_1 < R$. □

COROLLARY 4.2.6. *Power series can be differentiated and integrated any number of times inside their disk of convergence. Moreover, the radius of convergence of the differentiated (or integrated) series is equal to the radius of convergence of the original series.*

PROOF. This fact is a consequence of the second part of Weierstrass' Theorem 4.1.10. □

Exercises for Section 4.2

Find the radius and disk of convergence of each of the following power series.

1. $\displaystyle\sum_{n=1}^{\infty}\frac{(-1)^n}{n^2+1}\,z^n.$

2. $\displaystyle\sum_{n=1}^{\infty}\frac{2^n}{n!}\,(iz)^n.$

3. $\sum_{n=1}^{\infty}(n!)^2\,(z+1)^n$.

4. $\sum_{n=1}^{\infty}\frac{n^{2n}}{n!}\,(z-1)^{2n}$.

5. $\sum_{n=1}^{\infty}\sin(in)\,(z+2)^n$.

6. $\sum_{n=1}^{\infty}\frac{n^4}{n!}\,(z-i)^n$.

7. $\sum_{n=1}^{\infty}\left(1+\frac{\pi}{n}\right)^n\,(z+i)^n$.

8. $\sum_{n=1}^{\infty}n^n z^n$.

Suppose that the radii of convergence of the power series $\sum_{n=1}^{\infty}a_n z^n$ and $\sum_{n=1}^{\infty}b_n z^n$ are equal to R_1 and R_2, respectively, where $0 < R_1 < \infty$ and $0 < R_2 < \infty$. Estimate the radius of convergence, R, of each of the following power series.

9. $\sum_{n=1}^{\infty}(a_n + b_n)z^n$.

10. $\sum_{n=1}^{\infty}(a_n - b_n)z^n$.

11. $\sum_{n=1}^{\infty}a_n b_n z^n$.

12. $\sum_{n=1}^{\infty}\frac{a_n}{b_n}\,z^n$, $\qquad b_n \neq 0, \qquad \lim_{n\to\infty}\frac{a_n}{a_{n+1}}$ and $\lim_{n\to\infty}\frac{b_n}{b_{n+1}}$ exist.

13. $\sum_{n=1}^{\infty}n a_n z^n$.

14. $\sum_{n=1}^{\infty}n^k a_n z^n$, $\qquad k \in \mathbb{N}$.

15. $\sum_{n=1}^{\infty}\frac{a_n}{n}\,z^n$.

16. $\displaystyle\sum_{n=1}^{\infty} \frac{a_n}{n^k} z^n, \qquad k \in \mathbb{N}.$

17. Does there exist a power series in powers of z that converges at $z = 3+4i$ and diverges at $z = -3+3i$? Explain.

Find the sum of the following series.

18. $\displaystyle\sum_{n=1}^{\infty} n z^{n-1}.$

19. $\displaystyle\sum_{n=1}^{\infty} n^2 z^n.$

20. $\displaystyle\sum_{n=1}^{\infty} \frac{z^n}{n+1}.$

21. $\displaystyle\sum_{n=2}^{\infty} \frac{z^n}{n(n-1)}.$

4.3. Taylor series

We turn now to the Taylor series of an analytic function $f(z)$ and the relation between the radius of convergence of the series and the singularities of $f(z)$.

4.3.1. Taylor series and radius of convergence.

The following theorem is central in the theory of analytic functions.

THEOREM 4.3.1 (TAYLOR SERIES). *Let $f(z)$ be an analytic function in a domain D which contains the open disk $D_{z_0}^R : |z - z_0| < R$ and its boundary $C_R : |z - z_0| = R$. Then, at each point z in that disk, $f(z)$ has the convergent series representation*

$$f(z) = \sum_{n=0}^{\infty} c_n (z - z_0)^n, \tag{4.3.1}$$

where

$$c_n = \frac{f^{(n)}(z_0)}{n!}. \tag{4.3.2}$$

This series is called the Taylor series *of $f(z)$ with center at $z = z_0$.*

PROOF. Since C_R is in D, then Cauchy's integral formula,

$$f(z) = \frac{1}{2\pi i} \oint_{C_R} \frac{f(\zeta)}{\zeta - z}\, d\zeta, \tag{4.3.3}$$

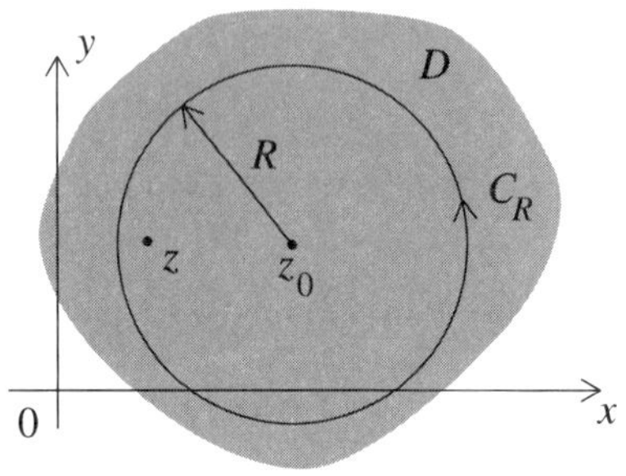

FIGURE 4.3. Shaded domain D containing the disk $D_{z_0}^R$ of convergence of a Taylor series centered at z_0.

is valid at any point z of $D_{z_0}^R$ (see Fig 4.3). We use the transformation

$$\frac{1}{\zeta - z} = \frac{1}{\zeta - z_0 - (z - z_0)} = \frac{1}{\zeta - z_0} \frac{1}{1 - \frac{z - z_0}{\zeta - z_0}} \tag{4.3.4}$$

to expand $1/(\zeta - z)$ in powers of $(z - z_0)/(\zeta - z_0)$ in a neighborhood of z_0. Since $z \in D_{z_0}^R$ and $\zeta \in C_R$, then $|z - z_0| < R$ and $|\zeta - z_0| = R$, so that

$$\left| \frac{z - z_0}{\zeta - z_0} \right| < 1.$$

Therefore, one can expand the right-hand side of (4.3.4) in a power series in $(z - z_0)/(\zeta - z_0)$:

$$\frac{1}{\zeta - z} = \sum_{n=0}^{\infty} \frac{(z - z_0)^n}{(\zeta - z_0)^{n+1}}. \tag{4.3.5}$$

We prove that series (4.3.5) is uniformly convergent with respect to ζ and z for all $\zeta \in C_R$ and all z strictly inside the disk $D_{z_0}^R$. Indeed, since z is an interior point, there exists $\rho > 0$ such that $|z - z_0| < \rho < R$ and

$$\left| \frac{z - z_0}{\zeta - z_0} \right| < \frac{\rho}{R} < 1.$$

Therefore, series (4.3.5) is majorized by the convergent series

$$\frac{1}{R} \sum_{n=0}^{\infty} \left(\frac{\rho}{R} \right)^n,$$

and thus is uniformly convergent by Weierstrass' M-Test (Theorem 4.1.8).

Substituting (4.3.5) into (4.3.3) and integrating termwise with respect to ζ (this integration is possible because the series is uniformly convergent with respect to $\zeta \in C_R$ and z strictly inside $D_{z_0}^R$) we obtain (4.3.1) where, by (3.4.8), the coefficients c_n are given by

$$c_n = \frac{1}{2\pi i} \oint_{C_R} \frac{f(\zeta)}{(\zeta - z_0)^{n+1}}\, d\zeta = \frac{f^{(n)}(z_0)}{n!}. \quad \square \tag{4.3.6}$$

We remark that the coefficients c_n given by (4.3.6) do not change if the radius R of the disk $D_{z_0}^R$ is increased as long as its boundary C_R does not cross any singularity of $f(z)$. But, as soon as at least one singular point of $f(z)$ is located inside the disk $D_{z_0}^R$, the Taylor series becomes divergent. Therefore the following theorem holds.

THEOREM 4.3.2. *The radius of convergence of the* Taylor series (4.3.1) *is equal to the distance from z_0 to the closest singular point of $f(z)$.*

This theorem explains why the radius of convergence of the series

$$\frac{1}{1+x^2} = \sum_{n=0}^{\infty} (-x^2)^n \tag{4.3.7}$$

is equal to 1, a fact that is not seen by considering the rational function $1/(1+x^2)$ of the real variable x. In the complex plane the series, with center $z_0 = 0$,

$$\frac{1}{1+z^2} = \sum_{n=0}^{\infty} (-z^2)^n \tag{4.3.8}$$

is convergent in the disk $|z| < 1$, that is, the radius of convergence of the series on the right-hand side of (4.3.8) is equal to 1 because the function $1/(1+z^2)$ on the left-hand side has two singular points, $z = \pm i$, in the complex plane at distance 1 from the center, $z_0 = 0$, of the series.

Note 4.3.1. The Taylor series (4.3.1) is a power series in $z - z_0$ with coefficients, c_n, given by (4.3.2). Suppose $f(z)$ is represented by another convergent power series in $z - z_0$,

$$f(z) = \sum_{n=0}^{\infty} a_n (z - z_0)^n. \tag{4.3.9}$$

Since this latter series can be differentiated any number of times in the disk of convergence $|z - z_0| < R$, it follows from (4.3.9) that

$$f(z_0) = a_0, \quad f'(z) = a_1 + 2a_2(z - z_0) + \dots, \qquad f'(z_0) = a_1 1!,$$

$$f''(z) = 2 \times 1 \times a_2 + 3 \times 2 \times a_3(z - z_0) + \dots, \qquad f''(z_0) = a_2 2!,$$

and, in general,

$$f^{(n)}(z_0) = a_n n!,$$

that is, the coefficients, a_n, of the power series (4.3.9) are equal to

$$a_n = \frac{f^{(n)}(z_0)}{n!}.$$

Hence, $a_n = c_n$, and the two series coincide. Therefore, there exists a deep link between the radius of convergence of the Taylor series of a function $f(z)$ in powers of $z - z_0$ and the distance from the point z_0 to the closest singular point of $f(z)$.

It follows from the previous results that if $f(z)$ is differentiable at $z = z_0$ and in some neighborhood, $|z - z_0| < \rho$, of z_0, then $f(z)$ can be represented by its Taylor series (4.3.1) in the same neighborhood. The converse statement is easily proved in the following theorem.

THEOREM 4.3.3. *If $f(z)$ is represented by the convergent power series*

$$f(z) = \sum_{n=0}^{\infty} a_n (z - z_0)^n \tag{4.3.10}$$

in the disk $|z - z_0| < \rho$, then it is differentiable at $z = z_0$.

PROOF. It follows from (4.3.10) that $f(z_0) = a_0$, and

$$\frac{f(z) - f(z_0)}{z - z_0} = a_1 + a_2(z - z_0) + \cdots + a_n(z - z_0)^{n-1} + \ldots. \tag{4.3.11}$$

Formula (4.3.11) shows that the limit

$$\lim_{z \to z_0} \frac{f(z) - f(z_0)}{z - z_0} = a_1 = f'(z_0)$$

exists and is finite, that is, $f(z)$ is differentiable at $z = z_0$. □

COROLLARY 4.3.1. *Let D be an open disk with center $z = z_0$. The following statements are equivalent:*

(1) $f(z)$ is differentiable in D;

(2) $f(z)$ is expandable in a power series in $z - z_0$ in D.

Remark 4.3.1. In the literature, a function $f(z)$ of a complex variable z is said to be *holomorphic* in a domain D if it is differentiable in D, and it is said to be *analytic* in D if it has a convergent power series about any point in D. Because of this equivalence, holomorphic functions are often called analytic.

EXAMPLE 4.3.1. *Find the radius of convergence of the Taylor series of* $\operatorname{Log}(1 - z)$ *about* $z = 0$,

$$\operatorname{Log}(1 - z) = -\sum_{n=1}^{\infty} \frac{z^n}{n}, \qquad |z| < 1. \tag{4.3.12}$$

SOLUTION. The function $\operatorname{Log}(1-z)$ is not defined at the branch point $z = 1$. Therefore, the radius of convergence of (4.3.12) is equal to the distance $|1 - 0| = 1$. □

4.3.2. Practical methods for obtaining Taylor series. We consider several examples of Taylor series expansions. In practice one tries to avoid computing the integral

$$\frac{f^{(n)}(z_0)}{n!} = \frac{1}{2\pi i} \oint_{C_R} \frac{f(\zeta)}{(\zeta - z_0)^{n+1}}\, d\zeta$$

in order to expand $f(z)$ in a Taylor series. It is often simpler to use some special methods. The geometric series,

$$\frac{1}{1-z} = \sum_{n=0}^{\infty} z^n, \qquad |z| < 1, \tag{4.3.13}$$

is often used for this purpose.

For example, differentiating (4.3.13) with respect to z, one obtains

$$\frac{1}{(1-z)^2} = \sum_{n=1}^{\infty} n z^{n-1}, \qquad \frac{2}{(1-z)^3} = \sum_{n=2}^{\infty} n(n-1) z^{n-2}, \qquad \text{for } |z| < 1, \tag{4.3.14}$$

and so on.

On the other hand, integrating (4.3.13) with respect to z from 0 to z, $|z| < 1$, we obtain

$$\operatorname{Log}(1-z) = -\sum_{n=0}^{\infty} \frac{z^{n+1}}{n+1}, \qquad |z| < 1. \tag{4.3.15}$$

Other useful expansions are

$$e^z = \sum_{n=0}^{\infty} \frac{z^n}{n!}, \qquad |z| < \infty, \tag{4.3.16}$$

$$\sin z = \sum_{n=0}^{\infty} (-1)^n \frac{z^{2n+1}}{(2n+1)!}, \qquad |z| < \infty, \tag{4.3.17}$$

$$\cos z = \sum_{n=0}^{\infty} (-1)^n \frac{z^{2n}}{(2n)!}, \qquad |z| < \infty. \tag{4.3.18}$$

It follows from the last three expansions that the radius of convergence of the corresponding series is equal to infinity, because the functions e^z,

$\sin z$, $\cos z$ do not have singular points in the finite part of the complex plane. Such functions are called *entire*. The only singular point of these functions is $z = \infty$ since, if $z = 1/z_1$, then e^{1/z_1}, $\sin(1/z_1)$, $\cos(1/z_1)$ have a singular point at $z_1 = 0$.

In order to expand a proper rational fraction $P_n(z)/Q_m(z)$, $n < m$, in a Taylor series, it suffices to represent this fraction as a sum of partial fractions and represent each of these fractions by a Taylor series using (4.3.13) and (4.3.14). We illustrate this technique by examples.

EXAMPLE 4.3.2. *Find the Taylor series expansion in powers of z of the function*

$$f(z) = \frac{1}{(z-2)(z-3)}. \tag{4.3.19}$$

SOLUTION. Since the singular points of $f(z)$ are $z = 2$ and $z = 3$, the radius of convergence of the Taylor series in powers of z is equal to 2. Using partial fractions, we have

$$f(z) = \frac{1}{z-3} - \frac{1}{z-2}. \tag{4.3.20}$$

We expand each of the fractions in (4.3.20) in a Taylor series in the disk $|z| < 2$ by means of (4.3.13). Thus, we have the series

$$\frac{1}{z-3} = -\frac{1}{3}\frac{1}{1-\frac{z}{3}} = -\frac{1}{3}\sum_{n=0}^{\infty}\left(\frac{z}{3}\right)^n,$$

which converges for

$$\left|\frac{z}{3}\right| < 1, \quad \text{that is,} \quad |z| < 3.$$

Similarly, we have the series

$$\frac{1}{z-2} = -\frac{1}{2}\frac{1}{1-\frac{z}{2}} = -\frac{1}{2}\sum_{n=0}^{\infty}\left(\frac{z}{2}\right)^n,$$

which converges for

$$\left|\frac{z}{2}\right| < 1, \quad \text{that is,} \quad |z| < 2.$$

Using these expansions, we obtain from (4.3.20) the Taylor series of $f(z)$ in the disk $|z| < 2$ in the form

$$f(z) = -\frac{1}{3}\sum_{n=0}^{\infty}\left(\frac{z}{3}\right)^n + \frac{1}{2}\sum_{n=0}^{\infty}\left(\frac{z}{2}\right)^n, \qquad |z| < 2,$$

or

$$f(z) = \sum_{n=0}^{\infty} \left(\frac{1}{2^{n+1}} - \frac{1}{3^{n+1}} \right) z^n, \qquad |z| < 2. \quad \square$$

However, there are cases where a Taylor series expansion (4.3.10) can be found only by using integrals.

EXAMPLE 4.3.3. *Find the Taylor series expansion of* $f(z) = e^{1/z}$ *with center at* $z_0 = 2$.

SOLUTION. Since $z = 0$ is the only singular point of $f(z)$, then the radius of convergence is equal to 2. Using (4.3.2) we obtain

$$\frac{f^{(n)}(2)}{n!} = \frac{1}{2\pi i} \oint_{C_R} \frac{e^{1/\zeta}}{(\zeta - 2)^{n+1}} \, d\zeta, \tag{4.3.21}$$

where C_R is the closed path $\zeta - 2 = \rho\, e^{i\theta}$, $-\pi \le \theta \le \pi$, $0 < \rho < 2$. Then $d\zeta = \rho e^{i\theta} i \, d\theta$ and (4.3.21) has the form

$$\begin{aligned} \frac{f^{(n)}(2)}{n!} &= \frac{1}{2\pi} \int_{-\pi}^{\pi} e^{\frac{1}{2+\rho e^{i\theta}}} \frac{\rho e^{i\theta}}{\left(\rho e^{i\theta}\right)^{n+1}} \, d\theta \\ &= \frac{1}{2\pi \rho^n} \int_{-\pi}^{\pi} e^{\frac{2+\rho\cos\theta - i\rho\sin\theta}{(2+\rho\cos\theta)^2 + \rho^2\sin^2\theta}} \, e^{-in\theta} \, d\theta \\ &= \frac{1}{2\pi \rho^n} \int_{-\pi}^{\pi} e^{\frac{2+\rho\cos\theta}{4+4\rho\cos\theta+\rho^2}} \, e^{-i\left[\frac{\rho\sin\theta}{4+4\rho\cos\theta+\rho^2} + n\theta\right]} \, d\theta \\ &= \frac{1}{2\pi \rho^n} \int_{-\pi}^{\pi} e^{\frac{2+\rho\cos\theta}{4+4\rho\cos\theta+\rho^2}} \left[\cos\beta_n - i\sin\beta_n\right] d\theta, \end{aligned} \tag{4.3.22}$$

where

$$\beta_n = \frac{\rho\sin\theta}{4 + 4\rho\cos\theta + \rho^2} + n\theta. \tag{4.3.23}$$

Since β_n is an odd function of θ, the integral of the imaginary part of the integrand on the right-hand side of (4.3.22) is equal to zero (as the integral of an odd function with symmetric limits of integration).

Therefore, it follows from (4.3.22) that

$$\frac{f^{(n)}(2)}{n!} = \frac{1}{2\pi\rho^n} \int_{-\pi}^{\pi} e^{\frac{2+\rho\cos\theta}{4+4\rho\cos\theta+\rho^2}} \cos\beta_n \, d\theta. \tag{4.3.24}$$

Hence the Taylor series expansion of $e^{1/z}$ about the point $z = 2$ is

$$e^{1/z} = \sum_{n=0}^{\infty} \left[\frac{1}{2\pi\rho^n} \int_{-\pi}^{\pi} e^{\frac{2+\rho\cos\theta}{4+4\rho\cos\theta+\rho^2}} \cos\beta_n \, d\theta \right] (z-2)^n, \tag{4.3.25}$$

where $|z-2|<2$ and $0<\rho<2$. □

Note 4.3.2. In fact, the integral in (4.3.25) does not depend on ρ, but this result is difficult to prove analytically.

Exercises for Section 4.3

Find the Taylor series for the following functions about the point z_0 and determine the radius of convergence.

1. $\cos z, \qquad z_0 = -\pi/2.$

2. $e^{3z}, \qquad z_0 = \pi i.$

3. $1/z, \qquad z_0 = -1.$

4. $1/(z-i), \qquad z_0 = -i.$

5. $\cos^2 z, \qquad z_0 = 0.$

6. $\cosh^2 z, \qquad z_0 = 0.$

7. $\dfrac{z}{z^2+4}, \qquad z_0 = 0.$

8. $\dfrac{z+2}{(z-1)^2}, \qquad z_0 = 0.$

9. $z^4 + 2z^3 - z + 1, \qquad z_0 = 2.$

10. $\dfrac{z}{(z+i)(z+3i)}, \qquad z_0 = 2i.$

11. $\dfrac{z-2}{(z+3)(z-1)}, \qquad z_0 = -1.$

12. $\dfrac{z^2}{(z-1)^2(z+2)}, \qquad z_0 = 3.$

Using Taylor series expansion for elementary functions given in Section 4.3, solve the following problems.

13. Prove that $(\sin z)' = \cos z.$

14. Prove that $(\cosh z)' = \sinh z.$

15. Show that $\tan z = z + \dfrac{z^3}{3} + 2\,\dfrac{z^5}{15} + \dots, \qquad |z| < \pi/2.$

16. Show that $\sec z = 1 + \dfrac{z^2}{2} + 5\,\dfrac{z^4}{24} + \dots, \qquad |z| < \pi/2.$

Find the Taylor series of the given functions about the given point, z_0, and determine their radii of convergence.

17. $\cos(3z-2), \qquad z_0 = 1.$

18. $\operatorname{Log}(3+z), \qquad z_0 = 0.$

19. $e^{z^2+2z}, \qquad z_0 = -1.$

20. $\sin(2z-5), \qquad z_0 = -2.$

Find the first three nonzero terms of the Taylor series about the given point, z_0, and determine the radius of convergence of the series.

21. $\dfrac{\cos^2 z}{1+z^2}, \qquad z_0 = 0.$

22. $\dfrac{z}{e^z-1}, \qquad z_0 = 0.$

23. $\operatorname{Log}(1+\cos z), \qquad z_0 = 0.$

24. $\dfrac{1}{1+\cos z}, \qquad z_0 = 0.$

25. $e^{1/z}, \qquad z_0 = 1.$

26. $\sin\dfrac{z}{1+z}, \qquad z_0 = 0.$

The series

$$J_n(x) = \sum_{m=0}^{\infty} \frac{(-1)^m x^{2m+n}}{2^{2m+n} m!(n+m)!} \tag{4.3.26}$$

defines the Bessel function of the first kind of order n for $n \in \mathbb{N}$. Using (4.3.26) and properties of power series, derive the following relations.

27. $[x^n J_n(x)]' = x^n J_{n-1}(x).$

28. $[x^{-n} J_n(x)]' = -x^{-n} J_{n+1}(x).$

Using Exercises 27 and 28, show that

29. $J_{n-1}(x) + J_{n+1}(x) = \dfrac{2n}{x} J_n(x).$

30. $J_{n-1}(x) - J_{n+1}(x) = 2J_n'(x).$

Use Exercises 27–30 to evaluate the integrals

31. $\displaystyle\int J_3(x)\, dx.$

32. $\displaystyle\int x^3 J_0(x)\, dx.$

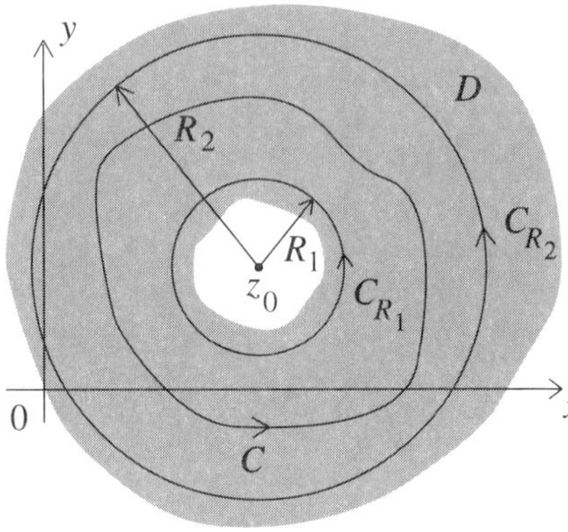

FIGURE 4.4. Shaded domain D, annulus and closed path C for a Laurent series.

4.4. Laurent series

An analytic function, $f(z)$, with a pole at z_0 can be expanded in a Laurent series with center at z_0, and whose domain of convergence is an annulus with center at z_0.

4.4.1. Laurent series and domain of convergence. The following theorem is central in the study of the local properties of meromorphic functions. See Definition 5.1.7 for the definition of a meromophic function.

THEOREM 4.4.1 (LAURENT SERIES). *Let the function $f(z)$ be analytic in a domain D containing the annulus*

$$R_1 \leq |z - z_0| \leq R_2, \tag{4.4.1}$$

(see Fig 4.4) bounded by the circles $C_{R_1} : |z| = R_1$ and $C_{R_2} : |z| = R_2$, and let C denote any positively oriented closed path around z_0 and lying inside the annulus. Then at each interior point z inside the annulus, $f(z)$ has the series expansion

$$f(z) = \sum_{n=-\infty}^{-1} c_n (z - z_0)^n + \sum_{n=0}^{\infty} c_n (z - z_0)^n, \tag{4.4.2}$$

where

$$c_n = \frac{1}{2\pi i} \oint_C \frac{f(\zeta)}{(\zeta - z_0)^{n+1}}\, d\zeta, \qquad n = 0, \pm 1, \pm 2, \ldots, \tag{4.4.3}$$

called the Laurent series *of $f(z)$ in the annulus* (4.4.1).

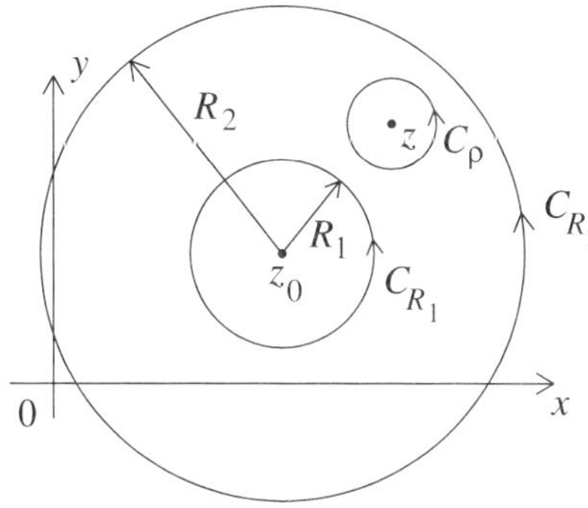

FIGURE 4.5. Annulus and circle C_ρ for the Laurent series.

PROOF. Take a circle C_ρ of radius ρ centered at z and lying entirely inside the annulus for ρ sufficiently small (see Fig 4.5). By Cauchy's Theorem for a multiply connected domain we have (see formula (3.4.14))

$$\oint_{C_{R_2}} \frac{f(\zeta)}{\zeta - z}\, d\zeta = \oint_{C_\rho} \frac{f(\zeta)}{\zeta - z}\, d\zeta + \oint_{C_{R_1}} \frac{f(\zeta)}{\zeta - z}\, d\zeta, \tag{4.4.4}$$

where the three circles C_{R_2}, C_ρ and C_{R_1} are taken counterclockwise. By Cauchy's integral formula (3.4.3), the integral along C_ρ is equal to $2\pi i f(z)$, so that from (4.4.4) we obtain

$$f(z) = \frac{1}{2\pi i} \oint_{C_{R_2}} \frac{f(\zeta)}{\zeta - z}\, d\zeta - \frac{1}{2\pi i} \oint_{C_{R_1}} \frac{f(\zeta)}{\zeta - z}\, d\zeta. \tag{4.4.5}$$

The integral along C_{R_2} can be transformed as in the previous subsection (see (4.3.1) and (4.3.2)):

$$\frac{1}{2\pi i} \oint_{C_{R_2}} \frac{f(\zeta)}{\zeta - z}\, d\zeta = \sum_{n=0}^{\infty} \left[\frac{1}{2\pi i} \oint_{C_{R_2}} \frac{f(\zeta)}{(\zeta - z_0)^{n+1}}\, d\zeta \right] (z - z_0)^n. \tag{4.4.6}$$

However, we cannot replace the expression in square brackets in (4.4.6) by $f^{(n)}(z_0)/n!$ since z_0 may be a singular point of $f(z)$.

We expand the expression $1/(\zeta - z)$ in the integral along C_{R_1} in (4.4.5) in negative powers of $z - z_0$ by means of the transformation

$$-\frac{1}{\zeta - z} = -\frac{1}{\zeta - z_0 - (z - z_0)} = \frac{1}{z - z_0} \frac{1}{1 - \frac{\zeta - z_0}{z - z_0}}. \tag{4.4.7}$$

Since $\zeta \in C_{R_1}$, $|\zeta - z_0| = R_1$. On the other hand, since z is an interior point of (4.4.1), $|z - z_0| > R_1$. Hence

$$\left|\frac{\zeta - z_0}{z - z_0}\right| < 1.$$

Therefore, the right-hand side of (4.4.7) can be expanded in a power series in $(\zeta - z_0)/(z - z_0)$ so that

$$-\frac{1}{\zeta - z} = \sum_{n=0}^{\infty} \frac{(\zeta - z_0)^n}{(z - z_0)^{n+1}}. \tag{4.4.8}$$

We show that the series (4.4.8) is uniformly convergent with respect to ζ and z for all $\zeta \in C_{R_1}$ and all z inside the annulus (4.4.1). Indeed, since z is an interior point of (4.4.1), there exists $\rho_1 > 0$ such that $R_1 < \rho_1 < |z - z_0|$. Then

$$\left|\frac{\zeta - z_0}{z - z_0}\right| < \frac{R_1}{\rho_1} < 1,$$

and the series (4.4.8) is majorized by the convergent series of positive numbers

$$\frac{1}{\rho_1}\sum_{n=0}^{\infty}\left(\frac{R_1}{\rho_1}\right)^n.$$

Therefore, by Weierstrass' M-test (Theorem 4.1.8), the series (4.4.8) is uniformly convergent.

Substituting (4.4.8) into the second term on the right-hand side of (4.4.5) and integrating termwise with respect to ζ (this is possible since the series is uniformly convergent with respect to ζ and z in the annulus (4.4.1)), we obtain

$$-\frac{1}{2\pi i}\oint_{C_{R_1}} \frac{f(\zeta)}{\zeta - z}\,d\zeta = \sum_{n=0}^{\infty}\left[\frac{1}{2\pi i}\oint_{C_{R_1}} \frac{f(\zeta)}{(\zeta - z_0)^{-n}}\,d\zeta\right](z - z_0)^{-n-1}$$

(and putting $n = -1 - k$)

$$= \sum_{k=-1}^{-\infty}\left[\frac{1}{2\pi i}\oint_{C_{R_1}} \frac{f(\zeta)}{(\zeta - z_0)^{k+1}}\,d\zeta\right](z - z_0)^{k}$$

$$= \sum_{n=-1}^{-\infty}\left[\frac{1}{2\pi i}\oint_{C_{R_1}} \frac{f(\zeta)}{(\zeta - z_0)^{n+1}}\,d\zeta\right](z - z_0)^{n}. \tag{4.4.9}$$

The integrands in (4.4.6) and (4.4.9) are the same. We show that the paths of integration, C_{R_1} and C_{R_2}, in these integrals can be replaced by an arbitrary closed path C lying entirely inside the annulus (4.4.1). Indeed, the function $f(\zeta)(\zeta - z_0)^{-n-1}$ is analytic in the region between C_{R_2} and C since its only singular point, $\zeta = z_0$, lies outside this domain. Therefore, by Cauchy's Theorem for multiply connected domains,

$$\oint_C \frac{f(\zeta)}{(\zeta - z_0)^{n+1}}\, d\zeta = \oint_{C_{R_2}} \frac{f(\zeta)}{(\zeta - z_0)^{n+1}}\, d\zeta. \tag{4.4.10}$$

Similarly, $f(\zeta)(\zeta - z_0)^{-n-1}$ is analytic in the region between C and C_{R_1} so that

$$\oint_C \frac{f(\zeta)}{(\zeta - z_0)^{n+1}}\, d\zeta = \oint_{C_{R_1}} \frac{f(\zeta)}{(\zeta - z_0)^{n+1}}\, d\zeta. \tag{4.4.11}$$

Substituting (4.4.10) and (4.4.11) into (4.4.6) and (4.4.9), respectively, and substituting the results into (4.4.5), we obtain

$$f(z) = \sum_{n=-\infty}^{-1} c_n (z - z_0)^n + \sum_{n=0}^{\infty} c_n (z - z_0)^n, \tag{4.4.12}$$

where, for any counterclockwise closed path C inside the annulus,

$$c_n = \frac{1}{2\pi i} \oint_C \frac{f(\zeta)}{(\zeta - z_0)^{n+1}}\, d\zeta, \qquad n = 0, \pm 1, \pm 2, \dots. \tag{4.4.13}$$

The two series in (4.4.12) can be combined into a single doubly infinite series

$$f(z) = \sum_{n=-\infty}^{\infty} c_n (z - z_0)^n. \quad \square \tag{4.4.14}$$

It follows from the derivation of (4.4.3) and (4.4.14) that the series (4.4.14) is absolutely and uniformly convergent in every closed annulus lying entirely inside the annulus (4.4.1).

Note 4.4.1. If $f(z)$ is analytic not only in the annulus (4.4.1) but in the whole disk $D : |z - z_0| < R_2$, then $f(z)/(z - z_0)^{n+1}$ is analytic in D for $n = -1, -2, -3, \dots$ Therefore, by Cauchy's Theorem for simply connected domains,

$$c_n = \frac{1}{2\pi i} \oint_C \frac{f(\zeta)}{(\zeta - z_0)^{n+1}}\, d\zeta = 0, \qquad \text{for} \quad n = -1, -2, -3, \dots.$$

In this case, the Laurent series (4.4.14) reduces to the Taylor series

$$f(z) = \sum_{n=0}^{\infty} c_n (z - z_0)^n, \qquad c_n = \frac{f^{(n)}(z_0)}{n!}.$$

DEFINITION 4.4.1. Let the coefficients c_n be determined by (4.4.3). Then the two series

$$\sum_{n=-\infty}^{-1} c_n(z-z_0)^n, \qquad \sum_{n=0}^{\infty} c_n(z-z_0)^n$$

are called the *principal* and *regular* parts, respectively, of Laurent series (4.4.14).

4.4.2. Practical methods of obtaining Laurent series. In practice, one tries to avoid computing the coefficients c_n of the Laurent series (4.4.3) by means of integration. Other practical methods are used instead. The background for these methods is the geometric series:

$$\frac{1}{1-z} = \sum_{n=0}^{\infty} z^n, \qquad |z| < 1. \tag{4.4.15}$$

Replacing z with $1/z$ in (4.4.15) we obtain

$$\frac{1}{1-(1/z)} = \sum_{n=0}^{\infty} \left(\frac{1}{z}\right)^n, \qquad \frac{1}{|z|} < 1,$$

so that

$$\frac{1}{1-z} = -\sum_{n=0}^{\infty} \frac{1}{z^{n+1}}, \qquad |z| > 1. \tag{4.4.16}$$

Differentiating (4.4.15) and (4.4.16), we obtain

$$\frac{1}{(1-z)^2} = \sum_{n=1}^{\infty} nz^{n-1}, \qquad |z| < 1, \tag{4.4.17}$$

$$\frac{1}{(1-z)^2} = \sum_{n=0}^{\infty} \frac{n+1}{z^{n+2}}, \qquad |z| > 1. \tag{4.4.18}$$

Similarly, one can get Laurent series expansions for $\log(1-1/z)$, $e^{1/z}$, $\sin 1/z$, $\cos 1/z$ by replacing z with $1/z$ in (4.3.15)–(4.3.18).

To find a Laurent series for a proper rational function $P_n(z)/Q_m(z)$, where $n < m$, it suffices to expand it into a sum of partial fractions and use the Taylor series (4.4.15), (4.4.17) or the Laurent series (4.4.16), (4.4.18) or consequences of these formulae which can be found by differentiation the necessary number of times.

EXAMPLE 4.4.1. *Find the Laurent series of the function*

$$f(z) = \frac{1}{(z-2)(z-3)}$$

(a) *in the annulus* $2 < |z| < 3$,
(b) *in the region* $3 < |z| < \infty$.

SOLUTION. First of all let us convince ourselves that it is possible to find the desired expansions. The only singular points of $f(z)$ are $z = 2$ and $z = 3$. These points do not lie inside the annulus $2 < |z| < 3$ or in the region $|z| > 3$. Therefore, in both cases (a) and (b), it is possible to find Laurent series.

In the case (a), we first expand $f(z)$ in partial fractions,

$$f(z) = \frac{1}{z-3} - \frac{1}{z-2}. \tag{4.4.19}$$

The fraction $1/(z-3)$ can be expanded either in a Taylor series by means of (4.4.15) or in a Laurent series by (4.4.16). If we use a Taylor series expansion in powers of z, then we obtain a series which is convergent in the disk $|z| < 3$ and therefore in the annulus $2 < |z| < 3$. This is what we need. Hence

$$\frac{1}{z-3} = -\frac{1}{3}\frac{1}{1-(z/3)} = -\frac{1}{3}\sum_{n=0}^{\infty}\left(\frac{z}{3}\right)^n, \qquad |z| < 3.$$

A similar approach can be used for the fraction $1/(z-2)$. If this fraction is expanded in a Taylor series by means of (4.4.15), then we obtain a series which is convergent in the disk $|z| < 2$, but we need a series which is convergent in the annulus $2 < |z| < 3$. Therefore we cannot use a Taylor series in this case. A Laurent series expansion by means of (4.4.16) gives a series which is convergent in the region $|z| > 2$ and, therefore, in the region $2 < |z| < 3$. This is what we need. Hence

$$\frac{1}{z-2} = \frac{1}{z}\frac{1}{1-(2/z)} = \sum_{n=0}^{\infty}\frac{2^n}{z^{n+1}}, \qquad |z| > 2.$$

Substituting these expansions into (4.4.19) we obtain the desired formula,

$$\begin{aligned} f(z) &= -\frac{1}{3}\sum_{n=0}^{\infty}\left(\frac{z}{3}\right)^n - \frac{1}{2}\sum_{n=0}^{\infty}\left(\frac{2}{z}\right)^{n+1} \\ &= -\sum_{n=0}^{\infty}\left(\frac{2^n}{z^{n+1}} + \frac{z^n}{3^{n+1}}\right), \qquad 2 < |z| < 3. \end{aligned}$$

The solution to the case (b) is left as an exercise to the reader. □

Note that, in general, a Laurent series expansion can be found only by means of formula (4.4.3), as shown in the following two examples taken from [**50**], pp. 101–102, Section 5.6, Examples 1 and 3, respectively.

EXAMPLE 4.4.2. *Prove that for all real* x

$$e^{[z-(1/z)]x/2} = \sum_{n=-\infty}^{\infty} J_n(x) z^n, \tag{4.4.20}$$

where

$$J_n(x) = \frac{1}{2\pi} \int_{-\pi}^{\pi} \cos(x \sin\theta - n\theta)\, d\theta. \tag{4.4.21}$$

The function $J_n(x)$ is called the Bessel function of the first kind of order n, and (4.4.21) is one of its integral representations. A power series representation of $J_n(x)$ is given in (4.3.26).

SOLUTION. Since $z = 0$ is the only singular point of the function on the left-hand side of (4.4.20), then its Laurent series expansion in powers of z is possible in every annulus $0 < \rho < |z| < R$. To determine the coefficients c_n, we use formula (4.4.3) and take the unit circle $|z| = 1$ as the path of integration. Then, $z = e^{i\theta}$, $dz = i\, e^{i\theta}\, d\theta$, and

$$\begin{aligned} c_n &= \frac{1}{2\pi i} \oint_{|z|=1} \frac{e^{[z-(1/z)]x/2}}{z^{n+1}}\, dz \\ &= \frac{1}{2\pi} \int_{-\pi}^{\pi} e^{-i(n+1)\theta} e^{x(e^{i\theta}-e^{-i\theta})/2} e^{i\theta}\, d\theta \\ &= \frac{1}{2\pi} \int_{-\pi}^{\pi} e^{i(x\sin\theta - n\theta)}\, d\theta \\ &= \frac{1}{2\pi} \int_{-\pi}^{\pi} \cos(x\sin\theta - n\theta)\, d\theta + \frac{i}{2\pi} \int_{-\pi}^{\pi} \sin(x\sin\theta - n\theta)\, d\theta. \end{aligned}$$

Since the last integral is equal to zero, then

$$c_n = \frac{1}{2\pi} \int_{-\pi}^{\pi} \cos(x\sin\theta - n\theta)\, d\theta = J_n(x). \quad \square$$

EXAMPLE 4.4.3. *Prove that*

$$e^{uz+(v/z)} = \sum_{n=-\infty}^{\infty} c_n z^n, \tag{4.4.22}$$

where

$$c_n = \frac{1}{2\pi} \int_{-\pi}^{\pi} e^{(u+v)\cos\theta} \cos[(u-v)\sin\theta - n\theta]\, d\theta. \tag{4.4.23}$$

SOLUTION. As in the previous example, we use (4.4.3) with the circle $z = e^{i\theta}$ as the path of integration:

$$\begin{aligned} c_n &= \frac{1}{2\pi i} \oint_{|z|=1} \frac{e^{uz+(v/z)}}{z^{n+1}}\, dz \\ &= \frac{1}{2\pi} \int_{-\pi}^{\pi} e^{-i(n+1)\theta} e^{ue^{i\theta}+ve^{-i\theta}} e^{i\theta}\, d\theta \\ &= \frac{1}{2\pi} \int_{-\pi}^{\pi} e^{-in\theta} e^{u(\cos\theta + i\sin\theta)+v(\cos\theta - i\sin\theta)}\, d\theta \\ &= \frac{1}{2\pi} \int_{-\pi}^{\pi} e^{(u+v)\cos\theta} e^{i[(u-v)\sin\theta - n\theta]}\, d\theta \\ &= \frac{1}{2\pi} \int_{-\pi}^{\pi} e^{(u+v)\cos\theta} \cos\left[(u-v)\sin\theta - n\theta\right] d\theta. \quad \square \end{aligned}$$

4.4.3. Cauchy's estimate for the coefficients of a Laurent series. We close this section with the following important theorem.

THEOREM 4.4.2. *Suppose that the function $f(z)$ is analytic in the annulus: $R_1 < |z - z_0| < R_2$. Then the coefficients of the Laurent series*

$$f(z) = \sum_{n=-\infty}^{\infty} c_n (z - z_0)^n$$

of $f(z)$ can be estimated, in absolute value, by the following inequalities:

$$|c_n| < \frac{M}{R^n}, \qquad n = 0, \pm 1, \pm 2, \dots, \tag{4.4.24}$$

where

$$M = \max_{z \in C_R} |f(z)|, \qquad C_R : |z - z_0| = R, \quad R_1 < R < R_2. \tag{4.4.25}$$

Inequality (4.4.24) is called *Cauchy's estimate* for the coefficients of a Laurent series.

PROOF. Using (4.4.3) we obtain

$$\begin{aligned} |c_n| &= \left| \frac{1}{2\pi i} \oint_{C_R} \frac{f(\zeta)}{(\zeta - z_0)^{n+1}}\, d\zeta \right| \\ &\le \frac{1}{2\pi} \int_{C_R} \frac{|f(\zeta)|}{|\zeta - z_0|^{n+1}}\, |d\zeta| \\ &\le \frac{M}{2\pi R^{n+1}} \int_0^{2\pi} R\, d\theta = \frac{M}{R^n}, \end{aligned}$$

where the last inequality follows from (4.4.25) and the following facts:

$$|\zeta - z_0| = R, \qquad \zeta - z_0 = R\,e^{i\theta}, \qquad |d\zeta| = |Ri\,e^{i\theta}\,d\theta| = R\,d\theta. \quad \square$$

Exercises for Section 4.4

Expand each of the following functions in a Laurent series about $z_0 = 0$. (Hints for Exercises 7 and 8: The functions

$$J_n(x) = \sum_{k=0}^{\infty} \frac{(-1)^k (x/2)^{2k+n}}{k!(k+n)!}, \qquad I_n(x) = \sum_{k=0}^{\infty} \frac{(x/2)^{2k+n}}{k!(k+n)!}$$

are the Bessel and modified Bessel functions, respectively, of the first kind of order n for $n \in \mathbb{N}$.)

1. $\dfrac{\sin z}{z^3}$.
2. $\dfrac{\cos^2 z}{z}$.
3. $\dfrac{e^z - 1 - z}{z^2}$.
4. $\dfrac{1 + z^2/2 - \cos z}{z^4}$.
5. $z^4 \sin(1/z)$.
6. $z\,e^{1/z}$.
7. $e^{z+1/z}$.
8. $\cos(1/z)\cos z$.

Expand each of the following functions in a Laurent series about z_0.

9. $\dfrac{z+2}{(z-3)^3}, \qquad z_0 = 3.$
10. $\dfrac{z-1}{(z+i)^2}, \qquad z_0 = -i.$
11. $(z+2)\sin[1/(z-i)], \qquad z_0 = i.$
12. $(z-1)e^{1/(z-2)}, \qquad z_0 = 2.$
13. $\dfrac{\cos z}{z+4}, \qquad z_0 = -4.$
14. $\dfrac{e^z}{(z+1)^3}, \qquad z_0 = -1.$

Expand each function in convergent Laurent series in the given domains.

15. $\dfrac{1}{(z+1)(z-2)}$, (a) $|z| < 1$, (b) $1 < |z| < 2$, (c) $2 < |z| < \infty$.

16. $\dfrac{1}{(z-3)(z+2)}$, (a) $|z| < 2$, (b) $2 < |z| < 3$, (c) $3 < |z| < \infty$.

17. $\dfrac{2z+1}{z^2+4z+3}$, (a) $|z| < 1$, (b) $1 < |z| < 3$, (c) $3 < |z| < \infty$.

18. $\dfrac{3z-5}{z^2+5z-6}$, (a) $|z| < 1$, (b) $1 < |z| < 6$, (c) $6 < |z| < \infty$.

19. $\dfrac{1}{(z+2)(z+3)}$, $3 < |z-1| < 4$.

20. $\dfrac{z}{z^2+7z-8}$, $3 < |z+2| < 6$.

21. $\dfrac{1}{(z^2+1)^2}$, $0 < |z+i| < 2$.

22. $\dfrac{1}{(z^2-9)^2}$, $0 < |z-3| < 3$.

CHAPTER 5

Singular Points and the Residue Theorem

5.1. Singular points of analytic functions

5.1.1. Zeros of analytic functions. In this subsection we define zeros of order m of an analytic function $f(z)$ and give a convenient representation of $f(z)$ in a neighborhood of a zero of order m.

DEFINITION 5.1.1. Let $f(z)$ be analytic in a neighborhood of $z = z_0$. The point z_0 is called a *zero of order* m of $f(z)$ if

$$f(z_0) = 0, \quad f'(z_0) = 0, \quad \ldots, \quad f^{(m-1)}(z_0) = 0,$$

but $f^{(m)}(z_0) \neq 0$.

If $z = z_0$ is a zero of order m of $f(z)$, the Taylor series of $f(z)$, centered at z_0, has the form

$$\begin{aligned} f(z) &= \sum_{n=m}^{\infty} \frac{f^{(n)}(z_0)}{n!}(z - z_0)^n \qquad \text{(and putting } n = m + k\text{)} \\ &= \sum_{k=0}^{\infty} \frac{f^{(m+k)}(z_0)}{(m+k)!}(z - z_0)^{m+k} \\ &= (z - z_0)^m \sum_{k=0}^{\infty} \frac{f^{(m+k)}(z_0)}{(m+k)!}(z - z_0)^k, \end{aligned} \tag{5.1.1}$$

that is,

$$f(z) = (z - z_0)^m \varphi(z), \tag{5.1.2}$$

where the function

$$\begin{aligned} \varphi(z) = \Bigg[\frac{f^{(m)}(z_0)}{m!} &+ \frac{f^{(m+1)}(z_0)}{(m+1)!}(z - z_0) + \ldots \\ &+ \frac{f^{(m+k)}(z_0)}{(m+k)!}(z - z_0)^k + \ldots \Bigg] \end{aligned} \tag{5.1.3}$$

is analytic at $z = z_0$ and, by definition,

$$\varphi(z_0) = \frac{f^{(m)}(z_0)}{m!} \neq 0.$$

Note that the series (5.1.1) and (5.1.3) have the same disk of convergence.

It follows that if z_0 is a zero of order m of $f(z)$, then $f(z)$ can be represented in the form (5.1.2) where $\varphi(z)$ is analytic at $z = z_0$, and $\varphi(z_0) \neq 0$ and $\varphi(z_0) \neq \infty$.

The converse statement is also true. If the function $f(z)$ is analytic at $z = z_0$ and is represented in the form (5.1.2), where $\varphi(z_0) \neq 0$, $\varphi(z_0) \neq \infty$ and $\varphi(z)$ is analytic, then z_0 is a zero of order m of $f(z)$.

EXAMPLE 5.1.1. *Determine the order of the zero of*

$$f(z) = (z-5)^{100} e^z. \tag{5.1.4}$$

SOLUTION. Comparing (5.1.2) and (5.1.4) and noting that $e^5 \neq 0$ and $e^5 \neq \infty$, we immediately see that $z = 5$ is a zero of order 100 of $f(z)$ since $f^{(100)}(5) \neq 0$. Moreover, it is the only zero of $f(z)$ since $e^z \neq 0$ for all $z \in \mathbb{C}$. □

EXAMPLE 5.1.2. *Determine the order of the zero $z = 0$ of*

$$f(z) = \frac{\sin^{10} z}{z^5}. \tag{5.1.5}$$

SOLUTION. To define the function $\varphi(z)$ we write

$$f(z) = z^5 \frac{\sin^{10} z}{z^{10}} =: z^5 \varphi(z),$$

where $\lim_{z\to 0} \varphi(z) = 1 \neq 0$. Therefore $z = 0$ is zero of order 5 of $f(z)$. □

5.1.2. Isolated singularities. A point $a \in \mathbb{C}$ is called a *singular point* of $f(z)$ if f is not defined at a.

For examples, the points $z = 1$, $z = 0$ and $z = \infty$ are singular points of $z/(z-1)^2$, $(\sin z)/z$ and $z + 1$, respectively.

By Liouville's Theorem 3.4.5, the only analytic functions which do not have any singular point in the extended complex plane are the constant functions, $f(z) = \text{constant}$.

DEFINITION 5.1.2. A singular point z_0 of $f(z)$ is called an *isolated singular point* if there exists $\delta > 0$ such that $f(z)$ is analytic in the punctured disk $0 < |z - z_0| < \delta$.

Not every singular point, z_0, is an isolated singular point, as can be seen from the following example taken from [**50**], p. 98, par. 5.501.

EXAMPLE 5.1.3. *Show that the function*

$$f(z) = z^2 + z^4 + z^8 + \cdots + z^{2^n} + \cdots = \sum_{n=1}^{\infty} z^{2^n} \tag{5.1.6}$$

has infinitely many singular points on the unit circle $|z| = 1$.

SOLUTION. It is seen by the ratio test that the radius of convergence of the series (5.1.6) is equal to 1, since

$$\left| \frac{z^{2^{n+1}}}{z^{2^n}} \right| = |z^{2^{n+1}-2^n}| = |z^{2^n}| < 1,$$

if $|z| < 1$. It is clear that $\lim_{z\to 1-0} f(z) = \infty$; hence, $z = 1$ is a singular point of $f(z)$. It follows from (5.1.6) that

$$\begin{aligned} f(z) &= z^2 + \sum_{n=2}^{\infty} z^{2^n} \qquad \text{(and putting } n = k+1\text{)} \\ &= z^2 + \sum_{k=1}^{\infty} z^{2^{k+1}} = z^2 + \sum_{k=1}^{\infty} \left(z^2\right)^{2^k} = z^2 + f(z^2). \end{aligned}$$

Furthermore, if $z^2 \to 1 - 0$ then $f(z^2) \to \infty$ and, therefore, $f(z) \to \infty$, since $f(z) = z^2 + f(z^2)$.

Hence, the points satisfying the equality $z^2 = 1$, that is, the points $z = \pm 1$, are singular points of $f(z)$. Similarly, it follows from (5.1.6) that

$$\begin{aligned} f(z) &= z^2 + z^4 + \sum_{n=3}^{\infty} z^{2^n} \qquad \text{(and putting } n = k+2\text{)} \\ &= z^2 + z^4 + \sum_{k=1}^{\infty} \left(z^2\right)^{k+2} = z^2 + z^4 + \sum_{k=1}^{\infty} \left(z^4\right)^{2^k} = z^2 + z^4 + f(z^4). \end{aligned}$$

It follows from the above formula that, if $z^4 \to 1 - 0$, then $f(z^4) \to \infty$ and, therefore, $f(z) \to \infty$. Similarly, one can show that

$$f(z) = z^2 + z^4 + \cdots + z^{2^n} + f(z^n).$$

Hence, for any positive integer n, the points satisfying the relation $z^n = 1$ are singular points of $f(z)$. Solving the equation $z^n = 1$ we obtain the n nth roots of unity,

$$z_k = e^{2\pi i k/n}, \qquad k = 0, 1, \ldots, n-1, \tag{5.1.7}$$

which are singular points of $f(z)$. These singular points are not isolated: any arc of the circle $|z| = 1$, no matter how small, contains infinitely many singular points since n in (5.1.7) can be taken as large as we please. □

In the sequel, we shall consider only isolated singular points unless stated otherwise.

If z_0 is an isolated singular point of $f(z)$, then $f(z)$ can be represented by the Laurent series

$$f(z) = \sum_{n=-\infty}^{-1} c_n(z - z_0)^n + \sum_{n=0}^{\infty} c_n(z - z_0)^n, \tag{5.1.8}$$

in the annulus $0 < \delta_1 < |z - z_0| < \delta$.

Isolated singularities are classified as follows.

An isolated singularity, z_0, is a *removable singularity*, a *pole* or an *essential singularity* of $f(z)$, if the principal part of the Laurent series (5.1.8) of $f(z)$ contains, respectively,

(a) no negative powers of $z - z_0$,
(b) a finite number of negative powers of $z - z_0$, or
(c) infinitely many negative powers of $z - z_0$.

5.1.3. Removable singularities.

DEFINITION 5.1.3. Let the function $f(z)$ be analytic in a punctured disk $0 < |z - z_0| < \delta$. If its Laurent series around $z = z_0$ has no principal part, that is,

$$\begin{aligned} f(z) &= \sum_{n=0}^{\infty} c_n(z - z_0)^n \\ &= c_0 + c_1(z - z_0) + c_2(z - z_0)^2 + \cdots \\ &\quad + c_n(z - z_0)^n + \ldots, \end{aligned} \tag{5.1.9}$$

for $0 < \delta_1 < |z - z_0| < \delta$, then z_0 is said to be a *removable singularity* of $f(z)$.

As can be seen from (5.1.9), if z_0 is a removable singularity then the limit

$$\lim_{z \to z_0} f(z) = c_0$$

exists and is finite. Therefore letting $f(z_0) = c_0$, we obtain that z_0 is a point of analyticity of $f(z)$, that is, the discontinuity is removed:

$$\lim_{z \to z_0} \frac{f(z) - f(z_0)}{z - z_0} = c_1 = f'(z_0).$$

It can easily be shown that the converse statement is correct, namely, if z_0 is a singular point of an analytic function $f(z)$ and the limit $c_0 = \lim_{z \to z_0} f(z)$ exists and is finite, then z_0 is a removable singularity, that is, the Laurent series of $f(z)$ has the form (5.1.9). Indeed, if, by contradiction, at least one of the coefficients c_n ($n = -1$ or $n = -2$, etc.) cannot be equal to zero in (5.1.8) then the limit $A = \lim_{z \to z_0} f(z)$ is not finite. Therefore we have proved the following theorem.

THEOREM 5.1.1. *A necessary and sufficient condition for a singular point z_0 of an analytic function in a punctured disk $0 < |z - z_0| < \delta$ to be a removable singularity is the existence of the finite limit*

$$\lim_{z \to z_0} f(z) = c_0.$$

Note 5.1.1. The above proof by contradiction is based on the following fact from mathematical logic: proposition "$A \to B$" is equivalent to proposition "not $B \to$ not A."

Similarly, one can prove the following theorem.

THEOREM 5.1.2. *If an analytic function $f(z)$ is bounded in the punctured disk $0 < |z - z_0| < \delta_2$, then either f is analytic at z_0 or z_0 is a removable singularity.*

5.1.4. Poles.

DEFINITION 5.1.4. Let the function $f(z)$ be analytic in the punctured disk $0 < |z - z_0| < \delta$. If the principal part of the Laurent series of $f(z)$ around $z = z_0$ contains a finite number of terms,

$$f(z) = \frac{c_{-m}}{(z - z_0)^m} + \frac{c_{-m+1}}{(z - z_0)^{m-1}} + \cdots + \frac{c_{-1}}{z - z_0} + \sum_{n=0}^{\infty} c_n (z - z_0)^n, \quad (5.1.10)$$

where $c_{-m} \neq 0$ and $m < \infty$, then z_0 is called a *pole of order* m of $f(z)$.

We can rewrite (5.1.10) in the form

$$f(z) = \frac{1}{(z - z_0)^m} \left[c_{-m} + c_{-m+1}(z - z_0) + \cdots + c_{-1}(z - z_0)^{m-1} + \sum_{n=0}^{\infty} c_n (z - z_0)^{n+m} \right], \quad (5.1.11)$$

or

$$f(z) = \frac{\varphi(z)}{(z - z_0)^m}, \quad (5.1.12)$$

where the function

$$\varphi(z) = c_{-m} + c_{-m+1}(z - z_0) + \cdots$$

$$+ c_{-1}(z - z_0)^{m-1} + \sum_{n=0}^{\infty} c_n (z - z_0)^{n+m} \quad (5.1.13)$$

is analytic in the disk $|z - z_0| < \delta$ and $\varphi(z_0) = c_{-m} \neq 0$.

Hence, if z_0 is a pole of order m of $f(z)$, then $f(z)$ can be represented in the form (5.1.12), where $\varphi(z)$ is analytic in some δ-neighborhood of z_0; moreover, $\varphi(z_0) \neq 0$ and $\varphi(z_0) \neq \infty$.

The converse statement is also true: if $f(z)$ is representable in the form (5.1.12) where $\varphi(z)$ is analytic in some δ-neighborhood of z_0 and $\varphi(z_0) \neq 0$, then z_0 is a pole of order m of $f(z)$. To prove this, it suffices to expand $\varphi(z)$ in a Taylor series in $z - z_0$ and divide every term of this series by $(z - z_0)^m$.

Therefore it is not necessary to find a Laurent series expansion (5.1.10) in order to determine that z_0 is is a pole of order m of $f(z)$; it suffices to transform $f(z)$ to the form (5.1.12).

In practice, $f(z)$ is often represented by a ratio of two analytic functions, $\varphi(z)$ and $\psi(z)$,

$$f(z) = \frac{\varphi(z)}{\psi(z)}. \quad (5.1.14)$$

Then z_0 is a pole of $f(z)$ if $\psi(z_0) = 0$ and $\varphi(z_0) \neq 0$. Let us assume, for example, that $\psi(z)$ has a zero of order m at z_0 and $\varphi(z_0) \neq 0$. Then $\psi(z)$ can be represented in the form (5.1.2):

$$\psi(z) = (z - z_0)^m \kappa(z),$$

where $\kappa(z)$ is analytic in some δ-neighborhood of z_0 and $\kappa(z_0) \neq 0$. Then $f(z)$ in (5.1.14) can be written in the form

$$f(z) = \frac{\frac{1}{\kappa(z)}\varphi(z)}{(z - z_0)^m}, \quad (5.1.15)$$

where $[\kappa(z)]^{-1}\varphi(z)$ is analytic at z_0 and $[\kappa(z_0)]^{-1}\varphi(z_0) \neq 0$. This means that $f(z)$ is of the form (5.1.12), that is, z_0 is a pole of order m of $f(z)$.

Hence if $f(z)$ is of the form (5.1.14), where $\varphi(z)$ and $\psi(z)$ are analytic, then each zero of order m of $\psi(z)$, which is not a zero of $\varphi(z)$, is a pole of order m of $f(z)$. Thus, we have the following theorem.

THEOREM 5.1.3. *A necessary and sufficient condition for a point z_0 to be a pole of a function $f(z)$ analytic in a puncture disk $0 < |z - z_0| < \delta$ is that*

$$\lim_{z \to z_0} |f(z)| = \infty \quad (5.1.16)$$

(independently of the direction of approach of z to z_0).

PROOF. The necessity follows from condition (5.1.12): if z_0 is a pole of order m, then $f(z)$ is of the form (5.1.12) and

$$\lim_{z\to z_0} |f(z)| = \lim_{z\to z_0} \frac{|\varphi(z)|}{|z-z_0|^m} = \infty,$$

since $\varphi(z_0) \neq 0$.

For the sufficiency, assume that (5.1.16) holds. Thus, if $g(z) = 1/f(z)$ is analytic in a punctured δ-neighborhood of z_0 and $\lim_{z\to z_0} g(z) = 0$, that is, the point z_0 is a zero of some integer order, m, of $g(z)$, then by (5.1.2) we have

$$g(z) = (z-z_0)^m \psi(z), \qquad \psi(z_0) \neq 0, \qquad \psi(z_0) \neq \infty, \tag{5.1.17}$$

or, equivalently,

$$f(z) = \frac{1}{\psi(z)(z-z_0)^m}, \qquad \psi(z_0) \neq 0, \qquad \psi(z_0) \neq \infty.$$

Thus, by (5.1.12), z_0 is a pole of order m of $f(z)$. □

We now consider a few examples.

EXAMPLE 5.1.4. *Find the singular points of*

$$f(z) = \frac{e^z}{(z-1)^{100}}$$

in the finite complex plane and determine their character.

SOLUTION. The only singular point is $z = 1$ and it is a zero of order 100 of the denominator. Moreover, $e^z|_{z=1} = e \neq 0$. Hence $z = 1$ is a pole of order 100. □

EXAMPLE 5.1.5. *Find the singular points of*

$$f(z) = \frac{z+1}{z(z+3)^5}$$

and determine their character.

SOLUTION. The zeros of the denominator are $z = 0$ (a simple zero) and $z = -3$ (a zero of order 5) and the numerator does not vanish at $z = 0$ or at $z = -3$. Hence $z = 0$ and $z = -3$ are poles of order 1 and 5, respectively. □

EXAMPLE 5.1.6. *Find the singular points of*

$$f(z) = \frac{\sin^2 z}{(z+1)^4 z^6}.$$

SOLUTION. We rewrite $f(z)$ in the form

$$f(z) = \frac{\frac{\sin^2 z}{z^2}}{(z+1)^4 z^4}.$$

The zeros of the denominator are $z = -1$ and $z = 0$, both of order 4. Moreover, the numerator is not equal to zero at $z = -1$ or at $z = 0$ since

$$\lim_{z\to 0} \frac{\sin^2 z}{z^2} = 1.$$

Hence $z = 0$ and $z = -1$ are poles of order 4. □

5.1.5. Essential singularities.

DEFINITION 5.1.5. Let the function $f(z)$ be analytic in a punctured disk $0 < |z - z_0| < \delta$. If the principal part of the Laurent series of $f(z)$ around $z = z_0$ contains an infinite number of terms:

$$f(z) = \sum_{n=-\infty}^{-1} c_n(z-z_0)^n + \sum_{n=0}^{\infty} c_n(z-z_0)^n, \tag{5.1.18}$$

that is, given any positive integer N there exists $c_{-n} \neq 0$ for infinitely many $n \geq N$, then the point z_0 is called an *essential singularity* of $f(z)$.

The behavior of an analytic function in a neighborhood of an essential singularity is described by the following theorem, which goes by the name Casorati–Weierstrass, Sokhotski, or simply Weierstrass' Theorem.

THEOREM 5.1.4 (WEIERSTRASS' THEOREM). *Let z_0 be an isolated essential singularity of a function $f(z)$ which is analytic in a punctured disk $0 < |z-z_0| < \delta$. Given any $\varepsilon > 0$ and $w \in \mathbb{C}$, then, in any punctured neighborhood of z_0, there exists at least one point z such that $|f(z) - w| < \varepsilon$.*

PROOF. Assume to the contrary, that is, given a complex number w and $\varepsilon > 0$, there exists $\delta > 0$ such that for all z such that $0 < |z - z_0| < \delta$ the inequality

$$|f(z) - w| > \varepsilon \tag{5.1.19}$$

is satisfied. Then, consider the auxiliary function

$$\psi(z) = \frac{1}{f(z) - w}. \tag{5.1.20}$$

By (5.1.19), $\psi(z)$ is analytic and bounded in a punctured η-neighborhood of z_0, that is (see Theorem 5.1.2), z_0 is a removable singularity of $\psi(z)$. This means that, in an η-neighborhood of z_0, $\psi(z)$ can be written in the form

$$\psi(z) = (z - z_0)^m \kappa(z), \qquad \kappa(z_0) \neq 0,$$

where $\kappa(z)$ is analytic in this neighborhood. Then it follows from (5.1.20) that, in a punctured η-neighborhood of z_0, $f(z)$ has the form

$$f(z) = \frac{1}{\kappa(z)(z - z_0)^m} + w, \tag{5.1.21}$$

where $1/\kappa(z)$ is analytic in $0 < |z - z_0| < \eta$ and $1/\kappa(z_0) \neq 0$. But this means that z_0 is either a pole of order m of $f(z)$ (if $m > 0$) or a point of analyticity of $f(z)$ (if $m = 0$). In both cases we have a contradiction with the assumption of the theorem. This contradiction proves the theorem. □

COROLLARY 5.1.1. *Suppose that a function $f(z)$ is analytic in the punctured disk $0 < |z - z_0| < \delta$ and has an isolated essential singularity at $z = z_0$. Then $f(z)$ approaches any value $w \in \mathbb{C}$ infinitely closely and infinitely often in any punctured $0 < |z - z_0| < \delta_1$, where $\delta_1 \leq \delta$.*

PROOF. By Theorem 5.1.4, for every $\varepsilon > 0$, in any sufficiently small punctured δ-neighborhood of z_0 there exists at least one point z_1 such that

$$|f(z_1) - w| < \varepsilon. \tag{5.1.22}$$

Then, by taking a nested sequence of shrinking punctured δ-neighborhood of z_0, we see that there exist infinitely many points z_1 for which (5.1.22) is satisfied. □

The point $z = 0$ is an essential singular point of the function $f(z) = e^{1/z}$ since its Laurent series around $z = 0$,

$$e^{1/z} = \sum_{n=0}^{\infty} \frac{1}{n! z^n},$$

contains infinitely many negative powers of z. For this function, we have the following example.

EXAMPLE 5.1.7. *Given an arbitrary complex number w, $w \neq 0$ and $w \neq \infty$, show that there exist infinitely many complex numbers z such that $e^{1/z} = w$ in any punctured neighborhood, $0 < |z| < \delta$, of 0, where δ can be taken as small as we please.*

SOLUTION. Taking the logarithm of $e^{1/z} = w$, we have

$$\frac{1}{z} = \log w = \operatorname{Log} w + 2k\pi i, \qquad k = 0, \pm 1, \pm 2, \ldots.$$

Thus

$$z_k = \frac{1}{\operatorname{Log} w + 2k\pi i}, \qquad k = 0, \pm 1, \pm 2, \ldots,$$

that is, for all $\delta > 0$ there exist infinitely many points z_k in the punctured neighborhood $0 < |z_k| < \delta$ for which the condition $e^{1/z_k} = w$ is satisfied for whatever number w, except $w = 0$ and $w = \infty$. □

Let us consider the behavior of $e^{1/z}$ in a neighborhood of the point $z = 0$.

If $z = x$ is real and approaches 0 from above, then

$$\lim_{x \to 0+} e^{1/x} = +\infty.$$

If $z = x$ is real and approaches 0 from below, then

$$\lim_{x \to 0-} e^{1/x} = 0.$$

If $z = iy$ is pure imaginary and y approaches 0 from above or below, then the limit

$$\lim_{y \to 0} e^{1/(iy)} = \lim_{y \to 0} \left(\cos\frac{1}{y} - i\sin\frac{1}{y} \right)$$

does not exist.

This strange behavior is typical of any analytic function in a neighborhood of an essential singularity.

COROLLARY 5.1.2. *If z_0 is an essential singularity of a function $f(z)$ analytic in the punctured disk $0 < |z - z_0| < \delta$, then $\lim_{z \to z_0} f(z)$ does not exist.*

PROOF. The statement follows from Theorem 5.1.4 since, depending on the choice of the sequence of points z_n approaching z_0 as $n \to \infty$, $f(z_n)$ can take any preassigned value $w \in \mathbb{C}$, except possibly one value. □

There is no reason to prove the following converse of Theorem 5.1.4: if no finite (or infinite) limit of $f(z)$ exists as $z \to z_0$ then, by Theorems 5.1.1 and 5.1.3, z_0 cannot be a removable singularity or a pole.

The following theorems are deeper than Theorem 5.1.4 and are stated without proofs. They are formulated in terms of entire functions which are analytic in the finite complex plane and meromorphic functions whose only singularities in the finite complex plane are poles. By definition, an entire function omits the value $z = \infty$ in the whole complex plane $\mathbb{C}$.

THEOREM 5.1.5 (LITTLE PICARD THEOREM). *If $f(z)$ is an entire function that omits two values in the finite plane, then it is a constant.*

PROOF. See [**2**], p. 307. □

It can be shown by an extension of Theorem 5.1.5, given in [**34**], Vol. 2, p. 268, that the entire functions $\cos z$ and $\sin z$ take every finite complex values in the complex plane.

THEOREM 5.1.6 (GREAT PICARD THEOREM). *Suppose an analytic (meromorphic) function, $f(z)$, has an essential singularity at $z = z_0$. Then in each neighborhood of z_0, $f(z)$ assumes each complex value, with one (two) possible exception(s), an infinite number of times.*

PROOF. See [**34**], Vol. 3, pp. 344–345. □

It can be shown that the meromorphic function $\tan z$ omits the values $\pm i$ in the complex plane.

We rephrase Theorem (5.1.6) in the following corollary.

COROLLARY 5.1.3. *If $f(z)$ has an isolated singularity at $z = z_0$ and if there are two complex numbers that are not assumed infinitely often by $f(z)$, then $z = z_0$ is either a pole or a removable singularity.*

In the above Example 5.1.7 the function $e^{1/z}$ takes any complex value w except $w = 0$ in any punctured δ-neighborhood of $z = 0$. The number $w = 0$ is called an *exceptional value* for the function $e^{1/z}$.

EXAMPLE 5.1.8. *The point $z = \infty$ is an essential singularity of the function $f(z) = \cos z$ and for any complex w the equation $\cos z = w$ has infinitely many solutions,*

$$z_k = \frac{1}{i} \operatorname{Log}\left(w + \sqrt{w^2 - 1}\right) + 2k\pi, \qquad k = 0, \pm 1, \pm 2, \ldots,$$

in any δ-neighborhood of the point $z = \infty$, that is, in the region $\delta < |z| < \infty$. Hence the function $\cos z$ does not have any exceptional values.

Note 5.1.2. It follows from Theorems 5.1.1–5.1.4 that, besides the characterization of the isolated singularities of an analytic function presented above, there exists another equivalent characterization (see [**42**]), namely, the point z_0 is said to be

(a) a *removable singularity* if $f(z)$ has a finite limit as $z \to z_0$,

(b) a *pole of order m* if $f(z) \to \infty$ as $z \to z_0$, and

(c) an *essential singularity* if $f(z)$ has no finite or infinite limit as $z \to z_0$.

5.1.6. Behavior of an analytic function near $z = \infty$. We now consider the behavior of an analytic function in a neighborhood of the point $z = \infty$.

DEFINITION 5.1.6. The point $z = \infty$ is said to be an *isolated singular point* of an analytic function $f(z)$ if there exists $R > 0$ such that there are no singular points in the region $R < |z| < \infty$.

For example, $z = \infty$ is not an isolated singular point for the function $f(z) = 1/\sin z$ since the singular points, $z_k = k\pi$, $k = 0, \pm 1, \ldots$, of this function tend to ∞ as $k \to \pm\infty$.

If $z = \infty$ is an isolated singular point of $f(z)$ in the region $R < |z| < \infty$ then it can be expanded in a Laurent series

$$f(z) = \sum_{n=-\infty}^{\infty} c_n z^n, \tag{5.1.23}$$

which is convergent in the region $R < |z| < \infty$.

Following Definition 4.4.1, the series

$$\sum_{n=-\infty}^{-1} c_n z^n \quad \text{and} \quad \sum_{n=0}^{\infty} c_n z^n,$$

valid in a punctured neighborhood of infinity, are called the *regular* and *principal* parts of the series (5.1.23), respectively.

As in the case of a finite isolated singular point z_0, there are three possible cases.

(1) The point $z = \infty$ is called a *removable singularity* of $f(z)$ if the limit

$$\lim_{z \to \infty} f(z) = c_0$$

exists, is finite and does not depend on the way z approaches infinity. The series (5.1.23) in this case does not contain positive powers of z. If, moreover, the coefficients $c_0, c_{-1}, \dots, c_{-m+1}$ are equal to zero in (5.1.23) but $c_{-m} \neq 0$, then the point $z = \infty$ is called a *zero of order* m of $f(z)$.

(2) The point $z = \infty$ is called a *pole of order* m of $f(z)$ if the series (5.1.23) contains a finite number of positive powers of z, that is,

$$f(z) = \sum_{n=-\infty}^{m} c_n z^n, \qquad c_m \neq 0, \quad m < \infty.$$

In this case, by (5.1.12), $f(z)$ can be represented in the form

$$f(z) = \varphi(z) z^m, \tag{5.1.24}$$

where $\varphi(z)$ is analytic in a neighborhood of $z = \infty$ and $\varphi(\infty) \neq 0$. In this case we see that $\lim_{z_k \to \infty} |f(z_k)| = \infty$, no matter how z_k approaches ∞.

(3) The point $z = \infty$ is called an *essential singular point* of $f(z)$ if the series (5.1.23) contains infinitely many positive powers of z, that is,

$$f(z) = \sum_{n=\infty}^{\infty} c_n z^n.$$

In this case, $f(z)$ has no finite nor infinite limit as $z \to \infty$.

In practice, to expand $f(z)$ in a Laurent series in a neighborhood of the isolated singular point $z = \infty$, one can use the inversion $z = 1/\zeta$ and expand the function $f(1/\zeta)$ in a Laurent series in a neighborhood of $\zeta = 0$.

EXAMPLE 5.1.9. *Represent the function*

$$f(z) = \frac{z^2}{\sqrt{1+z^2}} \tag{5.1.25}$$

in a Laurent series in a neighborhood of $z = \infty$.

SOLUTION. The points $z = \pm i$ are the branch points of $f(z)$. Joining these points by a cut we obtain two single-valued branches of $f(z)$. We select the branch of $f(z)$ for which $f(1) = 1/\sqrt{2}$. Thus, representing $f(z)$ in the form (5.1.24) we obtain

$$f(z) = \varphi(z)z, \qquad \varphi(z) = \sqrt{\frac{z^2}{1+z^2}}, \qquad \varphi(\infty) = 1.$$

Hence $z = \infty$ is a simple pole of $f(z)$. Letting $z = 1/\zeta$, we get

$$f\left(\frac{1}{\zeta}\right) = \frac{1}{\zeta\sqrt{1+\zeta^2}}, \tag{5.1.26}$$

and using the binomial series

$$(1+\xi)^\alpha = 1 + \frac{\alpha\xi}{1!} + \frac{\alpha(\alpha-1)\xi^2}{2!} + \dots + \frac{\alpha(\alpha-1)\cdots(\alpha-n+1)}{n!}\xi^n + \dots, \qquad |\xi| < 1, \tag{5.1.27}$$

we have

$$\begin{aligned}(1+\zeta^2)^{-1/2} &= 1 - \frac{1}{2}\zeta^2 + \frac{\left(-\frac{1}{2}\right)\left(-\frac{3}{2}\right)}{2!}(\zeta^2)^2 \\ &\quad + \dots + \frac{\left(-\frac{1}{2}\right)\left(-\frac{3}{2}\right)\cdots\left(-\frac{1}{2}-n+1\right)}{n!}(\zeta^2)^n + \dots \\ &= 1 - \frac{1}{2}\zeta^2 + \frac{3}{2^2 2!}\zeta^4 + \cdots \\ &\quad + (-1)^n \frac{1}{n! 2^n} 1\cdot 3\cdot 5\cdots(2n-1)\zeta^{2n} + \dots \\ &= \sum_{n=0}^{\infty}(-1)^n \frac{(2n-1)!!}{2^n n!}\zeta^{2n}, \qquad |\zeta| < 1, \end{aligned} \tag{5.1.28}$$

where $(2n-1)!! = 1\cdot 3\cdot 5\cdots(2n-1)$ and $(-1)!! = 1$.

Substituting (5.1.28) into (5.1.26), we obtain

$$f\left(\frac{1}{\zeta}\right) = \sum_{n=0}^{\infty}(-1)^n \frac{(2n-1)!!}{2^n n!}\zeta^{2n-1}, \qquad |\zeta| < 1. \tag{5.1.29}$$

Finally, letting $\zeta = 1/z$ in (5.1.29) we obtain the Laurent series of (5.1.25) in a neighborhood of $z = \infty$ in the form

$$\frac{z^2}{\sqrt{1+z^2}} = \sum_{n=0}^{\infty}(-1)^n \frac{(2n-1)!!}{2^n n!}\frac{1}{z^{2n-1}}, \qquad 1 < |z| < \infty. \quad \square$$

5.1.7. Generalized Liouville's Theorem. We recall that a function $f(z)$ which is analytic in the whole complex plane is called entire. This function can be represented by a Taylor series,

$$f(z) = \sum_{k=0}^{\infty} c_k z^k, \tag{5.1.30}$$

which has an infinite radius of convergence. By virtue of Liouville's Theorem 3.4.5 an entire function $f(z)$ (if it is not a constant) must have a singular point at $z = \infty$ which is either a pole of order n or an essential singularity.

THEOREM 5.1.7 (GENERALIZED LIOUVILLE'S THEOREM). *If there exist a nonnegative integer n and a positive number R_0 such that the entire function $f(z)$ satisfies the inequality*

$$|f(z)| \leq c|z|^n \tag{5.1.31}$$

in the region $|z| > R_0$, then $f(z)$ is a polynomial of degree not exceeding n.

PROOF. Since $k \geq 0$ in (5.1.30), one can use Cauchy's estimate (4.4.24) for the coefficients of the series (5.1.30) in the form

$$|c_k| \leq \frac{M}{R^k}, \qquad k = 0, 1, \ldots,$$

for all $R > R_0$, where, by (5.1.31),

$$M = \max_{|z|=R} |f(z)| \leq cR^n.$$

Therefore

$$|c_k| \leq \frac{cR^n}{R^k} = \frac{c}{R^{k-n}}. \tag{5.1.32}$$

If $k > n$, it follows from (5.1.32) that $c_k = 0$ since R can be taken as large as we please and the coefficients c_k are independent of R. Hence $c_{n+1} = c_{n+2} = \cdots = 0$ in (5.1.30), that is, $f(z)$ is a polynomial of degree not exceeding n. □

COROLLARY 5.1.4. *If $n = 0$ in (5.1.31) then $c_1 = c_2 = \cdots = 0$, that is, series (5.1.30) has the form*

$$f(z) = c_0 = \textit{constant},$$

so that we obtain Liouville's Theorem 3.4.5.

5.1.8. Expansion in partial fractions. In this subsection, the coefficients of the partial fraction expansion of a rational function are determined efficiently by means of Liouville's Theorem 3.4.5.

DEFINITION 5.1.7. An analytic function $f(z)$ is said to be a *meromorphic function* if its only singular points are poles, including possibly the point $z = \infty$.

The number of poles of a meromorphic function can be either finite (in the case of a rational function) or infinite (in the case of the transcendental functions $\tan z$, $\tanh z$, $\cot z$, $\coth z$, $\sec z$, etc.).

The following theorem holds.

THEOREM 5.1.8. *A meromorphic function $f(z)$ which has a finite number of poles, $z_1, z_2, \dots, z_s$, in the extended complex plane (the point $z = \infty$ can also be a pole) is a rational function.*

PROOF. Since each singular point z_k, $k = 1, 2, \dots, s$ (and perhaps $z = \infty$), is isolated, then $f(z)$ can be represented, in a punctured neighborhood of each z_k, by a convergent Laurent series which has a finite number of negative powers of $z - z_k$. Suppose that the singular point z_k is a pole of order μ_k, where $\mu_k \geq 1$, and the point $z = \infty$ is a pole of order m, where $m \geq 0$ (if $m = 0$, there is no pole at $z = \infty$). The Laurent series of $f(z)$ in a neighborhood of $z = \infty$ has the form

$$f(z) = \sum_{n=0}^{m} c_n^{(\infty)} z^n + \sum_{n=-\infty}^{-1} c_n^{(\infty)} z^n, \qquad R < |z| < \infty. \tag{5.1.33}$$

In a neighborhood of $z_k \neq \infty$, the Laurent series is

$$f(z) = \sum_{n=1}^{\mu_k} \frac{c_{-n}^{(k)}}{(z - z_k)^n} + \sum_{n=0}^{\infty} c_n^{(k)} (z - z_k)^n,$$
$$0 < \delta \leq |z - z_k| < R_k. \tag{5.1.34}$$

Note that, even if the series (5.1.34) has infinitely many positive powers of $z - z_k$, the outer radius R_k of convergence is finite and equal to the distance from z_k to the next closest pole of $f(z)$.

Consider an auxiliary function $\varphi(z)$ which is equal to the difference between $f(z)$ and the principal parts of the Laurent series (5.1.33) and (5.1.34):

$$\varphi(z) = f(z) - \sum_{n=0}^{m} c_n^{(\infty)} z^n - \sum_{k=1}^{s} \sum_{n=1}^{\mu_k} \frac{c_{-n}^{(k)}}{(z - z_k)^n}. \tag{5.1.35}$$

This function is analytic in the whole complex plane; therefore, it is equal to a constant by Liouville's Theorem 3.4.5:

$$\varphi(z) = a = \text{constant}.$$

It then follows from (5.1.35) that

$$f(z) = a + \sum_{n=0}^{m} c_n^{(\infty)} z^n + \sum_{k=1}^{s} \sum_{n=1}^{\mu_k} \frac{c_{-n}^{(k)}}{(z - z_k)^n}. \tag{5.1.36}$$

We infer from (5.1.36) that

$$a = \lim_{z \to \infty} \left[f(z) - \sum_{n=0}^{m} c_n^{(\infty)} z^n \right]. \tag{5.1.37}$$

The right-hand side of (5.1.36) is the *partial fraction expansion* of the given rational function; the term

$$a + \sum_{n=0}^{m} c_n^{(\infty)} z^n$$

is called the *regular part* of this function. □

Remark 5.1.1. The regular part of a proper rational function is equal to zero, so that (5.1.36) reduces to

$$f(z) = \frac{P_q(z)}{Q_r(z)} = \sum_{k=1}^{s} \sum_{n=1}^{\mu_k} \frac{c_{-n}^{(k)}}{(z - z_k)^n}, \tag{5.1.38}$$

where $P_q(z)$ and $Q_r(z)$ are polynomials of degrees q and r, respectively, with $r > q$. Formula (5.1.38) is, in fact, the partial fraction expansion of a proper rational fraction, and Theorem 5.1.8 gives a simple derivation of this formula.

In addition to formulae found in [**42**], we derive simple formulae for determining the coefficients $c_{-n}^{(k)}$ in (5.1.38). For this purpose, we rewrite (5.1.38) in the more explicit form

$$f(z) = \sum_{n=1}^{\mu_1} \frac{c_{-n}^{(1)}}{(z - z_1)^n} + \sum_{n=1}^{\mu_2} \frac{c_{-n}^{(2)}}{(z - z_2)^n} + \cdots + \sum_{n=1}^{\mu_k} \frac{c_{-n}^{(k)}}{(z - z_k)^n} + \cdots + \sum_{n=1}^{\mu_s} \frac{c_{-n}^{(s)}}{(z - z_s)^n}. \tag{5.1.39}$$

Multiplying (5.1.39) by $(z - z_k)^{\mu_k}$, we obtain

$$f(z)(z - z_k)^{\mu_k} = (z - z_k)^{\mu_k} \sum_{\substack{m=1 \\ m \neq k}}^{s} \sum_{n=1}^{\mu_m} \frac{c_{-n}^{(m)}}{(z - z_m)^n}$$

$$+ c_{-\mu_k}^{(k)} + c_{-(\mu_k - 1)}^{(k)}(z - z_k) + c_{-(\mu_k - 2)}^{(k)}(z - z_k)^2 + \dots$$

$$+ c_{-1}^{(k)}(z - z_k)^{\mu_k - 1}. \quad (5.1.40)$$

To determine the coefficients $c_{-(\mu_k - p)}^{(k)}$, $p = 0, 1, \dots, \mu_k - 1$, we differentiate (5.1.40) p times with respect to z and take the limit as $z \to z_k$. Thus,

$$c_{-(\mu_k - p)}^{(k)} = \frac{1}{p!} \lim_{z \to z_k} [f(z)(z - z_k)^{\mu_k}]^{(p)}, \qquad p = 0, 1, \dots, \mu_k - 1. \quad (5.1.41)$$

Setting $\mu_k - p = m$ in (5.1.41), we obtain

$$c_{-m}^{(k)} = \frac{1}{(\mu_k - m)!} \lim_{z \to z_k} [f(z)(z - z_k)^{\mu_k}]^{(\mu_k - m)},$$

$$m = 1, 2, \dots, \mu_k. \quad (5.1.42)$$

Formula (5.1.42) gives the partial fraction coefficient of the term with denominator $(z - z_k)^m$, where μ_k is the multiplicity of the root z_k; in fact, it gives the coefficients of all the terms in the partial fraction expansion (5.1.38).

Note 5.1.3. Using formula (5.2.12) of the next section to compute the residue at the pole of order $\mu_k - m$, we can rewrite (5.1.42) in the form

$$c_{-(m+1)}^{(k)} = \operatorname*{Res}_{z=z_k} \left[\frac{f(z)(z - z_k)^{\mu_k}}{(z - z_k)^{\mu_k - m}} \right] = \operatorname*{Res}_{z=z_k} \left[\frac{f(z)}{(z - z_k)^{-m}} \right]; \quad (5.1.43)$$

however, for practical purposes, it is more convenient to use (5.1.42).

EXAMPLE 5.1.10. *Expand the following proper rational function in partial fractions:*

$$f(x) = \frac{x + 1}{(x - 2)^2(x - 3)}.$$

SOLUTION. It follows from (5.1.38) and (5.1.42) that

$$f(x) = \frac{x + 1}{(x - 2)^2(x - 3)} = \frac{A}{(x - 2)^2} + \frac{B}{x - 2} + \frac{C}{x - 3}.$$

Then

$$A = \lim_{x \to 2} [f(x)(x - 2)^2] = \frac{2 + 1}{2 - 3} = -3,$$

$$B = \lim_{x\to 2}[f(x)(x-2)^2]'$$
$$= \lim_{x\to 2}\left(\frac{x+1}{x-3}\right)' = \lim_{x\to 2}\frac{x-3-(x+1)}{(x-3)^2} = -4,$$
$$C = \lim_{x\to 3}[f(x)(x-3)] = \frac{3+1}{(3-2)^2} = 4.$$

Thus

$$f(x) = \frac{x+1}{(x-2)^2(x-3)} = -\frac{3}{(x-2)^2} - \frac{4}{x-2} + \frac{4}{x-3}. \quad \square$$

We remark that the useful formula (5.1.42) seems to be absent from textbooks. It is especially useful for computing only one of the coefficients $c_{-m}^{(k)}$, without computing the others (see Example 6.1.6 in the next chapter).

Exercises for Section 5.1

Find the order of every zero of the given functions.

1. $z^2 + 16$.
2. $(z^2-1)^2(z^2+4)$.
3. $(1-\cos z)(z^2-9)^3$.
4. $z(e^z-1)^2$.
5. $z^2 \sin^3 z$.
6. $\dfrac{1-\cos z}{z^2(z-1)^2}$.
7. $\dfrac{1}{z}\sin^2 z$.
8. $\dfrac{1}{z}(e^{2z}-1)\sin z$.

Find the order of the zero at $z = 0$ of the following functions.

9. $z - \sin z$.
10. $\dfrac{1}{z}(1-\cos z)^2$.
11. $z^4(e^z-1)^2$.
12. $e^{\cos z} - e^z$.
13. $z\,\mathrm{Log}(1-z)$.

14. $\dfrac{z^4}{1 - z - e^{-z}}$.

Find the singular points of the given functions, including infinity.

15. $\dfrac{z+1}{(z^2+4)(z-1)^2}$.

16. $\dfrac{1}{(z-i)^2(z+2)}$.

17. $\dfrac{1}{e^z+1}$.

18. $\dfrac{\sin(z-1)}{z-1} + \dfrac{1}{z-1}$.

19. $\dfrac{z}{1-\cos z}$.

20. $\dfrac{1}{z^2} + e^{1/z}$.

21. $\cos \dfrac{1}{z-i}$.

22. $\tan z$.

23. $\dfrac{1}{z} + \cot^2 z$.

24. $\dfrac{1}{(z-1)^{10}} + \sin \dfrac{1}{z-1}$.

25. Let the functions $f(z)$ and $g(z)$ be analytic in a domain D except at the point z_0. Suppose that z_0 is a pole of order n and m of $f(z)$ and $g(z)$, respectively. Discuss the possible types of singularity of the function $f(z) + g(z)$ at z_0.

26. Let z_0 be a singular point of $f(z)$ and let $g(z)$ be analytic at z_0. Find the type of singularity of $f(z)g(z)$ if

(a) z_0 is a removable singularity.

(b) z_0 is a pole of order n.

(c) z_0 is an essential singularity.

27. Show that the function

$$\sum_{n=1}^{\infty} \frac{z^{2^n}}{n^2}$$

is continuous in and on the unit circle, but every point of the circle is a singularity.

5.2. The residue theorem

Let z_0 be an isolated singular point of an analytic function $f(z)$. Then $f(z)$ can be represented by a Laurent series in a neighborhood of z_0,

$$f(z) = \sum_{n=-\infty}^{\infty} c_n (z - z_0)^n, \tag{5.2.1}$$

where

$$c_n = \frac{1}{2\pi i} \oint_C \frac{f(\zeta)}{(\zeta - z_0)^{n+1}} \, d\zeta, \tag{5.2.2}$$

and C is any closed path which contains the only singular point z_0 inside and is taken in the positive direction. In particular,

$$c_{-1} = \frac{1}{2\pi i} \oint_C f(\zeta) \, d\zeta. \tag{5.2.3}$$

DEFINITION 5.2.1. The coefficient c_{-1} of a Laurent series in a neighborhood of an isolated singular point z_0 is called the *residue* of the analytic function $f(z)$ at z_0 and is denoted by $\operatorname{Res}_{z=z_0} f(z)$.

By (5.2.2), we have

$$\operatorname*{Res}_{z=z_0} f(z) = c_{-1} = \frac{1}{2\pi i} \oint_C f(\zeta) \, d\zeta. \tag{5.2.4}$$

If z_0 is a removable singularity of $f(z)$, then the Laurent series (5.2.1) does not contain negative powers of $z - z_0$ and therefore substituting (5.2.1) into (5.2.4) gives $c_{-1} = 0$.

Therefore, the residue of $f(z)$, in general, is not equal to zero if z_0 is a pole or an essential singularity. However, the residue can be equal to zero if the coefficient c_{-1} of the Laurent series is zero. For example, if $z = 0$ is a pole or an essential singularity and $f(z)$ is an even function (that is, $f(-z) = f(z)$), then its Laurent series contains only even (positive and negative) powers of z and

$$c_{-1} = \operatorname*{Res}_{z=0} f(z) = 0.$$

It is essential for the sequel to compute the coefficient c_{-1} not by using (5.2.4) but by either differentiating $f(z)$ at z_0 or computing c_{-1} by means of some special techniques for obtaining the Laurent series expansion of $f(z)$.

5.2.1. Computing residues. Let z_0 be a pole of order m of $f(z)$. Thus the Laurent series of $f(z)$ in a neighborhood of z_0 has the form

$$f(z) = \frac{c_{-m}}{(z-z_0)^m} + \frac{c_{-m+1}}{(z-z_0)^{-m+1}} + \cdots + \frac{c_{-1}}{z-z_0} + \sum_{n=0}^{\infty} c_n (z-z_0)^n, \quad (5.2.5)$$

where $c_{-m} \neq 0$. To determine c_{-1}, we proceed as follows:

(1) Multiply both sides of (5.2.5) by $(z-z_0)^m$:

$$(z-z_0)^m f(z) = c_{-m} + c_{-m+1}(z-z_0) + \cdots + c_{-1}(z-z_0)^{m-1} + \sum_{n=0}^{\infty} c_n (z-z_0)^{n+m}. \quad (5.2.6)$$

(2) Differentiate (5.2.6) $m-1$ times:

$$[(z-z_0)^m f(z)]^{(m-1)} = c_{-1}(m-1)! + \sum_{n=0}^{\infty} c_n (n+m)(n+m-1)\cdots(n+2)(z-z_0)^{n+1}. \quad (5.2.7)$$

(3) Take the limit in (5.2.7) as $z \to z_0$ and divide by $(m-1)!$:

$$c_{-1} = \operatorname*{Res}_{z=z_0} f(z) = \frac{1}{(m-1)!} \lim_{z \to z_0} [(z-z_0)^m f(z)]^{(m-1)}. \quad (5.2.8)$$

Formula (5.2.8) allows one to compute the residue at a pole, z_0, of order m by means of differentiation. In particular, if $m = 1$ (that is, z_0 is a simple pole) then, by (5.2.8), we have

$$\operatorname*{Res}_{z=z_0} f(z) = \lim_{z \to z_0} [(z-z_0) f(z)], \quad (5.2.9)$$

since $0! = 1$ and the derivative of order zero of $f(z)$ is $f(z)$.

Formula (5.2.9), which can be used to compute the residue at a simple pole, can be obtained directly by multiplying (5.2.5) by $z - z_0$, since $c_{-m} = c_{-m+1} = \cdots = c_{-2} = 0$, $c_{-1} \neq 0$, and by taking the limit as $z \to z_0$.

Suppose that $f(z)$ has the form

$$f(z) = \frac{\varphi(z)}{\psi(z)}, \quad (5.2.10)$$

where $\varphi(z)$ and $\psi(z)$ are analytic and z_0 is a pole of order 1, that is, $\psi(z_0) = 0$, $\psi'(z_0) \neq 0$ and $\varphi(z_0) \neq 0$. Since $\psi(z_0) = 0$, (5.2.9) can be transformed into another convenient form:

$$\begin{aligned}
\operatorname*{Res}_{z=z_0} \frac{\varphi(z)}{\psi(z)} &= \lim_{z\to z_0} (z-z_0)\frac{\varphi(z)}{\psi(z)} \\
&= \lim_{z\to z_0} \frac{\varphi(z)}{\frac{\psi(z)-\psi(z_0)}{z-z_0}} \\
&= \frac{\varphi(z_0)}{\lim_{z\to z_0} \frac{\psi(z)-\psi(z_0)}{z-z_0}} \\
&= \frac{\varphi(z_0)}{\psi'(z_0)}.
\end{aligned}$$

Hence, the formula

$$\operatorname*{Res}_{z=z_0} \frac{\varphi(z)}{\psi(z)} = \frac{\varphi(z_0)}{\psi'(z_0)}. \tag{5.2.11}$$

can be used to compute the residue at z_0 if z_0 is a simple pole.

Note 5.2.1. In practice, one can overestimate the order of a pole by overlooking the zeros of $\varphi(z)$ in (5.2.10). Thus, if z_0 is a zero of order l of $\varphi(z)$ and a zero of order $k > l$ of $\psi(z)$ then z_0 is a pole of order $m = k - l$ and not k. However, if such a mistake occurs, the result may still be correct when (5.2.8) is used. This fact can easily be justified if one multiplies both sides of (5.2.5) not by $(z-z_0)^m$ but by $(z-z_0)^k$, where $k \geq m$, differentiates the given expression $k-1$ times and takes the limit as $z \to z_0$. Thus, with $k \geq m$, we obtain

$$\begin{aligned}
c_{-1} &= \operatorname*{Res}_{z=z_0} f(z) \\
&= \frac{1}{(k-1)!} \lim_{z\to z_0} [(z-z_0)^k f(z)]^{(k-1)}.
\end{aligned} \tag{5.2.12}$$

Therefore, if the order of a pole is overestimated, the final result remains correct; but using formula (5.2.12) is less convenient since one has to compute derivatives of higher order than in (5.2.8).

EXAMPLE 5.2.1. *Find the residue of the function*

$$f(z) = \frac{z+1}{z}$$

at $z = 0$.

SOLUTION. The point $z = 0$ is a simple pole of $f(z)$. Using (5.2.9) we obtain

$$\operatorname*{Res}_{z=0} \frac{z+1}{z} = \lim_{z\to 0} \left[z\,\frac{z+1}{z} \right] = 1.$$

We obtain the same result by using (5.2.11):

$$\operatorname*{Res}_{z=0} \frac{z+1}{z} = \left.\frac{z+1}{1}\right|_{z=0} = 1.$$

Finally, formula (5.2.12) also gives the same result for any integer $k > 1$:

$$\begin{aligned}\operatorname*{Res}_{z=0} \frac{z+1}{z} &= \frac{1}{(k-1)!} \lim_{z\to 0} \left[z^k \frac{z+1}{z} \right]^{(k-1)} \\ &= \frac{1}{(k-1)!} \lim_{z\to 0} [(k-1)!(kz+1)] = 1. \quad \square\end{aligned}$$

5.2.2. Computing the residue at an essential singularity. There are no general formulae for computing residues at essential singularities. One may compute the integral (5.2.4) or use some special ways of getting the Laurent expansion and determining the coefficient c_{-1}. For example, if $f(z)$ is an even function with respect to $z - z_0$ then the Laurent series contains only even powers of $z - z_0$ and $\operatorname{Res}_{z=z_0} f(z) = 0$.

EXAMPLE 5.2.2. *Find the residue of the function*

$$f(z) = \frac{e^{-1/z^4}}{z^2+1}$$

at $z = 0$.

SOLUTION. In this case $z = 0$ is an essential singularity. Since $f(z)$ is an even function, then

$$\operatorname*{Res}_{z=0} \frac{e^{-1/z^4}}{z^2+1} = 0.$$

The residues at the simple poles $z = \pm i$ can easily be computed by formula (5.2.11). $\square$

EXAMPLE 5.2.3. *Find the residue of*

$$f(z) = z^2 e^{1/z}$$

at $z = 0$.

SOLUTION. The point $z = 0$ is an essential singularity. We expand $f(z)$ in a Laurent series:

$$f(z) = z^2 \left(1 + \frac{1}{1!z} + \frac{1}{2!z^2} + \frac{1}{3!z^3} + \dots \right). \tag{5.2.13}$$

It follows from (5.2.13) that the coefficient c_{-1} is equal to

$$c_{-1} = \operatorname*{Res}_{z=0} z^2 e^{1/z} = \frac{1}{3!} = \frac{1}{6}. \quad \square$$

EXAMPLE 5.2.4. *Find the residue of*

$$f(z) = \frac{1}{1-z} e^{1/z}$$

at $z = 0$.

SOLUTION. The point $z = 0$ is an essential singularity. We expand $1/(1-z)$ and $e^{1/z}$ in a Taylor and a Laurent series, respectively:

$$\frac{1}{1-z} e^{1/z} = \left(1 + z + z^2 + \dots\right)\left(1 + \frac{1}{1!z} + \frac{1}{2!z^2} + \dots\right). \tag{5.2.14}$$

To determine the coefficient at z^{-1}, one multiplies 1 by $1/z$, z by $1/(2!z^2)$, z^2 by $1/(3!z^3)$ and so on, and adds the results. Thus,

$$c_{-1} = \operatorname*{Res}_{z=0} \frac{1}{1-z}\, e^{1/z} = 1 + \frac{1}{2!} + \frac{1}{3!} + \dots + \frac{1}{n!} + \dots = e - 1. \quad \square$$

EXAMPLE 5.2.5. *Find the residue of*

$$f(z) = e^{z+1/z}$$

at $z = 0$.

SOLUTION. The point $z = 0$ is an essential singularity. We represent e^z by a Taylor series in powers of z and $e^{1/z}$ by a Laurent series in powers of z, and multiply both series, with the aim of determining the coefficient of z^{-1}. Thus,

$$\begin{aligned} e^{z+1/z} &= e^z\, e^{1/z} \\ &= \left(1 + \frac{z}{1!} + \frac{z^2}{2!} + \dots\right)\left(1 + \frac{1}{1!z} + \frac{1}{2!z^2} + \dots\right). \end{aligned}$$

The coefficient c_{-1} is equal to

$$\begin{aligned} c_{-1} &= \frac{1}{1!} + \frac{1}{1!2!} + \frac{1}{2!3!} + \dots + \frac{1}{n!(n+1)!} + \dots \\ &= \sum_{n=0}^{\infty} \frac{1}{n!(n+1)!}. \end{aligned} \tag{5.2.15}$$

The series in (5.2.15) is equal to the value, at $z = 2$, of the so-called *modified Bessel function,*

$$I_1(z) = \sum_{n=0}^{\infty} \frac{1}{n!(n+1)!} \left(\frac{z}{2}\right)^{2n+1}, \qquad |z| < \infty. \tag{5.2.16}$$

Thus,

$$c_{-1} = \operatorname*{Res}_{z=0} e^{z+1/z} = I_1(2). \quad \square$$

Many problems can be found, for example, in [**31**], pp. 79–80, numbers 314–336.

5.2.3. Residue of an analytic function at infinity.

DEFINITION 5.2.2. The function $f(z)$ is said to be *analytic at* $z = \infty$ if the function $\varphi(z) = f(1/z)$ is analytic at $z = 0$.

For example, $f(z) = \sin(1/z)$ is analytic at $z = \infty$ since $\varphi(z) = \sin z$ is analytic at $z = 0$.

Suppose that $f(z)$ is analytic in the infinite domain $D : R < |z| < \infty$, so that it can be represented there by a convergent Laurent series

$$f(z) = \sum_{n=-\infty}^{\infty} c_n z^n = \sum_{n=0}^{\infty} c_n z^n + \frac{c_{-1}}{z} + \frac{c_{-2}}{z^2} + \dots, \qquad R < |z| < \infty. \tag{5.2.17}$$

Let C be an arbitrary closed curve lying entirely in D and taken in the positive direction with respect to the bounded domain it encloses. Since the Laurent series in (5.2.17) is uniformly convergent in D, integrating the left-hand side and right-hand side termwise along C, we have

$$\oint_C f(z)\,dz = \sum_{n=0}^{\infty} c_n \oint_C z^n\,dz + c_{-1} \oint_C \frac{dz}{z} + c_{-2} \oint_C \frac{dz}{z^2} + \dots . \tag{5.2.18}$$

It is clear, by Cauchy's Theorem 3.3.4 for multiply connected domains, that the path C can be replaced by the circle $|z| = R_0 > R$, taken counterclockwise, in D. Hence, by the change of variable $z = R_0\, e^{i\theta}$, we have

$$\begin{aligned} \oint_C \frac{dz}{z^n} &= \oint_{|z|=R_0} \frac{dz}{z^n} \\ &= \frac{i}{R_0^{n-1}} \int_0^{2\pi} e^{-i(n-1)\theta}\,d\theta = \begin{cases} 2\pi i, & n = 1, \\ 0, & n \neq 1. \end{cases} \end{aligned} \tag{5.2.19}$$

Thus, by (5.2.18) we obtain

$$c_{-1} = \frac{1}{2\pi i} \oint_C f(z)\,dz. \tag{5.2.20}$$

DEFINITION 5.2.3. Let the function $f(z)$ be analytic in the infinite domain $D : R \le |z| < \infty$. The *residue of* $f(z)$ *at infinity*, denoted by $\operatorname*{Res}_{z=\infty} f(z)$, is the number $-c_{-1}$, where c_{-1} is the coefficient of $1/z$ in the

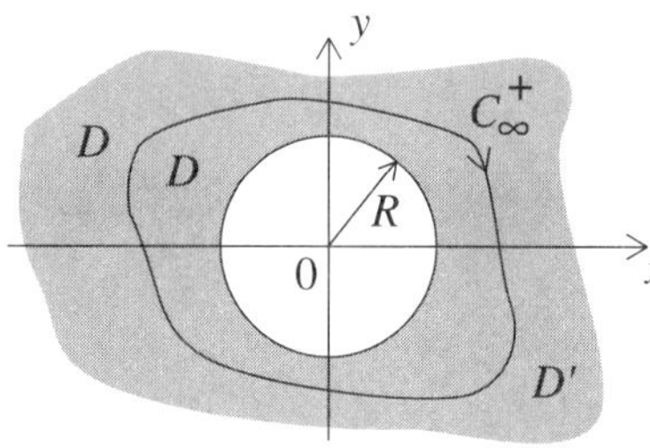

FIGURE 5.1. Positive direction of curve C_∞^+ bounding the infinite domain $D' \subset D$.

Laurent series of $f(z)$ convergent in D,

$$\operatorname*{Res}_{z=\infty} f(z) = -c_{-1} = -\frac{1}{2\pi i}\oint_C f(z)\,dz, \tag{5.2.21}$$

where C is any closed path lying in D and taken in the positive direction with respect to the bounded region it encloses.

Note 5.2.2. If we change the direction of C in (5.2.21) we obtain the path C_∞^+ which is traversed in the positive direction with respect to the infinite domain D' it encloses (see Fig 5.1).

In that case, (5.2.21) becomes

$$\operatorname*{Res}_{z=\infty} f(z) = \frac{1}{2\pi i}\oint_{C_\infty^+} f(z)\,dz. \tag{5.2.22}$$

Formulae (5.2.22) and (5.2.4) are identical since the paths C_∞^+ in (5.2.22) and C in (5.2.4) are both taken in the positive direction, and $z = \infty$ and $z = z_0$ are the only singular points of $f(z)$ inside C_∞^+ and C, respectively.

The coefficient c_k of the Laurent series (5.2.17) can be computed by the simple formula

$$c_k = \frac{1}{2\pi i}\oint_C \frac{f(z)}{z^{k+1}}\,dz, \qquad k = 0, \pm 1, \pm 2, \ldots, \tag{5.2.23}$$

where the closed path C lies entirely in the domain of analyticity, $D : R < |z| < \infty$, of $f(z)$ and is taken in the positive direction with respect to the bounded region it encloses.

If the point $z = \infty$ is a zero of order k of $f(z)$ then the Laurent series (5.2.17) has the form

$$f(z) = \frac{c_{-k}}{z^k} + c_{-(k-1)}z^{-k+1} + \ldots, \qquad c_{-k} \neq 0. \tag{5.2.24}$$

It follows from (5.2.24) that

$$f(z) = O\left(\frac{1}{z^k}\right), \qquad \text{as } k \to \infty.$$

If $k = 1$, then

$$\operatorname*{Res}_{z=\infty} f(z) = -c_{-1},$$

and if $k \geq 2$, then

$$\operatorname*{Res}_{z=\infty} f(z) = 0.$$

For example, in the case of a rational function of the form

$$f(z) = \frac{a_n z^n + a_{n-1} z^{n-1} + \cdots + a_0}{b_m z^m + b_{m-1} z^{m-1} + \cdots + b_0},$$

we have

$$f(z) \approx \frac{a_n}{b_m} \frac{1}{z^{m-n}}, \qquad \text{as } z \to \infty.$$

Thus,

$$\operatorname*{Res}_{z=\infty} f(z) = \begin{cases} -a_n/b_m, & \text{if } m = n+1, \\ 0, & \text{if } m > n+1. \end{cases}$$

Note that, for the function $f(z) = 1/z$, we have

$$\operatorname*{Res}_{z=\infty} \frac{1}{z} = -1 \neq 0,$$

despite the fact that $z = \infty$ is a point of analyticity of $1/z$. Hence we have proved the following important theorem.

THEOREM 5.2.1. *If $f(z)$ is analytic in an annulus $A : R < |z| < \infty$ and $f(z) = O(z^{-k})$ as $z \to \infty$, then*

$$\operatorname*{Res}_{z=\infty} f(z) = \begin{cases} -c_1, & \textit{if } k = 1, \\ 0, & \textit{if } k = 2, 3, \dots, \end{cases}$$

where c_1 is the coefficient of z^{-1} in the Laurent series of $f(z)$ convergent in A.

Note 5.2.3. Any attempt to use the substitution $z = 1/\zeta$ in the analytic function $f(z)$ and then compute the residue at $\zeta = 0$ does not give the same result for the residue at $z = \infty$, as can be seen from the function $f(z) = 1/z$.

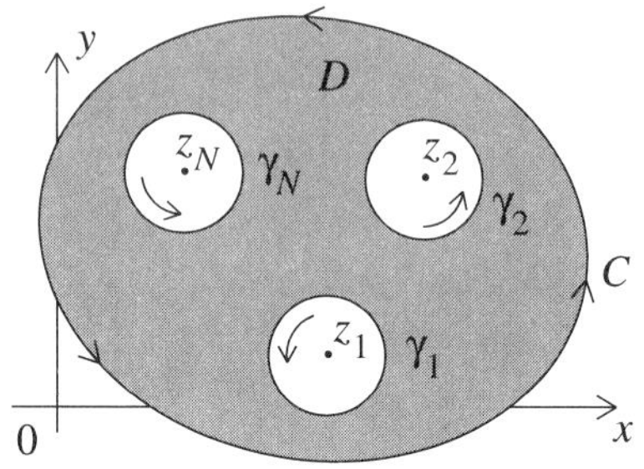

FIGURE 5.2. Closed region D and closed paths, γ_k, surrounding the singular points z_k, for $k = 1, 2, \ldots, N$.

5.2.4. The residue theorem. The following theorem can be used for evaluating integrals by means of the theory of residues.

THEOREM 5.2.2 (RESIDUE THEOREM). *Suppose the function $f(z)$ is analytic in a simply connected closed region D bounded by the path C taken in the positive direction, except for a finite number of isolated singularities, z_k, $k = 1, 2, \ldots, N$, located inside D. Then*

$$\oint_C f(z)\,dz = 2\pi i \sum_{k=1}^{N} \operatorname*{Res}_{z=z_k} f(z). \tag{5.2.25}$$

PROOF. We surround each singular point, z_k, $k = 1, 2, \ldots, N$, by a sufficiently small closed path γ_k, containing only the singular point z_k (see Fig 5.2). Then $f(z)$ is analytic in the region D' bounded by the paths $C, \gamma_1, \gamma_2, \ldots, \gamma_N$. Therefore by Cauchy's Theorem 3.3.4 for multiply connected domains, we have

$$\oint_C f(z)\,dz = \sum_{k=1}^{N} \oint_{\gamma_k} f(z)\,dz, \tag{5.2.26}$$

where the paths C and γ_k are taken counterclockwise. By definition of the residue at z_k we have (see formula (5.2.4)):

$$\oint_{\gamma_k} f(z)\,dz = 2\pi i \operatorname*{Res}_{z=z_k} f(z), \tag{5.2.27}$$

which, upon substitution in (5.2.26), yields (5.2.25). □

Using the last Theorem 5.2.2 and Definition 5.2.3 of the residue at $z = \infty$, one can prove the following theorem, which is useful for evaluating integrals.

THEOREM 5.2.3. *If an analytic function $f(z)$ has a finite number of singularities z_k, $k = 1, 2, \ldots, N$, in the complex plane, then the sum of all the residues of $f(z)$, including the residue at $z = \infty$, is equal to zero:*

$$\sum_{k=1}^{N} \operatorname*{Res}_{z=z_k} f(z) + \operatorname*{Res}_{z=\infty} f(z) = 0. \tag{5.2.28}$$

PROOF. Let C be a closed path which contains all N singular points z_k assumed to be situated at finite distance from $z = 0$. By the residue theorem we have

$$\frac{1}{2\pi i} \oint_C f(z)\, dz = \sum_{k=1}^{N} \operatorname*{Res}_{z=z_k} f(z). \tag{5.2.29}$$

On the other hand, it follows from (5.2.17) that

$$-\frac{1}{2\pi i} \oint_C f(z)\, dz = \operatorname*{Res}_{z=\infty} f(z). \tag{5.2.30}$$

Adding (5.2.29) and (5.2.30) we obtain that

$$\sum_{k=1}^{N} \operatorname*{Res}_{z=z_k} f(z) + \operatorname*{Res}_{z=\infty} f(z) = 0. \quad \square \tag{5.2.31}$$

EXAMPLE 5.2.6. *Evaluate the following integral counterclockwise:*

$$I_1 = \oint_{|z|=2} \frac{1 - \cos z}{z^2 - z}\, dz.$$

SOLUTION. The singular points of the integrand in the disk $|z| < 2$ are $z = 0$ and $z = 1$. At $z = 1$, the numerator is not zero and the denominator has a zero of order 1. Hence $z = 1$ is a pole of order 1 of the integrand. At $z = 0$, the numerator and denominator are equal to zero. However,

$$1 - \cos z = 1 - \left(1 - \frac{z^2}{2!} + \ldots\right) = O(z^2), \qquad \text{as } z \to 0;$$

thus

$$\lim_{z\to 0} \frac{1 - \cos z}{z(z-1)} = 0.$$

Hence $z = 0$ is a removable singularity and

$$\operatorname*{Res}_{z=0} \frac{1 - \cos z}{z(z-1)} = 0.$$

Thus, by the residue theorem, we obtain

$$\begin{aligned} I_1 &= 2\pi i \operatorname*{Res}_{z=1} \frac{1-\cos z}{z(z-1)} \\ &= 2\pi i \lim_{z\to 1} \frac{1-\cos z}{z} = 2\pi i(1-\cos 1). \quad \square \end{aligned}$$

EXAMPLE 5.2.7. *Evaluate the following integral counterclockwise:*

$$I_2 = \oint_{|z|=4} \frac{z^2}{\sin z}\, dz.$$

SOLUTION. The zeros of the denominator in the disk $|z| < 4$ are $z = 0$ and $z = \pm\pi$. The point $z = 0$ is a removable singularity since $z = 0$ is a zero of order 2 of the numerator and a zero of order 1 of the denominator. Thus

$$\lim_{z\to 0} \frac{z^2}{\sin z} = 0.$$

The points $z = \pm\pi$ are poles of order 1 of the integrand since $\sin(\pm\pi) = 0$, but $(\sin z)'\big|_{z=\pm\pi} \neq 0$. Therefore, by (5.2.11) and (5.2.25),

$$\begin{aligned} I_2 &= 2\pi i \left(\operatorname*{Res}_{z=\pi} + \operatorname*{Res}_{z=-\pi}\right)\left[\frac{z^2}{\sin z}\right] \\ &= 2\pi i \left[\frac{z^2}{\cos z}\bigg|_{z=\pi} + \frac{z^2}{\cos z}\bigg|_{z=-\pi}\right] \\ &= 4\pi i \frac{\pi^2}{\cos \pi} = -4\pi^3 i. \quad \square \end{aligned}$$

EXAMPLE 5.2.8. *Evaluate the following integral counterclockwise:*

$$I_3 = \oint_{|z|=2} \frac{e^{1/z^2}}{1-z}\, dz.$$

SOLUTION. There are two singular points in the region $|z| < 2$, namely, a pole, $z = 1$, of order 1 and an essential singularity, $z = 0$, since the Laurent series of e^{1/z^2} contains infinitely many negative powers of z. Therefore

$$I_3 = 2\pi i \left(\operatorname*{Res}_{z=1} + \operatorname*{Res}_{z=0}\right)\left[\frac{e^{1/z^2}}{1-z}\right]. \tag{5.2.32}$$

Using formula (5.2.11), we have

$$\operatorname*{Res}_{z=1} \frac{e^{1/z^2}}{1-z} = \frac{e^{1/1}}{-1} = -e. \tag{5.2.33}$$

In order to find the residue at $z = 0$ we expand $1/(1-z)$ in a Taylor series and e^{1/z^2} in a Laurent series in powers of z:

$$\frac{e^{1/z^2}}{1-z} = (1 + z + z^2 + z^3 + \dots)\left(1 + \frac{1}{1!z^2} + \frac{1}{2!z^4} + \frac{1}{3!z^6} + \dots\right).$$

To obtain the terms containing $1/z$ one has to take the terms with odd powers (say, z^{2k-1}) of the first series and multiply them by the terms of the second series which have one more power (that is, $1/z^{2k}$), and then add the results. Thus, we have

$$\operatorname*{Res}_{z=0} \frac{e^{1/z^2}}{1-z} = \frac{1}{1!} + \frac{1}{2!} + \frac{1}{3!} + \dots = e - 1. \tag{5.2.34}$$

Substituting (5.2.33) and (5.2.34) into (5.2.32) we obtain

$$I_3 = 2\pi i(-e + e - 1) = -2\pi i. \quad \square$$

If a closed path C surrounds all, or a large number of, the singular points of the integrand, then it is convenient to use Theorem 5.2.3, which says that, in the case of a finite number of singular points in the whole complex plane, the sum of all residues including the residue at $z = \infty$ is equal to zero.

Theorem 5.2.1 is also extremely useful if $f(z) = O(z^{-k})$ as $z \to \infty$, because the residue at $z = \infty$ is different from zero if and only if $k = 1$.

EXAMPLE 5.2.9. *Evaluate the following integral counterclockwise:*

$$I_4 = \oint_{|z|=2} \frac{dz}{1 + z^{10}}.$$

SOLUTION. The 10 singular points, $z_1, z_2, \dots, z_{10}$, of the integrand in the disk $|z| < 2$ are the roots of the equation $z^{10} = -1$. Therefore,

$$I_4 = 2\pi i \sum_{k=1}^{10} \operatorname*{Res}_{z=z_k} \frac{1}{1+z^{10}}.$$

By Theorem 5.2.3, we have

$$\sum_{k=1}^{10} \operatorname*{Res}_{z=z_k} \frac{1}{1+z^{10}} + \operatorname*{Res}_{z=\infty} \frac{1}{1+z^{10}} = 0.$$

Since

$$\frac{1}{1+z^{10}} = O\left(\frac{1}{z^{10}}\right) \qquad \text{as } z \to \infty,$$

then, by Theorem 5.2.1,

$$\operatorname*{Res}_{z=\infty} \frac{1}{1+z^{10}} = 0.$$

Hence, $I_4 = 0$. □

EXAMPLE 5.2.10. *Evaluate the given integral counterclockwise:*

$$I_5 = \oint_{|z|=5} \frac{z^{13}}{(3z^2+2)^4(z^3+3)^2}\, dz.$$

SOLUTION. The integrand has five singular points in the disk $|z| < 5$, namely, the two and three zeros of the first and second factors, respectively, in the denominator. Therefore it is more convenient to evaluate the integral by means of Theorem 5.2.3,

$$I_5 = -2\pi i \operatorname*{Res}_{z=\infty} \frac{z^{13}}{(3z^2+2)^4(z^3+3)^2}.$$

Since the order of the denominator as $z \to \infty$ is $O(z^{8+6}) = O(z^{14})$, and the order of the numerator is $O(z^{13})$, then the integrand is equivalent to $1/(3^4 z)$ as $z \to \infty$. Therefore by Theorem 5.2.1, the residue at $z = \infty$ is equal to $-1/3^4$ and $I_5 = 2\pi i/81$. □

5.2.5. Path of integration through poles of odd orders. The following theorem holds when the path of integration goes through poles of odd orders.

THEOREM 5.2.4. *Suppose that the function $f(z)$ is analytic in a closed region D bounded by the closed path C, except for a finite number of singular points, $z_1, z_2, \dots, z_N$, lying inside D, and a finite number of simple poles, $\tilde{z}_1, \tilde{z}_2, \dots, \tilde{z}_l$, lying on C at points where C is smooth. Then*

$$\text{p. v.} \oint_C f(z)\, dz = 2\pi i \sum_{k=1}^{N} \operatorname*{Res}_{z=z_k} f(z) + \pi i \sum_{k=1}^{l} \operatorname*{Res}_{z=\tilde{z}_k} f(z). \qquad (5.2.35)$$

PROOF. We bypass each singular point $\tilde{z}_k$ by a circular arc γ_k of radius δ and center at $\tilde{z}_k$, lying in D. We choose δ so small that the whole arc γ_k lies in the region of analyticity of $f(z)$. Then $f(z)$ is analytic on the closed path which consists of the arcs γ_k and the remaining part, $\widetilde{C}$ of C (see Fig 5.3). Therefore by the residue theorem

$$\int_{\widetilde{C}} f(z)\, dz + \sum_{k=1}^{l} \int_{\gamma_k} f(z)\, dz = 2\pi i \sum_{k=1}^{N} \operatorname*{Res}_{z=z_k} f(z). \qquad (5.2.36)$$

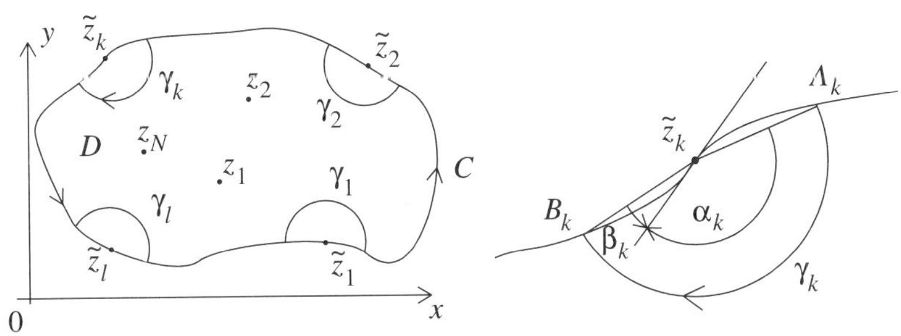

FIGURE 5.3. The path $\widetilde{C} + \gamma_1 + \cdots + \gamma_l$.

Expanding $f(z)$ in a Laurent series in a neighborhood of the simple pole $\tilde{z}_k$, we obtain

$$f(z) = \frac{c_{-1}}{z - \tilde{z}_k} + \sum_{n=0}^{\infty} c_n (z - \tilde{z}_k)^n. \tag{5.2.37}$$

Then

$$\int_{\gamma_k} f(z)\, dz = \int_{\gamma_k} \frac{c_{-1}}{z - \tilde{z}_k}\, dz + \sum_{n=0}^{\infty} c_n \int_{\gamma_k} (z - \tilde{z}_k)^n\, dz. \tag{5.2.38}$$

On the arc γ_k we have $z = \tilde{z}_k + \delta e^{i\theta}$, $\alpha_k \le \theta \le \beta_k$, where α_k is the angle between the secant joining the points A_k and $\tilde{z}_k$ and the tangent to $\widetilde{C}$ at $\tilde{z}_k$, and β_k is the angle between the secant joining the points B_k and $\tilde{z}_k$ and the same tangent (see the magnification of arc γ_k in Fig 5.3). With this notation, (5.2.38) becomes

$$\int_{\gamma_k} f(z)\, dz \;=\; c_{-1} \int_{\alpha_k}^{\beta_k} \frac{\delta e^{i\theta} i\, d\theta}{\delta e^{i\theta}} + \sum_{n=0}^{\infty} c_n \int_{\alpha_k}^{\beta_k} \left(\delta e^{i\theta}\right)^n \delta\, e^{i\theta} i\, d\theta. \tag{5.2.39}$$

In the limit, as $\delta \to 0$, we have $\alpha_k \to \pi$, $\beta_k \to 0$, and (5.2.39) becomes

$$\begin{aligned} \lim_{\delta \to 0} \int_{\gamma_k} f(z)\, dz &= i c_{-1} \int_{\pi}^{0} d\theta \\ &= -\pi i c_{-1} \\ &= -\pi i \operatorname*{Res}_{z = \tilde{z}_k} f(z). \end{aligned} \tag{5.2.40}$$

Hence, taking the limit of (5.2.36) as $\delta \to 0$ we obtain (5.2.35). □

Note 5.2.4. Formula (5.2.35) is true also in the case the points $\tilde{z}_k$ are poles of any odd order ($\tilde{z}_k$ and the principal part of the Laurent series contains only odd powers of $z - \tilde{z}_k$):

$$f(z) = \sum_{p=0}^{s} \frac{c_{-(2p+1)}}{(z - \tilde{z}_k)^{2p+1}} + \sum_{n=0}^{\infty} c_n (z - \tilde{z}_k)^n, \tag{5.2.41}$$

where $c_{-(2p+1)} \neq 0$.

Indeed, integrating each of the terms in the principal part of (5.2.41) along the arc γ_k from $\theta = \pi$ to $\theta = 0$ we obtain, as in the transition from (5.2.38) to (5.2.39), that the term containing c_{-1} is the only nonzero term. This term is

$$\begin{aligned}\int_{\gamma_k} \frac{dz}{(z - \tilde{z}_k)^{2p+1}} &= \int_{\pi}^{0} \frac{e^{i\theta} i d\theta}{(e^{i\theta})^{2p+1}} = i \int_{\pi}^{0} e^{-2pi\theta}\, d\theta \\ &= \begin{cases} -\pi i, & \text{if } \ p = 0, \\ 0, & \text{if } \ p = 1, 2, \dots, s. \end{cases}\end{aligned}$$

Note that simple poles of the integrand located on the path occur in diffraction problems (see [**49**]).

EXAMPLE 5.2.11. *Evaluate the following integral counterclockwise:*

$$I_6 = \text{p. v.} \oint_{|z|=1} \frac{\sin z}{(z^2 - 1)(z^2 + 1)}\, dz.$$

SOLUTION. The four singular points, $z = \pm 1$ and $z = \pm i$, of the integrand are simple poles. Moreover, all the singularities are located on the circle $|z| = 1$. Hence using (5.2.35) we obtain

$$\begin{aligned} I_6 &= \pi i \left(\operatorname*{Res}_{z=1} + \operatorname*{Res}_{z=-1} + \operatorname*{Res}_{z=i} + \operatorname*{Res}_{z=-i} \right) \left[\frac{\sin z}{(z^2 - 1)(z^2 + 1)} \right] \\ &= \pi i \left[\left. \frac{\sin z}{2z(z^2+1)} \right|_{z=1} + \left. \frac{\sin z}{2z(z^2+1)} \right|_{z=-1} \right. \\ &\quad \left. + \left. \frac{\sin z}{2z(z^2-1)} \right|_{z=i} + \left. \frac{\sin z}{2z(z^2-1)} \right|_{z=-i} \right] \\ &= \pi i \left[\frac{\sin 1}{4} + \frac{\sin 1}{4} + \frac{\sin i}{2i(-2)} + \frac{\sin(-i)}{2(-i)(-2)} \right] \\ &= \frac{\pi i}{2} (\sin 1 - \sinh 1). \quad \square \end{aligned}$$

Exercises for Section 5.2

Find the residue of the given functions at every singular point and at infinity.

1. $\dfrac{1}{z - z^3}$.

2. $\dfrac{z}{(z+1)(z-i)^2}$.

3. $\dfrac{z^2 + 4z + 1}{z^2(z+1)}$.

4. $\dfrac{z^3 + 1}{z(z-1)^2(z+i)^3}$.

5. $\dfrac{e^z}{(z-1)(z+3i)^2}$.

6. $\dfrac{\cos z}{(z-1)^2(z+4)}$.

Find the residue of the given functions at every finite singular point.

7. $\dfrac{1}{e^z - 1}$.

8. $\dfrac{\sin z}{z(z-1)^2}$.

9. $\dfrac{1 - \cos z}{z^2 \sin z}$.

10. $z^3 e^{1/z}$.

11. $z^2 \sin \dfrac{1}{z-1}$.

12. $\cos \dfrac{z}{z+2}$.

13. $\dfrac{\sin z}{z} + \dfrac{1}{z^3} + e^{1/z}$.

14. $\dfrac{1 - \cos z}{z^2} + \dfrac{1}{z^{25}} + \sin \dfrac{1}{z}$.

15. $e^{z/(z-1)}$.

16. $\cos\left(\dfrac{1}{z}\right) \cos z$.

17. Let $f(z) = \dfrac{\varphi(z)}{[\psi(z)]^2}$, $\varphi(z_0) \neq 0$, $\psi(z_0) = 0$, $\psi'(z_0) \neq 0$. Suppose that $\varphi(z)$ and $\psi(z)$ are analytic at $z = z_0$. Find the type of singularity of $f(z)$ at $z = z_0$ and $\operatorname*{Res}_{z=z_0} f(z)$.

18. Suppose that $z = z_0$ is a pole of order n of the function $f(z)$. Find $\operatorname*{Res}_{z=z_0} [f'(z)/f(z)]$.

Evaluate the following integrals counterclockwise along the given circles C.

19. $\displaystyle\oint_C \frac{z}{(z+1)^2(z-2)}\, dz, \qquad C: |z| = 3.$

20. $\displaystyle\oint_C \frac{\sin z}{z^2(z-1)}\, dz, \qquad C: |z| = 2.$

21. $\displaystyle\oint_C \frac{e^z}{(z+1)^2}\, dz, \qquad C: |z-4| = 1.$

22. $\displaystyle\oint_C \frac{\cos z}{(1-z)^2}\, dz, \qquad C: |z+2| = 1/2.$

23. $\displaystyle\oint_C \frac{1}{z \sin z}\, dz, \qquad C: |z| = 4.$

24. $\displaystyle\oint_C \frac{1}{z(e^z-1)^2}\, dz, \qquad C: |z| = 2.$

25. $\displaystyle\oint_C \frac{z}{z^3-1}\, dz, \qquad C: |z| = 3.$

26. $\displaystyle\oint_C \frac{z^2+1}{z^4-1}\, dz, \qquad C: |z| = 2.$

27. $\displaystyle\oint_C \frac{e^{1/z}}{z^2+4}\, dz, \qquad C: |z| = 3.$

28. $\displaystyle\oint_C \frac{\sin[1/(z-1)]}{z^2(z+2)}\, dz, \qquad C: |z+1| = 3/2.$

29. $\displaystyle\oint_C z \sin\frac{1}{z}\, dz, \qquad C: |z| = 1.$

30. $\displaystyle\oint_C \left(\frac{1}{z^2} + \sin\frac{1}{z^4}\right) dz, \qquad C: |z| = 1.$

Using the fact that the sum of the residues at all the singular points (including the point at infinity) is equal to zero, compute the given integrals counterclockwise along the given circles C.

31. $\displaystyle\oint_C \frac{z}{(z+2)(z^8+1)}\,dz, \qquad C: |z| = 5.$

32. $\displaystyle\oint_C \frac{1}{(z^6-64)(z-1)}\,dz, \qquad C: |z| = 100.$

33. $\displaystyle\oint_C \frac{1}{(z+1)(z+2)\cdots(z+100)}\,dz, \qquad C: |z| = 150.$

34. $\displaystyle\oint_C \frac{z(z+2)}{z^{24}-1}\,dz, \qquad C: |z| = 2.$

CHAPTER 6

Elementary Definite Integrals

The main idea in evaluating definite integrals over the real x-axis by means of Cauchy's Theorem and the theory of residues, in the simplest cases, is as follows. Instead of evaluating the integral of a function $f(x)$ of the real variable x from $-\infty$ to $+\infty$, one considers the integral of $f(z)$ of the complex variable z along a closed path, C, consisting of a segment, $[-R, R]$, of the real axis and a semicircle, C_R: $|z| = R$, $0 \leq \arg z \leq \pi$, in the upper half-plane. The residue theorem is applied to $f(z)$ over the region bounded by C and the limit is taken as $R \to \infty$. If $|f(z)| = O\left(1/|z|^2\right)$, the integral along C_R tends to zero as $R \to \infty$. Thus,

$$\int_{-\infty}^{\infty} f(x)\, dx = 2\pi i \sum \operatorname*{Res}_{\Im z>0} f(z).$$

In more complicated cases, other appropriate closed paths are chosen and the function $f(x)$ is replaced not by $f(z)$ but by some other functions.

There are many known variants of this simple method, which is far from being exhausted at the present time.

In this chapter, generally but not always,

$$a = \alpha + i\beta \in \mathbb{C}, \qquad \alpha,\, \alpha_k,\, \beta \in \mathbb{R}, \qquad 0 \leq \operatorname{Arg} z < 2\pi,$$

and branch cuts are taken along the positive real semi-axis.

6.1. Rational functions over $(-\infty, +\infty)$

In this section, we consider integrals of real rational functions of the form

$$\int_{-\infty}^{\infty} \frac{P_n(x)}{Q_m(x)}\, dx, \tag{6.1.1}$$

where $P_n(x)$ and $Q_m(x)$ are polynomials in x of degrees n and m, respectively, with real coefficients, and $m \geq n + 2$. This last condition ensures

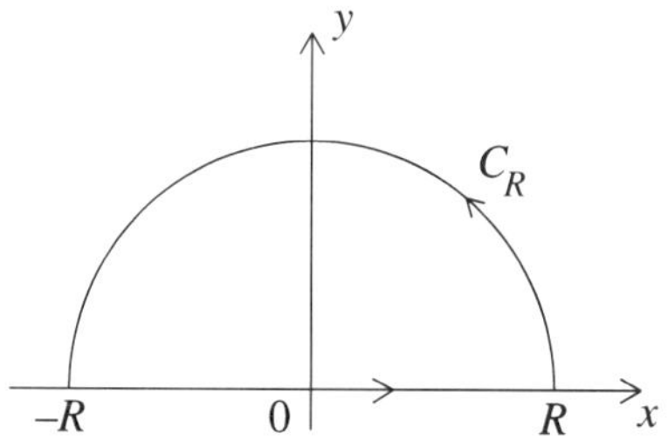

FIGURE 6.1. The path of integration for the integral (6.1.3) of $f(z)$ without any real poles.

the convergence of the integral in (6.1.1). We consider two cases.

6.1.1. The case of no real poles. Consider a rational function of a complex variable,

$$f(z) = \frac{P_n(z)}{Q_m(z)}, \tag{6.1.2}$$

where $Q_m(x) \neq 0$ for all real x. We take the closed path, C, consisting of the segment $[-R, R]$ of the x-axis and the semicircle C_R of radius R in the upper half-plane, as shown in Fig 6.1.

By the residue theorem, we have

$$\oint_C f(z)\,dz = 2\pi i \sum_k \operatorname*{Res}_{z=z_k} f(z), \tag{6.1.3}$$

where z_k are the singular points of $f(z)$ enclosed by C. Since $z = x$ on the segment $[-R, R]$, we have

$$\int_{-R}^{R} f(x)\,dx + \int_{C_R} f(z)\,dz = 2\pi i \sum_k \operatorname*{Res}_{z=z_k} f(z), \tag{6.1.4}$$

where $\Im z_k > 0$ since $Q_m(z)$ has no real zeros. We show that

$$\lim_{R\to\infty} \int_{C_R} f(z)\,dz = 0. \tag{6.1.5}$$

Indeed, on C_R one has

$$z = R\,e^{i\theta}, \qquad dz = iR\,e^{i\theta}\,d\theta, \qquad 0 \leq \theta \leq \pi.$$

Hence

$$\int_{C_R} \frac{P_n(z)}{Q_m(z)}\,dz = \int_0^{\pi} \frac{P_n(R\,e^{i\theta})}{Q_m(R\,e^{i\theta})}\,iR\,e^{i\theta}\,d\theta. \tag{6.1.6}$$

Since $m \geq n+2$, we have

$$\begin{aligned}\left|\frac{P_n(z)}{Q_m(z)}\, z\right|_{z=R\,e^{i\theta}} &= \left|\frac{(z^n + a_1 z^{n-1} + \cdots + a_n) z}{z^m + b_1 z^{m-1} + \cdots + b_m}\right|_{z=R\,e^{i\theta}} \\ &= \left|\frac{z^{-(m-n-1)} + a_1 z^{-(m-n)} + \cdots + a_n z^{-(m-1)}}{1 + b_1 z^{-1} + \cdots + b_m z^{-m}}\right|_{z=R\,e^{i\theta}} \\ &\to 0, \qquad \text{as } R \to \infty,\end{aligned}$$

because all the powers of z in the numerator and the denominator of the last fraction are negative. Hence, as $R \to \infty$, we get from (6.1.4) the formula

$$\int_{-\infty}^{\infty} \frac{P_n(x)}{Q_m(x)}\, dx = 2\pi i \sum_k \operatorname*{Res}_{z=z_k} \left[\frac{P_n(z)}{Q_m(z)}\right], \tag{6.1.7}$$

provided $\Im z_k > 0$, $m \geq n+2$ and $Q_m(x) \neq 0$.

EXAMPLE 6.1.1. *Evaluate the integral*

$$\int_{-\infty}^{\infty} \frac{dx}{x^2+1}.$$

SOLUTION. The conditions on (6.1.7) are satisfied because $P_n(x) = 1$ and $Q_m(x) = x^2 + 1$. Since the points $z = \pm i$ are poles of order 1 of the rational function $1/\left(z^2+1\right)$, then

$$\begin{aligned}\int_{-\infty}^{\infty} \frac{dx}{x^2+1} &= 2\pi i \operatorname*{Res}_{z=i} \left(\frac{1}{z^2+1}\right) \\ &= 2\pi i \left.\frac{1}{2z}\right|_{z=i} \\ &= \pi.\end{aligned}$$

This result can also be checked by direct evaluation. □

EXAMPLE 6.1.2. *Evaluate the integral*

$$\int_{-\infty}^{\infty} \frac{dx}{\left(x^2+1\right)^3}.$$

SOLUTION. The points $z = \pm i$ are poles of order 3 for the rational function $1/\left(z^2+1\right)^3$ because $\left(z^2+1\right)^3 = (z+i)^3(z-i)^3$; therefore, by formula (5.2.8)

$$\int_{-\infty}^{\infty} \frac{dx}{\left(x^2+1\right)^3} = 2\pi i \operatorname*{Res}_{z=i} \left[\frac{1}{\left(z^2+1\right)^3}\right]$$

$$
\begin{aligned}
&= 2\pi i \, \frac{1}{2!} \lim_{z\to i} \left[(z-i)^3 \frac{1}{(z^2+1)^3} \right]'' \\
&= \pi i \lim_{z\to i} \left[\frac{1}{(z+i)^3} \right]'' \\
&= \pi i \lim_{z\to i} \left[-\frac{3}{(z+i)^4} \right]' \\
&= \pi i \lim_{z\to i} \frac{12}{(z+i)^5} \\
&= \pi i \frac{12}{(2i)^5} = \frac{3\pi}{8}. \quad \square
\end{aligned}
$$

EXAMPLE 6.1.3. *Evaluate the integral*

$$\int_{-\infty}^{\infty} \frac{x^2+1}{x^4+1} \, dx.$$

SOLUTION. The poles of the function

$$f(z) = \frac{z^2+1}{z^4+1}$$

are the zeros of the denominator,

$$z^4 = -1 = e^{(\pi+2k\pi)i},$$

that is,

$$z_k = e^{(\pi+2k\pi)i/4}, \qquad k = 0, 1, 2, 3.$$

The function $f(z)$ has simple poles at these points because

$$Q(z_k) = z_k^4 + 1 = 0, \quad \text{but} \quad Q'(z_k) = 4z_k^3 \neq 0.$$

Since the first two poles, $z_0 = e^{\pi i/4}$ and $z_1 = e^{i3\pi/4}$, lie in the upper half-plane, then

$$
\begin{aligned}
\int_{-\infty}^{\infty} \frac{x^2+1}{x^4+1} \, dx &= 2\pi i \left(\operatorname*{Res}_{z=z_0} + \operatorname*{Res}_{z=z_1} \right) \left[\frac{z^2+1}{z^4+1} \right] \\
&= 2\pi i \left(\frac{e^{\pi i/2}+1}{4e^{3\pi i/4}} + \frac{e^{3\pi i/2}+1}{4e^{9\pi i/4}} \right) \\
&= \frac{\pi i}{2} \left(e^{-\pi i/4} + e^{-3\pi i/4} + e^{-3\pi i/4} + e^{-\pi i/4} \right) \\
&= \pi i \left[e^{-\pi i/4} + e^{-3\pi i/4} \right] \\
&= \pi i \left(-i\sqrt{2} \right)
\end{aligned}
$$

$$= \pi\sqrt{2}. \quad \square$$

EXAMPLE 6.1.4. *Evaluate the integral*

$$\int_{-\infty}^{\infty} \frac{dx}{x^6 + 1}.$$

SOLUTION. The poles of the function

$$f(z) = \frac{1}{z^6 + 1}$$

are the zeros of the denominator,

$$z^6 = -1 = e^{(\pi + 2k\pi)i},$$

that is,

$$z_k = e^{i(\pi + 2k\pi)/6}, \qquad k = 0, 1, \dots, 5.$$

Three of these roots,

$$z_0 = e^{\pi i/6}, \qquad z_1 = e^{3\pi i/6} = e^{\pi i/2} = i, \qquad z_2 = e^{5\pi i/6},$$

lie in the upper half-plane. Since $f(z)$ has simple poles at these points, then

$$\begin{aligned}
\int_{-\infty}^{\infty} \frac{dx}{x^6 + 1} &= 2\pi i \left(\operatorname*{Res}_{z=z_0} + \operatorname*{Res}_{z=z_1} + \operatorname*{Res}_{z=z_2} \right) \left[\frac{1}{z^6 + 1} \right] \\
&= 2\pi i \left(\left. \frac{1}{6z^5} \right|_{z=e^{\pi i/6}} + \left. \frac{1}{6z^5} \right|_{z=i} + \left. \frac{1}{6z^5} \right|_{z=e^{5\pi i/6}} \right) \\
&= \frac{2\pi i}{6} \left(\frac{1}{e^{5\pi i/6}} + \frac{1}{i} + \frac{1}{e^{25\pi i/6}} \right) \\
&= \frac{\pi i}{3} \left(e^{-5\pi i/6} - i + e^{-\pi i/6} \right) \\
&= \frac{\pi i}{3} \left(-2i\,\frac{1}{2} - i \right) = \frac{2\pi}{3}. \quad \square
\end{aligned}$$

Note 6.1.1. If the integrand is an even function, then an integral from 0 to ∞ can be replaced by half the integral from $-\infty$ to ∞ and formula (6.1.7) can be used, but this is not possible otherwise. It will be shown in Subsection 7.1.1 how to evaluate the integral of an arbitrary rational function of the form $P_n(x)/Q_m(x)$ from 0 to ∞ and even from a to b, which amounts to evaluating an indefinite integral, by means of the theory of residues.

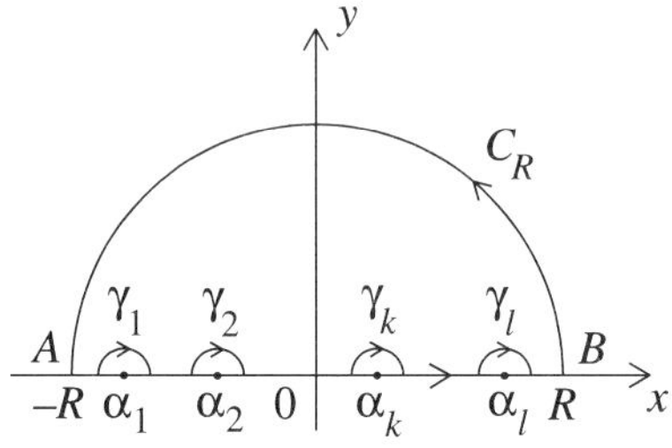

FIGURE 6.2. The path of integration bypassing real simple poles for integral (6.1.1).

6.1.2. The case of real poles. We consider the case $Q_m(x) = 0$ at the points $\alpha_1, \alpha_2, \dots, \alpha_l$, all of which are real simple poles of $P_n(x)/Q_m(x)$.

We bypass the real points $\alpha_1, \alpha_2, \dots, \alpha_l$ along semicircles $\gamma_1, \gamma_2, \dots, \gamma_l$ of small radii δ and centers α_k $(k = 1, \dots, l)$, lying in the upper half-plane, that is, we consider a closed path as shown in Fig 6.2.

By the residue theorem we have

$$\left(\int_{AB} + \sum_{k=1}^{l} \int_{\gamma_k} + \int_{C_R}\right) \frac{P_n(z)}{Q_m(z)}\, dz = 2\pi i \sum_k \operatorname*{Res}_{z=z_k} \frac{P_n(z)}{Q_m(z)}, \tag{6.1.8}$$

where $\Im z_k > 0$ and the first integral on the left-hand side of (6.1.8) is evaluated along the straight line subsegments from $-R$ to R, omitting the semicircles shown in Fig 6.2. Since $m \geq n + 2$, then, as before, the integral along the semicircle C_R approaches zero as $R \to \infty$. Moreover, since the points α_k $(k = 1, \dots, l)$ are poles of order 1, then a Laurent series expansion in a neighborhood of the point $z = \alpha_k$ has the form

$$\frac{P_n(z)}{Q_m(z)} = \frac{c_{-1}}{z - \alpha_k} + \sum_{\mu=0}^{\infty} c_\mu (z - \alpha_k)^\mu. \tag{6.1.9}$$

Hence

$$\int_{\gamma_k} \frac{P_n(z)}{Q_m(z)}\, dz = \int_{\gamma_k} \left[\frac{c_{-1}}{z - \alpha_k} + \sum_{\mu=0}^{\infty} c_\mu (z - \alpha_k)^\mu\right] dz. \tag{6.1.10}$$

Since $z \in \gamma_k$, then $z - \alpha_k = \delta\, e^{i\theta}$ and $dz = \delta i\, e^{i\theta}\, d\theta$, where θ varies clockwise from π to 0. Hence from (6.1.10) we obtain

$$\begin{aligned}\int_{\gamma_k} \frac{P_n(z)}{Q_m(z)}\, dz &= c_{-1} \int_{\pi}^{0} \frac{\delta i\, e^{i\theta}\, d\theta}{\delta\, e^{i\theta}} + \sum_{\mu=0}^{\infty} c_\mu \int_{\pi}^{0} \left(\delta\, e^{i\theta}\right)^\mu \delta i\, e^{i\theta}\, d\theta \\ &\to -c_{-1}\pi i + 0, \qquad \text{as} \quad \delta \to 0,\end{aligned}$$

where

$$c_{-1} = \operatorname*{Res}_{z=\alpha_k} \left[\frac{P_n(z)}{Q_m(z)} \right]. \tag{6.1.11}$$

Therefore, the left-hand side of (6.1.8) has a finite limit as $\delta \to 0$ and $R \to \infty$:

$$\text{p. v.} \int_{-\infty}^{\infty} \frac{P_n(x)}{Q_m(x)}\, dx = 2\pi i \sum_k \operatorname*{Res}_{z=z_k} \left[\frac{P_n(z)}{Q_m(z)} \right] + \pi i \sum_{k=1}^{l} \operatorname*{Res}_{z=\alpha_k} \left[\frac{P_n(z)}{Q_m(z)} \right], \tag{6.1.12}$$

where $\Im z_k > 0$. At the same time we have proved the convergence of integral (6.1.12) in the sense of the principal value.

Note 6.1.2. Formula (6.1.12) is valid also in the case where all the points α_k are real poles of odd order $2s+1$ for some $s = 1, 2, 3, \dots$, and the coefficients, c_{-2l}, of even order in the Laurent series expansion are all equal to zero.

Indeed, in this case,

$$\frac{P_n(z)}{Q_m(z)} = \frac{c_{-(2s+1)}}{(z-\alpha_k)^{2s+1}} + \frac{c_{-(2s-1)}}{(z-\alpha_k)^{2s-1}} + \cdots + \frac{c_{-1}}{(z-\alpha_k)} + \sum_{\mu=0}^{\infty} c_\mu (z-\alpha_k)^\mu, \tag{6.1.13}$$

and we have

$$\begin{aligned} \int_{\gamma_k} \frac{dz}{(z-\alpha_k)^{2s+1}} &= \int_{\pi}^{0} \frac{\delta\, e^{i\theta}}{(\delta\, e^{i\theta})^{2s+1}}\, i\, d\theta \\ &= \delta^{-2s} i \int_{\pi}^{0} e^{-2si\theta}\, d\theta \\ &= \begin{cases} -\pi i, & \text{if } s = 0, \\ 0, & \text{if } s = 1, 2, 3, \dots. \end{cases} \end{aligned}$$

EXAMPLE 6.1.5. *Evaluate the integral*

$$\text{p. v.} \int_{-\infty}^{\infty} \frac{dx}{(x-1)\,(x^2+1)}.$$

SOLUTION. We use formula (6.1.12):

$$\text{p. v.} \int_{-\infty}^{\infty} \frac{dx}{(x-1)\,(x^2+1)} = \left(2\pi i \operatorname*{Res}_{z=i} + \pi i \operatorname*{Res}_{z=1} \right) \left[\frac{1}{(z-1)\,(z^2+1)} \right]$$

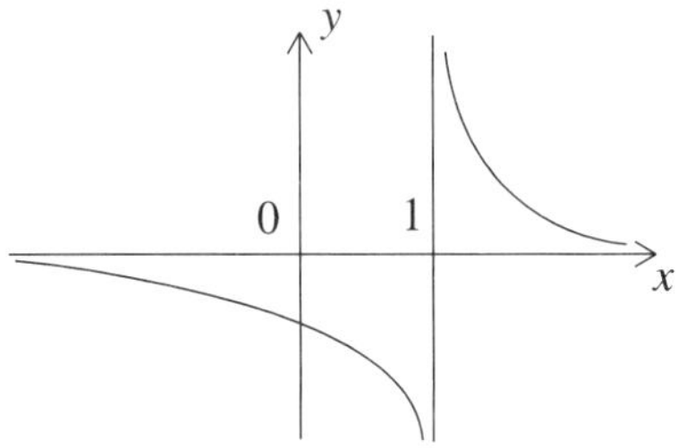

FIGURE 6.3. The graph of the integrand in Example 6.1.5.

$$
\begin{aligned}
&= 2\pi i \left.\frac{1}{2z(z-1)}\right|_{z=i} + \pi i \left.\frac{1}{z^2+1}\right|_{z=1} \\
&= \frac{\pi}{i-1} + \frac{\pi i}{2} = -\frac{\pi}{2}.
\end{aligned}
$$

A rough graph of the integrand is shown in Fig 6.3. □

EXAMPLE 6.1.6. *Find the values of the real parameters a and c, with $c > a^4 > 0$, for which the Cauchy principal value*

$$
I = \text{p. v.} \int_{-\infty}^{\infty} \frac{dx}{(x-a)^3(x^2+2a^2x+c)}
$$

is finite and evaluate I.

SOLUTION. We expand the integrand in partial fractions,

$$
\begin{aligned}
f(x) &= \frac{1}{(x-a)^3(x^2+2a^2x+c)} \\
&= \frac{A}{(x-a)^3} + \frac{B}{(x-a)^2} + \frac{C}{x-a} + \frac{Dx+E}{x^2+2a^2x+c}.
\end{aligned}
\tag{6.1.14}
$$

To have a finite principal value it is necessary that $B = 0$; thus the integrand is of the form (6.1.13). To use formula (5.1.42) to compute B, we multiply both sides of (6.1.14) by $(x-a)^3$, differentiate the resulting equation once with respect to x and consider the limit as $x \to a$:

$$
\begin{aligned}
B &= \lim_{x\to a} [f(x)(x-a)^3]' = \lim_{x\to a} \left(\frac{1}{x^2+2a^2x+c}\right)' \\
&= -\lim_{x\to a} \frac{2x+2a^2}{(x^2+2a^2x+c)^2} = -\frac{2a+2a^2}{(a^2+2a^3+c)^2} \\
&= 0,
\end{aligned}
$$

if $a = -1$. Hence I is finite if $a = -1$ and $c > 1$.

The singular points of the integrand are

$$z_1 = -1 \quad \text{(pole of order 3)}, \qquad z_{2,3} = -1 \pm i\sqrt{c-1} \quad \text{(poles of order 1)}.$$

Using formula (6.1.12) we have

$$\begin{aligned}
I &= \text{p. v.} \int_{-\infty}^{\infty} \frac{dx}{(x+1)^3(x^2+2x+c)} \\
&= \left(2\pi i \operatorname*{Res}_{z=-1+i\sqrt{c-1}} + \pi i \operatorname*{Res}_{z=-1} \right) \left[\frac{1}{(z+1)^3(z^2+2z+c)} \right] \\
&= 2\pi i \frac{1}{(z_2+1)^3 2(z_2+1)} + \frac{\pi i}{2} \lim_{z \to -1} \left(\frac{1}{z^2+2z+c} \right)'' \\
&= \frac{\pi i}{(c-1)^2} - \frac{\pi i}{2} \lim_{z \to -1} \left(\frac{2z+2}{(z^2+2z+c)^2} \right)' \\
&= \frac{\pi i}{(c-1)^2} - \pi i \lim_{z \to -1} \frac{(z^2+2z+c)^2 - 2(z^2+2z+c)(2z+2)(z+1)}{(z^2+2z+c)^4} \\
&= \frac{\pi i}{(c-1)^2} - \frac{\pi i}{(c-1)^2} = 0. \quad \square
\end{aligned}$$

6.2. Rational functions times sine or cosine

We consider integrals of the form

$$\int_{-\infty}^{\infty} \frac{P_n(x)}{Q_m(x)} \sin \alpha x \, dx, \qquad \int_{-\infty}^{\infty} \frac{P_n(x)}{Q_m(x)} \cos \alpha x \, dx, \qquad m \geq n+1. \tag{6.2.1}$$

The following lemma, due to Camille Jordan, will be used in the sequel.

LEMMA 6.2.1 (JORDAN'S LEMMA). *If a function $f(z)$ is continuous on a sequence of circular arcs*

$$C_{R_n}: \quad |z| = R_n, \quad \Im z \geq -a,$$

where $R_n \to \infty$, a is fixed and

$$M_n = \max_{z \in C_{R_n}} |f(z)| \to 0, \qquad \text{as } R_n \to \infty, \tag{6.2.2}$$

then, for any $\lambda > 0$,

$$\lim_{R_n \to \infty} \int_{C_{R_n}} f(z)\, e^{i\lambda z}\, dz = 0. \tag{6.2.3}$$

PROOF. Suppose that $a > 0$. Then, on the arc AB, $-\alpha_n \leq \arg z < 0$, where $\alpha_n > 0$ (see Fig 6.4). Clearly, $\alpha_n = \arcsin(a/R_n) \to 0$ as $R_n \to \infty$. Moreover, $\arcsin(a/R_n) \approx a/R_n$ for R_n large, so that $\alpha_n R_n \approx a = \text{constant}$ as $R_n \to \infty$. Since, on arc AB, $-\alpha_n \leq \theta \leq 0$, then

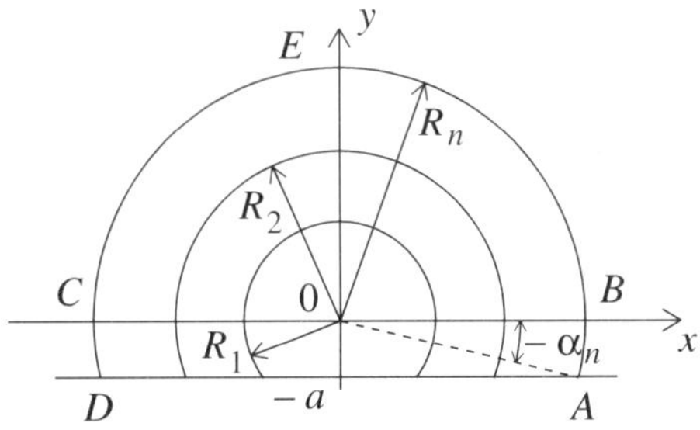

FIGURE 6.4. The sequence of circular arcs C_{R_n}.

$$-\sin\alpha_n \le \sin\theta \le 0, \qquad \text{that is,} \qquad 0 \le -\sin\theta \le \sin\alpha_n,$$

and

$$\begin{aligned}\left|e^{i\lambda z}\right| = \left|e^{i\lambda R_n(\cos\theta + i\sin\theta)}\right| &= e^{-\lambda R_n \sin\theta}\\ &\le e^{\lambda R_n \sin\alpha_n}\\ &\approx e^{\lambda a} = \text{constant}\end{aligned}$$

for large R_n. Hence, by (6.2.2),

$$\begin{aligned}\left|\int_{AB} f(z)\, e^{i\lambda z}\, dz\right| &\le \int_{-\alpha_n}^{0} |f(R_n\, e^{i\theta})|\, |e^{i\lambda z}| R_n |i|\, |e^{i\theta}|\, d\theta\\ &\le R_n M_n\, e^{a\lambda}\alpha_n\\ &\approx a M_n\, e^{a\lambda} \to 0, \qquad \text{as } R_n \to \infty.\end{aligned}$$

Similarly, it can be shown that $\int_{CD} \to 0$ as $R_n \to \infty$. Furthermore, on the arc BEC one has

$$z = R_n\, e^{i\theta}, \qquad 0 \le \theta \le \pi,$$

and

$$\left|e^{i\lambda z}\right| = \left|e^{i\lambda R_n(\cos\theta + i\sin\theta)}\right| = e^{-\lambda R_n \sin\theta}.$$

Then

$$\begin{aligned}\left|\int_{BEC} f(z)\, e^{i\lambda z}\, dz\right| &\le \int_0^{\pi} \left|f(R_n\, e^{i\theta})\right| \left|e^{i\lambda R_n e^{i\theta}}\right| R_n\, |i|\, \left|e^{i\theta}\right|\, d\theta\\ &\le M_n R_n \int_0^{\pi} e^{-\lambda R_n \sin\theta}\, d\theta\\ &= M_n R_n \left[\int_0^{\pi/2} + \int_{\pi/2}^{\pi}\right] e^{-\lambda R_n \sin\theta}\, d\theta\end{aligned}$$

(and substituting $\theta = \pi - t$ in the second integral)

$$= 2M_n R_n \int_0^{\pi/2} e^{-\lambda R_n \sin\theta}\, d\theta. \tag{6.2.4}$$

By the simple inequality (see Fig 6.5)

$$\sin x \geq \frac{2}{\pi}\, x, \qquad \text{where} \quad 0 \leq x \leq \frac{\pi}{2}, \tag{6.2.5}$$

the inequality (6.2.4) becomes

$$\begin{aligned} \left| \int_{BEC} f(z) e^{i\lambda z}\, dz \right| &\leq 2M_n R_n \int_0^{\pi/2} e^{-\lambda R_n 2\theta/\pi}\, d\theta \\ &= 2M_n R_n \left(-\frac{\pi}{2\lambda R_n} e^{-\lambda R_n 2\theta/\pi} \right) \Bigg|_0^{\pi/2} \\ &= M_n \frac{\pi}{\lambda} \left(1 - e^{-\lambda R_n}\right) \to 0, \qquad \text{as } R_n \to \infty. \quad \square \end{aligned}$$

Note 6.2.1. If $a < 0$, the proof of Jordan's Lemma is simpler since the estimates on the arcs AB and CD are not necessary.

As in Section 6.1, we consider integrals of the form (6.2.1) in the absence and in the presence of real poles.

6.2.1. The case of no real poles. Consider the function of a complex variable

$$f(z) = \frac{P_n(z)}{Q_m(z)}\, e^{i\lambda z}, \qquad \lambda \in \mathbb{R}, \tag{6.2.6}$$

where $Q_m(x) \neq 0$ for real x. We take the closed path C consisting of the segment $[-R, R]$ of the real axis and the semicircle C_R of radius R (see

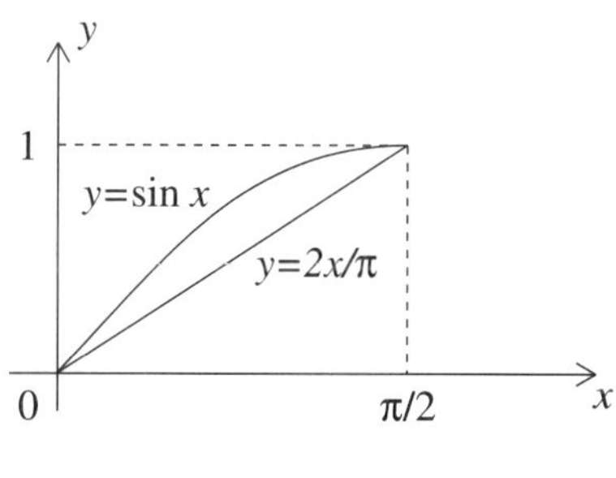

FIGURE 6.5. The inequality $\sin x \geq 2x/\pi$ on the interval $0 \leq x \leq \pi/2$.

Fig 6.1) in the upper half-plane. By the residue theorem we have

$$\oint_C \frac{P_n(z)}{Q_m(z)}\, e^{i\lambda z}\, dz = \int_{-R}^{R} \frac{P_n(x)}{Q_m(x)}\, e^{i\lambda x}\, dx + \int_{C_R} \frac{P_n(z)}{Q_m(z)}\, e^{i\lambda z}\, dz$$

$$= 2\pi i \sum_k \operatorname*{Res}_{z=z_k} \left[\frac{P_n(z)}{Q_m(z)}\, e^{i\lambda z}\right]. \tag{6.2.7}$$

Because $m \geq n+1$, then

$$\lim_{R\to\infty} \frac{P_n(z)}{Q_m(z)}\bigg|_{z\in C_R} = 0,$$

and, by Jordan's Lemma,

$$\lim_{R\to\infty} \int_{C_R} \frac{P_n(z)}{Q_m(z)}\, e^{i\lambda z}\, dz = 0.$$

Hence, from (6.2.7), as $R \to \infty$, we have

$$\int_{-\infty}^{\infty} \frac{P_n(x)}{Q_m(x)}\, e^{i\lambda x}\, dx - 2\pi i \sum_k \operatorname*{Res}_{z=z_k} \left[\frac{P_n(z)}{Q_m(z)}\, e^{i\lambda z}\right], \tag{6.2.8}$$

and, separating the real and imaginary parts in (6.2.8), we obtain

$$\int_{-\infty}^{\infty} \frac{P_n(x)}{Q_m(x)}\, \cos\lambda x\, dx = \Re\left\{ 2\pi i \sum_k \operatorname*{Res}_{z=z_k} \left[\frac{P_n(z)}{Q_m(z)}\, e^{i\lambda z}\right]\right\}, \tag{6.2.9}$$

$$\int_{-\infty}^{\infty} \frac{P_n(x)}{Q_m(x)}\, \sin\lambda x\, dx = \Im\left\{ 2\pi i \sum_k \operatorname*{Res}_{z=z_k} \left[\frac{P_n(z)}{Q_m(z)}\, e^{i\lambda z}\right]\right\}, \tag{6.2.10}$$

where

$$m \geq n+1, \qquad \Im z_k > 0, \qquad Q_m(x) \neq 0.$$

EXAMPLE 6.2.1. *Evaluate the integral*

$$I_1 = \int_{-\infty}^{\infty} \frac{\cos\alpha x}{x^2+\beta^2}\, dx, \qquad \alpha > 0, \quad \beta > 0.$$

SOLUTION. All the conditions are such that formula (6.2.9) is true, and hence to evaluate I_1 it suffices to find the residue of $e^{i\alpha z}/\left(z^2+\beta^2\right)$ at the sole and simple pole $z = \beta i$ in the upper half-plane:

$$I_1 = \Re\left\{ 2\pi i \operatorname*{Res}_{z=\beta i} \left[\frac{e^{i\alpha z}}{z^2+\beta^2}\right]\right\} = \Re\left[2\pi i \frac{e^{-\alpha\beta}}{2\beta i}\right]$$

$$= \frac{\pi}{\beta}\, e^{-\alpha\beta}. \tag{6.2.11}$$

If $\alpha < 0$, we let $\alpha = -\gamma$ in I_1. Then $\gamma > 0$ and one can use formula (6.2.9). □

The value of I_1, for arbitrary real α and β, is

$$\int_{-\infty}^{\infty} \frac{\cos \alpha x}{x^2 + \beta^2}\, dx = \frac{\pi}{|\beta|}\, e^{-|\alpha|\,|\beta|}. \tag{6.2.12}$$

6.2.2. The case of real poles. We consider the case $Q_m(x) = 0$ for real $x = \alpha_1, \dots, \alpha_l$. This case is similar to the one in Subsection 6.1.2. We assume that the function $[P_n(x)/Q_m(x)]\, e^{i\lambda x}$ has either simple poles at the points $\alpha_1, \alpha_2, \dots, \alpha_l$, or only poles of odd orders and that the Laurent series of the function $[P_n(z)/Q_m(z)]e^{i\lambda z}$ has only odd negative powers of $z - \alpha_k$ (k=1, 2, ..., l). Then, repeating the steps of Subsection 6.1.2 we obtain

$$\text{p. v.} \int_{-\infty}^{\infty} \frac{P_n(x)}{Q_m(x)} \sin \lambda x \, dx$$
$$= \Im\left\{ 2\pi i \sum_k \operatorname*{Res}_{z=z_k} \left[\frac{P_n(z)}{Q_m(z)}\, e^{i\lambda z} \right] + \pi i \sum_{k=1}^{l} \operatorname*{Res}_{z=\alpha_k} \left[\frac{P_n(z)}{Q_m(z)}\, e^{i\lambda z} \right] \right\}, \tag{6.2.13}$$

$$\text{p. v.} \int_{-\infty}^{\infty} \frac{P_n(x)}{Q_m(x)} \cos \lambda x \, dx$$
$$= \Re\left\{ 2\pi i \sum_k \operatorname*{Res}_{z=z_k} \left[\frac{P_n(z)}{Q_m(z)}\, e^{i\lambda z} \right] + \pi i \sum_{k=1}^{l} \operatorname*{Res}_{z=\alpha_k} \left[\frac{P_n(z)}{Q_m(z)}\, e^{i\lambda z} \right] \right\}. \tag{6.2.14}$$

EXAMPLE 6.2.2. *Evaluate the integral*

$$I_2 = \int_0^{\infty} \frac{\sin \alpha x}{x}\, dx, \qquad \alpha > 0.$$

Note that the symbol p.v. is not needed before the integral because $x = 0$ is a removable singularity of the integrand.

SOLUTION. Since the integrand is an even function, I_2 is equal to half the integral from $-\infty$ to ∞. The conditions $m = 1$, $n = 0$ and $\alpha > 0$ are such that (6.2.13) is true, and hence to evaluate I_2 this integral it suffices to find the residue at the sole and simple pole $z = 0$ of $e^{i\alpha z}/z$:

$$I_2 = \frac{1}{2} \int_{-\infty}^{\infty} \frac{\sin \alpha x}{x}\, dx = \frac{1}{2} \Im \left[\pi i \operatorname*{Res}_{z=0} \left(\frac{e^{i\alpha z}}{z} \right) \right] = \frac{\pi}{2}. \quad \square$$

If $\alpha < 0$ in I_2, we let $\alpha = -\gamma$ and get $I_2 = -\pi/2$. Thus, for arbitrary real α, we have

$$\int_0^\infty \frac{\sin \alpha x}{x} = \begin{cases} \pi/2, & \alpha > 0, \\ 0, & \alpha = 0, \\ -\pi/2, & \alpha < 0. \end{cases} \tag{6.2.15}$$

Formula (6.2.15) is known as the *Dirichlet discontinuous factor* (see, for example, [**32**], p. 602).

Note 6.2.2. Comparing formula (6.1.7) and the two formulae (6.2.13) and (6.2.14), we see that the first one is valid for $m \geq n+2$, that is, in the worst case, for $m = n+2$, while the second and the third ones are valid for $m \geq n+1$, that is, in the worst case, for $m = n+1$. This is because the necessary condition for the integral along the semicircle C_R to approach zero as $R \to \infty$ is $m \geq n+2$ in (6.1.7), while it suffices that $m \geq n+1$ in order to satisfy the same condition for (6.2.13) and (6.2.14), because Jordan's Lemma is used in the last two cases. We note that, in fact, conditionally convergent integrals are evaluated in (6.2.13) and (6.2.14) for the case $m = n+1$. In particular, using the Dirichlet Abel Test (see Theorem 4.2.2), one can prove that the integral in Example 6.2.2 is only conditionally convergent because the integral of the function $|\sin ax/x|$ from 0 to ∞ is divergent.

The integrals considered in Section 6.1 can converge in the sense of the principal value if $m = n+1$.

EXAMPLE 6.2.3. *Evaluate directly the integral*

$$I_3 = \text{p. v.} \int_{-\infty}^\infty \frac{dx}{x}. \tag{6.2.16}$$

SOLUTION. By definition of the principal value, we have

$$\begin{aligned} I_3 &= \lim_{b\to\infty} \lim_{\varepsilon\to 0+} \left(\int_{-b}^{-\varepsilon} + \int_\varepsilon^b \right) \left[\frac{dx}{x} \right] \\ &= \lim_{b\to\infty} \lim_{\varepsilon\to 0+} \left(\log|x| \Big|_{-b}^{-\varepsilon} + \log|x| \Big|_\varepsilon^b \right) \\ &= \lim_{b\to\infty} \lim_{\varepsilon\to 0+} \left(\log\frac{\varepsilon}{b} + \log\frac{b}{\varepsilon} \right) \\ &= \lim_{b\to\infty} \lim_{\varepsilon\to 0+} 0 = 0. \quad \square \end{aligned}$$

A formal application of (6.1.12) to (6.2.16) gives the incorrect answer

$$\pi i \operatorname*{Res}_{z=0} \frac{1}{z} = \pi i \neq 0,$$

because formula (6.1.12) is valid for $m \geq n+2$, and this condition is not satisfied by integral (6.2.16). In fact, the integral along C_R shown in Fig 6.1 $(R>0)$ is equal to πi:

$$\int_{C_R} \frac{dz}{z} = \int_0^\pi \frac{R\,e^{i\theta}}{R\,e^{i\theta}}\, i\, d\theta = i\int_0^\pi d\theta = \pi i, \qquad \text{for all } R>0.$$

One can get the correct answer if the point $z=0$ in Fig 6.1 is surrounded by a semicircle $C_\delta = \{z = \delta\, e^{i\theta}\}$ of radius $\delta < R$. Then, in addition to the integral on C_R which is already evaluated, one has to evaluate the integrals

$$\begin{aligned}\left(\int_{-R}^{-\delta} + \int_{C_\delta} + \int_\delta^R\right)\left[\frac{dz}{z}\right] &= \left(\int_{-R}^{-\delta} + \int_\delta^R\right)\left[\frac{dx}{x}\right] + \int_{C_\delta}\frac{dz}{z} \\ &= \log|x|\Big|_{-R}^{-\delta} + \log|x|\Big|_\delta^R + \int_\pi^0 \frac{\delta i\, e^{i\theta}}{\delta\, e^{i\theta}}\, d\theta \\ &= \log 1 + i\int_\pi^0 d\theta = -\pi i.\end{aligned}$$

Thus, we get $I_3 = \pi i - \pi i = 0$, which is the correct answer.

More complicated types of integrals will be considered in the remaining sections of this chapter and in the next two chapters, where we shall use closed paths that are different from those shown in Figs. 6.1 and 6.2, and the integrand $f(x)$ will be replaced either by $f(z)$ or by some other functions of z.

6.3. Rational functions times exponential functions

We consider integrals of the form

$$\int_{-\infty}^{\infty} \frac{P_n(e^x)}{Q_m(e^x)}\, e^{ax}\, dx, \qquad a = \alpha + i\beta \in \mathbb{C}, \tag{6.3.1}$$

where $P_n(e^x)$ and $Q_m(e^x)$ are polynomials in e^x of degrees n and m, respectively.

The integrand will approach zero as $x \to +\infty$ if $\alpha < m-n$. It will also approach zero as $x \to -\infty$ if $-k < \alpha$ and

$$\frac{P_n(e^x)}{Q_m(e^x)} = O\left(e^{-kx}\right), \qquad \text{as } x \to -\infty. \tag{6.3.2}$$

Thus, if $Q_m(e^x) \neq 0$ for all real x and $-k < \alpha < m-n$, it can be easily checked that the integral (6.3.1) is absolutely convergent since the integrand approaches zero exponentially both as $x \to -\infty$ and as $x \to \infty$.

As in Sections 6.1 and 6.2, we consider two cases:

(a) $Q_m(e^x) \neq 0$ for real x,

(b) $Q_m(e^x) = 0$ for real $x = \alpha_1, \alpha_2, \ldots, \alpha_l$.

6.3.1. The case of no real poles. We consider the case $Q_m(e^x) \neq 0$ for all real x. Since the function $f(x) = P_n(e^x)/Q_m(e^x)$ is periodic of period $2\pi i$, that is, $f(x+2\pi i) = f(x)$ for all x, because $e^{x+2k\pi i} = e^x$, then it is convenient to choose a closed rectangular path, C, described by the following inequalities (see Fig 6.6):

$$-R \leq x \leq R, \qquad 0 \leq y = \Im z \leq 2\pi.$$

By the residue theorem, we have

$$\oint_C \frac{P_n(e^z)}{Q_m(e^z)}\, e^{az}\, dz = 2\pi i \sum_k \operatorname*{Res}_{z=z_k} \left[\frac{P_n(e^z)}{Q_m(e^z)}\, e^{az}\right], \tag{6.3.3}$$

where, for sufficiently large R, all the singular points z_k, with $0 < \Im z_k < 2\pi$, lie in the strip $0 < \Im z < 2\pi$.

We consider the left-hand side of (6.3.3) in greater detail:

$$\left(\int_I + \int_{II} + \int_{III} + \int_{IV}\right) \left[\frac{P_n(e^z)}{Q_m(e^z)}\, e^{az}\right] dz$$
$$= 2\pi i \sum_k \operatorname*{Res}_{z=z_k} \left[\frac{P_n(e^z)}{Q_m(e^z)}\, e^{az}\right]. \tag{6.3.4}$$

We evaluate each of these four integrals along the corresponding side of the rectangle as $R \to \infty$.

On side I, $z = x$, and therefore

$$\int_I = \int_{-R}^{R} \frac{P_n(e^x)}{Q_m(e^x)}\, e^{ax}\, dx \to \int_{-\infty}^{\infty} \frac{P_n(e^x)}{Q_m(e^x)}\, e^{ax}\, dx, \qquad \text{as} \quad R \to \infty.$$

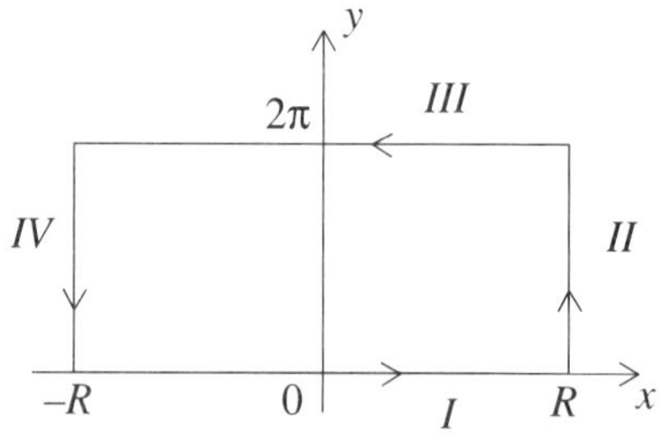

FIGURE 6.6. The path of integration for the integral (6.3.1).

On side II, since $z = R + iy$, $0 \leq y \leq 2\pi$, we have

$$\left|\frac{P_n\left(e^{R+iy}\right)}{Q_m\left(e^{R+iy}\right)}\right| \approx \left|\frac{e^{n(R+iy)}}{e^{m(R+iy)}}\right| = e^{(n-m)R}, \qquad \text{as } R \to \infty, \tag{6.3.5}$$

and

$$\left|e^{a(R+iy)}\right| = \left|e^{(\alpha+i\beta)(R+iy)}\right| = e^{\alpha R - \beta y}. \tag{6.3.6}$$

Because $\alpha < m - n$, by (6.3.5) and (6.3.6), we have the estimate

$$\begin{aligned} \left|\int_{II}\right| &\leq \int_0^{2\pi} \left|\frac{P_n\left(e^{R+iy}\right)}{Q_m\left(e^{R+iy}\right)}\right| \left|e^{a(R+iy)}\right| |i|\, dy \\ &\approx e^{-(m-n-\alpha)R} \int_0^{2\pi} e^{-\beta y}\, dy \to 0, \qquad \text{as} \quad R \to \infty. \end{aligned}$$

On side III, since $z = x + 2\pi i$, we have

$$\begin{aligned} \int_{III} &= \int_R^{-R} \frac{P_n\left(e^{x+2\pi i}\right)}{Q_m\left(e^{x+2\pi i}\right)} e^{a(x+2\pi i)}\, dx \\ &\to e^{2\pi a i} \int_{\infty}^{-\infty} \frac{P_n\left(e^x\right)}{Q_m\left(e^x\right)} e^{ax}\, dx, \qquad \text{as} \quad R \to \infty. \end{aligned}$$

On side IV, since $z = -R + iy$, we have

$$\left|\frac{P_n\left(e^{-R+iy}\right)}{Q_m\left(e^{-R+iy}\right)}\right| = Ce^{-kR}, \qquad C = \text{constant} \neq 0, \qquad \text{as} \quad R \to \infty,$$

because

$$\frac{P_n\left(e^x\right)}{Q_m\left(e^x\right)} = O\left(e^{-kx}\right), \qquad \text{as} \quad x \to -\infty.$$

Since $k + \alpha > 0$, we have the estimate

$$\begin{aligned} \left|\int_{IV}\right| &\leq \int_0^{2\pi} C\, e^{-kR}\, e^{-\alpha R - \beta y}\, dy \\ &= C\, e^{-(k+\alpha)R} \int_0^{2\pi} e^{-\beta y}\, dy \to 0, \qquad \text{as } R \to \infty. \end{aligned}$$

Therefore, we obtain from (6.3.4), in the limit as $R \to \infty$,

$$\begin{aligned} \int_{-\infty}^{\infty} \frac{P_n\left(e^x\right)}{Q_m\left(e^x\right)} e^{ax}\, dx - e^{2\pi a i} \int_{-\infty}^{\infty} \frac{P_n\left(e^x\right)}{Q_m\left(e^x\right)} e^{ax}\, dx \\ = 2\pi i \sum_k \operatorname*{Res}_{z=z_k} \left[\frac{P_n\left(e^z\right)}{Q_m\left(e^z\right)} e^{az}\right], \end{aligned}$$

or

$$\int_{-\infty}^{\infty} \frac{P_n(e^x)}{Q_m(e^x)} e^{ax}\, dx = \frac{2\pi i}{1 - e^{2\pi a i}} \sum_k \operatorname*{Res}_{z=z_k} \left[\frac{P_n(e^z)}{Q_m(e^z)} e^{az} \right], \tag{6.3.7}$$

where

$$Q_m(e^x) \neq 0, \qquad -k < \Re a < m - n, \qquad 0 < \Im z_k < 2\pi.$$

Note 6.3.1. Formula (6.3.7) can be obtained by using the closed path consisting of the segment $[-R, R]$ of the real axis and the semicircle C_R of radius R shown in Fig 6.1. In fact, if z_k is a singular point of the function $e^{az} P_n(e^z)/Q_m(e^z)$ in the strip $0 < \Im z < 2\pi$, then the points $z_k + 2p\pi i$, for $p = 0, 1, \ldots$, are also singular points of this function.

To establish the statement of this note, for simplicity, we assume that all the singular points, z_k, of the integrand of (6.3.1) in the strip $0 < \Im z < 2\pi$ are simple poles and we let $C_{k,p}$ denote the circle of radius p centered at $z_k + 2p\pi i$. In this case, the integral (6.3.1) is equal to the series

$$\begin{aligned}
S &= 2\pi i \sum_{p=0}^{\infty} \sum_k \operatorname*{Res}_{z=z_k+2p\pi i} \left[\frac{P_n(e^z)}{Q_m(e^z)} e^{az} \right] \\
&= 2\pi i \sum_{p=0}^{\infty} \sum_k \frac{1}{2\pi i} \oint_{C_{k,p}} \frac{P_n(e^\zeta)}{Q_m(e^\zeta)} e^{a\zeta}\, d\zeta \\
&= \sum_{p=0}^{\infty} \sum_k \int_0^{2\pi} \frac{P_n\left(e^{z_k + p\, e^{i\theta}}\right)}{Q_m\left(e^{z_k + p\, e^{i\theta}}\right)} e^{a(z_k + 2p\pi i + p\, e^{i\theta})}\, ip\, d\theta \\
&= \sum_{p=0}^{\infty} e^{2pa\pi i} \sum_k \int_0^{2\pi} \frac{P_n\left(e^{z_k + p\, e^{i\theta}}\right)}{Q_m\left(e^{z_k + p\, e^{i\theta}}\right)} e^{a(z_k + p\, e^{i\theta})}\, ip\, d\theta \\
&= \frac{2\pi i}{1 - e^{2\pi a i}} \sum_k \oint_{C_{k,0}} \frac{P_n(e^\zeta)}{Q_m(e^\zeta)} e^{a\zeta}\, d\zeta \\
&= \frac{2\pi i}{1 - e^{2\pi a i}} \sum_k \operatorname*{Res}_{z=z_k} \left[\frac{P_n(e^z)}{Q_m(e^z)} e^{az} \right], \qquad \text{if } \Im a > 0,
\end{aligned} \tag{6.3.8}$$

which is, in fact, formula (6.3.7).

If $\Im a = 0$, then the right-hand side of (6.3.8) can be obtained by assuming that

$$\sum_{p=0}^{\infty} e^{2p\pi a i} = \lim_{\varepsilon \to 0} \sum_{p=0}^{\infty} e^{2p(\alpha + \varepsilon i)\pi i} = \frac{1}{1 - e^{2\pi \alpha i}}.$$

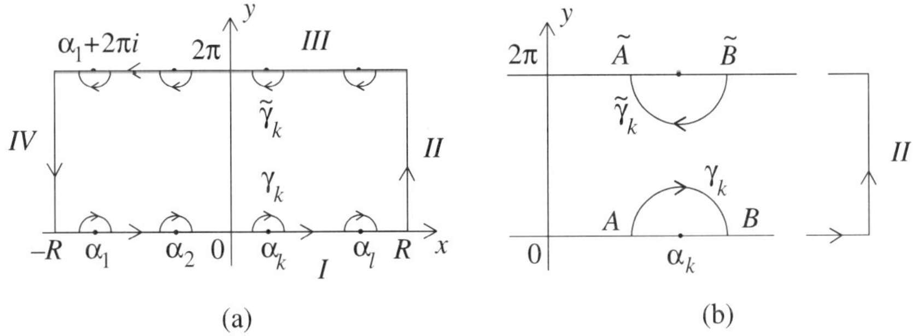

FIGURE 6.7. The path of integration in Subsection 6.3.2.

Finally, if $\Im a < 0$, to determine (6.3.7) one has to close the segment $[-R, R]$ of the real axis by a semicircle, C_R, in the lower half-plane and take into account the fact that the points $z_k - 2p\pi i$, for $p = 0, 1, \dots$, are simple poles.

6.3.2. The case of real poles. We assume that the integrand in (6.3.1) has real simple poles at $\alpha_1, \alpha_2, \dots, \alpha_l$.

Since the function

$$f(x) = \frac{P_n(e^x)}{Q_m(e^x)}$$

is periodic of period $2\pi i$, then the denominator vanishes at the points $\alpha_1, \alpha_2, \dots, \alpha_l$ on side I and also at the points $\alpha_1 + 2\pi i, \alpha_2 + 2\pi i, \dots, \alpha_l + 2\pi i$ on side III of the path, C, shown in Fig 6.7(a). We bypass these points along semicircles, $\gamma_1, \gamma_2, \dots, \gamma_l$, and $\widetilde{\gamma}_1, \widetilde{\gamma}_2, \dots, \widetilde{\gamma}_l$, of radius δ on sides I and III, respectively.

By the residue theorem, the value of the integral along C is

$$\left(\int_I + \int_{II} + \int_{III} + \int_{IV}\right) \left[\frac{P_n(e^z)}{Q_m(e^z)}\, e^{az}\right] dz = 2\pi i \sum_k \operatorname*{Res}_{z=z_k} \left[\frac{P_n(e^z)}{Q_m(e^z)}\, e^{az}\right], \tag{6.3.9}$$

where $0 < \Im z_k < 2\pi$.

As in Subsection 6.3.1, the integrals along sides II and IV approach zero as $R \to \infty$ if $-k < \Re a < m - n$, provided (6.3.2) holds.

On side I we have

$$\int_I \frac{P_n(e^z)}{Q_m(e^z)}\, e^{az}\, dz = \int_{-R}^{R} \frac{P_n(e^x)}{Q_m(e^x)}\, e^{ax}\, dx + \sum_{k=1}^{l} \int_{\gamma_k} \frac{P_n(e^z)}{Q_m(e^z)}\, e^{az}\, dz, \tag{6.3.10}$$

where the first integral on the right-hand side is evaluated along the straight line segments shown in Fig 6.7(a). On the arc γ_k, enlarged in Fig 6.7(b), we have

$$z-\alpha_k = \delta\, e^{i\theta}, \qquad \theta|_A = \arg(z-\alpha_k)|_A = \pi, \qquad \theta|_B = \arg(z-\alpha_k)|_B = 0,$$

because the arc γ_k is taken clockwise from A to B.

Consider the Laurent series expansion of the integrand in a neighborhood of the simple pole $z = \alpha_k$,

$$\frac{P_n(e^z)}{Q_m(e^z)}\, e^{az} = \frac{c_{-1}}{z-\alpha_k} + \sum_{m=0}^{\infty} c_m (z-\alpha_k)^m.$$

Then

$$\begin{aligned}\int_{\gamma_k} \frac{P_n(e^z)}{Q_m(e^z)}\, e^{az}\, dz &= \int_{\gamma_k} \frac{c_{-1}}{z-\alpha_k}\, dz + \sum_{m=0}^{\infty} c_m \int_{\gamma_k} (z-\alpha_k)^m\, dz \\ &= c_{-1} \int_{\pi}^{0} \frac{\delta\, e^{i\theta} i\, d\theta}{\delta\, e^{i\theta}} + \sum_{m=0}^{\infty} c_m \int_{\pi}^{0} \left(\delta e^{i\theta}\right)^m \delta\, e^{i\theta} i\, d\theta \\ &\to -c_{-1}\pi i + 0, \qquad \text{as } \delta \to 0,\end{aligned}$$

where

$$c_{-1} = \operatorname*{Res}_{z=\alpha_k} \left[\frac{P_n(e^z)}{Q_m(e^z)}\, e^{az} \right].$$

Therefore, it follows from (6.3.10), by taking the limits as $\delta \to 0$ and $R \to \infty$, that

$$\begin{aligned}\int_I \frac{P_n(e^z)}{Q_m(e^z)}\, e^{az}\, dz &= \int_{-\infty}^{\infty} \frac{P_n(e^x)}{Q_m(e^x)}\, e^{ax}\, dx \\ &\quad - \pi i \sum_{k=1}^{l} \operatorname*{Res}_{z=\alpha_k} \left[\frac{P_n(e^z)}{Q_m(e^z)}\, e^{az} \right]. \qquad (6.3.11)\end{aligned}$$

On side III, we have

$$\begin{aligned}\int_{III} \frac{P_n(e^z)}{Q_m(e^z)}\, e^{az}\, dz &= \int_{R}^{-R} \frac{P_n(e^x)}{Q_m(e^x)}\, e^{a(x+2\pi i)}\, dx \\ &\quad + \sum_{k=1}^{l} \int_{\widetilde{\gamma}_k} \frac{P_n(e^z)}{Q_m(e^z)}\, e^{az}\, dz. \qquad (6.3.12)\end{aligned}$$

The first term on the right-hand side is evaluated along the part of the segment $[-R, R]$, excluding the arcs $\widetilde{\gamma}_k$.

On the arcs $\widetilde{\gamma}_k$,

$$z - (\alpha_k + 2\pi i) = \delta\, e^{i\theta}.$$

Thus

$$\theta|_{\widetilde{B}} = \arg(z - \alpha_k - 2\pi i)|_{\widetilde{B}} = 0, \qquad \theta|_{\widetilde{A}} = \arg(z - \alpha_k - 2\pi i)|_{\widetilde{A}} = -\pi,$$

because the arc $\widetilde{\gamma}_k$ is taken clockwise from $\widetilde{B}$ to $\widetilde{A}$.

Consider the Laurent series expansion of the integrand in a neighborhood of the point $\alpha_k + 2\pi i$:

$$\frac{P_n(e^z)}{Q_m(e^z)} e^{az} = \frac{\tilde{c}_{-1}}{z - \alpha_k - 2\pi i} + \sum_{k=0}^{\infty} \tilde{c}_m (z - \alpha_k - 2\pi i)^m.$$

Then

$$\begin{aligned}\int_{\widetilde{\gamma}_k} \frac{P_n(e^z)}{Q_m(e^z)} e^{az}\, dz &= \tilde{c}_{-1} \int_0^{-\pi} \frac{\delta\, e^{i\theta}}{\delta\, e^{i\theta}}\, i\, d\theta + \sum_{m=0}^{\infty} \int_{-\pi}^{0} \tilde{c}_m \left(\delta\, e^{i\theta}\right)^m \delta\, e^{i\theta}\, i\, d\theta \\ &\to -\pi i \tilde{c}_{-1}, \qquad \text{as} \quad \delta \to 0,\end{aligned}$$

where

$$\tilde{c}_{-1} = \operatorname*{Res}_{z=\alpha_k+2\pi i} \left[\frac{P_n(e^z)}{Q_m(e^z)} e^{az}\right] = e^{2\pi a i} \operatorname*{Res}_{z=\alpha_k} \left[\frac{P_n(e^z)}{Q_m(e^z)} e^{az}\right].$$

Therefore, it follows from (6.3.12), as $\delta \to 0$ and $R \to \infty$, that

$$\begin{aligned}\int_{III} \frac{P_n(e^z)}{Q_m(e^z)} e^{az}\, dz = -e^{2\pi a i} \int_{-\infty}^{\infty} \frac{P_n(e^x)}{Q_m(e^x)} e^{ax}\, dx \\ - \pi i e^{2\pi a i} \sum_{k=1}^{l} \operatorname*{Res}_{z=\alpha_k} \left[\frac{P_n(e^z)}{Q_m(e^z)} e^{az}\right]. \qquad (6.3.13)\end{aligned}$$

Considering the limit of (6.3.9) as $\delta \to 0$ and $R \to \infty$ and using formulae (6.3.11) and (6.3.13), we obtain

$$\begin{aligned}\left(1 - e^{2\pi a i}\right) \int_{-\infty}^{\infty} \frac{P_n(e^x)}{Q_m(e^x)} e^{ax}\, dx - \pi i \left(1 + e^{2\pi a i}\right) \sum_{k=1}^{l} \operatorname*{Res}_{z=\alpha_k} \left[\frac{P_n(e^z)}{Q_m(e^z)} e^{az}\right] \\ = 2\pi i \sum_k \operatorname*{Res}_{z=z_k} \left[\frac{P_n(e^z)}{Q_m(e^z)} e^{az}\right].\end{aligned}$$

It follows from the last formula that

$$\begin{aligned}\text{p. v.} \int_{-\infty}^{\infty} \frac{P_n(e^x)}{Q_m(e^x)} e^{ax}\, dx = \frac{2\pi i}{1 - e^{2\pi a i}} \sum_k \operatorname*{Res}_{z=z_k} \left[\frac{P_n(e^z)}{Q_m(e^z)} e^{az}\right] \\ + \pi i \frac{1 + e^{2\pi a i}}{1 - e^{2\pi a i}} \sum_{k=1}^{l} \operatorname*{Res}_{z=\alpha_k} \left[\frac{P_n(e^z)}{Q_m(e^z)} e^{az}\right], \qquad (6.3.14)\end{aligned}$$

where α_k are poles of order 1 and $0 < \Im z_k < 2\pi$.

As in Sections 6.1 and 6.2, the real poles α_k in (6.3.14) may be of any odd order (if the Laurent series expansion of the integrand in a neighborhood of the points α_k does not contain any even negative powers of $(z - \alpha_k)$).

EXAMPLE 6.3.1. *Evaluate the integral*

$$I_4 = \int_{-\infty}^{\infty} \frac{e^{\alpha x}}{e^x + 1}\, dx, \qquad 0 < \alpha < 1. \tag{6.3.15}$$

SOLUTION. The function

$$f(z) = \frac{1}{e^z + 1}$$

has period $2\pi i$, and one can check that for $0 < \alpha < 1$ (or $0 < \Re\alpha < 1$) the integral I_4 converges. Since $e^x + 1$ has no real zeros, the singular points of $f(z)$ are the roots of the equation

$$e^z = -1 = e^{\pi i},$$

that is,

$$z_k = \pi i + 2k\pi i, \qquad \text{where } k = 0, \pm 1, \pm 2, \dots .$$

The only value of z_k which lies in the strip $0 < \Im z < 2\pi$ is $z = z_0 = e^{\pi i}$. Hence using formula (6.3.7) we obtain

$$\begin{aligned} I_4 &= \frac{2\pi i}{1 - e^{2\pi\alpha i}} \operatorname*{Res}_{z=\pi i} \left[\frac{e^{\alpha z}}{e^z + 1}\right] \\ &= \frac{2\pi i}{1 - e^{2\pi\alpha i}} \frac{e^{\alpha\pi i}}{e^{\pi i}} = \frac{2\pi i}{e^{\pi\alpha i} - e^{-\pi\alpha i}} \\ &= \frac{\pi}{\sin \pi\alpha}. \quad \square \end{aligned}$$

EXAMPLE 6.3.2. *Evaluate the integral*

$$I_5 = \int_{-\infty}^{\infty} \frac{e^{ax}}{1 + e^x + e^{2x}}\, dx, \qquad 0 < \Re a < 2. \tag{6.3.16}$$

SOLUTION. The integral I_5 is convergent for $0 < \Re a < 2$ because the integrand $f(x)$ is like $e^{-(2-a)x} \to 0$ as $x \to \infty$, since $\Re(2 - a) > 0$, and like $e^{ax} \to 0$ as $x \to -\infty$, since $\Re a > 0$. The poles of the function

$$f(z) = \frac{1}{1 + e^z + e^{2z}}$$

are the zeros of the equation

$$e^{2z} + e^z + 1 = 0,$$

that is,

$$e^z = -\frac{1}{2} \pm i\,\frac{\sqrt{3}}{2} = \begin{cases} e^{i\theta_1}, \\ e^{i\theta_2}, \end{cases} \tag{6.3.17}$$

where

$$\tan\theta_1 = -\sqrt{3} \implies \theta_1 = \frac{2\pi}{3}, \qquad \tan\theta_2 = \sqrt{3} \implies \theta_2 = \frac{4\pi}{3}.$$

Thus, there are two simple poles,

$$z_1 = 2\pi i/3 \quad \text{and} \quad z_2 = 4\pi i/3,$$

in the strip $0 < \Im z < 2\pi$. Therefore, using (6.3.7), we have

$$\begin{aligned} I_5 &= \frac{2\pi i}{1 - e^{2\pi a i}} \left(\operatorname*{Res}_{z=2\pi i/3} + \operatorname*{Res}_{z=4\pi i/3} \right) \left[\frac{e^{az}}{1 + e^z + e^{2z}} \right] \\ &= \frac{2\pi i}{1 - e^{2\pi a i}} \left[\frac{e^{2\pi a i/3}}{2e^{4\pi i/3} + e^{2\pi i/3}} + \frac{e^{4\pi a i/3}}{2e^{2\pi i/3} + e^{4\pi i/3}} \right] \\ &= -\frac{\pi}{\sin\pi a} \left[\frac{e^{-\pi a i/3}}{-1 - i\sqrt{3} - 1/2 + i\sqrt{3}/2} + \frac{e^{\pi a i/3}}{-1 + i\sqrt{3} - 1/2 - i\sqrt{3}/2} \right], \end{aligned}$$

where the last term is obtained by multiplying the numerator and denominator of the previous term by $e^{-\pi a i}$.

Finally, noting that the second fraction inside the square brackets is the complex conjugate of the first one, we obtain

$$\begin{aligned} I_5 &= \frac{\pi}{\sin\pi a} 2\Re \left[\frac{e^{-\pi a i/3}}{3/2 + i\sqrt{3}/2} \right] \\ &= \frac{4\pi}{12\sin\pi a} \Re \left[\left(\cos\frac{\pi a}{3} - i\sin\frac{\pi a}{3} \right) \left(3 - i\sqrt{3} \right) \right] \\ &= \frac{\pi}{3\sin\pi a} \left[3\cos\frac{\pi a}{3} - \sqrt{3}\sin\frac{\pi a}{3} \right] \\ &= \frac{2\pi}{\sqrt{3}\,\sin\pi a} \left[\frac{\sqrt{3}}{2}\cos\frac{\pi a}{3} - \frac{1}{2}\sin\frac{\pi a}{3} \right] \\ &= \frac{2\pi}{\sqrt{3}\,\sin\pi a} \left[\sin\frac{\pi}{3}\cos\frac{\pi a}{3} - \cos\frac{\pi}{3}\sin\frac{\pi a}{3} \right] \\ &= \frac{2\pi}{\sqrt{3}\,\sin\pi a} \sin\frac{\pi(1-a)}{3}. \quad \square \end{aligned}$$

EXAMPLE 6.3.3. *Evaluate the integral*

$$I_6 = \int_{-\infty}^{\infty} \frac{\cos ax}{\cosh x}\,dx = \Re\left[\int_{-\infty}^{\infty} \frac{e^{iax}}{\cosh x}\,dx \right], \qquad |\Im a| < 1. \tag{6.3.18}$$

SOLUTION. If $a = \alpha \pm i$, the integrand

$$\frac{\cos(\alpha + i)x}{\cosh x} = \frac{\cos \alpha x \cosh x - i \sin \alpha x \sinh x}{\cosh x}$$

does not approach zero as $x \to \infty$ and I_6 diverges; but it converges if $|\Im a| < 1$. The zeros of $\cosh z$ are

$$z_k = (2k+1)\frac{\pi}{2}i, \qquad k = 0, \pm 1, \pm 2, \dots .$$

The two points $z_0 = \pi i/2$ and $z_1 = 3\pi i/2$ are located in the horizontal strip $0 \le \Im z \le 2\pi$, and these are poles of order 1. Hence, using formula (6.3.7) we obtain

$$\begin{aligned} I &= \Re\left\{\frac{2\pi i}{1 - e^{-2\pi a}}\left(\operatorname*{Res}_{z=\pi i/2} + \operatorname*{Res}_{z=3\pi i/2}\right)\left[\frac{e^{iaz}}{\cosh z}\right]\right\} \\ &= \Re\left\{\frac{2\pi i}{1 - e^{-2\pi a}}\left[\frac{e^{-\pi a/2}}{i} + \frac{e^{-3\pi a/2}}{-i}\right]\right\} \\ &= \frac{2\pi\, e^{-\pi a/2}\left(1 - e^{-\pi a}\right)}{\left(1 - e^{-\pi a}\right)\left(1 + e^{-\pi a}\right)} \\ &= \frac{2\pi\, e^{-\pi a/2}}{1 + e^{-\pi a}} = \frac{2\pi}{e^{\pi a/2} + e^{-\pi a/2}} \\ &= \frac{\pi}{\cosh(\pi a/2)}. \quad \square \end{aligned}$$

EXAMPLE 6.3.4. *Evaluate the integral*

$$I = \text{p. v.} \int_{-\infty}^{\infty} \frac{e^{ax}}{e^{2x} - 1}\, dx, \qquad 0 < \Re a < 2. \tag{6.3.19}$$

SOLUTION. If we write the function $e^{2z} - 1$ in the form $e^{2z} = e^{2k\pi i}$, we see that its zeros are $z_k = k\pi i$ and are all simple. Hence in the strip $0 \le \Im z < 2\pi$, the simple poles of the integrand are $z_0 = 0$ and $z_1 = \pi i$. Moreover, z_0 is a real zero. Therefore, by formula (6.3.14), we have

$$\begin{aligned} I &= \frac{2\pi i}{1 - e^{2\pi a i}} \operatorname*{Res}_{z=\pi i}\left[\frac{e^{az}}{e^{2z} - 1}\right] + \pi i\, \frac{1 + e^{2\pi a i}}{1 - e^{2\pi a i}} \operatorname*{Res}_{z=0}\left[\frac{e^{az}}{e^{2z} - 1}\right] \\ &= \frac{2\pi i}{1 - e^{2\pi a i}}\, \frac{e^{\pi a i}}{2e^{2\pi i}} + \pi i\, \frac{1 + e^{2\pi a i}}{2\left(1 - e^{2\pi a i}\right)} \\ &= -\frac{\pi i}{e^{\pi a i} - e^{-\pi a i}} + \pi i\, \frac{e^{-\pi a i} + e^{\pi a i}}{2\left(e^{-\pi a i} - e^{\pi a i}\right)} \\ &= -\frac{\pi}{2 \sin \pi a} - \pi\, \frac{\cos \pi a}{2 \sin \pi a} \end{aligned}$$

$$= -\frac{\pi}{2}\,\frac{2\cos^2(\pi a/2)}{2\sin(\pi a/2)\cos(\pi a/2)}$$
$$= -\frac{\pi}{2}\cot\frac{\pi a}{2}. \quad \square$$

6.4. Rational functions times a power of x

We consider integrals of the form

$$\int_0^\infty \frac{P_n(x)}{Q_m(x)}\,x^{\alpha-1}\,dx, \qquad \alpha \in \mathbb{R}\setminus\mathbb{Z}. \tag{6.4.1}$$

These integrals can be reduced to integrals considered in Section 6.3 by the substitution $x = e^t$. But they can also be reduced directly to line integrals.

Consider first the conditions of convergence of the integral (6.4.1). Assuming that $P_n(0) \neq 0$ and $Q_m(0) \neq 0$, we see that $\alpha > 0$ is a necessary condition for its convergence as $x \to 0$. A necessary condition for its convergence as $x \to \infty$ is that $m - n - \alpha > 0$. Hence the integral (6.4.1) converges if

$$0 < \alpha < m - n. \tag{6.4.2}$$

As in the previous sections we consider two cases:

(a) $Q_m(x) \neq 0$ for $x > 0$,

(b) $Q_m(x)$ has positive zeros of order 1 at the points $x = \alpha_1, \alpha_2, \dots, \alpha_l$, distinct from zero.

6.4.1. The case of no real poles. Consider the case $Q_m(x) \neq 0$ for $x > 0$. We take a closed path C (see Fig 6.8) consisting of the segments of a lower and an upper cut along the positive x-axis and the circles C_R and C_δ of radii R and δ, respectively, and centers at the origin.

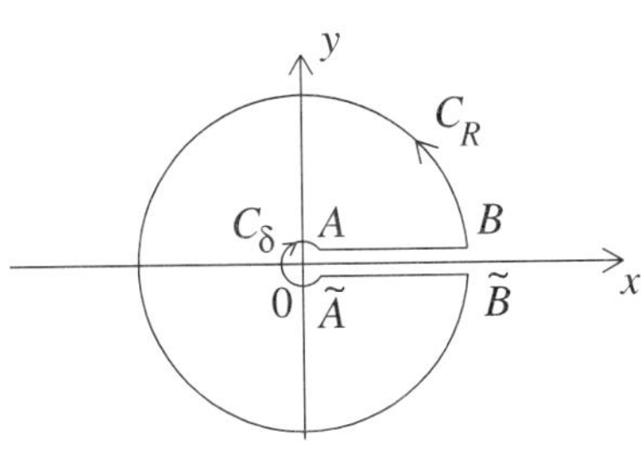

FIGURE 6.8. The path of integration in Subsection 6.4.1

The function of a complex variable

$$f(z) = \frac{P_n(z)}{Q_m(z)}\, z^{\alpha-1}, \tag{6.4.3}$$

where $z^{\alpha-1} = e^{(\alpha-1)\log z}$, is single-valued and analytic in the region bounded by C, except at the poles. By the residue theorem,

$$\oint_C \frac{P_n(z)}{Q_m(z)}\, z^{\alpha-1}\, dz = 2\pi i \sum_k \operatorname*{Res}_{z=z_k} \left[\frac{P_n(z)}{Q_m(z)}\, z^{\alpha-1}\right],$$

that is,

$$\left(\int_{AB} + \int_{C_R} + \int_{\widetilde{B}\widetilde{A}} + \int_{C_\delta}\right) \left[\frac{P_n(z)}{Q_m(z)}\, z^{\alpha-1}\, dz\right] = 2\pi i \sum_k \operatorname*{Res}_{z=z_k} \left[\frac{P_n(z)}{Q_m(z)}\, z^{\alpha-1}\right]. \tag{6.4.4}$$

On the segment AB, $z = x$, and we have

$$\int_{AB} = \int_\delta^R \frac{P_n(x)}{Q_m(x)}\, x^{\alpha-1}\, dx \quad \to \quad \int_0^\infty \frac{P_n(x)}{Q_m(x)}\, x^{\alpha-1}\, dx \tag{6.4.5}$$

as $R \to \infty$, $\delta \to 0$.

On the circle C_R,

$$z = Re^{i\theta}, \qquad 0 \le \theta \le 2\pi,$$

and we have

$$\int_{C_R} = \int_0^{2\pi} \frac{P_n\left(R\, e^{i\theta}\right)}{Q_m\left(R\, e^{i\theta}\right)} \left(R\, e^{i\theta}\right)^{\alpha-1} R\, e^{i\theta}\, i\, d\theta \to 0 \tag{6.4.6}$$

as $R \to \infty$, because, by (6.4.2), $m - n - \alpha > 0$.

On the segment $\widetilde{B}\widetilde{A}$,

$$z = x\, e^{2\pi i}, \qquad z^{\alpha-1} = \left(x\, e^{2\pi i}\right)^{\alpha-1} = e^{2\pi\alpha i}\, x^{\alpha-1},$$

and we have

$$\int_{\widetilde{B}\widetilde{A}} = e^{2\pi\alpha i} \int_R^\delta \frac{P_n(x)}{Q_m(x)}\, x^{\alpha-1}\, dx \to e^{2\pi\alpha i} \int_\infty^0 \frac{P_n(x)}{Q_m(x)}\, x^{\alpha-1}\, dx, \tag{6.4.7}$$

as $\delta \to 0$ and $R \to \infty$.

On the circle C_δ, we have $z = \delta\, e^{i\theta}$; thus

$$\int_{C_\delta} = \int_{2\pi}^0 \frac{P_n\left(\delta\, e^{i\theta}\right)}{Q_m\left(\delta\, e^{i\theta}\right)} \left(\delta\, e^{i\theta}\right)^{\alpha-1} \delta\, e^{i\theta} i\, d\theta \to 0 \tag{6.4.8}$$

as $\delta \to 0$ because $P_n(0) \neq 0$, $Q_m(0) \neq 0$ and $\alpha > 0$.

Hence, considering the limit as $\delta \to 0$ and $R \to \infty$ in (6.4.4) and using (6.4.5)–(6.4.8), we obtain

$$\left(1 - e^{2\pi\alpha i}\right) \int_0^\infty \frac{P_n(x)}{Q_m(x)}\, x^{\alpha-1}\, dx = 2\pi i \sum_k \operatorname*{Res}_{z=z_k} \left[\frac{P_n(z)}{Q_m(z)}\, z^{\alpha-1}\right], \tag{6.4.9}$$

where the residues are evaluated at all the poles z_k in the complex plane (we recall that the cut contains no singular points since $Q_m(x) \neq 0$ there). It follows from (6.4.9) that

$$\int_0^\infty \frac{P_n(x)}{Q_m(x)}\, x^{\alpha-1}\, dx = \frac{2\pi i}{1 - e^{2\pi\alpha i}} \sum_k \operatorname*{Res}_{z=z_k} \left[\frac{P_n(z)}{Q_m(z)}\, z^{\alpha-1}\right], \tag{6.4.10}$$

provided

$$Q_m(0)P_n(0) \neq 0, \qquad 0 < \alpha < m - n, \qquad Q_m(x) \neq 0 \quad \text{for} \quad x > 0.$$

EXAMPLE 6.4.1. *Evaluate the integral*

$$I_7 = \int_0^\infty \frac{x^{\alpha-1}}{x+1}\, dx, \qquad 0 < \alpha < 1. \tag{6.4.11}$$

SOLUTION. The conditions $P_n(x) = 1$ and $Q_m(x) = x+1$ are such that (6.4.10) is true, and hence the value of I_7 is found by evaluating the residue of $z^{\alpha-1}/(z+1)$ at the only pole $z = -1 = e^{\pi i}$ (we take $0 \leq \arg z \leq 2\pi$). Thus

$$\begin{aligned} I_7 &= \frac{2\pi i}{1 - e^{2\pi\alpha i}} \operatorname*{Res}_{z=-1} \left[\frac{z^{\alpha-1}}{z+1}\right] \\ &= \frac{2\pi i}{1 - e^{2\pi\alpha i}} \frac{\left(e^{\pi i}\right)^{\alpha-1}}{1} = 2\pi i\, \frac{e^{\pi\alpha i}}{e^{2\pi\alpha i} - 1} \\ &= 2\pi i\, \frac{1}{e^{\pi\alpha i} - e^{-\pi\alpha i}} = 2\pi i\, \frac{1}{2i \sin \pi\alpha} \\ &= \frac{\pi}{\sin \pi\alpha}. \quad \square \end{aligned}$$

6.4.2. The case of real poles. We suppose that the l strictly positive real numbers, $\alpha_1 < \alpha_2 < \cdots < \alpha_l$, are simple zeros of $Q_m(x)$.

We replace the path shown in Fig 6.8 by a closed path where the singular points, $\alpha_1, \alpha_2, \ldots, \alpha_l$, are bypassed along the semicircles, $\gamma_1, \gamma_2, \ldots, \gamma_l$, of radius δ on the upper part of the cut and similarly along the semicircles, $\widetilde{\gamma}_1, \widetilde{\gamma}_2, \ldots, \widetilde{\gamma}_l$, of radius δ on the lower part of the cut (see Fig 6.9(a)).

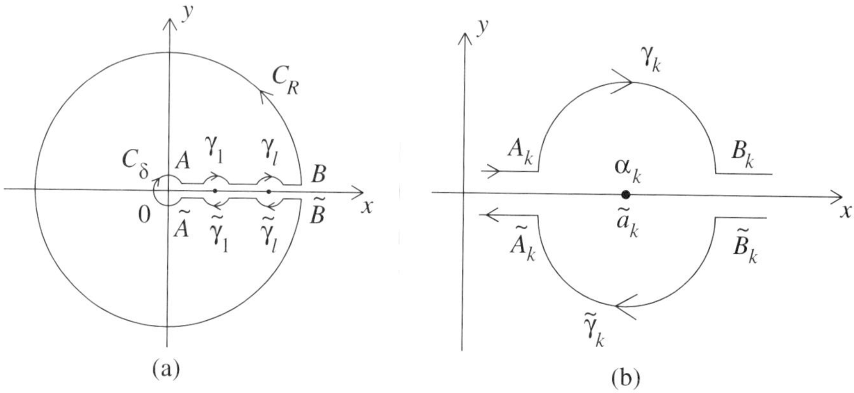

FIGURE 6.9. The path of integration and the points $\tilde{a}_k = \alpha_k e^{2\pi i}$, $k = 1, 2, \dots, l$, in Subsection 6.4.2.

By the residue theorem,

$$\left(\int_{AB} + \int_{C_R} + \int_{\widetilde{B}\widetilde{A}} + \int_{C_\delta} \right) \frac{P_n(z)}{Q_m(z)} z^{\alpha-1} \, dz = 2\pi i \sum_k \operatorname*{Res}_{z=z_k} \left[\frac{P_n(z)}{Q_m(z)} z^{\alpha-1} \right]. \tag{6.4.12}$$

As in Subsection 6.4.1, the integrals along the circles C_R and C_δ approach zero as $R \to \infty$ and $\delta \to 0$.

On the curve AB, we have

$$\int_{AB} = \int_\delta^R \frac{P_n(x)}{Q_m(x)} x^{\alpha-1} \, dx + \sum_{k=1}^{l} \int_{\gamma_k} \frac{P_n(z)}{Q_m(z)} z^{\alpha-1} \, dz. \tag{6.4.13}$$

The first integral on the right-hand side of (6.4.13) is evaluated along the straight line segments on $[\delta, R]$, excluding the arcs γ_k. On the arc γ_k taken clockwise, we have $z - \alpha_k = \delta \, e^{i\theta}$ where

$$\theta_{A_k} = \arg(z - \alpha_k)|_{A_k} = \pi, \qquad \theta|_{B_k} = \arg(z - \alpha_k)|_{B_k} = 0.$$

Expanding the integrand in a Laurent series in a neighborhood of the simple pole $z = \alpha_k$, we have

$$\frac{P_n(z)}{Q_m(z)} z^{\alpha-1} = \frac{c_{-1}}{z - \alpha_k} + \sum_{\mu=0}^{\infty} c_\mu (z - \alpha_k)^\mu.$$

Thus,

$$\int_{\gamma_k} \frac{P_n(z)}{Q_m(z)} z^{\alpha-1}\,dz = \int_{\gamma_k} \frac{c_{-1}}{z-\alpha_k}\,dz + \sum_{\mu=0}^{\infty} c_\mu \int_{\gamma_k} (z-\alpha_k)^\mu\,dz$$
$$= c_{-1}\int_\pi^0 \frac{\delta\,e^{i\theta}}{\delta\,e^{i\theta}}\,i\,d\theta + \sum_{\mu=0}^{\infty} c_\mu \int_\pi^0 \left(\delta\,e^{i\theta}\right)^\mu \delta\,e^{i\theta}\,i\,d\theta$$
$$\to -c_{-1}\pi i + 0, \qquad \text{as} \quad \delta \to 0,$$

where

$$c_{-1} = \operatorname*{Res}_{z=\alpha_k}\left[\frac{P_n(z)}{Q_m(z)} z^{\alpha-1}\right].$$

It follows from (6.4.13) that

$$\int_{AB} = \int_0^\infty \frac{P_n(x)}{Q_m(x)} x^{\alpha-1}\,dx - \pi i \sum_{k=1}^{l} \operatorname*{Res}_{z=\alpha_k}\left[\frac{P_n(z)}{Q_m(z)} z^{\alpha-1}\right], \tag{6.4.14}$$

as $\delta \to 0$ and $R \to \infty$.

Similarly, on $\widetilde{B}\widetilde{A}$ we have

$$\int_{\widetilde{B}\widetilde{A}} = \int_R^\delta \frac{P_n(x)}{Q_m(x)} \left(x\,e^{2\pi i}\right)^{\alpha-1}\,dx + \sum_{k=1}^{l} \int_{\widetilde{\gamma}_k} \frac{P_n(z)}{Q_m(z)} z^{\alpha-1}\,dz. \tag{6.4.15}$$

The first integral on the right-hand side of (6.4.15) is evaluated along the straight line segments excluding the curves $\widetilde{\gamma}_k$. On the arc $\widetilde{\gamma}_k$ taken clockwise, with center $\tilde{a}_k = \alpha_k\,e^{2\pi i}$, we have $z - \tilde{a}_k = \delta\,e^{i\theta}$ where

$$\theta|_{\widetilde{B}_k} = \arg(z-\tilde{a}_k)|_{\widetilde{B}_k} = 2\pi, \qquad \theta|_{\widetilde{A}_k} = \arg(z-\tilde{a}_k)|_{\widetilde{A}_k} = \pi.$$

Expanding the integrand in a Laurent series in a neighborhood of the simple pole $z = \tilde{a}_k$, we have

$$\frac{P_n(z)}{Q_m(z)} z^{\alpha-1} = \frac{c_{-1}}{z-\tilde{a}_k} + \sum_{\mu=0}^{\infty} c_\mu (z-\tilde{a}_k)^\mu.$$

Thus,

$$\int_{\widetilde{\gamma}_k} \frac{P_n(z)}{Q_m(z)} z^{\alpha-1}\,dz = \int_{\widetilde{\gamma}_k} \frac{c_{-1}}{z-\tilde{a}_k}\,dz + \sum_{\mu=0}^{\infty} c_\mu \int_{\widetilde{\gamma}_k} (z-\tilde{a}_k)^\mu\,dz$$
$$= c_{-1}\int_{2\pi}^{\pi} \frac{\delta\,e^{i\theta} i\,d\theta}{\delta\,e^{i\theta}} + \sum_{\mu=0}^{\infty} c_\mu \int_{2\pi}^{\pi} \left(\delta\,e^{i\theta}\right)^\mu \delta\,e^{i\theta} i\,d\theta$$
$$\to -c_{-1}\pi i + 0, \qquad \text{as } \delta \to 0,$$

where, with $z = \xi\, e^{2\pi i}$,

$$\begin{aligned} c_{-1} &= \operatorname*{Res}_{z=\alpha_k\, e^{2\pi i}} \left[\frac{P_n(z)}{Q_m(z)}\, z^{\alpha-1} \right] \\ &= \operatorname*{Res}_{\xi=\alpha_k} \left[\frac{P_n(\xi)}{Q_m(\xi)}\, \left(\xi\, e^{2\pi i}\right)^{\alpha-1} \right] \\ &= e^{2\pi\alpha i} \operatorname*{Res}_{z=\alpha_k} \left[\frac{P_n(z)}{Q_m(z)}\, z^{\alpha-1} \right]. \end{aligned} \tag{6.4.16}$$

Therefore, it follows from (6.4.15) that

$$\begin{aligned} \int_{\widetilde{B}\widetilde{A}} = e^{2\pi\alpha i} \int_{\infty}^{0} \frac{P_n(x)}{Q_m(x)}\, x^{\alpha-1}\, dx \\ - \pi i\, e^{2\pi\alpha i} \sum_{k=1}^{l} \operatorname*{Res}_{z=\alpha_k} \left[\frac{P_n(z)}{Q_m(z)}\, z^{\alpha-1} \right], \end{aligned} \tag{6.4.17}$$

as $R \to \infty$ and $\delta \to 0$.

Hence, considering the limit in (6.4.12) as $R \to \infty$ and $\delta \to 0$, and using (6.4.13)–(6.4.17), we obtain

$$\begin{aligned} \left(1 - e^{2\pi\alpha i}\right) \text{p. v.} \int_0^{\infty} \frac{P_n(x)}{Q_m(x)}\, x^{\alpha-1}\, dx \\ - \pi i \left(1 + e^{2\pi\alpha i}\right) \sum_{k=1}^{l} \operatorname*{Res}_{z=\alpha_k} \left[\frac{P_n(z)}{Q_m(z)}\, z^{\alpha-1} \right] \\ = 2\pi i \sum_{k} \operatorname*{Res}_{z=z_k} \left[\frac{P_n(z)}{Q_m(z)}\, z^{\alpha-1} \right]. \end{aligned} \tag{6.4.18}$$

We thus obtain, from (6.4.18), the following formula for the evaluation of the integral:

$$\begin{aligned} \text{p. v.} \int_0^{\infty} \frac{P_n(x)}{Q_m(x)}\, x^{\alpha-1}\, dx = \frac{2\pi i}{1 - e^{2\pi\alpha i}} \sum_{k} \operatorname*{Res}_{z=z_k} \left[\frac{P_n(z)}{Q_m(z)}\, z^{\alpha-1} \right] \\ + \pi i\, \frac{1 + e^{2\pi\alpha i}}{1 - e^{2\pi\alpha i}} \sum_{k=1}^{l} \operatorname*{Res}_{z=\alpha_k} \left[\frac{P_n(z)}{Q_m(z)}\, z^{\alpha-1} \right], \end{aligned} \tag{6.4.19}$$

provided

$$z_k \notin (0, +\infty), \qquad \alpha_k > 0, \qquad 0 < \alpha < m - n.$$

Note 6.4.1. The authors have not seen formula (6.4.19) in the literature for the evaluation of the previous integral.

In Problem 28.21 of [**21**] one finds the formula

$$\text{p. v.} \int_{-\infty}^{\infty} R(x)\, x^{a-1}\, dx = \frac{\pi}{\sin \pi a} \sum_{k} \operatorname*{Res}_{z=z_k} \left[R(z)(-z)^{a-1}\right] + \pi \cot \pi a \sum_{k=1}^{l} \operatorname*{Res}_{z=a_k} \left[R(z)\, z^{a-1}\right], \qquad (6.4.20)$$

where $R(z)$ is a rational function such that

$$R(z) = O\left(z^{-p}\right) \quad \text{as} \quad z \to 0,$$
$$R(z) = O\left(z^{-q}\right) \quad \text{as} \quad z \to \infty,$$

and

$$p < \Re a < q, \qquad (-z)^{a-1} = e^{(a-1)[\log|z| + i \arg(-z)]}, \qquad -\pi < \arg(-z) < \pi.$$

From our point of view, formula (6.4.20) is less convenient for practical evaluations because it is not clear from the inequality $-\pi < \arg(-z) < \pi$ whether $\arg(-1)$ is equal to π or $-\pi$.

EXAMPLE 6.4.2. *Evaluate the integral*

$$I_8 = \text{p. v.} \int_{0}^{\infty} \frac{dx}{x^{\alpha}(x-\beta)}, \qquad 0 < \alpha < 1,\ \beta > 0. \qquad (6.4.21)$$

SOLUTION. The conditions $m = 1$, $n = 0$, $x^{-\alpha} = x^{-\alpha+1-1}$ are such that formula (6.4.19) is true. Moreover, since

$$0 < \alpha < 1 \implies 0 < -\alpha + 1 < 1 = m - n,$$

in evaluating I_8 it suffices to use formula (6.4.19) and evaluate the residue of the function $[z^{\alpha}(z-\beta)]^{-1}$ at $z = \beta$:

$$\begin{aligned} I_8 &= \pi i \frac{1 + e^{2\pi i(1-\alpha)}}{1 - e^{2\pi i(1-\alpha)}} \operatorname*{Res}_{z=\beta} \left[\frac{1}{z^{\alpha}(z-\beta)}\right] \\ &= \pi i \frac{1 + e^{-2\pi \alpha i}}{1 - e^{-2\pi \alpha i}} \frac{1}{\beta^{\alpha}} = \pi i \frac{e^{i\pi\alpha} + e^{-i\pi\alpha}}{e^{i\pi\alpha} - e^{-i\pi\alpha}} \frac{1}{\beta^{\alpha}} \\ &= \pi \beta^{-\alpha} \cot \pi\alpha. \quad \square \end{aligned}$$

EXAMPLE 6.4.3. *Evaluate the integral*

$$I_9 = \text{p. v.} \int_{0}^{\infty} \frac{\sin(\alpha \log x)}{x^2 - 1}\, dx. \qquad (6.4.22)$$

SOLUTION. We first transform I_9 in the form

$$\begin{aligned} I_9 &= \Im\left[\text{p. v.}\int_0^\infty \frac{e^{i\alpha\log x}}{x^2-1}\,dx\right] \\ &= \Im\left[\text{p. v.}\int_0^\infty \frac{x^{(i\alpha+1)-1}}{x^2-1}\,dx\right]. \end{aligned}$$

It can easily be checked that the conditions are such that formula (6.4.19) is true, and hence, by evaluating the residues of the function $z^{i\alpha}/(z^2-1)$ at $z=1$ and $z=-1$, we have

$$\begin{aligned} I_9 &= \Im\left[\frac{2\pi i}{1-e^{2\pi i(i\alpha+1)}}\frac{\left(e^{\pi i}\right)^{i\alpha}}{2(-1)} + \pi i\,\frac{1+e^{2\pi i(i\alpha+1)}}{1-e^{2\pi i(i\alpha+1)}}\frac{1}{2}\right] \\ &= \Im\left[-\frac{\pi i e^{-\pi\alpha}}{1-e^{-2\pi\alpha}} + \frac{1}{2}\pi i\,\frac{1+e^{-2\pi\alpha}}{1-e^{-2\pi\alpha}}\right] \\ &= \pi\,\frac{-2e^{-\pi\alpha}+1+e^{-2\pi\alpha}}{2\left(1-e^{-2\pi\alpha}\right)} \\ &= \pi\frac{\left(1-e^{-\pi\alpha}\right)^2}{2\left(1-e^{-\pi\alpha}\right)\left(1+e^{-\pi\alpha}\right)} \\ &= \frac{\pi}{2}\,\frac{1-e^{-\pi\alpha}}{1+e^{-\pi\alpha}} = \frac{\pi}{2}\,\frac{e^{\pi\alpha/2}-e^{-\pi\alpha/2}}{e^{\pi\alpha/2}+e^{-\pi\alpha/2}} \\ &= \frac{\pi}{2}\tanh\frac{\pi\alpha}{2}. \quad \square \end{aligned}$$

Exercises for Chapter 6

Evaluate the following integrals.

1. $\displaystyle\int_{-\infty}^{\infty} \frac{x^2}{x^4+1}\,dx.$

2. $\displaystyle\int_0^{\infty} \frac{x^2}{x^4+x^2+1}\,dx.$

3. $\displaystyle\int_0^{\infty} \frac{1-\cos x}{x^2}\,dx.$

4. $\displaystyle\int_0^{\infty} \frac{\sin^2 x}{x^2}\,dx.$

5. $\displaystyle\int_{-\infty}^{\infty} \frac{x^2}{(x^2+1)^2(x^2+2x+2)}\,dx.$

Verify the following formulae.

6. $\displaystyle\int_0^\infty \frac{\cos ax}{x^2+1}\,dx = \frac{\pi}{2}\,e^{-a}, \qquad a > 0.$

7. $\displaystyle\int_0^\infty \frac{dx}{(x^2+1)^2} = \frac{\pi}{4}.$

8. $\displaystyle\int_{-\infty}^\infty \frac{x-\sin x}{x^2}\,dx = \frac{\pi}{2}.$

9. $\displaystyle\int_0^\infty \frac{\cos ax}{(x^2+1)^2}\,dx = \frac{\pi(a+1)e^{-a}}{4}, \qquad a > 0.$

10. $\displaystyle\int_0^\infty \frac{x^{\alpha-1}}{1+x^2}\,dx = \frac{\pi}{2\sin\left(\frac{\pi\alpha}{2}\right)}, \qquad 0 < \alpha < 2.$

11. $\displaystyle\int_0^\infty \frac{dx}{(x^2+a^2)^2} = \frac{\pi}{4a^3}, \qquad a > 0.$

12. $\displaystyle\int_{-\infty}^\infty \frac{e^{ax}}{1+e^x}\,dx = \frac{\pi}{\sin a\pi}, \qquad 0 < a < 1.$

13. $\displaystyle\int_0^\infty \frac{x^{\alpha-1}}{1+x^{2n}}\,dx = \frac{\pi}{2n\sin\left(\frac{\pi\alpha}{2n}\right)}, \qquad n = 1, 2, 3, \ldots, \quad 0 < \alpha < 2n.$

CHAPTER 7

Intermediate Definite Integrals

7.1. Rational functions over $(0, +\infty)$

Let $P_n(x)$ and $Q_m(x)$ be polynomials of degrees n and m, respectively. We consider integrals of the form

$$\int_0^\infty \frac{P_n(x)}{Q_m(x)}\,dx, \tag{7.1.1}$$

where the rational function $P_n(x)/Q_m(x)$ is not even and $m \geq n+2$. These integrals can be evaluated by taking the limit in (6.4.10) or (6.4.18) as $\alpha \to 1$. But this procedure leads to an indefinite form $0/0$ and, in general, the limit cannot be easily found. On the other hand, these integrals can be evaluated directly by the theory of residues.

7.1.1. The case of no real poles. Suppose that $Q_m(x) \neq 0$ for $x > 0$. We consider the auxiliary function

$$f(z) = \frac{P_n(z)}{Q_m(z)}\,\mathrm{Log}\,z \tag{7.1.2}$$

and the closed path shown in Fig 6.8. By the residue theorem 5.2.2 we have

$$\begin{aligned}\left(\int_{AB} + \int_{C_R} + \int_{\widetilde{B}\widetilde{A}} + \int_{C_\delta}\right)&\left[\frac{P_n(z)}{Q_m(z)}\,\mathrm{Log}\,z\right] dz \\ &= 2\pi i \sum_k \operatorname*{Res}_{z=z_k}\left[\frac{P_n(z)}{Q_m(z)}\,\mathrm{Log}\,z\right]. \end{aligned}\tag{7.1.3}$$

As in Subsection 6.4.1, the integrals along the circles C_R and C_δ approach zero as $R \to \infty$ and $\delta \to 0$, respectively. Since $z = x$ on the segment AB, we have

$$\int_{AB} = \int_\delta^R \frac{P_n(x)}{Q_m(x)}\ln x\,dx \to \int_0^\infty \frac{P_n(x)}{Q_m(x)}\ln x\,dx, \tag{7.1.4}$$

as $R \to \infty$ and $\delta \to 0$. Since, on the segment $\widetilde{B}\widetilde{A}$,

$$z = x\, e^{2\pi i} \quad \text{and} \quad \operatorname{Log} z = \ln x + 2\pi i,$$

we have

$$\begin{aligned} \int_{\widetilde{B}\widetilde{A}} &= \int_R^\delta \frac{P_n(x)}{Q_m(x)} (\ln x + 2\pi i)\, dx \\ &\to \int_\infty^0 \frac{P_n(x)}{Q_m(x)} (\ln x + 2\pi i)\, dx, \end{aligned} \tag{7.1.5}$$

as $R \to \infty$ and $\delta \to 0$. Hence, by taking the limit in (7.1.3) as $R \to \infty$ and $\delta \to 0$, and using (7.1.4) and (7.1.5), we obtain

$$\begin{aligned} \int_0^\infty \frac{P_n(x)}{Q_m(x)} \ln x\, dx - \int_0^\infty \frac{P_n(x)}{Q_m(x)} (\ln x + 2\pi i)\, dx \\ = 2\pi i \sum_k \operatorname*{Res}_{z=z_k} \left[\frac{P_n(z)}{Q_m(z)} \operatorname{Log} z \right]. \end{aligned}$$

Thus, the formula

$$\int_0^\infty \frac{P_n(x)}{Q_m(x)}\, dx = -\sum_k \operatorname*{Res}_{z=z_k} \left[\frac{P_n(z)}{Q_m(z)} \operatorname{Log} z \right] \tag{7.1.6}$$

is valid if $m \geq n+2$ and $Q_m(x) \neq 0$ for $x > 0$.

Note 7.1.1. There is an interesting extension of formula (7.1.6). Consider the integral

$$I = \int_0^b \frac{P_n(x)}{Q_m(x)}\, dx, \tag{7.1.7}$$

where $m \geq n+2$ and $Q_m(x) \neq 0$ for $x > 0$. By the linear fractional transformation

$$t = \frac{x}{b-x}, \quad \text{that is} \quad x = \frac{b\,t}{t+1} \quad \text{and} \quad dx = \frac{b}{(t+1)^2}\, dt, \tag{7.1.8}$$

the previous integral becomes

$$\begin{aligned} I &= b \int_0^\infty \frac{P_n\big(bt/(t+1)\big)}{Q_m\big(bt/(t+1)\big)} \frac{dt}{(t+1)^2} \\ &= -b \sum_k \operatorname*{Res}_{z=z_k} \left[\frac{P_n\big(bz/(z+1)\big)}{Q_m\big(bz/(z+1)\big)} \frac{\operatorname{Log} z}{(z+1)^2} \right], \end{aligned} \tag{7.1.9}$$

provided

$$m \geq n+2 \quad \text{and} \quad Q_m\left(\frac{bx}{x+1}\right) \neq 0 \quad \text{for} \quad \frac{bx}{x+1} > 0.$$

Formula (7.1.7) can be considered as an indefinite integral of the function $P_n(x)/Q_m(x)$. Therefore indefinite integrals of rational functions can be evaluated by the theory of residues by means of (7.1.9).

EXAMPLE 7.1.1. *Use (7.1.9) to derive the formula*

$$\int_0^b \frac{dx}{1+x^2} = \arctan b. \tag{7.1.10}$$

SOLUTION. Using (7.1.8), we have

$$\begin{aligned} \int_0^b \frac{dx}{1+x^2} &= b\int_0^\infty \frac{1}{1+[bt/(t+1)]^2}\frac{dt}{(1+t)^2} \\ &= b\int_0^\infty \frac{dt}{(b^2+1)t^2+2t+1}. \end{aligned} \tag{7.1.11}$$

The singular points of the integrand are the zeros of the denominator,

$$t_{1,2} = \frac{-1 \pm \sqrt{1-(b^2+1)}}{b^2+1} = \frac{-1 \pm bi}{b^2+1}. \tag{7.1.12}$$

Hence, by (7.1.9) and (7.1.11), we have

$$\begin{aligned} \int_0^b \frac{dx}{1+x^2} &= -b\left(\operatorname*{Res}_{z=t_1} + \operatorname*{Res}_{z=t_2}\right)\left[\frac{\operatorname{Log} z}{(b^2+1)z^2+2z+1}\right] \\ &= -\frac{b}{2}\left[\frac{\operatorname{Log} t_1}{(b^2+1)t_1+1} + \frac{\operatorname{Log} t_2}{(b^2+1)t_2+1}\right] \\ &= -\frac{b}{2}\left[\frac{\operatorname{Log} t_1}{bi} - \frac{\operatorname{Log} t_2}{bi}\right] \\ &= -\frac{b}{2}\frac{1}{bi}\operatorname{Log}\frac{t_1}{t_2} \\ &= \frac{1}{2i}\operatorname{Log}\frac{1+bi}{1-bi} \\ &= \arctan b. \quad \square \end{aligned}$$

We may raise the following question: is it possible to evaluate by the theory of residues all the known types of indefinite integrals that can be evaluated in closed form? This is possible, at least, for the indefinite integrals that can be reduced to indefinite integrals of rational functions by a change of variable, that is,

(a) Integrals of the form $\int R(\sin x, \cos x)\,dx$, where $R(x,y)$ is a rational function of two variables.

(b) Integrals of the form $\int R\left(x, \sqrt{ax^2+bx+c}\right)\,dx$, that can be reduced to integral (7.1.7) by means of one of the three Euler's substitutions (see [**11**], Vol. 1, p. 190).

7.1.2. The case of positive real poles. Suppose that the real zeros of Q_m are positive and simple and are ordered as follows:

$$0 < \alpha_1 < \alpha_2 < \cdots < \alpha_l.$$

We consider the path shown in Fig 6.9 and the auxiliary function (7.1.2). The only difference between the present case and the one considered in the previous subsection is in the evaluation of the integrals along the semicircles γ_k and $\widetilde{\gamma}_k$ attached to the upper and lower parts, AB and $\widetilde{B}\widetilde{A}$, respectively, of the cut $[0,+\infty]$. Hence, the sums

$$\sum_{k=1}^{l}\int_{\gamma_k}\frac{P_n(z)}{Q_m(z)}\,\operatorname{Log} z\,dz, \qquad \sum_{k=1}^{l}\int_{\widetilde{\gamma}_k}\frac{P_n(z)}{Q_m(z)}\,\operatorname{Log} z\,dz, \tag{7.1.13}$$

are to be added to the integrals (7.1.4) and (7.1.5) along AB and $\widetilde{B}\widetilde{A}$, respectively. The limit of the integral along γ_k, as $\delta \to 0$, is

$$-c_{-1}\pi i = \int_{\gamma_k}\frac{P_n(z)}{Q_m(z)}\,\operatorname{Log} z\,dz$$

where

$$c_{-1} = \operatorname*{Res}_{z=\alpha_k}\left[\frac{P_n(z)}{Q_m(z)}\,\operatorname{Log} z\right]. \tag{7.1.14}$$

The limit of the integral along $\widetilde{\gamma}_k$, as $\delta \to 0$, is

$$-c_{-1}\pi i = \int_{\widetilde{\gamma}_k}\frac{P_n(z)}{Q_m(z)}\,\operatorname{Log} z\,dz,$$

where, by letting $z = \zeta\, e^{2\pi i}$, we have

$$\begin{aligned} c_{-1} &= \operatorname*{Res}_{z=\alpha_k\, e^{2\pi i}}\left[\frac{P_n(z)}{Q_m(z)}\,\operatorname{Log} z\right] \\ &= \operatorname*{Res}_{\zeta=\alpha_k}\left[\frac{P_n(\zeta)}{Q_m(\zeta)}\,\operatorname{Log}\left(\zeta\, e^{2\pi i}\right)\right] \\ &= \operatorname*{Res}_{z=\alpha_k}\left[\frac{P_n(z)}{Q_m(z)}\,\operatorname{Log} z\right] + 2\pi i\,\operatorname*{Res}_{z=\alpha_k}\left[\frac{P_n(z)}{Q_m(z)}\right]. \end{aligned} \tag{7.1.15}$$

Therefore, as $R \to \infty$ and $\delta \to 0$, the sum

$$-\pi i \sum_{k=1}^{l} \operatorname*{Res}_{z=\alpha_k} \left[\frac{P_n(z)}{Q_m(z)} \operatorname{Log} z \right]$$

is to be added to integral (7.1.4) along AB, while the sum

$$-\pi i \sum_{k=1}^{l} \left\{ \operatorname*{Res}_{z=\alpha_k} \left[\frac{P_n(z)}{Q_m(z)} \operatorname{Log} z \right] + 2\pi i \operatorname*{Res}_{z=\alpha_k} \left[\frac{P_n(z)}{Q_m(z)} \right] \right\}$$

is to be added to (7.1.5) along $\widetilde{B}\widetilde{A}$.

It then follows from (7.1.3), as $R \to \infty$ and $\delta \to 0$, that

$$\begin{aligned} \text{p. v.} \int_0^\infty \frac{P_n(x)}{Q_m(x)} \ln x \, dx &- \pi i \sum_{k=1}^{l} \operatorname*{Res}_{z=\alpha_k} \left[\frac{P_n(z)}{Q_m(z)} \operatorname{Log} z \right] \\ &- \text{p. v.} \int_0^\infty \frac{P_n(x)}{Q_m(x)} (\ln x + 2\pi i) \, dx \\ &- \pi i \sum_{k=1}^{l} \left\{ \operatorname*{Res}_{z=\alpha_k} \left[\frac{P_n(z)}{Q_m(z)} \operatorname{Log} z \right] + 2\pi i \operatorname*{Res}_{z=\alpha_k} \left[\frac{P_n(z)}{Q_m(z)} \right] \right\} \\ = 2\pi i \sum_{k} \operatorname*{Res}_{z=z_k} &\left[\frac{P_n(z)}{Q_m(z)} \operatorname{Log} z \right]. \end{aligned}$$

The formula for the evaluation of integral (7.1.1) over the positive real axis follows from this last relation,

$$\begin{aligned} \text{p. v.} \int_0^\infty \frac{P_n(x)}{Q_m(x)} \, dx &= -\sum_{k} \operatorname*{Res}_{z=z_k} \left[\frac{P_n(z)}{Q_m(z)} \operatorname{Log} z \right] \\ &- \sum_{k=1}^{l} \operatorname*{Res}_{z=\alpha_k} \left[\frac{P_n(z)}{Q_m(z)} \operatorname{Log} z \right] - \pi i \sum_{k=1}^{l} \operatorname*{Res}_{z=\alpha_k} \left[\frac{P_n(z)}{Q_m(z)} \right], \end{aligned} \tag{7.1.16}$$

where

$$\alpha_k > 0, \qquad \Im z_k \neq 0 \text{ if } \Re z_k > 0, \qquad m \geq n + 2, \qquad 0 \leq \arg z < 2\pi.$$

To the authors' knowledge, formula (7.1.16) is not found explicitly in the literature; however it can be obtained by the recurrence relation (7.2.2) given in Problems 29.03 and 29.05 of [**21**] and derived in Subsection 7.2.1.

If all $\alpha_k = 0$, then (7.1.16) reduces to (7.1.6), which is given in [**21**], Problem 29.01.

EXAMPLE 7.1.2. *Evaluate the integral*

$$I = \text{p. v.} \int_0^\infty \frac{dx}{(x-1)(x^2+1)}.$$

SOLUTION. Using formula (7.1.16), we have

$$\begin{aligned}
I &= -\left(\operatorname*{Res}_{z=i} + \operatorname*{Res}_{z=-i} + \operatorname*{Res}_{z=1}\right)\left[\frac{\operatorname{Log} z}{(z-1)(z^2+1)}\right] - \pi i \operatorname*{Res}_{z=1}\left[\frac{1}{(z-1)(z^2+1)}\right] \\
&= -\left[\frac{\operatorname{Log} i}{2i(i-1)} + \frac{\operatorname{Log}(-i)}{(-i-1)(-2i)}\right] - \pi i \frac{1}{1^2+1} \\
&= \left[\frac{\operatorname{Log}\left(e^{i\pi/2}\right)}{2i+2} + \frac{\operatorname{Log}\left(e^{i3\pi/2}\right)}{-2i+2}\right] - \frac{\pi i}{2} \\
&= \frac{1}{2}\left[\frac{1-i}{2}\, i\, \frac{\pi}{2} + \frac{1+i}{2}\, i\, \frac{3\pi}{2}\right] - \frac{\pi i}{2} \\
&= -\frac{\pi}{4}. \quad \square
\end{aligned}$$

Note 7.1.2. Since $0 \le \operatorname{Arg} z < 2\pi$ in (7.1.16), then in the previous example, $\operatorname{Arg}(-i) = 3\pi/2$ and not $-\pi/2$. Therefore

$$\frac{\operatorname{Log} i}{2i(i-1)} + \frac{\operatorname{Log}(-i)}{(-i-1)(-2i)} \neq 2\Re\left[\frac{\operatorname{Log} i}{2i(i-1)}\right],$$

although, at first glance, the second term on the left-hand side appears to be the complex conjugate of the first one.

7.2. Forms containing $(\ln x)^p$ in the numerator

We consider integrals of the form

$$I_p = \int_0^\infty \frac{P_n(x)}{Q_m(x)} (\ln x)^p \, dx, \tag{7.2.1}$$

where $P_n(x)$ and $Q_m(x)$ are real polynomials of degrees n and m, respectively, with $m \ge n+2$ and $p = 1, 2, \dots$. The case $p = 0$ was considered in the previous subsection.

7.2.1. $Q_m(x) \neq 0$ **for all** $x \ge 0$**.** Suppose that $Q_m(x) \neq 0$ for $x \ge 0$. We first prove the recurrence relation

$$\sum_{s=0}^{p-1} C_p^s (2\pi i)^{p-1-s} I_s = -\sum_k \operatorname*{Res}_{z=z_k}\left[\frac{P_n(z)}{Q_m(z)} (\operatorname{Log} z)^p\right], \tag{7.2.2}$$

where the numbers

$$C_p^s = \frac{p!}{s!(p-s)!}, \qquad C_p^0 = 1, \tag{7.2.3}$$

are the binomial coefficients, the branch cut of the logarithm is taken along the positive real axis (see Fig 6.8)

$$\operatorname{Log} z = \ln|z| + i \operatorname{Arg} z, \qquad 0 \le \operatorname{Arg} z < 2\pi, \tag{7.2.4}$$

the numbers z_k are the zeros of $Q_m(z)$, and $\Im(z_k) \neq 0$ if $\Re(z_k) \geq 0$.

Consider the auxiliary function

$$f(z) = \frac{P_n(z)}{Q_m(z)} (\operatorname{Log} z)^p \tag{7.2.5}$$

and the closed path shown in Fig 6.8. By the residue theorem 5.2.2 we have

$$\begin{aligned}\left(\int_{AB} + \int_{C_R} + \int_{\widetilde{B}\widetilde{A}} + \int_{C_\delta}\right) &\left[\frac{P_n(z)}{Q_m(z)} (\operatorname{Log} z)^p\right] dz \\ &= 2\pi i \sum_k \operatorname*{Res}_{z=z_k} \left[\frac{P_n(z)}{Q_m(z)} (\operatorname{Log} z)^p\right].\end{aligned} \tag{7.2.6}$$

As in the previous subsection, the integrals along the semicircles C_R and C_δ approach zero as $R \to \infty$ and $\delta \to 0$.

Since $z = x$ on AB, then

$$\begin{aligned}\int_{AB} \frac{P_n(z)}{Q_m(z)} (\operatorname{Log} z)^p \, dz &= \int_\delta^R \frac{P_n(x)}{Q_m(x)} (\ln x)^p \, dx \\ &\to \int_0^\infty \frac{P_n(x)}{Q_m(x)} (\ln x)^p \, dx\end{aligned} \tag{7.2.7}$$

as $R \to \infty$ and $\delta \to 0$.

Since

$$z = x\, e^{2\pi i} \quad \text{and} \quad \operatorname{Log} z = \ln x + 2\pi i$$

on $\widetilde{B}\widetilde{A}$, we have

$$\begin{aligned}\int_{\widetilde{B}\widetilde{A}} \frac{P_n(z)}{Q_m(z)} (\operatorname{Log} z)^p \, dz &= \int_R^\delta \frac{P_n(x)}{Q_m(x)} (\ln x + 2\pi i)^p \, dx \\ &\to -\int_0^\infty \frac{P_n(x)}{Q_m(x)} \sum_{s=0}^{p} C_p^s (\ln x)^s (2\pi i)^{p-s} \, dx.\end{aligned} \tag{7.2.8}$$

Therefore, taking the limit in (7.2.6) as $R \to \infty$ and $\delta \to 0$ and using (7.2.7) and (7.2.8), we obtain

$$\int_0^\infty \frac{P_n(x)}{Q_m(x)} \left[(\ln x)^p - \sum_{s=0}^{p} C_p^s (\ln x)^s (2\pi i)^{p-s} \right] dx = 2\pi i \sum_k \operatorname*{Res}_{z=z_k} \left[\frac{P_n(z)}{Q_m(z)} (\operatorname{Log} z)^p \right]$$

where $\Im(z_k) \neq 0$ if $\Re(z_k) \geq 0$. It follows from the last formula that

$$-\sum_{s=0}^{p-1} C_p^s (2\pi i)^{p-s-1} I_s = \sum_k \operatorname*{Res}_{z=z_k} \left[\frac{P_n(z)}{Q_m(z)} (\operatorname{Log} z)^p \right],$$

which coincides with (7.2.2). □

We now consider the cases $p = 2$ and $p = 3$.

(a) The case $p = 2$. In this case we obtain from (7.2.2) and (7.2.3) that

$$C_2^0 2\pi i I_0 + C_2^1 (2\pi i)^0 I_1 = -\sum_k \operatorname*{Res}_{z=z_k} \left[\frac{P_n(z)}{Q_m(z)} (\operatorname{Log} z)^2 \right], \tag{7.2.9}$$

that is,

$$2\pi i \int_0^\infty \frac{P_n(x)}{Q_m(x)}\, dx + 2 \int_0^\infty \frac{P_n(x)}{Q_m(x)} \ln x\, dx = -\sum_k \operatorname*{Res}_{z=z_k} \left[\frac{P_n(z)}{Q_m(z)} (\operatorname{Log} z)^2 \right]. \tag{7.2.10}$$

Equating the real parts on the left- and right-hand sides in (7.2.10), we obtain the following formula for I_1:

$$\begin{aligned} I_1 &= \int_0^\infty \frac{P_n(x)}{Q_m(x)} \ln x\, dx \\ &= -\frac{1}{2} \Re \left\{ \sum_k \operatorname*{Res}_{z=z_k} \left[\frac{P_n(z)}{Q_m(z)} (\operatorname{Log} z)^2 \right] \right\}, \end{aligned} \tag{7.2.11}$$

where $m \geq n+2$, $Q_m(x) \neq 0$ for $x > 0$ and $\Im(z_k) \neq 0$ if $\Re(z_k) \geq 0$. If (7.1.6) is used to evaluate I_0, then one can derive from (7.2.10) a rather bulky formula for I_1 given in [**21**], p. 295:

$$I_1 = -\sum_k \operatorname*{Res}_{z=z_k} \left\{ \frac{P_n(z)}{Q_m(z)} \left[\frac{1}{2} (\operatorname{Log} z)^2 - \pi i \operatorname{Log} z \right] \right\}. \tag{7.2.12}$$

If we equate the imaginary parts on the left- and right-hand sides of (7.2.10), we obtain another formula for I_0 (compare with (7.1.6)):

$$\begin{aligned} I_0 &= \int_0^\infty \frac{P_n(x)}{Q_m(x)}\,dx \\ &= -\frac{1}{2\pi}\Im\left\{\sum_k \operatorname*{Res}_{z=z_k}\left[\frac{P_n(z)}{Q_m(z)}\,(\operatorname{Log} z)^2\right]\right\}. \end{aligned} \tag{7.2.13}$$

(b) The case $p = 3$. In this case, by (7.2.2) we have

$$C_3^0(2\pi i)^2 I_0 + C_3^1(2\pi i)^1 I_1 + C_3^2(2\pi i)^0 I_2 = -\sum_k \operatorname*{Res}_{z=z_k}\left[\frac{P_n(z)}{Q_m(z)}\,(\operatorname{Log} z)^3\right],$$

that is,

$$-4\pi^2 I_0 + 6\pi i I_1 + 3I_2 = -\sum_k \operatorname*{Res}_{z=z_k}\left[\frac{P_n(z)}{Q_m(z)}\,(\operatorname{Log} z)^3\right], \tag{7.2.14}$$

where I_0, I_1 and I_2 are real integrals. Equating the real parts on the left- and right-hand sides of (7.2.14) and using (7.1.6) we obtain a simple formula for I_2:

$$\begin{aligned} I_2 &= \int_0^\infty \frac{P_n(x)}{Q_m(x)}\,(\ln x)^2\,dx \\ &= -\frac{1}{3}\Re\left\{\sum_k \operatorname*{Res}_{z=z_k}\left[\frac{P_n(z)}{Q_m(z)}\,((\operatorname{Log} z)^3 + 4\pi^2 \operatorname{Log} z)\right]\right\}, \end{aligned} \tag{7.2.15}$$

where $m \geq n+2$, $Q_m(x) \neq 0$ if $x \geq 0$ and $\Im(z_k) \neq 0$ if $\Re(z_k) \geq 0$.

One can obtain a bulkier formula for I_2 by using(7.1.6) and (7.2.11) (this formula is given in [**21**], p. 295]).

7.2.2. $Q_m(x) = 0$ has simple positive roots. We suppose that $Q_m(x)$ has simple real zeros at the points $x_j = \alpha_j$, $j = 1, \ldots, s$, ordered as $0 < \alpha_1 < \alpha_2 < \ldots < \alpha_s$. In this case, we use the closed path shown in Fig 6.9 and the auxiliary function (7.2.5).

The only difference from the previous subsection (corresponding to Fig 6.8) is that the integrals

$$\sum_{k=1}^{s} \int_{\gamma_k} \frac{P_n(z)}{Q_m(z)}\,(\operatorname{Log} z)^p\,dz \tag{7.2.16}$$

and

$$\sum_{k=1}^{s} \int_{\widetilde{\gamma}_k} \frac{P_n(z)}{Q_m(z)} (\operatorname{Log} z)^p \, dz \tag{7.2.17}$$

along the semicircles γ_k and $\widetilde{\gamma}_k$ on the upper and lower parts, AB and $\widetilde{B}\widetilde{A}$, respectively, of the cut, are added to the integral along $\widetilde{B}\widetilde{A}$ in formula (7.2.8).

As in Subsection 7.1.2, the limit of the integral along γ_k, as $\delta \to 0$, is

$$\int_{\gamma_k} \frac{P_n(z)}{Q_m(z)} (\operatorname{Log} z)^p \, dz = -c_{-1}\pi i,$$

where

$$c_{-1} = \operatorname*{Res}_{z=\alpha_k} \left[\frac{P_n(z)}{Q_m(z)} (\operatorname{Log} z)^p \right]. \tag{7.2.18}$$

The limit of the integral along $\widetilde{\gamma}_k$, as $\delta \to 0$, is

$$\int_{\widetilde{\gamma}_k} \frac{P_n(z)}{Q_m(z)} (\operatorname{Log} z)^p \, dz = -c_{-1}\pi i,$$

where

$$\begin{aligned} c_{-1} &= \operatorname*{Res}_{z=\alpha_k e^{2\pi i}} \left[\frac{P_n(z)}{Q_m(z)} (\operatorname{Log} z)^p \right] \\ &\qquad \text{(and letting } z = \zeta\, e^{2\pi i}) \\ &= \operatorname*{Res}_{\zeta=\alpha_k} \left\{ \frac{P_n(\zeta)}{Q_m(\zeta)} \left[\operatorname{Log} \left(\zeta\, e^{2\pi i} \right)^p \right] \right\} \\ &= \operatorname*{Res}_{z=\alpha_k} \left[\frac{P_n(z)}{Q_m(z)} (\operatorname{Log} z + 2\pi i)^p \right]. \end{aligned} \tag{7.2.19}$$

Hence, as $R \to \infty$ and $\delta \to 0$, one has to add the sum

$$-\pi i \sum_{k=1}^{s} \operatorname*{Res}_{z=\alpha_k} \left[\frac{P_n(z)}{Q_m(z)} (\operatorname{Log} z)^p \right]$$

to the integral (7.2.7) along AB, while the sum

$$-\pi i \sum_{k=1}^{s} \operatorname*{Res}_{z=\alpha_k} \left[\frac{P_n(z)}{Q_m(z)} (\operatorname{Log} z + 2\pi i)^p \right]$$

is added to the integral (7.2.8) along $\widetilde{B}\widetilde{A}$. Thus, as $R \to \infty$ and $\delta \to 0$, we obtain from formula (7.2.6) that

$$\sum_{r=0}^{p-1} C_p^r (2\pi i)^{p-r-1} I_r = -\sum_k \operatorname*{Res}_{z=z_k} \left[\frac{P_n(z)}{Q_m(z)} (\operatorname{Log} z)^p \right]$$
$$- \frac{1}{2} \sum_{k=1}^{s} \operatorname*{Res}_{z=\alpha_k} \left\{ \frac{P_n(z)}{Q_m(z)} \left[(\operatorname{Log} z)^p + (\operatorname{Log} z + 2\pi i)^p \right] \right\}, \quad (7.2.20)$$

where $m \geq n+2$, $\alpha_k > 0$ and $\Im(z_k) \neq 0$ if $\Re(z_k) \geq 0$.

Note 7.2.1. It was shown in Subsection 7.1.1 that any indefinite integral of a rational function of the form $P_n(x)/Q_m(x)$, $m \geq n+2$, can be computed by means of the theory of residues. It is known that any such integral can also be computed directly. However, an indefinite integral of the form $f(x) = [P_n(x)/Q_n(x)] \ln x$, in general, cannot be expressed in terms of a finite number of elementary functions. Therefore, the definite integral of $f(x)$ from 0 to b cannot be computed by means of the theory of residues.

Let us consider the difficulties in evaluating the integral

$$I = \int_0^b \frac{\ln x}{1+x^2}\, dx, \quad (7.2.21)$$

which cannot be evaluated directly. The change of variable, $t = -1 + b/x$, reduces (7.2.21) to the form

$$I = b \int_0^\infty \frac{\ln b - \ln(t+1)}{(t+1)^2 + b^2}\, dt, \quad (7.2.22)$$

which cannot be computed by means of the residue theory (see formula (7.2.11)) since the integrand contains the term $\ln(t+1)$ (and not $\ln t$). In other words, because the interval of integration, $0 \leq t < +\infty$, does not coincide with the upper cut $-1 \leq t < +\infty$ of the function $\operatorname{Log}(t+1)$, it is impossible to make a change of variable, $t = \varphi(\xi)$, such that, simultaneously, the integration interval becomes $0 \leq \xi < +\infty$ and the function $\operatorname{Log}(t+1)$ is transformed to $\operatorname{Log} \xi$.

EXAMPLE 7.2.1. *Evaluate the integral*

$$J_1 = \int_0^\infty \frac{\ln x}{(x+1)^2}\, dx.$$

SOLUTION. The conditions are such that formula (7.2.11) is true, and hence

$$J_1 = -\frac{1}{2} \Re \left[\operatorname*{Res}_{z=-1} \frac{(\operatorname{Log} z)^2}{(z+1)^2} \right] = -\frac{1}{2} \Re \lim_{z \to -1} \left[(\operatorname{Log} z)^2 \right]'$$

$$
\begin{aligned}
&= -\frac{1}{2}\Re \lim_{z\to -1} \frac{2\,\mathrm{Log}\, z}{z} = \Re\,\mathrm{Log}(-1) \\
&= \Re(i\pi) = 0. \quad \square
\end{aligned}
$$

EXAMPLE 7.2.2. *Evaluate the integral*

$$
J_2 = \int_0^\infty \frac{\ln x}{x^2 + 2x + 2}\, dx.
$$

SOLUTION. We use formula (7.2.11). The zeros, $z = -1 \pm i$, of the denominator, $z^2 + 2z + 2$, are the simple poles of the integrand. Hence, by (7.2.11) we have

$$
\begin{aligned}
J_2 &= -\frac{1}{2}\Re\left\{\left(\operatorname*{Res}_{z=-1+i} + \operatorname*{Res}_{z=-1-i}\right)\left[\frac{(\mathrm{Log}\, z)^2}{z^2+2z+2}\right]\right\} \\
&= -\frac{1}{2}\Re\left\{\left.\frac{[\mathrm{Log}(-1+i)]^2}{2z+2}\right|_{z=-1+i} + \left.\frac{[\mathrm{Log}(-1-i)]^2}{2z+2}\right|_{z=-1-i}\right\} \\
&= -\frac{1}{4}\Re\left\{\frac{1}{i}\left[[\mathrm{Log}(-1+i)]^2 - [\mathrm{Log}(-1-i)]^2\right]\right\}.
\end{aligned}
$$

Since $0 \le \mathrm{Arg}\, z < 2\pi$, then

$$
\begin{aligned}
\mathrm{Log}(-1+i) &= \mathrm{Log}\left(\sqrt{2}\, e^{i3\pi/4}\right) = \ln\sqrt{2} + \frac{3\pi}{4}\, i, \\
\mathrm{Log}(-1-i) &= \mathrm{Log}\left(\sqrt{2}\, e^{i5\pi/4}\right) = \ln\sqrt{2} + \frac{5\pi}{4}\, i.
\end{aligned}
$$

Thus

$$
\begin{aligned}
J_2 &= \frac{1}{4}\Re\left\{i\left[\left(\ln\sqrt{2}\right)^2 + 2i\,\frac{3\pi}{4}\,\ln\sqrt{2} - \frac{9\pi^2}{16} - \left(\ln\sqrt{2}\right)^2\right.\right. \\
&\qquad \left.\left. - 2i\,\frac{5\pi}{4}\,\ln\sqrt{2} + \frac{25\pi^2}{16}\right]\right\} \\
&= -\frac{1}{2}\Re\left\{\left[\frac{3\pi}{4} - \frac{5\pi}{4}\right]\ln\sqrt{2} - \frac{1}{2}\, i\,\pi^2\right\} \\
&= \frac{\pi}{8}\ln 2. \quad \square
\end{aligned}
$$

EXAMPLE 7.2.3. *Evaluate the integral*

$$
J_3 = \int_0^\infty \frac{(\ln x)^2}{x^2 + a^2}\, dx, \qquad a > 0.
$$

SOLUTION. We use formula (7.2.15). The zeros, $z = \pm ai$, of the denominator, $z^2 + a^2$, are the simple poles of the integrand. Therefore

$$
\begin{aligned}
J_3 &= -\frac{1}{3}\Re\left(\operatorname*{Res}_{z=ai} + \operatorname*{Res}_{z=-ai}\right)\left[\frac{1}{z^2+a^2}((\operatorname{Log} z)^3 + 4\pi^2 \operatorname{Log} z)\right] \\
&= -\frac{1}{3a}\Re\left\{\frac{1}{2i}\left[(\operatorname{Log}(ai))^3 + 4\pi^2\operatorname{Log}(ai) - (\operatorname{Log}(-ai))^3 - 4\pi^2\operatorname{Log}(-ai)\right]\right\} \\
&= -\frac{1}{6a}\Im\left[\left(\ln a + \frac{\pi}{2}i\right)^3 + 4\pi^2\left(\ln a + \frac{\pi}{2}i\right) - \left(\ln a + \frac{3\pi}{2}i\right)^3\right. \\
&\quad \left. - 4\pi^2\left(\ln a + \frac{3\pi}{2}i\right)\right] \\
&= -\frac{1}{6a}\Im\left\{(\ln a)^3 + 3(\ln a)^2\frac{\pi}{2}i + 3(\ln a)\left(\frac{-\pi^2}{4}\right) - \frac{\pi^3}{8}i + 2\pi^3 i\right. \\
&\quad \left. - \left[(\ln a)^3 + 3(\ln a)^2\frac{3\pi}{2}i + 3(\ln a)\left(-\frac{9\pi^2}{4}\right) - \frac{27\pi^3 i}{8}\right] - 6\pi^3 i\right\} \\
&= -\frac{1}{6a}\left[3(\ln a)^2\left(\frac{\pi}{2} - \frac{3\pi}{2}\right) + \frac{26\pi^3}{8} - 4\pi^3\right] \\
&= \frac{\pi}{2a}(\ln a)^2 + \frac{\pi^3}{8a}. \quad \square
\end{aligned}
$$

7.3. Forms containing $\ln g(x)$ or $\arctan g(x)$

In this section, we consider integrals of the form

$$
\begin{aligned}
I_l &= \int_{-\infty}^{\infty} \frac{P_n(x)}{Q_m(x)} \ln|x-a|\,dx, & I_{ll} &= \int_{-\infty}^{\infty} \frac{P_n(x)}{Q_m(x)} \ln|x-a|\,\ln|x-b|\,dx, \\
A &= \int_{-\infty}^{\infty} \frac{P_n(x}{Q_m(x)} \ln|x^2-a^2|\,dx, & B &= \int_{-\infty}^{\infty} \frac{P_n(x}{Q_m(x)} \ln|x^2+a^2|\,dx, \\
C &= \int_{-\infty}^{\infty} \frac{P_n(x)}{Q_m(x)} \operatorname{Arctan}\frac{a}{x}\,dx, & D &= \int_{-\infty}^{\infty} \frac{P_n(x)}{Q_m(x)} \operatorname{Arctan} x\,dx.
\end{aligned}
$$

These integrals are computed by separating the real and imaginary parts of specially chosen analytic functions. Integrals I_l, A and B are computed by this method in [**21**], Subsection 29.12, Examples 1, 3, 4 (under the assumption that $Q_m(x) \neq 0$ for $x > 0$). Integral I_{ll}, to the authors' knowledge, is absent from the literature.

7.3.1. Integral I_l. We first prove the following lemma.

LEMMA 7.3.1. *Suppose $Q_m(x) \neq 0$ for real x and z_k are the zeros of $Q_m(z)$ in the upper half-plane; then*

$$I_l = \Re\left\{2\pi i \sum_k \operatorname*{Res}_{z=z_k}\left[\frac{P_n(z)}{Q_m(z)}\operatorname{Log}(z-a)\right]\right\}, \tag{7.3.1}$$

where $\operatorname{Log}(z-a)$ *is the principal value of* $\log(z-a)$ *with branch cut along the half-line $[a,+\infty)$, and $m \geq n+2$.*

PROOF. The function $\operatorname{Log}(z-a)$ is analytic in the upper half-plane if we make a cut along the positive real axis joining the branch points $z=a$ and $z=\infty$ and assume that $\operatorname{Arg}(z-a)=0$ on the upper part of the cut, $\operatorname{Arg}(z-a)=2\pi$ on the lower part of the cut and $\operatorname{Arg}(z-a)=\pi$ if $z=x$ is any point on the real axis such that $x<a$. Consider a closed path consisting of the interval $[-R,R]$ $(R>|a|)$ of the x-axis, a semicircle γ_a of radius δ around the branch point $z=a$ and a semicircle C_R of radius R (see Fig 7.1). The function $P_n(z)\operatorname{Log}(z-a)/Q_m(z)$ is analytic inside the path; therefore, by the residue theorem 5.2.2 we have

$$\left(\int_{-R}^{a-\delta}+\int_{\gamma_a}+\int_{a+\delta}^{R}+\int_{C_R}\right)\left[\frac{P_n(z)}{Q_m(z)}\operatorname{Log}(z-a)\right]dz$$
$$= 2\pi i \sum_k \operatorname*{Res}_{z=z_k}\left[\frac{P_n(z)}{Q_m(z)}\operatorname{Log}(z-a)\right]. \tag{7.3.2}$$

Letting $z=R\,e^{i\theta}$ on C_R and $z-a=\delta\,e^{i\theta}$ on γ_a and taking the inequality $m\geq n+2$ into account, one can easily verify that the integrals along C_R and γ_a tend to zero as $R\to\infty$ and $\delta\to 0$.

On the interval $-R\leq x\leq a-\delta$,

$$z-a=|x-a|\,e^{i\pi} \quad \text{and} \quad \operatorname{Log}(z-a)=\ln|x-a|+i\pi.$$

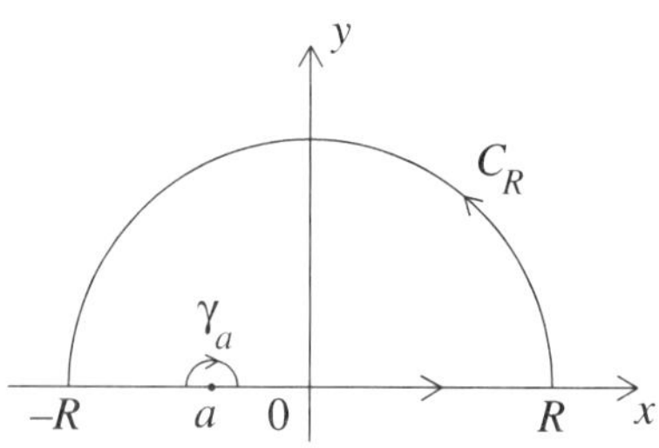

FIGURE 7.1. The closed path for the evaluation of integral I_l.

Thus,

$$\int_{-R}^{a-\delta} \frac{P_n(x)}{Q_m(x)} [\ln|x-a| + i\pi]\, dx \to \int_{-\infty}^{a} \frac{P_n(x)}{Q_m(x)} [\ln|x-a| + i\pi]\, dx$$

as $R \to \infty$ and $\delta \to 0$.

Since $z - a = |x - a|$ on the interval $a + \delta \le x \le R$, we have the limit

$$\int_{a+\delta}^{R} \frac{P_n(x)}{Q_m(x)} \ln|x-a|\, dx \to \int_{a}^{\infty} \frac{P_n(x)}{Q_m(x)} \ln|x-a|\, dx$$

as $R \to \infty$ and $\delta \to 0$. Hence, from (7.3.2) we have the formula

$$\int_{-\infty}^{\infty} \frac{P_n(x)}{Q_m(x)} \ln|x-a|\, dx + i\pi \int_{-\infty}^{a} \frac{P_n(x)}{Q_m(x)}\, dx = 2\pi i \sum_k \operatorname*{Res}_{z=z_k} \left[\frac{P_n(z)}{Q_m(z)} \operatorname{Log}(z-a) \right], \tag{7.3.3}$$

as $R \to \infty$ and $\delta \to 0$. Equating the real parts in (7.3.3) we obtain (7.3.1). □

Note 7.3.1. Equating the imaginary parts in (7.3.3), we obtain a formula similar to (7.1.6),

$$\int_{-\infty}^{a} \frac{P_n(x)}{Q_m(x)}\, dx = 2\Im \left[i \sum_k \operatorname*{Res}_{z=z_k} \frac{P_n(z)}{Q_m(z)} \operatorname{Log}(z-a) \right]. \tag{7.3.4}$$

Note 7.3.2. If $Q_m(x)$ has simple zeros at $x = a_k$, $k = 1, \dots, s$, then the terms

$$\Re\left\{ \pi i \sum_{k=1}^{s} \operatorname*{Res}_{z=\alpha_k} \left[\frac{P_n(z)}{Q_m(z)} \operatorname{Log}(z-a) \right] \right\} \tag{7.3.5}$$

and

$$\Im\left\{ i \sum_{k=1}^{s} \operatorname*{Res}_{z=\alpha_k} \left[\frac{P_n(z)}{Q_m(z)} \operatorname{Log}(z-a) \right] \right\} \tag{7.3.6}$$

are to be added to the right-hand sides of (7.3.1) and (7.3.4), respectively. These integrals are to be understood in the sense of the Cauchy principal value.

7.3.2. Integral I_{ll}. Consider the integral of the form

$$I_{ll} = \int_{-\infty}^{\infty} \frac{P_n(x)}{Q_m(x)} \ln|x-a| \ln|x-b|\, dx,$$

where a and b are real numbers and $a < b$.

We prove the following lemma.

LEMMA 7.3.2. *If $m \geq n+2$ and $Q_m(x) \neq 0$ for real x, then*

$$\int_{-\infty}^{\infty} \frac{P_n(x)}{Q_m(x)} \ln|x-a| \ln|x-b|\, dx = \Re\Bigg\{ 2\pi i \times \sum_k \operatorname*{Res}_{z=z_k} \left[\frac{P_n(z)}{Q_m(z)} \Big(\operatorname{Log}(z-a)\operatorname{Log}(z-b) + \frac{\pi}{i}\operatorname{Log}(z-a) \Big) \right] \Bigg\}, \quad (7.3.7)$$

where $\Im(z_k) > 0$.

PROOF. We use the closed path shown in Fig 7.1 with one additional semicircle γ_b centered at b with radius δ, where $a < b < R$. By the residue theorem 5.2.2 we have

$$\left(\int_{-R}^{a-\delta} + \int_{\gamma_a} + \int_{a+\delta}^{b-\delta} + \int_{\gamma_b} + \int_{b+\delta}^{R} + \int_{C_R} \right) \left[\frac{P_n(z)}{Q_m(z)} \operatorname{Log}(z-a)\operatorname{Log}(z-b) \right] dz = 2\pi i \sum_k \operatorname*{Res}_{z=z_k} \left[\frac{P_n(z)}{Q_m(z)} \operatorname{Log}(z-a)\operatorname{Log}(z-b) \right]. \quad (7.3.8)$$

The integrals along γ_a, γ_b and C_R approach zero as $R \to \infty$ and $\delta \to 0$.

When $z = x$, the function $\operatorname{Log}(z-a)\operatorname{Log}(z-b)$ has the form

$$\operatorname{Log}(z-a)\operatorname{Log}(z-b) = \begin{cases} (\ln|x-a| + i\pi)(\ln|z-b| + i\pi), & x < a, \\ \ln|x-a|(\ln|z-b| + i\pi), & a < x < b, \\ \ln|x-a| \ln|x-b|, & x > b. \end{cases}$$

Therefore, as $R \to \infty$ and $\delta \to 0$, we obtain from (7.3.8) that

$$\int_{-\infty}^{a} \frac{P_n(x)}{Q_m(x)} \big[\ln|x-a| + i\pi\big]\big[\ln|x-b| + i\pi\big]\, dx + \int_{a}^{b} \frac{P_n(x)}{Q_m(x)} \ln|x-a| \big[\ln|x-b| + i\pi\big]\, dx + \int_{b}^{\infty} \frac{P_n(x)}{Q_m(x)} \ln|x-a| \ln|x-b|\, dx$$

$$= 2\pi i \sum_k \operatorname*{Res}_{z=z_k} \left[\frac{P_n(z)}{Q_m(z)} \operatorname{Log}(z-a) \operatorname{Log}(z-b) \right]. \qquad (7.3.9)$$

Equating the real parts in (7.3.9) we obtain

$$\int_{-\infty}^{\infty} \frac{P_n(x)}{Q_m(x)} \ln|x-a| \ln|x-b| \, dx - \pi^2 \int_{-\infty}^{a} \frac{P_n(x)}{Q_m(x)} \, dx$$
$$= \Re\left\{ 2\pi i \sum_k \operatorname*{Res}_{z=z_k} \left[\frac{P_n(z)}{Q_m(z)} \operatorname{Log}(z-a) \operatorname{Log}(z-b) \right] \right\}. \qquad (7.3.10)$$

Substituting the value of the integral (7.3.4) into (7.3.10) and using the relation $\Im[if(z)] = \Re f(z)$, we obtain (7.3.7). □

7.3.3. Integrals A, B, C, D. We consider integrals of the form

$$A = \int_{-\infty}^{\infty} \frac{P_n(x)}{Q_m(x)} \ln|x^2 - a^2| \, dx, \qquad B = \int_{-\infty}^{\infty} \frac{P_n(x)}{Q_m(x)} \ln|x^2 + a^2| \, dx,$$
$$C = \int_{-\infty}^{\infty} \frac{P_n(x)}{Q_m(x)} \operatorname{Arctan} \frac{a}{x} \, dx, \qquad D = \int_{-\infty}^{\infty} \frac{P_n(x)}{Q_m(x)} \operatorname{Arctan} x \, dx.$$

Since the evaluation of these integrals make use of the principal values of a few functions, for simplicity we shall assume that $a > 0$.

To evaluate integral A it is sufficient to replace a by $-a$ in (7.3.1) and add the resulting formula to (7.3.1). As a result, we obtain the formula

$$\int_{-\infty}^{\infty} \frac{P_n(x)}{Q_m(x)} \ln|x^2 - a^2| \, dx$$
$$= \Re\left\{ 2\pi i \sum_k \operatorname*{Res}_{z=z_k} \left[\frac{P_n(z)}{Q_m(z)} \operatorname{Log}(z^2 - a^2) \right] \right\}, \qquad (7.3.11)$$

where $\Im(z_k) > 0$ and $m \geq n + 2$.

To evaluate integral B (see [**21**], Problem 29.12, Example 4]) it is sufficient to compute the integral of the function

$$f(z) = \frac{P_n(z)}{Q_m(z)} \operatorname{Log}(z + ai)$$

along the whole real axis, where $m \geq n + 2$ and $Q_m(x) \neq 0$ for real x. To select a branch $\operatorname{Log}(z + ai)$ of $\log(z + ai)$ it is sufficient to join the branch points, $z = -ai$ and $z = -\infty i$, by a cut along the negative y-axis and use a closed path which consists of the interval $[-R, R]$ of the real axis and the semicircle C_R of radius R in the upper half-plane. By the residue theorem 5.2.2

$$\left(\int_{-R}^{R} + \int_{C_R}\right) \left[\frac{P_n(z)}{Q_m(z)} \operatorname{Log}(z+ai)\right] dz = 2\pi i \sum_k \operatorname*{Res}_{z=z_k} \left[\frac{P_n(z)}{Q_m(z)} \operatorname{Log}(z+ai)\right]. \qquad (7.3.12)$$

The integral along C_R tends to zero as $R \to \infty$ since $m \geq n+2$. On the interval $[-R, R]$, we have $z = x$ and

$$\operatorname{Log}(x+ai) = \operatorname{Log}\left[\sqrt{x^2+a^2}\, e^{i \operatorname{Arg}(x+ai)}\right], \qquad (7.3.13)$$

where

$$\operatorname{Arg}(x+ai) = \begin{cases} \operatorname{Arctan}(a/x), & x \geq 0, \\ \operatorname{Arctan}(a/x) + \pi, & x < 0. \end{cases}$$

Using (7.3.13) and letting $R \to \infty$, we obtain from (7.3.12) that

$$\int_{-\infty}^{\infty} \frac{P_n(x)}{Q_m(x)} \ln\sqrt{x^2+a^2}\, dx + i \int_{-\infty}^{\infty} \frac{P_n(x)}{Q_m(x)} \operatorname{Arg}(x+ai)\, dx = 2\pi i \sum_k \operatorname*{Res}_{z=z_k} \left[\frac{P_n(z)}{Q_m(z)} \operatorname{Log}(z+ai)\right]. \qquad (7.3.14)$$

Equating the real parts in (7.3.14), we obtain the following formula for evaluating integral B:

$$\int_{-\infty}^{\infty} \frac{P_n(x)}{Q_m(x)} \ln(x^2+a^2)\, dx = \Re\left\{4\pi i \sum_k \operatorname*{Res}_{z=z_k} \left[\frac{P_n(z)}{Q_m(z)} \operatorname{Log}(z+ai)\right]\right\}, \qquad (7.3.15)$$

where $\Im z_k > 0$.

Equating the imaginary parts in (7.3.14), we obtain the following formula for evaluating integral C:

$$\int_{-\infty}^{\infty} \frac{P_n(x)}{Q_m(x)} \operatorname{Arg}(x+ai)\, dx = \Im\left\{2\pi i \sum_k \operatorname*{Res}_{z=z_k} \left[\frac{P_n(z)}{Q_m(z)} \operatorname{Log}(z+ai)\right]\right\}.$$

Using (7.3.13) and (7.3.4) and assuming that the upper limit a in (7.3.4) is equal to zero, we can rewrite the last formula in the form

$$\int_{-\infty}^{\infty} \frac{P_n(x)}{Q_m(x)} \operatorname{Arctan}\frac{a}{x}\, dx = 2\pi \Im\left\{i \sum_k \operatorname*{Res}_{z=z_k} \left[\frac{P_n(z)}{Q_m(z)} \operatorname{Log}(z+ai)\right]\right\}$$

$$- 2\pi\Im\left\{ i \sum_k \operatorname*{Res}_{z=z_k} \left[\frac{P_n(z)}{Q_m(z)} \operatorname{Log} z \right] \right\}, \qquad (7.3.16)$$

where $Q_m(x) \neq 0$ for real x, $\Im z_k > 0$.

Similarly, for evaluating integral D (see [**21**], Problem 29.12, Example 5]) it is sufficient to compute the integral of the function

$$f(z) = \frac{P_n(z)}{Q_m(z)} \operatorname{Log}(1 - iz),$$

along the whole real axis, where $m \geq n + 2$ and $Q_m(x) \neq 0$ for real x. To select a branch, $\operatorname{Log}(1 - iz)$, of $\log(1 - iz)$ it suffices to join the branch point $z = -i$ with the point $z = -\infty - i$ by a cut parallel to the negative real axis. Then

$$\begin{aligned} \operatorname{Log}(1 - iz)|_{y=0} &= \operatorname{Log}(1 - ix) \\ &= \ln\sqrt{1 + x^2} - i \operatorname{Arctan} x \end{aligned}$$

for all $-\infty < x < +\infty$. Using the residue theorem 5.2.2, we obtain

$$\begin{aligned} &\int_{-\infty}^{\infty} \frac{P_n(x)}{Q_m(x)} \left[\ln\sqrt{1 + x^2} - i \operatorname{Arctan} x \right] dx \\ &\qquad = 2\pi i \sum_k \operatorname*{Res}_{z=z_k} \left[\frac{P_n(z)}{Q_m(z)} \operatorname{Log}(1 - iz) \right]. \qquad (7.3.17) \end{aligned}$$

Finally, equating the imaginary parts in (7.3.17), we obtain the following formula for evaluating integral D:

$$\begin{aligned} &\int_{-\infty}^{\infty} \frac{P_n(x)}{Q_m(x)} \operatorname{Arctan} x \, dx \\ &\qquad = -2\pi\Re\left\{ \sum_k \operatorname*{Res}_{z=z_k} \left[\frac{P_n(z)}{Q_m(z)} \operatorname{Log}(1 - iz) \right] \right\}, \qquad (7.3.18) \end{aligned}$$

where $\Im z_k > 0$.

Note 7.3.3. If $Q_m(x)$ has simple zeros at the points

$$x_j = a_j, \qquad j = 1, \dots, s,$$

then one has to replace $\sum_k \operatorname*{Res}_{z=z_k}$ with

$$\sum_k \operatorname*{Res}_{z=z_k} + \frac{1}{2} \sum_{j=1}^{s} \operatorname*{Res}_{z=a_j}$$

on the right-hand sides of (7.3.11), (7.3.15), (7.3.16) and (7.3.18). Moreover, the same function is used for computing the residues at the points $x_j = a_j$ for $j = 1, \dots, s$ and at z_k.

7.4. Forms containing ln in the denominator

Consider integrals of the form

$$\int_0^\infty \frac{P_n(x)}{Q_m(x)} \frac{dx}{(\ln x)^2 + \pi^2},$$

where $P_n(x)$ and $Q_m(x)$ are polynomials, $m \geq n+2$, $Q_m(-1) \neq 0$ and $Q_m(x) \neq 0$ for $x \geq 0$.

We cut the complex plane along the positive real axis and use the closed path C shown in Fig 6.8 for the function

$$f(z) = \frac{P_n(z)}{Q_m(z)} \frac{1}{\operatorname{Log} z - \pi i}.$$

By the residue theorem 5.2.2,

$$\oint_C f(z)\,dz = 2\pi i \sum_k \operatorname*{Res}_{z=z_k} \left[\frac{P_n(z)}{Q_m(z)} \frac{1}{\operatorname{Log} z - \pi i}\right] + 2\pi i \operatorname*{Res}_{z=e^{i\pi}} \frac{P_n(z)}{Q_m(z)} \frac{1}{\operatorname{Log} z - \pi i},$$

that is,

$$\left(\int_{AB} + \int_{C_R} + \int_{\widetilde{B}\widetilde{A}} + \int_{C_\delta}\right) \left[\frac{P_n(z)}{Q_m(z)} \frac{1}{\operatorname{Log} z - \pi i}\right] dz = 2\pi i \sum_k \operatorname*{Res}_{z=z_k} \left[\frac{P_n(z)}{Q_m(z)} \frac{1}{\operatorname{Log} z - \pi i}\right] + 2\pi i \frac{P_n(-1)}{Q_m(-1)}\, e^{i\pi}. \quad (7.4.1)$$

It can easily be shown that the integrals along the circles C_R and C_δ approach zero as $R \to \infty$ and $\delta \to 0$. On the segment AB, $z = x$ and we have

$$\int_{AB} = \int_\delta^R \frac{P_n(x)}{Q_m(x)} \frac{1}{\ln x - \pi i}\,dx \to \int_0^\infty \frac{P_n(x)}{Q_m(x)} \frac{1}{\ln x - \pi i}\,dx$$

as $R \to \infty$ and $\delta \to 0$. On the segment $\widetilde{B}\widetilde{A}$, $z = x\,e^{2\pi i}$ and we have

$$\int = \int_R^\delta \frac{P_n(x)}{Q_m(x)} \frac{1}{\ln x + 2\pi i - \pi i}\,dx \to -\int_0^\infty \frac{P_n(x}{Q_m(x)} \frac{dx}{\ln x + \pi i}$$

as $R \to \infty$ and $\delta \to 0$. Therefore, it follows from (7.4.1) that

$$\int_0^\infty \frac{P_n(x)}{Q_m(x)} \left[\frac{1}{\ln x - \pi i} - \frac{1}{\ln x + \pi i}\right] dx$$
$$= 2\pi i \sum_k \operatorname*{Res}_{z=z_k} \left[\frac{P_n(z)}{Q_m(z)} \frac{1}{\operatorname{Log} z - \pi i}\right] - 2\pi i \frac{P_n(-1)}{Q_m(-1)}$$

as $R \to \infty$ and $\delta \to 0$, and, after obvious transformations,

$$\int_0^\infty \frac{P_n(x)}{Q_m(x)} \frac{dx}{(\ln x)^2 + \pi^2}$$
$$= \sum_k \operatorname*{Res}_{z=z_k} \left[\frac{P_n(z)}{Q_m(z)} \frac{1}{\operatorname{Log} z - \pi i}\right] - \frac{P_n(-1)}{Q_m(-1)}, \qquad (7.4.2)$$

where $\Im z_k \neq 0$, $Q_m(x) \neq 0$ for $x \geq 0$.

7.5. Forms containing $P_n(e^x)/Q_m(e^x)$

In this section, we consider integrals of the form

$$\int_{-\infty}^\infty \frac{P_n(e^x)}{Q_m(e^x)} \frac{dx}{x^2 + (2s+1)^2\pi^2},$$

where $s = 0, 1, 2, \dots$, $m \geq n$, $Q_m(e^x) \neq 0$ for real x, and

$$Q_m(e^{\pi i}) = Q_m(-1) \neq 0, \qquad Q_m(e^{-\infty}) = Q_m(0) \neq 0.$$

EXAMPLE 7.5.1. *Letting z_k be the zeros of $Q_m(e^z)$ lying in the strip $0 < \Im z < 2\pi$, we prove the following formula:*

$$\int_{-\infty}^\infty \frac{P_n(e^x)}{Q_m(e^x)} \frac{dx}{x^2 + (2s+1)^2\pi^2} = \frac{1}{2s+1} \frac{P_n(-1)}{Q_m(-1)}$$
$$+ \frac{1}{2s+1} \sum_k \operatorname*{Res}_{z=z_k} \left[\frac{P_n(e^z)}{Q_m(e^z)} \sum_{k=-s}^{s} \frac{1}{z + (2k-1)\pi i}\right]. \qquad (7.5.1)$$

PROOF. We use the closed rectangular path C shown in Fig 6.6, and consider the auxiliary function

$$F(z) = \frac{P_n(e^z)}{Q_m(e^z)} \sum_{k=-s}^{s} \frac{1}{z + (2k-1)\pi i}. \qquad (7.5.2)$$

In the rectangle of height 2π shown in Fig 6.6, the function $F(z)$ has poles at the zeros, z_k, of $Q_m(e^z)$ and at the singular point $z = \pi i$ corresponding to $k = 0$ in (7.5.2). The other singular points of (7.5.2), namely,

$$z_k = -(2k-1)\pi i, \qquad k = -s, -s+1, \dots, s, \quad k \neq 0,$$

lie outside the rectangle. Therefore by the residue theorem 5.2.2 we have

$$\oint_C \frac{P_n(e^z)}{Q_m(e^z)} \sum_{k=-s}^{s} \frac{1}{z+(2k-1)\pi i}\,dz$$

$$= 2\pi i \frac{P_n(e^{\pi i})}{Q_m(e^{\pi i})} + 2\pi i \sum_k \operatorname*{Res}_{z=z_k}\left[\frac{P_n(e^z)}{Q_m(e^z)} \sum_{k=-s}^{s} \frac{1}{z+(2k-1)\pi i}\right],$$

that is (see Fig 6.6),

$$\left(\int_I + \int_{II} + \int_{III} + \int_{IV}\right)\left[\frac{P_n(e^z)}{Q_m(e^z)} \sum_{k=-s}^{s} \frac{1}{z+(2k-1)\pi i}\right] dz$$

$$= 2\pi i \frac{P_n(-1)}{Q_m(-1)} + 2\pi i \sum_k \operatorname*{Res}_{z=z_k}\left[\frac{P_n(e^z)}{Q_m(e^z)} \sum_{k=-s}^{s} \frac{1}{z+(2k-1)\pi i}\right]. \quad (7.5.3)$$

It can easily be shown that the integrals along sides II and IV tend to zero as $R \to \infty$. On side I, $z = x$ and we have (see (7.5.2))

$$\int_I = \int_{-R}^{R} F(x)\,dx \to \int_{-\infty}^{\infty} F(x)\,dx.$$

On side III, $z = x + 2\pi i$ and we have

$$\int_{III} = \int_{R}^{-R} F(x+2\pi i)\,dx \to -\int_{-\infty}^{\infty} F(x+2\pi i)\,dx.$$

Therefore, as $R \to \infty$, (7.5.3) can be written in the form

$$\int_{-\infty}^{\infty} [F(x) - F(x+2\pi i)]\,dx = 2\pi i \frac{P_n(-1)}{Q_m(-1)} + 2\pi i \sum_k \operatorname*{Res}_{z=z_k} F(z). \quad (7.5.4)$$

Using (7.5.2) we obtain

$$F(x) - F(x+2\pi i)$$

$$= \frac{P_n(e^x)}{Q_m(e^x)}\left[\sum_{k=-s}^{s} \frac{1}{x+(2k-1)\pi i} - \sum_{k=-s}^{s} \frac{1}{x+(2k+1)\pi i}\right]$$

(and letting $k = r-1$ in the second sum)

$$= \frac{P_n(e^x)}{Q_m(e^x)}\left[\sum_{k=-s}^{s} \frac{1}{x+(2k-1)\pi i} - \sum_{r=-s+1}^{s+1} \frac{1}{x+(2r-1)\pi i}\right]$$

$$= \frac{P_n(e^x)}{Q_m(e^x)}\left[\frac{1}{x+(-2s-1)\pi i} - \frac{1}{x+[2(s+1)-1]\pi i}\right]$$

$$= \frac{P_n(e^x)}{Q_m(e^x)} \frac{(2s+1)2\pi i}{x^2 + (2s+1)^2\pi^2}.$$

Hence by (7.5.1), the integral (7.5.4) has the form

$$(2s+1)2\pi i \int_{-\infty}^{\infty} \frac{P_n(e^x)}{Q_m(e^x)} \frac{dx}{x^2 + (2s+1)^2\pi^2}$$
$$= 2\pi i \frac{P_n(-1)}{Q_m(-1)} + 2\pi i \sum_k \operatorname*{Res}_{z=z_k} \left[\frac{P_n(e^z)}{Q_m(e^z)} \sum_{k=-s}^{s} \frac{1}{z + (2k-1)\pi i} \right]. \quad \square$$

Note 7.5.1. If $Q_m(e^x)$ has simple zeros at the points $x = a_1, a_2, \dots, a_p$, then $\sum\limits_k \operatorname*{Res}\limits_{z=z_k}$ in (7.5.1) has to be replaced by

$$\sum_k \operatorname*{Res}_{z=z_k} + \frac{1}{2} \sum_{s=1}^{p} \operatorname*{Res}_{z=a_s}.$$

Note 7.5.2. If $s = 0$ and $0 < \Im z_k < 2\pi$, then formula (7.5.1) becomes

$$\int_{-\infty}^{\infty} \frac{P_n(e^x)}{Q_m(e^x)} \frac{dx}{x^2 + \pi^2} = \frac{P_n(-1)}{Q_m(-1)} + \sum_k \operatorname*{Res}_{z=z_k} \left[\frac{P_n(e^z)}{Q_m(e^z)} \frac{1}{z - \pi i} \right]. \quad (7.5.5)$$

EXAMPLE 7.5.2. *If $a > 0$, derive the formula (found in* [**4**]*)*

$$J = \int_{-\infty}^{\infty} \frac{du}{(\pi^2/4 + u^2)(1 + a \tanh^2 u)} = \frac{2\sqrt{a}}{(1+a) \arctan \sqrt{a}}. \quad (7.5.6)$$

SOLUTION. The integral in (7.5.6) was computed as the sum of all the residues of the integrand in the upper half-plane and by a subsequent summation of the resulting series.

Let $u = x/2$ and $1/a = b^2$ in (7.5.6). Using the formula

$$\tanh \frac{x}{2} = \frac{e^x - 1}{e^x + 1}$$

we obtain from (7.5.6) that

$$J = 2b^2 \int_{-\infty}^{\infty} \left\{ \left[b^2 + \left(\frac{e^x - 1}{e^x + 1} \right)^2 \right] (\pi^2 + x^2) \right\}^{-1} dx. \quad (7.5.7)$$

Therefore, one can use formula (7.5.5) to compute (7.5.7). For this purpose, we have to find the roots of the equation

$$\tanh^2 \frac{z}{2} = \left(\frac{e^z - 1}{e^z + 1} \right)^2$$

$$= -b^2 \tag{7.5.8}$$

which are located in the strip $0 < \Im z < 2\pi$. Taking square roots on both sides of the previous equation we have

$$\tanh\frac{z}{2} = \pm bi, \qquad b > 0, \tag{7.5.9}$$

and setting $z = i\xi$ we obtain

$$\frac{\xi}{2} = \operatorname{Arctan}(\pm b) + k\pi, \qquad k = 0, \pm 1, \dots,$$

which we rewrite in the form

$$\xi = 2\big[\operatorname{Arctan}(\pm b) + k\pi\big], \qquad k = 0, \pm 1, \pm 2, \dots. \tag{7.5.10}$$

The only roots of (7.5.10) in the strip $0 < \xi < 2\pi$ are

$$\xi_1 = 2\operatorname{Arctan} b := 2\theta \quad \text{with } k = 0 \text{ and the plus sign,}$$

and

$$\xi_2 = 2[\pi - \operatorname{Arctan} b] := 2(\pi - \theta) \quad \text{with } k = 1 \text{ and the minus sign.}$$

Therefore we have to use (7.5.5) for the cases $z_1 = i\xi_1$ and $z_2 = i\xi_2$. Since $P_n(-1)/Q_m(-1) = 0$ in the present example, we obtain

$$\begin{aligned}\left[\left(\frac{e^z - 1}{e^z + 1}\right)^2\right]' &= 2\frac{e^z - 1}{e^z + 1}\,\frac{e^z(e^z + 1) - e^z(e^z - 1)}{(e^z + 1)^2}\\ &= 4\frac{e^z(e^z - 1)}{(e^z + 1)^3}.\end{aligned}$$

Thus we have

$$\begin{aligned}J &= \left(\operatorname*{Res}_{z=i\xi_1} + \operatorname*{Res}_{z=i\xi_2}\right)\left[\frac{2b^2}{[b^2 + [(e^z - 1)/(e^z + 1)]^2]\,(z - \pi i)}\right]\\ &= \frac{2b^2}{4(2i\theta - \pi i)}\,\frac{\left(e^{2i\theta} + 1\right)^3}{e^{2i\theta}\left(e^{2i\theta} - 1\right)} + \frac{2b^2}{4i(\pi - 2\theta)}\,\frac{\left(e^{-2i\theta} + 1\right)^3}{e^{-2i\theta}\left(e^{-2i\theta} - 1\right)}\\ &= \frac{b^2}{i(2\theta - \pi)}\left[\frac{\left(e^{2i\theta} + 1\right)^3}{e^{2i\theta}\left(e^{2i\theta} - 1\right)}\right]\\ &= \frac{b^2}{i(2\theta - \pi)}\,\frac{\left(e^{i\theta} + e^{-i\theta}\right)^3}{e^{i\theta} - e^{-i\theta}}\\ &= \frac{4b^2}{2i^2(\theta - \pi/2)}\,\frac{\cos^2\theta}{\tan\theta} \qquad (\text{since } \theta = \operatorname{Arctan} b)\end{aligned}$$

$$
\begin{aligned}
&= \frac{2b^2}{\pi/2 - \operatorname{Arctan} b} \frac{1}{\tan\theta(1+\tan^2\theta)} \\
&= \frac{2}{a \operatorname{Arctan}\sqrt{a}} \frac{\sqrt{a}}{1+1/a} \\
&= \frac{2\sqrt{a}}{(1+a)\operatorname{Arctan}\sqrt{a}}. \quad \square
\end{aligned}
$$

7.6. Poisson's integral

To derive Poisson's integral in example 7.6.1, we make use of the well-known formula:

$$\int_{-\infty}^{\infty} e^{-x^2}\,dx = \sqrt{\pi}, \tag{7.6.1}$$

which is easily obtained by considering the double integral

$$
\begin{aligned}
\int_0^\infty e^{-y^2}\,dy \int_0^\infty e^{-x^2}\,dx &= \int_0^\infty \int_0^\infty e^{-(x^2+y^2)}\,dx\,dy \\
&= \int_0^\infty \int_0^{\pi/2} e^{-r2)} r\,dr\,d\theta.
\end{aligned}
$$

EXAMPLE 7.6.1. *Derive Poisson's integral,*

$$P = \int_{-\infty}^{\infty} e^{-\alpha x^2} \cos\beta x\,dx = \sqrt{\frac{\pi}{\alpha}}\, e^{-\beta^2/(4\alpha)}, \tag{7.6.2}$$

for $\alpha > 0$ and real β.

PROOF. We have

$$
\begin{aligned}
P &= \Re \int_{-\infty}^{\infty} e^{-\alpha x^2 - i\beta x}\,dx \\
&= \Re \left[e^{-\beta^2/(4\alpha)} \int_{-\infty}^{\infty} e^{-\alpha[x+\beta i/(2\alpha)]^2}\,dx \right] \\
&\qquad \left(\text{letting } \sqrt{\alpha}\,[x+\beta i/(2\alpha)] = t, \quad dx = 1/\sqrt{\alpha}\,dt\right) \\
&= \Re \left[e^{-\beta^2/(4\alpha)} \frac{1}{\sqrt{\alpha}} \int_{-\infty+\beta i/(2\sqrt{\alpha})}^{+\infty+\beta i/(2\sqrt{\alpha})} e^{-t^2}\,dt \right].
\end{aligned} \tag{7.6.3}
$$

To complete the proof of (7.6.2) one has to show that $\beta i/(2\sqrt{\alpha}\,)$ can be discarded in (7.6.3); then, using (7.6.1), we obtain (7.6.2).

Let us consider a closed rectangular path in the complex plane with base $[-R, R]$ on the x-axis and height $\beta/(2\sqrt{\alpha}\,)$ (see Fig 7.2).

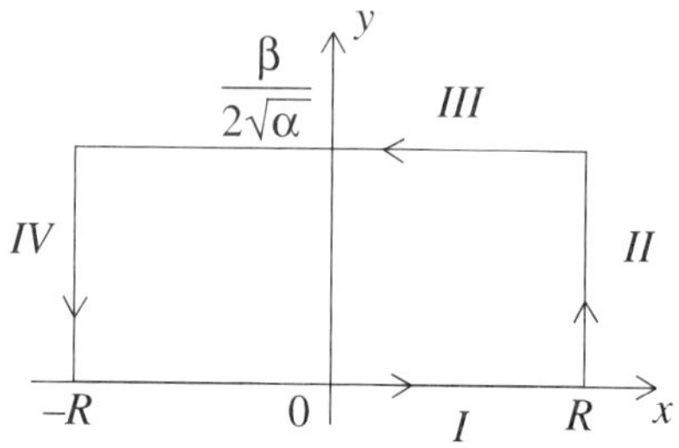

FIGURE 7.2. Rectangular region for Poisson's integral.

Since the function $\exp\left(-z^2\right)$ has no singular points inside the rectangle, then by the residue theorem 5.2.2 we have

$$\oint_C e^{-z^2}\,dz = \left(\int_I + \int_{II} + \int_{III} + \int_{IV}\right) e^{-z^2}\,dz = 0. \tag{7.6.4}$$

On side I, $z = x$; thus we have

$$\int_I = \int_{-R}^{R} e^{-x^2}\,dx \to \int_{-\infty}^{\infty} e^{-x^2}\,dx$$

as $R \to \infty$. On side II, $z = R + iy$; thus we have

$$\int_{II} = i \int_0^{\beta/(2\sqrt{\alpha}\,)} e^{-(R+iy)^2}\,dy = i\,e^{-R^2} \int_0^{\beta/(2\sqrt{\alpha}\,)} e^{-2iRy+y^2}\,dy \to 0$$

as $R \to \infty$ since $|e^{-2iRy}|$ is bounded. Similarly, one can show that

$$\int_{IV} \to 0, \qquad \text{as } R \to \infty.$$

On side III, $z = x + i\beta/\sqrt{\alpha}$; hence we have

$$\begin{aligned}
\int_{III} &= \int_R^{-R} e^{-[x+i\beta/(2\sqrt{\alpha}\,)]^2}\,dx \\
&\quad \text{(and letting } x + i\beta/(2\sqrt{\alpha}\,) = t\text{)} \\
&= -\int_{-R+i\beta/(2\sqrt{\alpha}\,)}^{R+i\beta/(2\sqrt{\alpha}\,)} e^{-t^2}\,dt \\
&\to -\int_{-\infty+i\beta/(2\sqrt{\alpha}\,)}^{+\infty+i\beta/(2\sqrt{\alpha}\,)} e^{-t^2}\,dt, \qquad \text{as } R \to \infty.
\end{aligned}$$

Therefore, as $R \to \infty$, from (7.6.4) we obtain that

$$\int_{-\infty}^{\infty} e^{-x^2}\,dx - \int_{-\infty+i\beta/(2\sqrt{\alpha}\,)}^{+\infty+i\beta/(2\sqrt{\alpha}\,)} e^{-t^2}\,dt = 0. \tag{7.6.5}$$

It follows from (7.6.5) that the constant $i\beta/(2\sqrt{\alpha}\,)$ can be discarded from the limits of the integral on the right-hand side of (7.6.3). Thus, (7.6.2) follows from (7.6.1) and (7.6.3). □

Note 7.6.1. Equation (7.6.5) implies that the horizontal line of integration

$$\left(-\infty + i\,\frac{\beta}{2\sqrt{\alpha}}\,, +\infty + i\,\frac{\beta}{2\sqrt{\alpha}}\right)$$

can be translated parallel to the real axis. Such an operation is a particular case of deformation of the path of integration. Another deformation will be seen in the next section.

7.7. Fresnel integrals

In this section, we use the calculus of residues to derive the *Fresnel integrals* from more general formulae. These integrals first appeared in the theory of diffraction of waves. More recently they have been applied to designing highways for high-speed automobiles.

EXAMPLE 7.7.1. *Derive Fresnel integrals*

$$\int_0^{\infty} \cos(x^2)\,dx = \frac{1}{2}\sqrt{\frac{\pi}{2}}, \qquad \int_0^{\infty} \sin(x^2)\,dx = \frac{1}{2}\sqrt{\frac{\pi}{2}}. \tag{7.7.1}$$

PROOF. We consider a closed path, C, in the complex plane consisting of the segment $[0, R]$ of the real axis, the arc C_R of radius R and angle $0 \le \theta \le \theta_0$, and the ray $z = r\,e^{i\theta_0}$ where $0 \le r \le R$ and $\theta_0 =$ constant (see Fig 7.3).

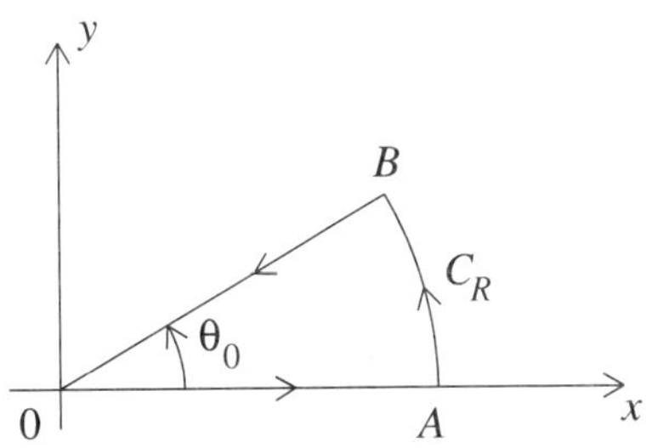

FIGURE 7.3. The closed path of integration $OABO$ for the derivation of (7.7.10) and (7.7.11).

Since $\exp(-z^2)$ does not have singular points inside C, then by the residue theorem 5.2.2 we have

$$\left(\int_{OA} + \int_{AB} + \int_{BO}\right) e^{-z^2}\, dz = 0. \tag{7.7.2}$$

On the segment OA, $z = x$; hence we have

$$\begin{aligned}\int_{OA} e^{-z^2}\, dz &= \int_0^R e^{-x^2}\, dx \\ &\to \int_0^\infty e^{-x^2}\, dx \\ &= \frac{\sqrt{\pi}}{2}, \qquad \text{as } R \to \infty.\end{aligned}$$

We show that the integral along BO has a finite limit as $R \to \infty$ if θ_0 lies in the interval $[0, \pi/4]$. Since on BO $z = r\, e^{i\theta_0}$, we have

$$\begin{aligned}\int_{BO} e^{-z^2}\, dz &= \int_R^0 e^{-r^2 \exp(2i\theta_0)}\, e^{i\theta_0}\, dr \\ &= -\int_0^R e^{-r^2(\cos 2\theta_0 + i \sin 2\theta_0)}\, e^{i\theta_0}\, dr. \qquad (7.7.3)\end{aligned}$$

The integrand in (7.7.3) remains bounded for $0 \leq r \leq R$ as $R \to \infty$ if $\cos 2\theta_0 \geq 0$, that is for $0 \leq \theta_0 \leq \pi/4$. In this case,

$$\begin{aligned}&\int_{BO} e^{-z^2}\, dz \\ &\quad \to -\int_0^\infty e^{-r^2 \cos 2\theta_0} \left[\cos(r^2 \sin 2\theta_0) - i \sin(r^2 \sin 2\theta_0)\right] e^{i\theta_0}\, dr. \qquad (7.7.4)\end{aligned}$$

We prove that the integral along AB in (7.7.2) approaches zero as $R \to \infty$. On the arc AB, $z = R\, e^{i\theta}$ for $0 \leq \theta \leq \theta_0$; hence we have

$$\begin{aligned}\left|\int_{AB} e^{-z^2}\, dz\right| &\leq \int_0^{\theta_0} \left|e^{-R^2 \exp 2i\theta}\right| \left|R\, e^{i\theta}\, i\right| d\theta \\ &= R \int_0^{\theta_0} e^{-R^2 \cos 2\theta}\, d\theta. \qquad (7.7.5)\end{aligned}$$

First, consider θ_0 in the interval $0 < \theta_0 < \pi/4$. Since $\cos 2\theta > 0$ for all $\theta \in [0, \theta_0]$, it follows immediately from (7.7.5) that the integral along AB approaches zero as $R \to \infty$. In the case $\theta_0 = \pi/4$, the integrand in (7.7.5)

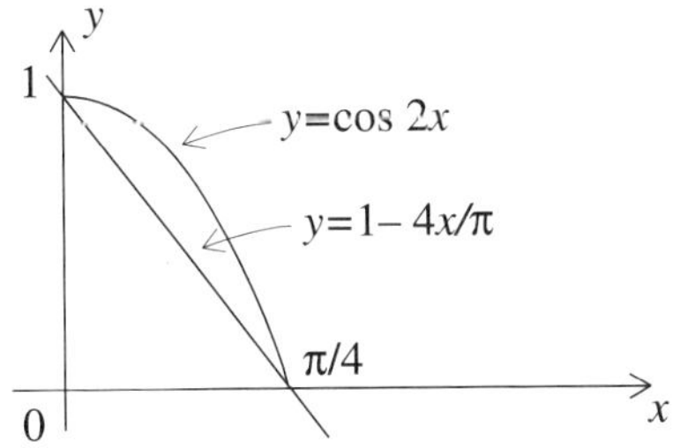

FIGURE 7.4. The inequality $\cos 2x \geq 1 - 4x/\pi$ over the interval $0 \leq x \leq \pi/4$.

is equal to 1 at the upper limit; thus we need a finer investigation. Using the inequality

$$\cos 2\theta \geq 1 - \frac{4}{\pi}\theta, \qquad \text{if } 0 \leq \theta \leq \frac{\pi}{4} \tag{7.7.6}$$

(see Fig 7.4), we obtain from (7.7.5) that

$$\begin{aligned}\left| \int_{AB} e^{-z^2}\, dz \right| &\leq R \int_0^{\theta_0} e^{-R^2(1-4/\pi\theta)}\, d\theta \\ &= R\, e^{-R^2} \frac{\pi}{4R^2} \left(e^{R^2 4/\pi\theta_0} - 1 \right) \to 0\end{aligned}$$

as $R \to \infty$, provided $0 \leq \theta_0 \leq \pi/4$. Therefore, it follows from (7.7.2), as $R \to \infty$, that

$$\begin{aligned}\int_0^\infty e^{-r^2 \cos 2\theta_0} &\left[\cos(r^2 \sin 2\theta_0) - i \sin(r^2 \sin 2\theta_0)\right] \\ &\times \left[\cos\theta_0 + i \sin\theta_0\right] dr = \frac{\sqrt{\pi}}{2}.\end{aligned} \tag{7.7.7}$$

Equating the real and imaginary parts in (7.7.7) we obtain

$$\begin{aligned}\int_0^\infty e^{-r^2 \cos 2\theta_0} &\left[\cos(r^2 \sin 2\theta_0) \cos\theta_0 \right. \\ &\left. + \sin(r^2 \sin 2\theta_0) \sin\theta_0\right] dr = \frac{\sqrt{\pi}}{2}\end{aligned} \tag{7.7.8}$$

and

$$\begin{aligned}\int_0^\infty e^{-r^2 \cos 2\theta_0} &\left[\cos(r^2 \sin 2\theta_0) \sin\theta_0 \right. \\ &\left. - \sin(r^2 \sin 2\theta_0) \cos\theta_0\right] dr = 0,\end{aligned} \tag{7.7.9}$$

which is a system of two linear equations in the unknown integrals

$$J_1(\theta_0) = \int_0^\infty e^{-r^2 \cos 2\theta_0} \cos(r^2 \sin 2\theta_0)\, dr, \tag{7.7.10}$$

$$J_2(\theta_0) = \int_0^\infty e^{-r^2 \cos 2\theta_0} \sin(r^2 \sin 2\theta_0)\, dr. \tag{7.7.11}$$

Written more concisely this system becomes

$$\cos\theta_0 J_1(\theta_0) + \sin\theta_0 J_2(\theta_0) = \frac{\sqrt{\pi}}{2}, \tag{7.7.12}$$

$$\sin\theta_0 J_1(\theta_0) - \cos\theta_0 J_2(\theta_0) = 0. \tag{7.7.13}$$

It follows from (7.7.9) and (7.7.13) that

$$J_1(\theta_0) = \frac{\sqrt{\pi}}{2}\cos\theta_0, \quad J_2(\theta_0) = \frac{\sqrt{\pi}}{2}\sin\theta_0, \qquad 0 \le \theta_0 \le \frac{\pi}{4}. \tag{7.7.14}$$

Finally, letting $\theta_0 = \pi/4$ in (7.7.10), (7.7.11) and (7.7.14), we have the formulae

$$\int_0^\infty \cos r^2\, dr = \frac{1}{2}\sqrt{\frac{\pi}{2}}, \qquad \int_0^\infty \sin r^2\, dr = \frac{1}{2}\sqrt{\frac{\pi}{2}},$$

which coincide with formulae (7.7.1). □

Exercises for Chapter 7

Evaluate the following integrals.

1. $\displaystyle\int_0^\infty \frac{1}{x^2+x+1}\, dx.$

2. $\displaystyle\int_0^\infty \frac{x+4}{x^4+x^2+1}\, dx.$

3. $\displaystyle\int_0^\infty \frac{x+1}{x^4+1}\, dx.$

4. $\displaystyle\int_0^\infty \frac{1}{x^2+2x+2}\, dx.$

5. $\displaystyle\int_0^\infty \frac{\ln x}{(x^2+a^2)(1+b^2x^2)}\, dx, \qquad a>0, \quad b>0.$

6. $\displaystyle\int_0^\infty \frac{(1-x^2)\ln x}{(1+x^2)^2}\, dx.$

7. $\displaystyle\int_0^\infty \frac{x^2 \ln x}{(a^2+b^2x^2)(1+x^2)}\,dx, \qquad ab>0.$

8. $\displaystyle\int_0^\infty \frac{(\ln x)^2}{(x-1)(x+a)}\,dx.$

9. $\displaystyle\int_0^\infty \frac{(\ln x)^2}{x^2+x+1}\,dx.$

10. $\displaystyle\int_0^\infty \frac{(1+x^2)\,(\ln x)^2}{1+x^4}\,dx.$

11. $\displaystyle\int_0^\infty \frac{1}{(x^2+a^2)[(\ln x)^2+\pi^2]}\,dx.$

Prove the following formulae.

12. $\displaystyle\int_0^\infty \frac{(\ln x)^3}{x^2+1}\,dx = 0.$

13. $\displaystyle\int_0^\infty \frac{\ln x}{(x^2+1)^2}\,dx = -\frac{\pi}{4}.$

14. $\displaystyle\int_0^\infty \frac{\ln(x^2+1)}{x^2+1}\,dx = \pi\ln 2.$

Evaluate the following integrals.

15. $\displaystyle\text{p. v.}\int_0^\infty \frac{1}{(x-2)(x^2+4)}\,dx.$

16. $\displaystyle\text{p. v.}\int_0^\infty \frac{1}{(x-4)(x^2+2x+2)}\,dx.$

17. $\displaystyle\int_{-\infty}^\infty \frac{\ln|x-2|}{(x^2+4)(x^2+9)}\,dx.$

18. $\displaystyle\int_{-\infty}^\infty \frac{\ln|x-1|}{x^2+1}\,dx.$

19. $\displaystyle\int_{-\infty}^\infty \frac{x\,\ln|x-1|\,\ln|x-5|}{(x^2+1)(x^2+4)}\,dx.$

20. $\displaystyle\int_{-\infty}^\infty \frac{\ln|x-2|\,\ln|x-4|}{x^2+2x+10}\,dx.$

21. $\displaystyle\int_{-\infty}^\infty \frac{x\,\operatorname{Arctan} x}{(x^2+4x+20)(x^2+1)}\,dx.$

22. $\displaystyle\int_{-\infty}^\infty \frac{\operatorname{Arctan} x}{x^2+3x+8.5}\,dx.$

CHAPTER 8

Advanced Definite Integrals

8.1. Rational functions times trigonometric functions

In this section, new classes of integrals over the real line, announced in [**6**], are evaluated in closed form by means of the calculus of residue. The integrands are a combination of rational and trigonometric functions. Some known tabulated formulae are easily derived from, corrected or completed by means of the general formulae obtained here.

8.1.1. Introduction. We consider the Cauchy principal value of integrals of the form

$$\begin{aligned} I_s &= \text{p. v.} \int_{-\infty}^{\infty} \frac{P_n(x)}{Q_m(x)} \frac{dx}{\sin ax}, \\ I_c &= \text{p. v.} \int_{-\infty}^{\infty} \frac{P_n(x)}{Q_m(x)} \frac{dx}{\cos ax}, \end{aligned} \tag{8.1.1}$$

$$\begin{aligned} I_s^s &= \text{p. v.} \int_{-\infty}^{\infty} \frac{P_n(x)}{Q_m(x)} \frac{\sin bx}{\sin ax}\, dx, \\ I_s^c &= \text{p. v.} \int_{-\infty}^{\infty} \frac{P_n(x)}{Q_m(x)} \frac{\cos bx}{\sin ax}\, dx, \end{aligned} \tag{8.1.2}$$

$$\begin{aligned} I_c^c &= \text{p. v.} \int_{-\infty}^{\infty} \frac{P_n(x)}{Q_m(x)} \frac{\cos bx}{\cos ax}\, dx, \\ I_c^s &= \text{p. v.} \int_{-\infty}^{\infty} \frac{P_n(x)}{Q_m(x)} \frac{\sin bx}{\cos ax}\, dx, \end{aligned} \tag{8.1.3}$$

where $Q_m(x)$ and $P_n(x)$ are real polynomials of the real variable x, of degrees m and n, respectively. We assume that the real zeros of Q_m are simple. We also assume that $m \geq n+1$ and remark that if $P_n(x)/Q_m(x)$ is even, then necessarily $m \geq n+2$.

NOTATION 8.1.1. When suitable, the following notation will be used:

- $\mathcal{A}$ and $\widetilde{\mathcal{A}}$ denote the sets of simple real zeros of $Q_m(x)$ and $Q_m(-x)$, respectively:

$$\mathcal{A} = \{a_k \in \mathbb{R};\ Q_m(a_k) = 0,\ Q'_m(a_k) \neq 0\},$$
$$\widetilde{\mathcal{A}} = \{\tilde{a}_k \in \mathbb{R};\ Q_m(-\tilde{a}_k) = 0, Q'_m(-\tilde{a}_k) \neq 0\};$$

- $\mathcal{Z}$ and $\widetilde{\mathcal{Z}}$ denote the sets of complex zeros of $Q_m(z)$ and $Q_m(-z)$ in the upper half-plane, respectively:

$$\mathcal{Z} = \{z_k \in \mathbb{C},\ \Im z_k > 0;\ Q_m(z_k) = 0\},$$
$$\widetilde{\mathcal{Z}} = \{\tilde{z}_k \in \mathbb{C},\ \Im \tilde{z}_k > 0;\ Q_m(-\tilde{z}_k) = 0\};$$

- $\mathcal{B}_0$, $\mathcal{B}_1$, $\widetilde{\mathcal{B}}_0$, $\widetilde{\mathcal{B}}_1$ denote the sets of admissible values of $a \in \mathbb{R}$:

$$\mathcal{B}_0 = \{a \in \mathbb{R};\ \forall\, a_k \in \mathcal{A}, \sin a a_k \neq 0\},$$
$$\mathcal{B}_1 = \{a \in \mathbb{R};\ \forall\, a_k \in \mathcal{A}, \cos a a_k \neq 0\},$$
$$\widetilde{\mathcal{B}}_0 = \{a \in \mathbb{R},\ \forall\, \tilde{a}_k \in \widetilde{\mathcal{A}}, \sin a\tilde{a}_k \neq 0\}$$
$$\widetilde{\mathcal{B}}_1 = \{a \in \mathbb{R},\ \forall\, \tilde{a}_k \in \widetilde{\mathcal{A}}, \cos a\tilde{a}_k \neq 0\}.$$

The general idea for the evaluation of these integrals is clear: one sums residues at the zeros of $Q_m(x)$ and $\sin ax$ or $\cos ax$. However, general formulae to evaluate these integrals seem to be missing in the literature.

When $|b| \leq |a|$, some particular cases can be found (see [**23**], Sections 3.743–3.749, with references to older handbooks). But in such examples (see, for example, [**18**], pp. 81–82, formulae 30–32, and p. 23, formulae 36–37), it is impossible to take the inverse Fourier sine and cosine transforms because, in these transforms, the parameter y (here denoted b) in I_s^s, I_s^c, I_c^s and I_c^c varies over the interval $[0, +\infty)$.

When $|b| > |a|$, even particular cases of the last four integrals (8.1.2), (8.1.3) seem to be absent from the literature.

The main idea of this section is that, although the number of singular points in these integrals is equal to infinity, these integrals can be expressed by means of a finite number of terms, namely, by the sum of the residues at the zeros of $Q_m(x)$. For the first two integrals (8.1.1), the sum of the residues at the zeros of $\sin ax$ and $\cos ax$, respectively, is equal to zero, and the same holds in the case of the last four integrals (8.1.2), (8.1.3) if $|b| < |a|$. Moreover, if $|b| > |a|$, the corresponding series for the last four integrals can be expressed by a finite sum of residues at the zeros of $Q_m(x)$.

It is found that the last four integrals are equal to the sum of some function of a and b and a $2a$-periodic function of b.

As a by-product, the following four series, denoted by S_1, S_2. S_3 and S_4 in (8.1.45), (8.1.60), (8.1.68) and (8.1.69), respectively,

$$\sum_{k=-\infty}^{\infty} (-1)^k \frac{P_n(k\pi/a)}{Q_m(k\pi/a)} \cos\left(\frac{b\pi k}{a}\right), \qquad \sum_{k=-\infty}^{\infty} (-1)^{k+1} \frac{P_n(k\pi/a)}{Q_m(k\pi/a)} \sin\left(\frac{b\pi k}{a}\right),$$

$$\sum_{k=-\infty}^{\infty} (-1)^k \frac{P_n(\gamma_k)}{Q_m(\gamma_k)} \sin(b\gamma_k), \qquad \sum_{k=-\infty}^{\infty} (-1)^{k+1} \frac{P_n(\gamma_k)}{Q_m(\gamma_k)} \cos(b\gamma_k),$$

where $\gamma_k = (2k+1)\pi/(2a)$, will be evaluated in closed form as a finite sum of residues at the zeros of $Q_m(x)$.

8.1.2. The integral I_s. We derive several formulae for the integral

$$I_s = \text{p. v.} \int_{-\infty}^{\infty} \frac{P_n(x)}{Q_m(x)} \frac{dx}{\sin ax}, \qquad m \geq n+1. \tag{8.1.4}$$

We first consider the case where $P_n(x)/Q_m(x)$ is odd and Q_m has no real zeros.

FORMULA 8.1.1. *If Q_m has no real zeros and $P_n(x)/Q_m(x)$ is odd, then*

$$\text{p. v.} \int_{-\infty}^{\infty} \frac{P_n(x)}{Q_m(x)} \frac{dx}{\sin ax} = 2\pi i \sum_k \operatorname*{Res}_{z_k \in \mathcal{Z}} \left[\frac{P_n(z)}{Q_m(z)} \frac{1}{\sin az}\right], \qquad m \geq n+1. \tag{8.1.5}$$

PROOF. Let

$$f(z) = \frac{P_n(z)}{Q_m(z)} \frac{1}{\sin az} \tag{8.1.6}$$

be a function of the complex variable z and let C be a closed path that consists of parts of the segment $[-R_k, R_k]$ of the real axis, shown in Fig 8.1, with $|a|R_k = (2k+1)\pi/2$, $k = 0, 1, \ldots$, where the zeros of $\sin az$, that is, the points $ax_l = l\pi$, $l = 0, \pm 1, \pm 2, \ldots, \pm k$, are bypassed along the semicircles γ_l of radius δ in the upper half-plane, and the semicircle C_{R_k} of radius R_k. By the residue theorem we have

$$\left(\int_{C_{R_k}} + \int_{-R_k}^{R_k} + \sum_{l=-k}^{k} \int_{\gamma_l}\right) \left[\frac{P_n(z)}{Q_m(z)} \frac{1}{\sin az}\, dz\right] = 2\pi i \sum_k \operatorname*{Res}_{z=z_k} \left[\frac{P_n(z)}{Q_m(z)} \frac{1}{\sin az}\right], \tag{8.1.7}$$

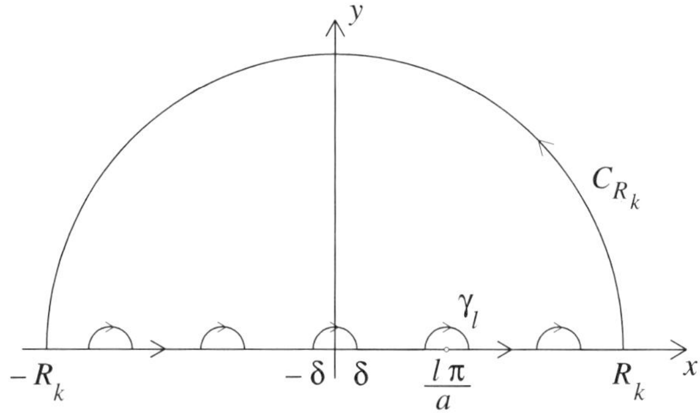

FIGURE 8.1. The path of integration for the integral (8.1.7).

where z_k are the zeros of $Q_m(z)$ that lie inside C and the integral from $-R_k$ to R_k is evaluated along line segments of the x-axis excluding the arcs γ_l.

It is shown in Lemma 8.1.1 that the integral along the arc C_{R_k} approaches zero as $R_k \to \infty$. Since $x_l = l\pi/a$ is a simple pole of $f(z)$, then, by using a Laurent series in a neighborhood of the point x_l, we obtain the formula

$$\int_{\gamma_l} \frac{P_n(z)}{Q_m(z)} \frac{dz}{\sin az} \to -\pi i \operatorname*{Res}_{x_l = l\pi/a} \frac{P_n(z)}{Q_m(z)} \frac{1}{\sin az}$$

$$= -\frac{\pi i}{a} \frac{P_n(x_l)}{Q_m(x_l)} \frac{1}{(-1)^l}, \tag{8.1.8}$$

as $\delta \to 0$. Since $P_n(x)/Q_m(x)$ is odd and continuous,

$$\sum_{l=-k}^{k} \frac{P_n(x_l)}{Q_m(x_l)} (-1)^l = \frac{P_n(0)}{Q_m(0)} = 0. \tag{8.1.9}$$

Therefore, taking the limit in (8.1.7), as $R_k \to \infty$ and $\delta \to 0$, we obtain (8.1.5). □

We note that in (8.1.9) P_n is odd, Q_m is even and $Q_m(0) \neq 0$ since $Q_m(x)$ has no real zeros. The case where P_n is even and Q_m is odd is impossible, because $Q_m(0) = 0$ contradicts the assumptions of Formula 8.1.1.

LEMMA 8.1.1. *The first integral along the arc C_{R_k} in* (8.1.7) *approaches zero as $R_k \to \infty$:*

$$\lim_{R_k \to \infty} \int_{C_{R_k}} \frac{P_n(z)}{Q_m(z)} \frac{1}{\sin az}\, dz = 0, \qquad m \geq n + 1. \tag{8.1.10}$$

PROOF. Since $\sin z = \sin(x+iy) = \sin x \cosh y + i \cos x \sinh y$, one has

$$|\sin az| = \sqrt{\sinh^2 ay + \sin^2 ax}. \tag{8.1.11}$$

Then, a heuristic argument gives

$$\begin{aligned}
\left| \int_{C_{R_k}} \frac{P_n(z)}{Q_m(z)} \frac{1}{\sin az} dz \right| &\leq \int_{C_{R_k}} \left| \frac{P_n(z)}{Q_m(z)} \right| \frac{|dz|}{|\sin az|} \\
&= \int_0^\pi \left| \frac{P_n\left(R_k e^{i\theta}\right)}{Q_m\left(R_k e^{i\theta}\right)} \right| \frac{R_k |e^{i\theta}|\,|i|\,d\theta}{\sqrt{\sinh^2\left(aR_k \sin\theta\right) + \sin^2\left(aR_k \cos\theta\right)}} \\
&\leq C \int_0^\pi \frac{d\theta}{\sqrt{\sinh^2\left(aR_k \sin\theta\right) + \sin^2\left(aR_k \cos\theta\right)}} \\
&\to 0, \qquad \text{as} \quad R_k \to \infty,
\end{aligned} \tag{8.1.12}$$

since $m \geq n+1$ and the integrand approaches zero as $R_k \to \infty$ in the sector $0 < \theta < \pi$ and is equal to 1 if $\theta = 0$ or $\theta = \pi$ for all R_k (C = constant > 0).

To supply a rigorous proof of (8.1.10) we choose an arbitrary $\varepsilon > 0$ and divide the interval of integration in the last integral in (8.1.12) into three parts:

$$0 \leq \theta \leq \theta_0, \qquad \theta_0 \leq \theta \leq \pi - \theta_0, \qquad \pi - \theta_0 \leq \theta \leq \pi.$$

Since $|\sin az| = 1$ for $\theta = 0$ and $\theta = \pi$, then, by taking θ_0 sufficiently small, we obtain (by the continuity of the function (8.1.11)) that the following inequality is satisfied in the intervals $0 \leq \theta \leq \theta_0$ and $\pi - \theta_0 \leq \theta \leq \pi$:

$$|\sin az|\big|_{z \in C_{R_k}} \geq \frac{1}{2}, \qquad \text{that is,} \qquad \frac{1}{|\sin az|\big|_{z \in C_{R_k}}} \leq 2. \tag{8.1.13}$$

Then the moduli of the integrals with respect to the first and third intervals are smaller than $2C\theta_0$; therefore, with θ_0 sufficiently small, one can satisfy the inequality

$$2C\theta_0 \leq \frac{1}{3}\varepsilon \qquad \text{if} \qquad \theta_0 \leq \frac{1}{6C}\varepsilon. \tag{8.1.14}$$

Since, by the nonheuristic part of the heuristic argument, the integral over the interval $\theta_0 \leq \theta \leq \pi - \theta_0$ approaches zero as $R_k \to \infty$, there exists a constant K such that for all $k \geq K$ the following inequality is satisfied:

$$C \int_{\theta_0}^{\pi - \theta_0} \frac{1}{|\sin az|_{z \in C_{R_k}}} d\theta < \frac{\varepsilon}{3}. \tag{8.1.15}$$

Therefore, for all $\varepsilon > 0$, there exists a constant K such that, for all $k \geq K$,

$$C\int_0^{\pi} \frac{1}{|\sin az|_{z\in C_{R_k}}}\, d\theta \;<\; \varepsilon \quad \text{or} \quad \left|\int_{C_{R_k}} \frac{P_n(z)}{Q_m(z)}\,\frac{dz}{\sin az}\right| \;<\; \varepsilon. \qquad (8.1.16)$$

The last two inequalities imply that the limit on the left-hand side in (8.1.10) exists and is equal to zero. □

EXAMPLE 8.1.1. *Derive Formula* 3.747(3) *in* [**23**]:

$$I_1 = \text{p. v.} \int_{-\infty}^{\infty} \frac{x}{x^2+\beta^2}\,\frac{dx}{\sin ax} = \frac{\pi}{\sinh(a\beta)}, \qquad \Re\beta > 0. \quad (8.1.17)$$

SOLUTION. The formula follows from (8.1.5) with $n = 1$, $P_n(x) = x$, $m = 2$, $Q_m(x) = x^2 + \beta^2 \neq 0$ for real x. Since $z = \beta i$ is the only pole of $P_n(z)/Q_m(z)$ in the upper half-plane, then

$$I_1 = 2\pi i \operatorname*{Res}_{z=\beta i}\left[\frac{z}{z^2+\beta^2}\,\frac{1}{\sin az}\right] = 2\pi i\frac{\beta i}{2\beta i}\,\frac{1}{\sin(a\beta i)} = \frac{\pi}{\sinh(a\beta)}. \quad \Box$$

Formula 8.1.1 is easily generalized to the following formula.

FORMULA 8.1.2. *If Q_m has no real zeros and $m \geq n+l+2$, then*

$$\begin{aligned}\text{p. v.} \int_{-\infty}^{\infty} &\frac{P_n(x)}{Q_m(x)}\,\frac{x^l\,dx}{\prod_{k=1}^{l}\sin a_k x}\\ &= 2\pi i \sum_k \operatorname*{Res}_{z_k\in\mathcal{Z}}\left[\frac{P_n(z)}{Q_m(z)}\,\frac{z^l}{\prod_{k=1}^{l}\sin a_k z}\right], \qquad (8.1.18)\end{aligned}$$

provided $P_n(x)/Q_m(x)$ is even and a_i/a_j is not equal to a rational number (in other words, the zeros of $\sin a_i x$ and $\sin a_j x$ do not coincide if $i \neq j$). □

Integrals of the form (8.1.18), for the case $l > 1$, seem to be absent from handbooks, even in the form of examples. An instance of such a formula is the integral

$$\text{p. v.} \int_{-\infty}^{\infty} \frac{P_n(x)}{Q_m(x)}\,\frac{x^3\,dx}{\sin a_1x\sin a_2x\sin a_3x}, \qquad m \geq n+5, \quad (8.1.19)$$

where $P_n(x)/Q_m(x)$ is even. In this case, the finite sums, of the form (8.1.9), of the residues at the zeros of $\sin a_1 x$, $\sin a_2 x$ and $\sin a_3 x$ are equal to zero since the functions

$$\frac{x^3P_n(x)}{Q_m(x)\sin a_2x\sin a_3x}, \quad \frac{x^3P_n(x)}{Q_m(x)\sin a_1x\sin a_3x}, \quad \frac{x^3P_n(x)}{Q_m(x)\sin a_1x\sin a_2x}$$

are odd.

EXAMPLE 8.1.2. *Derive the formula*

$$\begin{aligned} I_2 &= \text{p. v.} \int_{-\infty}^{\infty} \frac{1}{(x^2+\alpha^2)(x^2+\beta^2)} \frac{x^2\,dx}{\sin ax \sin bx} \\ &= \frac{\pi}{-\alpha^2+\beta^2}\left[\frac{\alpha}{\sinh a\alpha \sinh b\alpha} - \frac{\beta}{\sinh a\beta \sinh b\beta}\right], \end{aligned}$$

where a, b, α, $\beta > 0$, *and* $a/b \notin \mathbb{Q}$.

SOLUTION. The formula follows from (8.1.18). In fact, we have

$$\begin{aligned} I_2 &= 2\pi i \left(\operatorname*{Res}_{z=\alpha i} + \operatorname*{Res}_{z=\beta i}\right)\left[\frac{z^2}{(z^2+\alpha^2)(z^2+\beta^2)\sin az \sin bz}\right] \\ &= 2\pi i \left[\frac{-\alpha^2}{2\alpha i(\beta^2-\alpha^2)(-\sinh a\alpha \sinh b\alpha)}\right. \\ &\qquad \left. + \frac{-\beta^2}{2\beta i(\alpha^2-\beta^2)(-\sinh a\beta \sinh b\beta)}\right] \\ &= \frac{\pi}{\beta^2-\alpha^2}\left[\frac{\alpha}{\sinh a\alpha \sinh b\alpha} - \frac{\beta}{\sinh a\beta \sinh b\beta}\right]. \quad \square \end{aligned}$$

Note 8.1.1. In formula (8.1.18), instead of $P_n(x)x^l$ and $P_n(z)z^l$, one may have

$$P_n(x)\prod_{k=1}^{p}\sin b_k x \prod_{k=1}^{q}\cos c_k x \quad \text{and} \quad P_n(z)\prod_{k=1}^{p}\sin b_k z \prod_{k=1}^{q}\cos c_k z,$$

respectively, if

$$p \geq l, \qquad m \geq n+1, \qquad \left|\sum_{k=1}^{p} b_k + \sum_{k=1}^{q} c_k\right| < \left|\sum_{k=1}^{l} a_k\right|,$$

$P_n(x)/Q_m(x)$ is even for $p-l$ even, and $P_n(x)/Q_m(x)$ is odd for $p-l$ odd.

Second, we consider the case where the function $P_n(x)/Q_m(x)$ is neither even nor odd and Q_m has no real zeros.

FORMULA 8.1.3. *If* Q_m *has no real zeros,* $P_n(x)/Q_m(x)$ *is neither even nor odd and* $m \geq n+1$, *then*

$$\begin{aligned} &\text{p. v.} \int_{-\infty}^{\infty} \frac{P_n(x)}{Q_m(x)\sin ax}\,dx \\ &\quad = \pi i \sum_k \left\{\operatorname*{Res}_{z_k\in\mathcal{Z}}\left[\frac{P_n(z)}{Q_m(z)\sin az}\right] - \operatorname*{Res}_{\tilde{z}_k\in\widetilde{\mathcal{Z}}}\left[\frac{P_n(-z)}{Q_m(-z)\sin az}\right]\right\}, \qquad (8.1.20) \end{aligned}$$

where $z_k \in \mathcal{Z}$ *and* $\tilde{z}_k \in \widetilde{\mathcal{Z}}$.

PROOF. If we represent $f(x) = P_n(x)/Q_m(x)$ as the sum of an odd and an even function,

$$f(x) = \frac{1}{2}[f(x) - f(-x)] + \frac{1}{2}[f(x) + f(-x)]$$
$$=: \frac{\widetilde{P}_n(x)}{\widetilde{Q}_m(x)} + \frac{\widehat{P}_n(x)}{\widehat{Q}_m(x)}, \tag{8.1.21}$$

then

$$\frac{\widetilde{P}_n(x)}{\widetilde{Q}_m(x)\sin ax} \quad \text{and} \quad \frac{\widehat{P}_n(x)}{\widehat{Q}_m(x)\sin ax}$$

are even and odd, respectively. Therefore

$$\text{p. v.} \int_{-\infty}^{\infty} \frac{\widehat{P}_n(x)}{\widehat{Q}_m(x)\sin ax}\, dx = 0. \quad \square$$

Formula (8.1.20), in contrast with (8.1.5), allows one to evaluate the integral in the case $P_n(x)/Q_m(x)$ is not odd.

EXAMPLE 8.1.3. *Evaluate the integral*

$$I_3 = \text{p. v.} \int_{-\infty}^{\infty} \frac{dx}{(x^2 + 2x + 2)\sin ax}. \tag{8.1.22}$$

SOLUTION. By means of (8.1.20) with $Q_m(z) = z^2 + 2z + 2$, we have

$$Q_m(z) = 0 \quad \Rightarrow \quad z^2 + 2z + 2 = 0 \quad \Rightarrow \quad z_1 = -1 + i, \quad z_2 = -1 - i,$$
$$Q_m(-z) = 0 \quad \Rightarrow \quad z^2 - 2z + 2 = 0 \quad \Rightarrow \quad \tilde{z}_1 = 1 + i, \quad \tilde{z}_2 = 1 - i.$$

The zeros of $Q_m(z)$ and $Q_m(-z)$ in the upper half-plane are z_1 and $\tilde{z}_1$, respectively. Thus,

$$\begin{aligned}
I_3 &= \pi i \left\{ \operatorname*{Res}_{z=-1+i} \left[\frac{1}{(z^2 + 2z + 2)\sin az} \right] - \operatorname*{Res}_{z=1+i} \left[\frac{1}{(z^2 - 2z + 2)\sin az} \right] \right\} \\
&= \pi i \left[\frac{1}{2(z+1)\sin az}\bigg|_{z=-1+i} - \frac{1}{2(z-1)\sin az}\bigg|_{z=1+i} \right] \\
&= \frac{\pi}{2\sin[a(-1+i)]} - \frac{\pi}{2\sin[a(1+i)]} \\
&= -2\Re \left\{ \frac{\pi}{2\sin[a(1+i)]} \right\} \\
&= -\Re \left[\frac{1}{\sin a \cosh a + i \cos a \sinh a} \right]
\end{aligned}$$

$$= -\Re\left[\frac{\sin a \cosh a - i \cos a \sinh a}{\sin^2 a \cosh^2 a + \cos^2 a \sinh^2 a}\right]$$
$$= -\frac{\sin a \cosh a}{\sinh^2 a + \sin^2 a}. \quad \square$$

Thirdly, we show that formulae (8.1.5) and (8.1.20) are still valid if Q_m has real zeros.

We suppose that Q_m has real zeros $a_k \in \mathcal{A}$ for $k = 1, 2, \dots, l$ and $a \in \mathcal{B}_0$. Bypassing the singular points a_k on the segment $[-R_k, R_k]$ along semicircles of radius δ in the upper half-plane, we find that the term

$$A = \pi i \sum_{k=1}^{l} \operatorname*{Res}_{a_k \in \mathcal{A}} \left[\frac{P_n(z)}{Q_m(z) \sin az}\right] \tag{8.1.23}$$

has to be added to the right-hand side of (8.1.5), and the term

$$B = \frac{\pi i}{2} \sum_{k=1}^{l} \left\{ \operatorname*{Res}_{a_k \in \mathcal{A}} \left[\frac{P_n(z)}{Q_m(z) \sin az}\right] - \operatorname*{Res}_{\tilde{a}_k \in \widetilde{\mathcal{A}}} \left[\frac{P_n(-z)}{Q_m(-z) \sin az}\right] \right\} \tag{8.1.24}$$

has to be added to the right-hand side of (8.1.20), where $\tilde{a}_k \in \widetilde{\mathcal{A}}$ and $a \in \mathcal{B}_0 \cap \widetilde{\mathcal{B}}_0$. But these two terms are zero, as proven in the following lemma.

LEMMA 8.1.2. *If Q_m has real zeros $a_k \in \mathcal{A}$ for $k = 1, 2, \dots, l$, and $a \in \mathcal{B}_0$, then the finite sums A in* (8.1.23) *and B in* (8.1.24) *are both equal to zero.*

PROOF. We first show that $A = 0$. In (8.1.23), $P_n(z)/Q_m(z)$ and $P_n(z)$ are odd, $Q_m(z)$ is even, and $Q_m(0) \neq 0$ because $\sin(aa_k) \neq 0$. Hence, if $Q_m(a_k) = 0$, $k = 1, 2, \dots, l$, then $Q_m(-a_k) = 0$, that is, $l = 2p$ is even. Thus the zeros of $Q_m(x)$ are $a_{-p}, a_{-p+1}, \dots, a_{-1}, a_1, a_2, \dots, a_p$, where $a_{-r} = -a_r$ for $r = 1, 2, \dots, p$. It then follows from (8.1.23) that

$$\begin{aligned}
A &= \pi i \left(\sum_{k=-p}^{-1} + \sum_{k=1}^{p} \right) \operatorname*{Res}_{a_k \in \mathcal{A}} \left[\frac{P_n(z)}{Q_m(z) \sin az}\right] \\
&= \pi i \left(\sum_{k=-p}^{-1} + \sum_{k=1}^{p} \right) \left[\frac{P_n(a_k)}{Q'_m(a_k) \sin aa_k}\right] \\
&= 0,
\end{aligned}$$

because $Q'_m(z)$ is odd, so that $P_n(z)/[Q'_m(z) \sin az]$ is odd.

Next, to show that $B = 0$, we consider the auxiliary odd function

$$\begin{aligned} f(z) &= \frac{P_n(z)}{Q_m(z)} - \frac{P_n(-z)}{Q_m(-z)} \\ &= \frac{P_n(z)Q_m(-z) - P_n(-z)Q_m(z)}{Q_m(z)Q_m(-z)}. \end{aligned} \tag{8.1.25}$$

If $Q_m(a_k) = 0$ and

$$|a_i| \neq |a_j|, \qquad \text{if} \quad i \neq j, \tag{8.1.26}$$

then $Q_m(-a_k) \neq 0$ since $Q_m(z)$ is neither odd nor even. Thus, if a_k are the simple real zeros of $Q_m(z)$, then the even function

$$\psi(z) = Q_m(z)Q_m(-z)$$

also has only simple real zeros and $\psi(0) = Q_m^2(0) \neq 0$ because $\sin(aa_k) \neq 0$ and $\sin(a\tilde{a}_k) \neq 0$, where $\tilde{a}_k$ are the simple real zeros of $Q_m(-z)$. If we let the zeros of $\psi(z)$ be $\hat{a}_{-p}, \hat{a}_{-p+1}, \dots, \hat{a}_{-1}, \hat{a}_1, \hat{a}_2, \dots, \hat{a}_p$, where $\hat{a}_{-r} = -a_r$ for $r = 1, 2, \dots, p$, and consider the odd function

$$\phi(z) = P_n(z)Q_m(-z) - P_n(-z)Q_m(z),$$

then from (8.1.23) we have

$$\begin{aligned} B &= \left(\sum_{k=-p}^{-1} + \sum_{k=1}^{p} \right) \operatorname*{Res}_{z=\hat{a}_k} \left[\frac{\phi(z)}{\psi(z) \sin az} \right] \\ &= \left(\sum_{k=-p}^{-1} + \sum_{k=1}^{p} \right) \left[\frac{\phi(\hat{a}_k)}{\psi'(\hat{a}_k) \sin a\hat{a}_k} \right] \\ &= 0, \end{aligned}$$

because $\psi'(z)$ is odd, and hence $\phi(z)/[\psi'(z)\sin az]$ is odd.

Suppose now that condition (8.1.26) does not hold. For instance, let $a_1 = -a_2$, hence $|a_1| = |a_2|$, but for the remaining values of i and j (8.1.26) holds. Then

$$Q_m(z) = \left(z^2 - a_1^2\right) \psi_{m-2}(z),$$

where (8.1.26) holds for the polynomial $\psi_{m-2}(z)$. Then the even function

$$Q_m(z)Q_m(-z) = \left(z^2 - a_1^2\right)^2 \psi_{m-2}(z)\psi_{m-2}(-z)$$

has a pair of double zeros at $z = a_1$ and $z = -a_1$. However, in this case, the function $\phi(z)$ contains the factor $z^2 - a_1^2$ and the function $f(z)$ is of the form

$$f(z) = \frac{P_n(z)\psi_{m-2}(-z) - P_n(-z)\psi_{m-2}(z)}{\left(z^2 - a_1^2\right)\psi_{m-2}(z)\psi_{m-2}(-z)},$$

that is, $f(z)$ is an odd function with only simple real poles. Therefore the equality $B = 0$ is still valid. □

COROLLARY 8.1.1. *Formulae* (8.1.5) *and* (8.1.20) *still hold if* Q_m *has real zeros* $a_k \in \mathcal{A}$ *for* $k = 1, 2, \dots, l$, *and* $a \in \mathcal{B}_0$.

PROOF. The corollary follows from Lemma 8.1.2. □

Note 8.1.2. Let $f(z)$ denote any of the three functions

$$\frac{P_n(z)}{Q_m(z)\cos az}, \qquad \frac{P_n(z)}{Q_m(z)}\,\frac{\sin bz}{\sin az}, \qquad \frac{P_n(z)}{Q_m(z)}\,\frac{\cos bz}{\cos az}, \tag{8.1.27}$$

where $P_n(z)/Q_m(z)$ is even, or any of the functions

$$\frac{P_n(z)}{Q_m(z)}\,\frac{\cos bz}{\sin az}, \qquad \frac{P_n(z)}{Q_m(z)}\,\frac{\sin bz}{\cos az}, \tag{8.1.28}$$

where $P_n(z)/Q_m(z)$ is odd. Then it can be shown, as in the proof of Lemma 8.1.2, that the finite sum of residues of f at $a_k \in \mathcal{A}$ is zero:

$$\sum_{k=-p}^{p} \operatorname*{Res}_{a_k \in \mathcal{A}} f(z) = 0.$$

In (8.1.27), since $Q_m(x)$ is even, $Q_m(0) \neq 0$ because an even function cannot have the simple zero $a_k = 0$. For the first function of (8.1.28), $Q_m(0) \neq 0$ because $\sin(a_k a) \neq 0$; however, for the second function, $Q_m(x)$ may be odd and then $Q_m(0) = 0$, but $Q_m'(0) \neq 0$. In this last case

$$\operatorname*{Res}_{z=0}\left[\frac{P_n(z)}{Q_m(z)}\,\frac{\sin bz}{\cos az}\right] = 0$$

because $\sin 0 = 0$.

8.1.3. The integral I_c. In this subsection, we derive several formulae for the integral

$$I_c = \text{p. v.} \int_{-\infty}^{\infty} \frac{P_n(x)}{Q_m(x)\cos ax}\,dx, \qquad m \geq n+1. \tag{8.1.29}$$

We first consider the case where $P_n(x)/Q_m(x)$ is even and Q_m has no real zeros.

FORMULA 8.1.4. *If* Q_m *has no real zeros and* $P_n(x)/Q_m(x)$ *is even, then*

$$\text{p. v.} \int_{-\infty}^{\infty} \frac{P_n(x)}{Q_m(x)}\,\frac{dx}{\cos ax} = 2\pi i \sum_k \operatorname*{Res}_{z_k \in \mathcal{Z}} \frac{P_n(z)}{Q_m(z)}\,\frac{1}{\cos az},$$
$$m \geq n+2. \tag{8.1.30}$$

PROOF. If $P_n(x)/Q_m(x)$ is an even function and $m \geq n+1$, then $m \geq n+2$ because $P_n(x)$ and $Q_m(x)$ are both even (if $P_n(x)$ and $Q_m(x)$ were both odd, then one power of x would cancel out). We show that, if $Q_m(x) \neq 0$ for real x, (8.1.30) is obtained from (8.1.5) by replacing $\sin ax$ with $\cos ax$. To prove this, it is sufficient to show that the series, S, of residues at the zeros of $\cos ax$ is equal to zero. Then the reader need only verify that the rest of the derivation is as in Subsection 8.1.2, with the appropriate modifications to Fig 8.1. The series S is

$$\begin{aligned} S &:= \sum_{k=-\infty}^{\infty} \operatorname*{Res}_{z=(2k+1)\pi/(2a)} \left[\frac{P_n(z)}{Q_m(z)} \frac{1}{\cos az} \right] \\ &= \frac{1}{a} \sum_{k=-\infty}^{\infty} (-1)^{k+1} \frac{P_n(x_{2k+1})}{Q_m(x_{2k+1})}, \end{aligned} \tag{8.1.31}$$

where $x_{2k+1} = (2k+1)\pi/(2a)$ are the zeros of $\cos ax$. Since $m \geq n+2$, the series (8.1.31) is absolutely convergent. Moreover, since $P_n(x)/Q_m(x)$ is an even function of x, we can use the notation

$$\frac{1}{a} \frac{P_n(x_{2k+1})}{Q_m(x_{2k+1})} = F\left(x_{2k+1}^2\right). \tag{8.1.32}$$

Inserting (8.1.32) into (8.1.31), we have

$$\begin{aligned} S &= \sum_{k=-\infty}^{\infty} (-1)^{k+1} F\left(x_{2k+1}^2\right) \\ &= \sum_{k=-\infty}^{-1} (-1)^{k+1} F\left(x_{2k+1}^2\right) + \sum_{k=0}^{\infty} (-1)^{k+1} F\left(x_{2k+1}^2\right). \end{aligned} \tag{8.1.33}$$

Now putting $k = -l-1$ in the first term on the right-hand side and changing the summation from 0 to ∞, as k changes from $-\infty$ to -1, we have

$$S = \sum_{l=0}^{\infty} (-1)^{-l} F\left(x_{2(-l-1)+1}^2\right) + \sum_{k=0}^{\infty} (-1)^{k+1} F\left(x_{2k+1}^2\right) = 0,$$

since

$$x_{2(-l-1)+1}^2 = x_{-2l-1}^2 = \left[\frac{(-2l-1)\pi}{2a}\right]^2 = x_{2l+1}^2$$

and $(-1)^{-l} = (-1)^l$. This implies (8.1.30). □

EXAMPLE 8.1.4. *From* (8.1.30) *we have the formula*

$$\begin{aligned} I_4 = \text{p. v.} \int_{-\infty}^{\infty} \frac{1}{x^2+\beta^2} \frac{dx}{\cos bx} &= 2\pi i \operatorname*{Res}_{z=\beta i} \left[\frac{1}{z^2+\beta^2} \frac{1}{\cos bz} \right] \\ &= 2\pi i \frac{1}{2\beta i} \frac{1}{\cos b\beta i} \\ &= \frac{\pi}{\beta \cosh b\beta}. \end{aligned}$$

This is a particular case of formula 3.743(4) *in* [**23**], *p. 416, with* $a = 0$ *in the integrand*

$$\frac{\cos(ax)}{\cos(bx)} \frac{1}{x^2+\beta^2}. \quad \square$$

Using a technique similar to the one used for the derivation of formula (8.1.18), we can easily generalize formula (8.1.30) to the following formula.

FORMULA 8.1.5. *If* Q_m *has no real zeros,* $P_n(x)/Q_m(x)$ *is even and* $m \geq n+2$, *then*

$$\begin{aligned} \text{p. v.} \int_{-\infty}^{\infty} \frac{P_n(x)}{Q_m(x)} \frac{1}{\prod_{k=1}^{l} \cos a_k x}\, dx \\ = 2\pi i \sum_k \operatorname*{Res}_{z_k \in \mathcal{Z}} \left[\frac{P_n(z)}{Q_m(z)} \frac{1}{\prod_{k=1}^{l} \cos a_k z} \right], \end{aligned} \tag{8.1.34}$$

provided a_i/a_j *is not equal to a rational number for* $i \neq j$.

Note 8.1.3. In formula (8.1.34), instead of $P_n(x)$ and $P_n(z)$, one may have

$$P_n(x) \prod_{k=1}^{p} \sin b_k x \prod_{k=1}^{q} \cos c_k x \quad \text{and} \quad P_n(z) \prod_{k=1}^{p} \sin b_k z \prod_{k=1}^{q} \cos c_k z,$$

respectively, if

$$m \geq n+1, \qquad \left| \sum_{k=1}^{p} b_k + \sum_{k=1}^{q} c_k \right| < \left| \sum_{k=1}^{l} a_k \right|,$$

$P_n(x)/Q_m(x)$ is even for p even, and $P_n(x)/Q_m(x)$ is odd for p odd.

A similar formula holds for the integral

$$\text{p. v.} \int_{-\infty}^{\infty} \frac{P_n(x)}{Q_m(x)} \frac{x^l}{\prod_{k=1}^{l} \sin a_k x \prod_{r=1}^{p} \cos b_r x}\, dx, \tag{8.1.35}$$

if $P_n(x)/Q_m(x)$ is even, a_i/a_j and b_i/b_j are not equal to rational numbers for $i \neq j$ and $m \geq n+l+2$.

Note 8.1.4. In formula (8.1.35), instead of $P_n(x)x^l$ one may have

$$P_n(x)\prod_{k=1}^{q}\sin c_k x\prod_{k=1}^{s}\cos d_k x,$$

if

$$q\ge l,\qquad m\ge n+1,\qquad \left|\sum_{k=1}^{q}c_k+\sum_{k=1}^{s}d_k\right|<\left|\sum_{k=1}^{l}a_k+\sum_{k=1}^{p}b_k\right|,$$

$P_n(x)/Q_m(x)$ is even for $l+q$ even, and $P_n(x)/Q_m(x)$ is odd for $l+q$ odd.

Second, we consider the case where the function $P_n(x)/Q_m(x)$ is neither even nor odd and Q_m has no real zeros.

FORMULA 8.1.6. *If $Q_m(x)\neq 0$, for real x, $m\ge n+1$ and $a\in\mathcal{B}_1\cap\widetilde{\mathcal{B}}_1$, then*

$$\begin{aligned}\text{p. v.}&\int_{-\infty}^{\infty}\frac{P_n(x)}{Q_m(x)}\frac{dx}{\cos ax}\\&=\pi i\sum_k\left\{\operatorname*{Res}_{z_k\in\mathcal{Z}}\left[\frac{P_n(z)}{Q_m(z)\cos az}\right]+\operatorname*{Res}_{\tilde z_k\in\widetilde{\mathcal{Z}}}\left[\frac{P_n(-z)}{Q_m(-z)\cos az}\right]\right\},\end{aligned}\tag{8.1.36}$$

where $z_k\in\mathcal{Z}$ and $\tilde z_k\in\widetilde{\mathcal{Z}}$.

PROOF. If $P_n(x)/Q_m(x)$ is neither even nor odd, $m\ge n+1$ and Q_m has no real zeros, then the value of the integral (8.1.29) is obtained by representing $P_n(x)/Q_m(x)$ as the sum of an even and an odd functions. Since the function

$$\frac{1}{\cos ax}\left[\frac{P_n(x)}{Q_m(x)}-\frac{P_n(-x)}{Q_m(-x)}\right]$$

is odd, then its integral from $-\infty$ to $+\infty$ is equal to zero, so that the value of (8.1.29) is given by (8.1.36). □

EXAMPLE 8.1.5. *Obtain the following formula (cf. (8.1.22)):*

$$I_5=\text{p. v.}\int_{-\infty}^{\infty}\frac{dx}{(x^2+2x+2)\cos ax}=\frac{\cos a\cosh a}{\sinh^2 a+\cos^2 a}.$$

SOLUTION. By (8.1.36) we have

$$\begin{aligned}I_5&=\pi i\left\{\operatorname*{Res}_{z=-1+i}\left[\frac{1}{(z^2+2z+2)\cos az}\right]+\operatorname*{Res}_{z=1+i}\left[\frac{1}{(z^2-2z+2)\cos az}\right]\right\}\\&=\pi i\left[\frac{1}{2(z+1)\cos az|_{z=-1+i}}+\frac{1}{2(z-1)\cos az|_{z=1+i}}\right]\\&=\frac{\pi}{2\cos[a(-1+i)]}+\frac{\pi}{2\cos[a(1+i)]}\end{aligned}$$

$$
\begin{aligned}
&= 2\Re\left\{\frac{\pi}{2\cos[a(1+i)]}\right\} \\
&= \Re\left[\frac{1}{\cos a\cosh a - i\sin a\sinh a}\right] \\
&= \Re\left[\frac{\cos a\cosh a + i\sin a\sinh a}{\cos^2 a\cosh^2 a + \sin^2 a\sinh^2 a}\right] \\
&= \frac{\cos a\cosh a}{\sinh^2 a + \cos^2 a}. \qquad \square
\end{aligned}
$$

Thirdly, we show that formulae (8.1.30) and (8.1.36) are still valid if Q_m has real zeros.

If $a_k \in \mathcal{A}$ for $k = 1, 2, \ldots, l$, and $a \in \mathcal{B}_1$, then, as in Subsection 8.1.2, the term

$$\pi i \sum_{k=1}^{l} \operatorname*{Res}_{a_k \in \mathcal{A}} \left[\frac{P_n(z)}{Q_m(z)\cos az}\right] \tag{8.1.37}$$

should be added to the right-hand side of (8.1.30), and, if $a \in \mathcal{B}_1 \cap \widetilde{\mathcal{B}}_1$, the term

$$\frac{\pi i}{2} \sum_{k=1}^{l} \left\{\operatorname*{Res}_{a_k \in \mathcal{A}} \left[\frac{P_n(z)}{Q_m(z)\cos az}\right] + \operatorname*{Res}_{\tilde{a}_k \in \widetilde{\mathcal{A}}} \left[\frac{P_n(-z)}{Q_m(-z)\cos az}\right]\right\} \tag{8.1.38}$$

should be added to the right-hand side of (8.1.36), where $\tilde{a}_k \in \widetilde{\mathcal{A}}$. But these terms are equal to zero by Lemma 8.1.2. Thus formulae (8.1.30) and (8.1.36) also hold if $a_k \in \mathcal{A}$ and $a \in \mathcal{B}_1$. We then have the following corollary.

COROLLARY 8.1.2. *Formulae* (8.1.30) *and* (8.1.36) *still hold if Q_m has real zeros $a_k \in \mathcal{A}$ for $k = 1, 2, \ldots, l$, and $a \in \mathcal{B}_1$.*

8.1.4. The integrals I_s^s and I_s^c. In this subsection we consider the integrals

$$
\begin{aligned}
I_s^s &= \text{p. v.} \int_{-\infty}^{\infty} \frac{P_n(x)}{Q_m(x)} \frac{\sin bx}{\sin ax}\, dx, \\
I_s^c &= \text{p. v.} \int_{-\infty}^{\infty} \frac{P_n(x)}{Q_m(x)} \frac{\cos bx}{\sin ax}\, dx,
\end{aligned} \tag{8.1.39}
$$

where $m \geq n + 2$ and $m \geq n + 1$ for the first and second integrals, respectively.

We begin with the first integral, I_s^s. There are two cases to be considered:

(1) $|b| \le |a|$,

(2) $|b| > |a|$.

In the **first case**, $|b| \le |a|$, the following condition is satisfied on the arc C_{R_k} as $R_k \to \infty$ (see Fig 8.1 and formula (8.1.11)):

$$\lim_{R_k\to\infty} \frac{P_n(z)}{Q_m(z)} \frac{\sin bz}{\sin az}\bigg|_{z\in C_{R_k}} = 0. \tag{8.1.40}$$

Therefore the derivation procedure that has led to formulae (8.1.5) and (8.1.20) is valid, but for the case which has led to formula (8.1.5), the function $P_n(x)/Q_m(x)$ must be even for the first integral in (8.1.39) and odd for the second one. Hence, we have the following pair of formulae.

FORMULA 8.1.7. *If $|b| \le |a|$, $P_n(x)/Q_m(x)$ is even and $m \ge n+2$, then*

$$\text{p. v.}\int_{-\infty}^{\infty} \frac{P_n(x)}{Q_m(x)} \frac{\sin bx}{\sin ax}\,dx = 2\pi i \sum_k \operatorname*{Res}_{z_k\in\mathcal{Z}} \left[\frac{P_n(z)}{Q_m(z)} \frac{\sin bz}{\sin az}\right], \tag{8.1.41}$$

where $z_k \in \mathcal{Z}$, $a_k \in \mathcal{A}$ for $k = 1, 2, \dots, l$, and $a \in \mathcal{B}_0$.
Similarly, if $|b| < |a|$, $P_n(x)/Q_m(x)$ is odd and $m \ge n+1$, then

$$\text{p. v.}\int_{-\infty}^{\infty} \frac{P_n(x)}{Q_m(x)} \frac{\cos bx}{\sin ax}\,dx = 2\pi i \sum_k \operatorname*{Res}_{z_k\in\mathcal{Z}} \left[\frac{P_n(z)}{Q_m(z)} \frac{\cos bz}{\sin az}\right]. \tag{8.1.42}$$

In formulae (8.1.41) and (8.1.42), the sum of the residues at the points $a_k \in \mathcal{A}$ is equal to zero by Lemma 8.1.2.

If the function $P_n(x)/Q_m(x)$ is neither even nor odd, the term

$$\operatorname*{Res}_{z_k\in\mathcal{Z}} \left[\frac{P_n(z)}{Q_m(z)} \frac{\sin bz}{\sin az}\right]$$

on the right-hand side of (8.1.41) has to be replaced with

$$\frac{1}{2}\left\{\operatorname*{Res}_{z_k\in\mathcal{Z}} \left[\frac{P_n(z)}{Q_m(z)} \frac{\sin bz}{\sin az}\right] + \operatorname*{Res}_{\tilde z_k\in\widetilde{\mathcal{Z}}} \left[\frac{P_n(-z)}{Q_m(-z)} \frac{\sin bz}{\sin az}\right]\right\}, \tag{8.1.43}$$

and in (8.1.42) one has to make a substitution similar to (8.1.43) where $\sin bz$ is replaced with $\cos bz$.

It seems that the **second case**, $|a| < |b|$, has not been treated in the literature (not even in particular examples), despite the fact that if the integrals (8.1.39) are convergent for $|b| \le |a|$, they are also convergent for $|b| > |a|$.

In this case, however, the situation is more complicated because condition (8.1.40) is not satisfied — the left-hand side of (8.1.40) exponentially approaches infinity since $|b| > |a|$. But, if one uses the substitutions $\sin bx = \Im e^{ibx}$ and $\cos bx = \Re e^{ibx}$ in (8.1.39), then, for $m \geq n+1$, by Jordan Lemma 6.2.1 we have

$$\lim_{R_k \to \infty} \int_{C_{R_k}} \frac{P_n(z)}{Q_m(z)} \frac{e^{ibz}}{\sin az}\, dz = 0, \qquad \text{if } b > 0. \tag{8.1.44}$$

Therefore the derivation procedure which has led to formulae (8.1.5) and (8.1.20) can be used here. However, in this case, the symmetry breaks down because of the factor e^{ibz} and because the series of residues at the zeros, $z_n = n\pi/a$, $n = 0, \pm1, \pm2, \dots$, of $\sin az$ is not zero for the first integral in (8.1.39). Hence, the following term is added to the right-hand side of (8.1.41):

$$\begin{aligned} S_1 &:= \Im\left[\pi i \sum_{k=-\infty}^{\infty} \frac{P_n(k\pi/a)}{Q_m(k\pi/a)} \frac{e^{ibk\pi/a}}{a(-1)^k}\right] \\ &= \frac{\pi}{a} \sum_{k=-\infty}^{\infty} (-1)^k \frac{P_n(k\pi/a)}{Q_m(k\pi/a)} \cos\left(\frac{bk\pi}{a}\right) \end{aligned} \tag{8.1.45}$$

and, instead of (8.1.41), we obtain the following formula:

FORMULA 8.1.8. *If $a > 0$, $b > 0$, $P_n(x)/Q_m(x)$ is even and $m \geq n+2$, then*

$$\begin{aligned} \text{p. v.} \int_{-\infty}^{\infty} \frac{P_n(x)}{Q_m(x)} \frac{\sin bx}{\sin ax}\, dx &= \Im\Bigg\{ 2\pi i \sum_k \operatorname*{Res}_{z_k \in \mathcal{Z}} \left[\frac{P_n(z)}{Q_m(z)} \frac{e^{ibz}}{\sin az}\right] \\ &\quad + \pi i \sum_{k=1}^{l} \operatorname*{Res}_{a_k \in \mathcal{A}} \left[\frac{P_n(z)}{Q_m(z)} \frac{e^{ibz}}{\sin az}\right]\Bigg\} + S_1, \end{aligned} \tag{8.1.46}$$

where S_1 is given by (8.1.45), $z_k \in \mathcal{Z}$, $a_k \in \mathcal{A}$ for $k = 1, 2, \dots, l$, and $a \in \mathcal{B}_0$.

We have written $a > 0$, $b > 0$ in (8.1.46) since (8.1.44) is valid for all $b > 0$, and, therefore, we may have $a < b$ or $a \geq b$.

Since the function $P_n(x)/Q_m(x)$ is even, then the series (8.1.45) can always be expressed in closed form by expanding $P_n(x)/Q_m(x)$ in partial fractions and using formula 1.445(3) from [**23**], p. 40:

$$\sum_{k=1}^{\infty} \frac{(-1)^k \cos kx}{k^2 + \alpha} = \frac{\pi}{2\sqrt{\alpha}} \frac{\cosh x\sqrt{\alpha}}{\sinh \pi\sqrt{\alpha}} - \frac{1}{2\alpha}, \qquad -\pi \leq x \leq \pi. \tag{8.1.47}$$

Differentiating (8.1.47) m times with respect to α, we obtain

$$(-1)^m \sum_{k=1}^{\infty} \frac{(-1)^k \cos kx}{(k^2+\alpha)^{m+1}} = \frac{d^m}{d\alpha^m}\left[\frac{\pi}{2\sqrt{\alpha}}\frac{\cosh x\sqrt{\alpha}}{\sinh \pi\sqrt{\alpha}} - \frac{1}{2\alpha}\right],$$
$$-\pi \le x \le \pi. \quad (8.1.48)$$

We shall use the values of the series (8.1.48) outside the interval $[-\pi, \pi]$. Since each term of the series (8.1.47) is a 2π-periodic function because $\cos k(x+2\pi) = \cos kx$, then its sum $S(x)$, which is equal to the right-hand side of (8.1.47) in the interval $-\pi \le x \le \pi$, must be 2π-periodic, that is,

$$\sum_{k=1}^{\infty} \frac{(-1)^k \cos kx}{k^2+\alpha} = \frac{\pi}{2\sqrt{\alpha}}\frac{\cosh(x-2p\pi)\sqrt{\alpha}}{\sinh \pi\sqrt{\alpha}} - \frac{1}{2\alpha}, \quad (8.1.49)$$

with

$$-\pi \le x - 2p\pi \le \pi, \qquad p = 0, \pm 1, \pm 2, \dots.$$

EXAMPLE 8.1.6. *Derive Formula* 3.743(1) *in* [**23**], *p. 416:*

$$I_6 = \text{p.v.} \int_{-\infty}^{\infty} \frac{\sin bx}{\sin ax}\frac{dx}{x^2+\beta^2} = \frac{\pi}{\beta}\frac{\sinh(b\beta)}{\sinh(a\beta)},$$
$$0 < b \le a, \quad \Re\beta > 0. \quad (8.1.50)$$

SOLUTION. The formula follows from formula (8.1.41) since the integrand satisfies the condition of validity of this formula. Thus,

$$\begin{aligned} I_6 &= 2\pi i \operatorname*{Res}_{z=\beta i}\left[\frac{\sin bz}{\sin az}\frac{1}{z^2+\beta^2}\right] \\ &= 2\pi i \frac{\sin(b\beta i)}{\sin(a\beta i)}\frac{1}{2\beta i} \\ &= \frac{\pi}{\beta}\frac{\sinh(b\beta)}{\sinh(a\beta)}, \qquad 0 < b \le a, \qquad \Re\beta > 0. \quad \square \end{aligned}$$

Formula (8.1.41) cannot be used for the case $0 < a < b$, but, for all $a > 0$ and $b > 0$, one can use formula (8.1.46). Thus

$$\begin{aligned} I_6 &= \Im\left\{2\pi i \operatorname*{Res}_{z=\beta i}\left[\frac{e^{ibz}}{\sin az}\frac{1}{z^2+\beta^2}\right]\right\} + S_1 \\ &= \Im\left(2\pi i \frac{e^{-b\beta}}{i\sinh a\beta}\frac{1}{2\beta i}\right) + S_1 \\ &= -\frac{\pi}{\beta}\frac{e^{-b\beta}}{\sinh a\beta} + S_1, \end{aligned} \quad (8.1.51)$$

where

$$
\begin{aligned}
S_1 &= \frac{\pi}{a} \sum_{k=-\infty}^{\infty} \frac{(-1)^k}{\left(\frac{k\pi}{a}\right)^2 + \beta^2} \cos\left(\frac{bk\pi}{a}\right) \\
&= \frac{a}{\pi} \sum_{k=-\infty}^{\infty} \frac{(-1)^k}{k^2 + \left(\frac{a\beta}{\pi}\right)^2} \cos\left(\frac{bk\pi}{a}\right) \\
&= \begin{cases} \frac{a}{\pi}\left[\frac{1}{\alpha} + \frac{\pi}{\sqrt{\alpha}} \frac{\cosh b\pi\sqrt{\alpha}/a}{\sinh \pi\sqrt{\alpha}} - \frac{1}{\alpha}\right], & -\pi \le b\pi/a \le \pi, \\ \frac{a}{\pi}\left[\frac{1}{\alpha} + \frac{\pi}{\sqrt{\alpha}} \frac{\cosh(b\pi\sqrt{\alpha}/a - 2\pi\sqrt{\alpha})}{\sinh \pi\sqrt{\alpha}} - \frac{1}{\alpha}\right], & \pi \le b\pi/a \le 3\pi \end{cases}
\end{aligned}
$$

by (8.1.47) and (8.1.49) with $p = 1$ and $\sqrt{\alpha} = a\beta/\pi$. Thus

$$
S_1 = \begin{cases} \frac{\pi}{\beta} \frac{\cosh b\beta}{\sinh a\beta}, & -a \le b \le a, \\ \frac{\pi}{\beta} \frac{\cosh(b-2a)\beta}{\sinh a\beta}, & a \le b \le 3a. \end{cases} \tag{8.1.52}
$$

It follows from (8.1.51) and (8.1.52) that

$$
I_6 = \begin{cases} \frac{\pi}{\beta} \frac{1}{\sinh a\beta}\left(\cosh b\beta - e^{-b\beta}\right) = \frac{\pi}{\beta} \frac{\sinh b\beta}{\sinh a\beta}, & -a \le b \le a, \\ \frac{\pi}{\beta} \frac{1}{\sinh a\beta}\left[\cosh(b-2a)\beta - e^{-b\beta}\right], & a \le b \le 3a. \end{cases} \tag{8.1.53}
$$

That is, for $0 < b \le a$, the values given by (8.1.41) and (8.1.46) coincide.

Now we show that, using different values for the integral (8.1.39), that is, formulae (8.1.41) and (8.1.46), one can express the sum S_1 of the series (8.1.45) in terms of a finite sum of residues at the zeros of $Q_m(x)$. We assume that $|b| \le |a|$ and equate the right-hand sides of (8.1.41) and (8.1.46):

$$
\begin{aligned}
2\pi i \sum_k \operatorname*{Res}_{z_k \in \mathcal{Z}} \left[\frac{P_n(z)}{Q_m(z)} \frac{\sin bz}{\sin az}\right] &= \Im\left\{2\pi i \sum_k \operatorname*{Res}_{z_k \in \mathcal{Z}} \left[\frac{P_n(z)}{Q_m(z)} \frac{e^{ibz}}{\sin az}\right]\right. \\
&\quad \left. + \pi i \sum_{k=1}^{l} \operatorname*{Res}_{a_k \in \mathcal{A}} \left[\frac{P_n(z)}{Q_m(z)} \frac{e^{ibz}}{\sin az}\right]\right\} + S_1,
\end{aligned} \tag{8.1.54}
$$

so that

$$
\begin{aligned}
S_1 &= \frac{\pi}{a} \sum_{k=-\infty}^{\infty} (-1)^k \frac{P_n(k\pi/a)}{Q_m(k\pi/a)} \cos\left(\frac{bk\pi}{a}\right) \\
&= 2\pi i \sum_k \operatorname*{Res}_{z_k \in \mathcal{Z}} \left[\frac{P_n(z)}{Q_m(z)} \frac{\sin bz}{\sin az}\right] - \Im\left\{2\pi i \sum_k \operatorname*{Res}_{z_k \in \mathcal{Z}} \left[\frac{P_n(z)}{Q_m(z)} \frac{e^{ibz}}{\sin az}\right]\right.
\end{aligned}
$$

$$+\pi i \sum_{k=1}^{l} \operatorname*{Res}_{a_k \in \mathcal{A}} \left[\frac{P_n(z)}{Q_m(z)} \frac{e^{ibz}}{\sin az} \right] \Bigg\}, \tag{8.1.55}$$

with

$$-\pi \le \frac{b\pi}{a} \le \pi, \quad \text{that is,} \quad -a \le b \le a, \qquad m \ge n+2.$$

The series on the left-hand side of (8.1.55) does not change if we replace $b\pi/a$ with $b\pi/a - 2p\pi$ for $p = 0, \pm 1, \dots$, that is, S_1 does not change under the substitution of b with $b - 2pa$. Hence, the right-hand side of (8.1.55) also remains unchanged under this substitution. Therefore,

$$S_1 = 2\pi i \sum_k \operatorname*{Res}_{z_k \in \mathcal{Z}} \left[\frac{P_n(z)}{Q_m(z)} \frac{\sin(b-2pa)z}{\sin az} \right]$$
$$- \Im \left\{ \sum_k \left(2\pi i \operatorname*{Res}_{z_k \in \mathcal{Z}} + \pi i \operatorname*{Res}_{a_k \in \mathcal{A}} \right) \left[\frac{P_n(z)}{Q_m(z)} \frac{e^{i(b-2pa)z}}{\sin az} \right] \right\},$$
$$-a \le b - 2pa \le a. \tag{8.1.56}$$

Hence, (8.1.56) gives the formula for the sum, S_1, of the series.

Substituting the value for S_1 from (8.1.56) into (8.1.46), we obtain a formula to evaluate the first integral in (8.1.39), which is valid for any relation between a and b ($b > 0$).

FORMULA 8.1.9. *If $(2p-1)a \le b \le (2p+1)a$, $P_n(x)/Q_m(x)$ is even and $m \ge n+2$, then*

$$\text{p. v.} \int_{-\infty}^{\infty} \frac{P_n(x)}{Q_m(x)} \frac{\sin bx}{\sin ax}\, dx = 2\pi i \sum_k \operatorname*{Res}_{z_k \in \mathcal{Z}} \left[\frac{P_n(z)}{Q_m(z)} \frac{\sin(b-2pa)z}{\sin az} \right]$$
$$+ \Im \left\{ \sum_k \left(2\pi i \operatorname*{Res}_{z_k \in \mathcal{Z}} + \pi i \operatorname*{Res}_{a_k \in \mathcal{A}} \right) \right.$$
$$\left. \left[\frac{P_n(z)}{Q_m(z)} \frac{1}{\sin az} \left(e^{ibz} - e^{i(b-2pa)z} \right) \right] \right\}, \tag{8.1.57}$$

where $z_k \in \mathcal{Z}$, $a_k \in \mathcal{A}$ and $a \in \mathcal{B}_0$.

Returning to Example 8.1.6, let us evaluate the integral (8.1.50) by means of formula (8.1.57):

$$I_6 = \text{p. v.} \int_{-\infty}^{\infty} \frac{1}{x^2+\beta^2} \frac{\sin bx}{\sin ax}\, dx$$
$$= 2\pi i \frac{1}{2\beta i} \frac{\sinh(b-2pa)\beta}{\sinh a\beta}$$

$$+\Im\left\{2\pi i\,\frac{1}{2\beta i}\,\frac{1}{i\sinh a\beta}\left[e^{-b\beta}-e^{-(b-2pa)\beta}\right]\right\}, \tag{8.1.58}$$

with $(2p-1)a \le b \le (2p+1)a$. It follows from (8.1.58) that, if $p=0$, then

$$I_6=\frac{\pi}{\beta}\,\frac{\sinh b\beta}{\sinh a\beta}, \qquad 0\le b\le a,$$

which coincides with the value (8.1.50) found before. It follows from (8.1.58), in the case $p=1$, that

$$\begin{aligned} I_6 &= \frac{\pi}{\beta}\,\frac{\sinh{(b-2a)\beta}}{\sinh a\beta}-\frac{\pi}{\beta}\,\frac{e^{-b\beta}-e^{-(b-2a)\beta}}{\sinh a\beta} \\ &= \frac{\pi}{\beta}\,\frac{1}{\sinh a\beta}\left[\cosh{(b-2a)\beta}-e^{-b\beta}\right], \qquad a\le b\le 3a, \end{aligned}$$

which coincides with the value (8.1.53) found before. □

The second integral in (8.1.39) can be evaluated in a similar way. This integral is evaluated by (8.1.42), if $|b|<|a|$, and by the following formula for arbitrary values of a and b:

FORMULA 8.1.10. *If $(2p-1)a<b<(2p+1)a$ for $p=0,\pm1,\pm2,\dots$, $P_n(x)/Q_m(x)$ is odd and $m\ge n+1$, then*

$$\begin{aligned} \text{p. v.}\int_{-\infty}^{\infty}\frac{P_n(x)}{Q_m(x)}\,\frac{\cos bx}{\sin ax}\,dx &= 2\pi i\sum_k \operatorname*{Res}_{z_k\in\mathcal{Z}}\left[\frac{P_n(z)}{Q_m(z)}\,\frac{\cos{(b-2pa)z}}{\sin az}\right] \\ &\quad+\Re\left\{\sum_k\left(2\pi i\operatorname*{Res}_{z_k\in\mathcal{Z}}+\pi i\operatorname*{Res}_{a_k\in\mathcal{A}}\right)\right. \\ &\qquad\left.\left[\frac{P_n(z)}{Q_m(z)}\,\frac{1}{\sin az}\left(e^{ibz}-e^{i(b-2pa)z}\right)\right]\right\}, \end{aligned} \tag{8.1.59}$$

where $z_k\in\mathcal{Z}$, $a_k\in\mathcal{A}$ for $k=1,2,\dots,l$, and $a\in\mathcal{B}_0$.

At the same time, the sum of the following series is found:

$$\begin{aligned} S_2 &= \frac{\pi}{a}\sum_{k=-\infty}^{\infty}(-1)^{k+1}\frac{P_n(k\pi/a)}{Q_m(k\pi/a)}\,\sin\left(\frac{bk\pi}{a}\right) \\ &= 2\pi i\sum_k \operatorname*{Res}_{z_k\in\mathcal{Z}}\left[\frac{P_n(z)}{Q_m(z)}\,\frac{\cos{(b-2pa)z}}{\sin az}\right] \\ &\quad-\Re\left\{\sum_k\left(2\pi i\operatorname*{Res}_{z_k\in\mathcal{Z}}+\pi i\operatorname*{Res}_{a_k\in\mathcal{A}}\right)\left[\frac{P_n(z)}{Q_m(z)}\,\frac{e^{i(b-2pa)z}}{\sin az}\right]\right\}, \end{aligned} \tag{8.1.60}$$

provided $-a < b - 2pa < a$ and $p = 0, \pm 1, \pm 2, \dots$.

EXAMPLE 8.1.7. *If* $(2p-1)a < b < (2p+1)a$, *derive the following formula:*

$$\begin{aligned}
I_7 &= \text{p. v.} \int_{-\infty}^{\infty} \frac{x}{x^2+\beta^2} \frac{\cos bx}{\sin ax}\, dx \\
&= 2\pi i \operatorname*{Res}_{z=\beta i} \left[\frac{z}{z^2+\beta^2} \frac{\cos (b-2pa)z}{\sin az} \right] \\
&\quad + \Re \left\{ 2\pi i \operatorname*{Res}_{z=\beta i} \left[\frac{z}{z^2+\beta^2} \frac{1}{\sin az} \left(e^{ibz} - e^{i(b-2pa)z} \right) \right] \right\} \\
&= \pi \frac{\cosh (b-2pa)\beta}{\sinh a\beta} + \pi \frac{1}{\sinh a\beta} \left[e^{-b\beta} - e^{-(b-2pa)\beta} \right].
\end{aligned} \tag{8.1.61}$$

SOLUTION. The formula follows from (8.1.59). In fact, if $p = 0$, we have

$$I_7 = \pi \frac{\cosh b\beta}{\sinh a\beta}, \qquad -a < b < a,$$

which coincides with formula 3.743(3) in [**23**], p. 416. If $p = 1$,

$$I_7 = \frac{\pi}{\sinh a\beta} \left[\sinh (b-2a)\beta + e^{-b\beta} \right], \qquad a < b < 3a. \quad \square$$

We remark that, in this example, the integral I_7 is discontinuous as $b \to a$:

$$I_7\big|_{b\to a+0} \neq I_7\big|_{b\to a-0}.$$

We can see from formulae (8.1.59) and (8.1.60) and from the last Example 8.1.7 that these formulae do not allow us to evaluate the integral in (8.1.59) if $b = (2p \pm 1)a$, for $p = 0, \pm 1, \pm 2, \dots$. But this evaluation is easy if we remark that $S_2 = 0$ in formula (8.1.60) when $b = (2p \pm 1)a$, because $\sin[(2p \pm 1)\pi k)] = 0$. Since the series in (8.1.60) is equal to the sum of the residues at the zeros of $\sin ax$, then $S_2 = 0$ and we obtain the following simple formula for the evaluation of this integral in the form of a finite sum of residues at the zeros of $Q_m(x)$.

FORMULA 8.1.11. *If* $p = 0, \pm 1, \pm 2, \dots$, $P_n(x)/Q_m(x)$ *is an odd function of* x *and* $m \geq n+1$, *then*

$$\begin{aligned}
&\text{p. v.} \int_{-\infty}^{\infty} \frac{P_n(x)}{Q_m(x)} \frac{\cos[(2p \pm 1)ax]}{\sin ax}\, dx \\
&\quad = \Re \left\{ \sum_k \left(2\pi i \operatorname*{Res}_{z_k \in \mathcal{Z}} + \pi i \operatorname*{Res}_{a_k \in \mathcal{A}} \right) \left[\frac{P_n(z)}{Q_m(z)} \frac{e^{i(2p\pm 1)az}}{\sin az} \right] \right\},
\end{aligned} \tag{8.1.62}$$

where $z_k \in \mathcal{Z}$, $a_k \in \mathcal{A}$ *for* $k = 1, 2, \dots, l$, *and* $a \in \mathcal{B}_0$.

EXAMPLE 8.1.8. *Obtain Formula 3.749(2) in* [**23**], *p. 418:*

$$I_8 = \text{p. v.} \int_{-\infty}^{\infty} \frac{x \cot ax}{x^2 + \beta^2}\, dx = \frac{2\pi}{e^{2a\beta} - 1}, \qquad a > 0, \qquad \beta > 0.$$

SOLUTION. This formula follows from formula (8.1.62), with $p = 0$ and the plus sign in the term ± 1. In fact we have

$$\begin{aligned} I_8 &= \Re \left\{ 2\pi i \operatorname*{Res}_{z=\beta i} \left[\frac{z\, e^{iaz}}{(z^2 + \beta^2) \sin az} \right] \right\} = \Re \left[2\pi i \, \frac{e^{-a\beta}}{2 \sin a\beta i} \right] \\ &= \frac{\pi\, e^{-a\beta}}{\sinh a\beta} = \frac{2\pi}{e^{2a\beta} - 1}. \quad \square \end{aligned}$$

8.1.5. The integrals I_c^c and I_c^s. Lastly, we consider the integrals

$$\begin{aligned} I_c^c &= \text{p. v.} \int_{-\infty}^{\infty} \frac{P_n(x)}{Q_m(x)} \, \frac{\cos bx}{\cos ax}\, dx, \\ I_c^s &= \text{p. v.} \int_{-\infty}^{\infty} \frac{P_n(x)}{Q_m(x)} \, \frac{\sin bx}{\cos ax}\, dx, \end{aligned} \tag{8.1.63}$$

where, in the first integral, the function $P_n(x)/Q_m(x)$ is even and $m \geq n+2$, and, in the second integral, this function is odd and $m \geq n+1$. These integrals are evaluated as the integrals I_s^s and I_s^c (8.1.39) in the previous Subsection 8.1.4.

If $|b| < |a|$, we have the following formulae for I_c^c and I_c^s.

FORMULA 8.1.12. *If* $|b| \leq |a|$, $P_n(x)/Q_m(x)$ *is even and* $m \geq n+2$, *then*

$$\text{p. v.} \int_{-\infty}^{\infty} \frac{P_n(x)}{Q_m(x)} \, \frac{\cos bx}{\cos ax}\, dx = 2\pi i \sum_k \operatorname*{Res}_{z_k \in \mathcal{Z}} \left[\frac{P_n(z)}{Q_m(z)} \, \frac{\cos bz}{\cos az} \right], \tag{8.1.64}$$

where $z_k \in \mathcal{Z}$, $a_k \in \mathcal{A}$ *for* $k = 1, 2, \dots, l$, *and* $a \in \mathcal{B}_1$.
Similarly, if $|b| < |a|$, $P_n(x)/Q_m(x)$ *is odd and* $m \geq n+1$, *then*

$$\text{p. v.} \int_{-\infty}^{\infty} \frac{P_n(x)}{Q_m(x)} \, \frac{\sin bx}{\cos ax}\, dx = 2\pi i \sum_k \operatorname*{Res}_{z_k \in \mathcal{Z}} \left[\frac{P_n(z)}{Q_m(z)} \, \frac{\sin bz}{\cos az} \right], \tag{8.1.65}$$

and, if $m \geq n+3$, *we may have* $|b| \leq |a|$.

In formulae (8.1.64) and (8.1.65), the sum of residues at the points $a_k \in \mathcal{A}$ is equal to zero by Lemma 8.1.2.

The strict inequality $|b| < |a|$, when $m = n+1$, is needed because the integrals are conditionally convergent and have a discontinuity as $|b| \to |a|$ (see Example 8.1.7).

The formulae for the evaluation of (8.1.63) in the case of arbitrary values of a and b are as follows.

FORMULA 8.1.13. *If* $(2p-1)a \le b \le (2p+1)a$ *for* $p = 0, \pm 1, \pm 2, \dots$, $P_n(x)/Q_m(x)$ *is even and* $m \ge n+2$, *then*

$$\text{p. v.} \int_{-\infty}^{\infty} \frac{P_n(x)}{Q_m(x)} \frac{\cos bx}{\cos ax}\, dx = (-1)^p 2\pi i \sum_k \operatorname*{Res}_{z_k \in \mathcal{Z}} \left[\frac{P_n(z)}{Q_m(z)} \frac{\cos (b-2pa)z}{\cos az} \right]$$
$$+ \Re \left\{ \sum_k \left(2\pi i \operatorname*{Res}_{z_k \in \mathcal{Z}} + \pi i \operatorname*{Res}_{a_k \in \mathcal{A}} \right) \left[\frac{P_n(z)}{Q_m(z)} \frac{1}{\cos az} \left(e^{ibz} - (-1)^p e^{i(b-2pa)z} \right) \right] \right\}, \quad (8.1.66)$$

where $z_k \in \mathcal{Z}$, $a_k \in \mathcal{A}$ *for* $k = 1, 2, \dots, l$, *and* $a \in \mathcal{B}_1$.
Similarly, if $(2p-1)a < b < (2p+1)a$ *for* $p = 0, \pm 1, \pm 2, \dots$, $P_n(x)/Q_m(x)$ *is odd and* $m \ge n+1$, *then*

$$\text{p. v.} \int_{-\infty}^{\infty} \frac{P_n(x)}{Q_m(x)} \frac{\sin bx}{\cos ax}\, dx = (-1)^p 2\pi i \sum_k \operatorname*{Res}_{z_k \in \mathcal{Z}} \left[\frac{P_n(z)}{Q_m(z)} \frac{\sin (b-2pa)z}{\cos az} \right]$$
$$+ \Im \left\{ \sum_k \left(2\pi i \operatorname*{Res}_{z_k \in \mathcal{Z}} + \pi i \operatorname*{Res}_{a_k \in \mathcal{A}} \right) \left[\frac{P_n(z)}{Q_m(z)} \frac{1}{\cos az} \left(e^{ibz} - (-1)^p e^{i(b-2pa)z} \right) \right] \right\}. \quad (8.1.67)$$

While deriving (8.1.66) one finds the sum of the following series:

$$S_3 = \frac{\pi}{a} \sum_{k=-\infty}^{\infty} (-1)^k \frac{P_n((2k+1)\pi/(2a))}{Q_m((2k+1)\pi/(2a))} \sin\left(\frac{b(2k+1)\pi}{2a} \right)$$
$$= (-1)^p 2\pi i \sum_k \operatorname*{Res}_{z_k \in \mathcal{Z}} \left[\frac{P_n(z)}{Q_m(z)} \frac{\cos (b-2pa)z}{\cos az} \right] - (-1)^p$$
$$\times \Re \left\{ \sum_k \left(2\pi i \operatorname*{Res}_{z_k \in \mathcal{Z}} + \pi i \operatorname*{Res}_{a_k \in \mathcal{A}} \right) \left[\frac{P_n(z)}{Q_m(z)} \frac{e^{i(b-2pa)z}}{\cos az} \right] \right\}, \quad (8.1.68)$$

provided $-a \le b - 2pa \le a$ for $p = 0, \pm 1, \pm 2, \dots$.

Similarly, while deriving (8.1.67) one finds the sum of the following series:

$$
\begin{aligned}
S_4 &= \frac{\pi}{a} \sum_{k=-\infty}^{\infty} (-1)^{k+1} \frac{P_n\big((2k+1)\pi/(2a)\big)}{Q_m\big((2k+1)\pi/(2a)\big)} \cos\left(\frac{b(2k+1)\pi}{2a}\right) \\
&= (-1)^p 2\pi i \sum_k \operatorname*{Res}_{z_k \in \mathcal{Z}} \left[\frac{P_n(z)}{Q_m(z)} \frac{\sin(b-2pa)z}{\cos az}\right] - (-1)^p \\
&\quad \times \Im\left\{\sum_k \left(2\pi i \operatorname*{Res}_{z_k\in\mathcal{Z}} + \pi i \operatorname*{Res}_{a_k\in\mathcal{A}}\right)\left[\frac{P_n(z)}{Q_m(z)} \frac{e^{i(b-2pa)z}}{\cos az}\right]\right\}, \qquad (8.1.69)
\end{aligned}
$$

provided $-a < b - 2pa < a$ for $p = 0, \pm 1, \pm 2, \dots$.

Formula (8.1.67) is not valid if $b = (2p \pm 1)a$, for $p = 0, \pm 1, \pm 2, \dots$; but, as with formula (8.1.62), this integral can be expressed as a finite sum of residues at the zeros of $Q_m(x)$ by the following formula.

FORMULA 8.1.14. *If $p = 0, \pm 1, \pm 2, \dots$, $P_n(x)/Q_m(x)$ is an odd function of x, and $m \geq n+1$, then*

$$
\begin{aligned}
&\text{p. v.} \int_{-\infty}^{\infty} \frac{P_n(x)}{Q_m(x)} \frac{\sin[(2p\pm 1)ax]}{\cos ax}\, dx \\
&\quad = \Im\left\{\sum_k \left(2\pi i \operatorname*{Res}_{z_k\in\mathcal{Z}} + \pi i \operatorname*{Res}_{a_k\in\mathcal{A}}\right)\left[\frac{P_n(x)}{Q_m(x)} \frac{e^{i(2p\pm 1)az}}{\cos az}\right]\right\}, \qquad (8.1.70)
\end{aligned}
$$

where $z_k \in \mathcal{Z}$, $a_k \in \mathcal{A}$ for $k = 1, 2, \dots, l$, and $a \in \mathcal{B}_1$.

EXAMPLE 8.1.9. *Derive Formula* 3.747(10) *in* [**23**], *p.!418:*

$$
I_9 = \text{p. v.} \int_{-\infty}^{\infty} \tan ax \,\frac{dx}{x} = \pi, \qquad a > 0.
$$

SOLUTION. The formula follows from formula (8.1.70), with $p = 0$. In fact, we have

$$
I_9 = \Im\left\{\pi i \operatorname*{Res}_{z=0}\left[\frac{1}{z} \frac{e^{iaz}}{\cos az}\right]\right\} = \Im[\pi i] = \pi. \quad \square
$$

EXAMPLE 8.1.10. *Derive Formula* 3.749(1) *in* [**23**], *p. 418:*

$$
I_{10} = \text{p. v.} \int_{-\infty}^{\infty} \frac{x \tan ax}{x^2 + \beta^2}\, dx = \frac{2\pi}{e^{2a\beta} + 1}, \qquad a > 0, \qquad \beta > 0.
$$

SOLUTION. The formula follows from formula (8.1.70), with $p = 0$. In fact we have

$$
I_{10} = \Im\left\{2\pi i \operatorname*{Res}_{z=\beta i}\left[\frac{z\, e^{iaz}}{(z^2+\beta^2)\cos az}\right]\right\}
$$

$$= \Im\left[2\pi i \frac{e^{-a\beta}}{2\cos(a\beta i)}\right] = \frac{\pi\, e^{-a\beta}}{\cosh a\beta} = \frac{2\pi}{e^{2a\beta}+1}. \quad \square$$

EXAMPLE 8.1.11. *Compute the Fourier cosine transform of*

$$\frac{x}{x^2+\beta^2}\tan ax,$$

that is,

$$I_{11}(y) = \text{p. v.} \int_0^\infty \frac{x}{x^2+\beta^2}\, \frac{\sin ax}{\cos ax}\, \cos xy\, dx, \qquad y > 0.$$

SOLUTION. We have

$$\begin{aligned} I_{11}(y) &= \frac{1}{4}\,\text{p. v.} \int_{-\infty}^\infty \frac{x\sin{(y+a)x}}{(x^2+\beta^2)\cos ax}\, dx - \frac{1}{4}\,\text{p. v.} \int_{-\infty}^\infty \frac{x\sin{(y-a)x}}{(x^2+\beta^2)\cos ax}\, dx \\ &=: \frac{1}{4}(A_1 - A_2), \end{aligned} \tag{8.1.71}$$

which defines A_1 and A_2. To evaluate A_1 we use formula (8.1.67) with $b = y + a$:

$$\begin{aligned} A_1 &= (-1)^p 2\pi i \,\frac{\beta i}{2\beta i}\, \frac{\sin[(y+a-2pa)\beta i]}{\cos(a\beta i)} \\ &\quad + \Im\left\{2\pi i\, \frac{\beta i}{2\beta i}\, \frac{1}{\cos(a\beta i)} \left[e^{i(y+a)\beta i} - (-1)^p\, e^{i(y+a-2pa)\beta i}\right]\right\}, \end{aligned}$$

where $2(p-1)a < y < 2pa$ for $p = 1, 2, 3, \dots$, or

$$\begin{aligned} A_1 = \frac{\pi}{\cosh a\beta}\Big\{&(-1)^{p+1}\sinh[(y+a-2pa)\beta] \\ &+ e^{-(y+a)\beta} + (-1)^{p+1}\, e^{-(y+a-2pa)\beta}\Big\}. \end{aligned} \tag{8.1.72}$$

To evaluate A_2 it suffices to replace $y + a$ by $y - a$ (or y by $y - 2a$) and p by $p - 1$ in (8.1.72):

$$\begin{aligned} A_2 = \frac{\pi}{\cosh a\beta}\Big\{&(-1)^{p}\sinh[(y+a-2pa)\beta] \\ &+ e^{-(y-a)\beta} + (-1)^{p}\, e^{-(y+a-2pa)\beta}\Big\}, \end{aligned} \tag{8.1.73}$$

where $2(p-1)a < y < 2pa$ for $p = 1, 2, 3, \dots$. Combining (8.1.71), (8.1.72) and (8.1.73), we have

$$\begin{aligned} I_{11}(y) = \frac{\pi}{4\cosh a\beta}\Big\{&2(-1)^{p+1}\cosh[(y+a-2pa)\beta] \\ &+ e^{-(y+a)\beta} - e^{-(y-a)\beta}\Big\}, \end{aligned} \tag{8.1.74}$$

where $2(p-1)a < y < 2pa$ for $p = 1, 2, 3, \ldots$. If we set $p = 1$ in (8.1.74) then we obtain

$$\begin{aligned} I_{11}(y)\Big|_{p=1} &= \frac{\pi}{4\cosh a\beta}\left[e^{(y-a)\beta} + e^{-(y-a)\beta} + e^{-(y+a)\beta} - e^{-(y-a)\beta}\right] \\ &= \frac{\pi}{2\left(e^{a\beta} + e^{-a\beta}\right)}\, 2\, e^{-a\beta}\cosh\beta y \\ &= \frac{\pi\cosh\beta y}{e^{2a\beta} + 1}, \qquad 0 < y < 2a. \end{aligned} \tag{8.1.75}$$

This result is the same as in [**18**], p. 23, Sect. 1.6, formula (34), provided we add the restrictive inequality $0 < y < 2a$ to this formula. □

It is easy to verify that the inverse Fourier cosine transform of $I_{11}(y)$ of (8.1.74) is

$$\frac{2}{\pi}\int_0^\infty I_{11}(y)\cos xy\, dy = \frac{x}{x^2+\beta^2}\tan ax,$$

as it should be.

EXAMPLE 8.1.12. *Compute the Fourier cosine transform of*

$$\frac{x}{x^2+\beta^2}\cot ax,$$

that is,

$$I_{12}(y) = \text{p. v.}\int_0^\infty \frac{x}{x^2+\beta^2}\,\frac{\cos ax}{\sin ax}\cos xy\, dx, \qquad y > 0.$$

SOLUTION. We have

$$\begin{aligned} I_{12}(y) &= \frac{1}{4}\,\text{p. v.}\int_{-\infty}^\infty \frac{x\cos(y+a)x}{(x^2+\beta^2)\sin ax}\, dx + \frac{1}{4}\,\text{p. v.}\int_{-\infty}^\infty \frac{x\cos(y-a)x}{(x^2+\beta^2)\sin ax}\, dx \\ &=: \frac{1}{4}(B_1 - B_2), \end{aligned} \tag{8.1.76}$$

which defines B_1 and B_2. To evaluate B_1 we use formula (8.1.59) with $b = y + a$:

$$\begin{aligned} B_1 &= \pi i\,\frac{\cos[(b-2pa)\beta i]}{\sin(a\beta i)} + \Re\left\{\pi i\,\frac{1}{\sin(a\beta i)}\left[e^{ib\beta i} - e^{i(b-2pa)\beta i}\right]\right\} \\ &= \frac{\pi}{\sinh a\beta}\left\{\cosh[(b-2pa)\beta] + e^{-b\beta} - e^{-(b-2pa)\beta}\right\}, \end{aligned}$$

where $2(p-1)a < y < 2pa$ for $p = 1, 2, 3, \ldots$, or

$$B_1 = \frac{\pi}{\sinh a\beta}\left\{\sinh[(y+a-2pa)\beta] + e^{-(y+a)\beta}\right\}. \tag{8.1.77}$$

To evaluate B_2 it suffices to replace $y+a$ by $y-a$ (or y by $y-2a$) and p by $p-1$ in (8.1.77):

$$B_2 = \frac{\pi}{\sinh a\beta}\left\{\sinh[(y+a-2pa)\beta] + e^{-(y-a)\beta}\right\}, \tag{8.1.78}$$

where $2(p-1)a < y < 2pa$ for $p = 1, 2, 3, \ldots$. Combining (8.1.76), (8.1.77) and (8.1.78), we have

$$I_{12}(y) = \frac{\pi}{4\sinh a\beta}\Big\{2\sinh[(y+a-2pa)\beta] \\ + e^{-(y+a)\beta} + e^{-(y-a)\beta}\Big\}, \tag{8.1.79}$$

where $2(p-1)a < y < 2pa$ for $p = 1, 2, 3, \ldots$. If we set $p = 1$ in (8.1.79), then we obtain

$$\begin{aligned} I_{12}(y)\Big|_{p=1} &= \frac{\pi}{4\sinh a\beta}\left\{2\sinh[(y-a)\beta] + e^{-(y+a)\beta} + e^{-(y-a)\beta}\right\} \\ &= \frac{\pi}{4\sinh a\beta}\left[e^{(y-a)\beta} + e^{-(y+a)\beta}\right] \\ &= \frac{\pi\, e^{-a\beta}\cosh\beta y}{2\sinh a\beta} \\ &= \frac{\pi\cosh\beta y}{e^{2a\beta}-1}, \qquad 0 < y < 2a. \end{aligned} \tag{8.1.80}$$

This result is the same as in [**18**], p. 23, Sect. 1.6, formula (35), provided we add the restrictive inequality $0 < y < 2a$ to this formula. □

Note 8.1.5. We remark that, in [**23**], the integrals of formulae 3.745(1,2) diverge at $x = b$, and formulae 3.743(5) and 3.749(3) (taken from Tables 191 and 161 of [**9**]), related to the forms discussed in this section, are incorrect. So do formula 18 of Table 191 and formulae 7–9 of Table 161 of [**9**]). In Table 191 of [**9**], the values of all the integrals of the types discussed here, namely, formulae 1–9 and 12–29, are incorrect. Some of these formulae are given in a correct form in [**18**], namely, formulae 21 and 31–32 of Section 2.6, pp. 80–82, and formulae 36 and 37 of Section 1.6, p. 23. However, in [**18**], formulae 34 and 35 of Section 1.6, p. 23 (see Examples 8.1.11 and 8.1.12) and formula 30 of Section 2.6, p. 80, are also incorrect. The correct formulae 3.749 (1 and 2) in [**23**] are taken from Table 333, formulae 79a and 79b in [**24**] (see Examples 8.1.8 and 8.1.9).

8.2. Forms containing $(x^2 - 2a\sin x + a^2)^{-1}$

We consider integrals of the form

$$I_q^p = \int_{-\infty}^{\infty} \frac{P_n(x)}{Q_m(x)} \frac{dx}{x^2 - 2ax\sin x + a^2}, \tag{8.2.1}$$

where $0 < a < \pi/2$, $P_n(x)$ and $Q_m(x)$ are real polynomials of the real variable x, of degrees n and m, respectively, $m \geq n$, $P_n(x)/Q_m(x)$ is an even function of x and $Q_m(x) \neq 0$ for real x.

If $P_n(x)/Q_m(x)$ is neither even nor odd, it suffices to consider its even part since the integral of the odd part is equal to zero.

If $0 < a < \pi/2$, the function $x^2 - 2ax\sin x + a^2$ has no real zeros since $|(x^2 + a^2)/(2ax)| \geq 1$, but if $a = \pi/2$, its only real zero is $x = \pi/2$.

8.2.1. The particular case $P_n(x)/Q_m(x) \equiv 1$. We first consider the simple case $P_n(x)/Q_m(x) \equiv 1$ and obtain the following formula.

FORMULA 8.2.1. *Derive the formula*

$$I_1^1 = \int_{-\infty}^{\infty} \frac{dx}{x^2 - 2ax\sin x + a^2} = \frac{\pi}{a}\,\frac{1 + \sin a}{\cos a}, \qquad 0 < a < \frac{\pi}{2}. \tag{8.2.2}$$

This integral was computed for the first time in the monograph [**47**], p. 181, formula (2.4.41), in a roundabout way by determining the minimal eigenvalue of a boundary-value problem in two ways.

PROOF. Considering the transformation

$$\begin{aligned}\frac{1}{x^2 - 2ax\sin x + a^2} &= \frac{1}{a^2\cos^2 x + (x - a\sin x)^2} \\ &= \frac{1}{2a\cos x}\left[\frac{1}{a\cos x - i(x - a\sin x)} + \frac{1}{a\cos x + i(x - a\sin x)}\right],\end{aligned}$$

we have

$$\begin{aligned}I_1^1 &= \frac{1}{2a}\,\text{p. v.} \int_{-\infty}^{\infty} \frac{dx}{(a\,e^{ix} - ix)\cos x} + \frac{1}{2a}\,\text{p. v.} \int_{-\infty}^{\infty} \frac{dx}{(a\,e^{-ix} + ix)\cos x} \\ &\qquad \text{(and putting } x = -t \text{ in the second integral)} \\ &= \frac{1}{a}\,\text{p. v.} \int_{-\infty}^{\infty} \frac{dx}{(a\,e^{ix} - ix)\cos x}. \end{aligned} \tag{8.2.3}$$

Set

$$f(z) = \frac{1}{(a\,e^{iz} - iz)\cos z}, \qquad z \in \mathbb{C}. \tag{8.2.4}$$

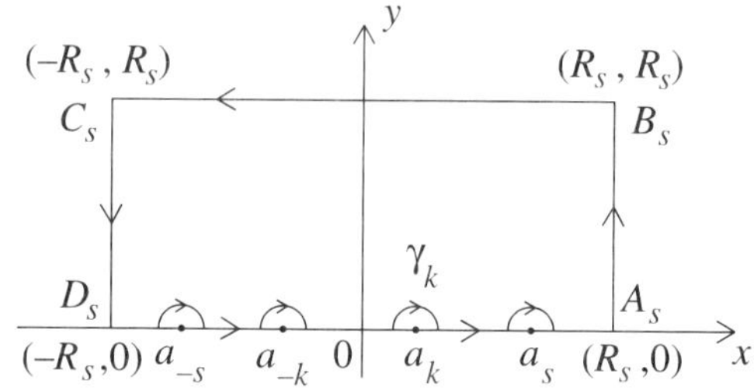

FIGURE 8.2. The path of integration in Subsection 8.2.2.

Let $R_s = (s+1)\pi$ for $s \in \mathbb{N}$ and consider the "rectangular" closed path, C, consisting of those segments of the real interval $[-R_s, R_s]$ where the points $a_k = (2k+1)\pi/2$ for $k = 0, \pm 1, \pm 2, \dots, \pm s$, are bypassed along the semicircles γ_k of radii δ in the upper half-plane, and the sides of the rectangle $A_sB_sC_sD_s$, with vertices

$$A_s = (R_s, 0), \qquad B_s = (R_s, R_s), \qquad C_s = (-R_s, R_s), \qquad D_s = (-R_s, 0)$$

shown in Fig 8.2. By the residue theorem we have

$$\left(\int_{\eta_s} + \int_{-R_s}^{R_s} + \sum_{k=-s}^{s} \int_{\gamma_k}\right) \left[\frac{1}{(a\,e^{iz} - iz)\cos z}\,dz\right] = 2\pi i \sum_p \operatorname*{Res}_{z=z_p} \frac{1}{(a\,e^{iz} - iz)\cos z}, \tag{8.2.5}$$

where the path η_s consists of the union

$$\eta_s = A_sB_s \cup B_sC_s \cup C_sD_s, \tag{8.2.6}$$

z_p are the zeros of the function $\varphi(z) = ae^{iz} - iz$ inside C, and the integral from $-R_s$ to R_s is evaluated along the line segments of the x-axis excluding the arcs γ_k.

It will be shown in Lemmas 8.2.1 and 8.2.2 that the integral along η_s in (8.2.5) approaches zero as $R_s = (s+1)\pi \to \infty$ and that there are no zeros of $\varphi(z) = a\,e^{iz} - iz$ in the region $\Im z \geq 0$. Consequently, the sum on the right-hand side of (8.2.5) is equal to zero.

Since a_k is a simple pole of $f(z)$, using a Laurent series in a neighborhood of a_k we obtain

$$\lim_{\delta \to 0} \int_{\gamma_k} f(z)\,dz = -\pi i \operatorname*{Res}_{z=a_k} f(z) \tag{8.2.7}$$

(for a similar calculation, see formula (6.1.9) in Subsection 6.1.2). Therefore, taking the limit in (8.2.5) as $s \to \infty$, $\delta \to 0$ and using (8.2.7) we obtain

$$\begin{aligned}
I_1^1 &= \lim_{\substack{s\to\infty \\ \delta\to 0}} \frac{1}{a}\int_{-R_s}^{R_s} \frac{dx}{(a\,e^{ix} - ix)\cos x} \\
&= \frac{\pi i}{a}\sum_{k=0}^{\infty}\left(\operatorname*{Res}_{z=a_k} + \operatorname*{Res}_{z=-a_k}\right)\left[\frac{1}{(a\,e^{iz} - iz)\cos z}\right] \\
&= \frac{\pi i}{a}\sum_{k=0}^{\infty}\left[\frac{(-1)^{k+1}}{a\,e^{ia_k} - ia_k} - \frac{(-1)^{k+1}}{a\,e^{-ia_k} + ia_k}\right] \\
&\qquad\qquad \left(\text{since } e^{\pm ia_k} = \pm i\sin a_k = \pm i(-1)^k\right) \\
&= \frac{\pi}{a}\sum_{k=0}^{\infty}(-1)^{k+1}\left[\frac{1}{a(-1)^k - a_k} + \frac{1}{a(-1)^k - a_k}\right] \\
&= \frac{2\pi}{a}\sum_{k=0}^{\infty}\frac{(-1)^k}{a_k - a(-1)^k} = \frac{2\pi}{a}\sum_{k=0}^{\infty}\frac{(-1)^k a_k + a}{a_k^2 - a^2} \\
&= \frac{\pi}{a}\left(\frac{1}{\cos a} + \tan a\right),
\end{aligned} \tag{8.2.8}$$

where $a_k = (2k+1)\pi/2$. In order to derive (8.2.7) we have used Formulae 1.421(1) and 1.422(1) in [**23**], p. 36, namely,

$$\tan\frac{\pi x}{2} = \frac{4x}{\pi}\sum_{k=1}^{\infty}\frac{1}{(2k-1)^2 - x^2} \tag{8.2.9}$$

and

$$\frac{1}{\cos(\pi x/2)} = \frac{4}{\pi}\sum_{k=1}^{\infty}\frac{(-1)^{k+1}(2k-1)}{(2k-1)^2 - x^2}. \tag{8.2.10}$$

This completes the proof of (8.2.2). □

LEMMA 8.2.1. *The following integral along the path η_s given by* (8.2.6) *approaches zero as $s \to \infty$:*

$$\lim_{s\to\infty}\int_{\eta_s}\frac{dz}{(a\,e^{iz} - iz)\cos z} = 0, \qquad 0 < a < \frac{\pi}{2}. \tag{8.2.11}$$

PROOF. Using (8.2.4) we write

$$F_s = \int_{\eta_s} f(z)\,dz = \int_{\eta_s}\frac{dz}{(a\,e^{iz} - iz)\cos z}.$$

Thus

$$|F_s| = \left| \int_{\eta_s} f(z)\, dz \right| \le \int_{\eta_s} \frac{|dz|}{|a\, e^{iz} - iz|\, |\cos z|}. \tag{8.2.12}$$

Next, we obtain an upper bound for the integral along η_s on the right-hand side of (8.2.12). We have

$$|\cos z|\Big|_{z\in\eta_s} = \sqrt{\sinh^2 y + \cos^2 x}\,\Big|_{(x,y)\in\eta_s}, \tag{8.2.13}$$

and

$$\begin{aligned} |a\, e^{iz} - iz|\Big|_{z\in\eta_s} &\ge \left(|iz| - |a|\, |e^{iz}|\right)\Big|_{z\in\eta_s} \\ &\ge R_s - \frac{\pi}{2}\, e^{-y} \ge (s+1)\pi - \frac{\pi}{2} = \left(s + \frac{1}{2}\right)\pi. \end{aligned}$$

Consequently,

$$|f(z)|\Big|_{z\in\eta_s} \le \frac{1}{(s+1/2)\pi}\, \frac{1}{|\cos z|}.$$

Using (8.2.12) we obtain

$$|F_s| \le \frac{1}{(s+1/2)\pi} \left(\int_{A_sB_s} + \int_{B_sC_s} + \int_{C_sD_s} \right) \left[\frac{|dz|}{|\cos z|} \right]. \tag{8.2.14}$$

On the segment A_sB_s,

$$z = (s+1)\pi + iy, \qquad 0 \le y \le (s+1)\pi, \qquad |dz| = dy;$$

thus we have

$$\begin{aligned} \int_{A_sB_s} \frac{|dz|}{|\cos z|} &= \int_0^{(s+1)\pi} \frac{dy}{\sqrt{\sinh^2 y + 1}} \\ &= \int_0^{(s+1)\pi} \frac{dy}{\cosh y} \\ &= 2\arctan\left(e^y\right)\Big|_0^{(s+1)\pi} \\ &\to 2\left(\frac{\pi}{2} - \frac{\pi}{4}\right) = \frac{\pi}{2}, \qquad \text{as } s \to \infty. \end{aligned} \tag{8.2.15}$$

Similarly, it can be shown that the integral along C_sD_s approaches $-\pi/2$ as $s \to \infty$. On the segment B_sC_s,

$$z = x + i(s+1)\pi, \qquad |dz| = dx;$$

hence, we have

$$\int_{B_s C_s} \frac{|dz|}{|\cos z|} = \int_{(s+1)\pi}^{-(s+1)\pi} \frac{dx}{\sqrt{\sinh^2(s+1)\pi + \cos^2 x}}$$

(and putting $x = (s+1)\pi t$)

$$= (s+1)\pi \int_1^{-1} \frac{dt}{\sqrt{\sinh^2(s+1)\pi + \cos^2(s+1)\pi t}}$$

$$\to 0, \qquad \text{as } s \to \infty. \tag{8.2.16}$$

Using (8.2.14)–(8.2.16) we obtain that $F_s \to 0$ as $s \to \infty$. □

LEMMA 8.2.2. *If $0 < a < \pi/2$ and $z = x + iy$, then the function*

$$\varphi(z) = a\, e^{iz} - iz = y + a\, e^{-y} \cos x + i \left(a\, e^{-y} \sin x - x\right)$$

has no zeros in the upper half-plane $\Im z \geq 0$.

PROOF. The equation $\varphi(z) = 0$ is equivalent to the following system of equations:

$$y + a\, e^{-y} \cos x = 0, \tag{8.2.17}$$

$$a\, e^{-y} \sin x - x = 0. \tag{8.2.18}$$

Let $y \geq 0$. Then $e^{-y} \leq 1$ and it follows from (8.2.17) that $\cos x \leq 0$ so that

$$\frac{\pi}{2} + 2k\pi < x < \frac{3\pi}{2} + 2k\pi, \qquad k = 0, \pm 1, \dots .$$

From (8.2.18) we have

$$|x| = a\, e^{-y} |\sin x|. \tag{8.2.19}$$

If $y \geq 0$, equation (8.2.19) does not have a solution because on the left-hand side $|x|$ satisfies the inequality $|x| \geq \pi/2$, while on the right-hand side we have $a\, e^{-y}|\sin x| < \pi/2$ since $0 < a < \pi/2$, $e^{-y} \leq 1$, and $|\sin x| \leq 1$. □

8.2.2. The case for general $P_n(x)/Q_m(x)$. We now turn to the general case of integral I_q^p. If $m = n$ then

$$\frac{P_n(x)}{Q_m(x)} = A + B\, \frac{P_{n-2}(x)}{Q_m(x)}$$

is an even function, where A and B are constants, and the integral for the case $P_n(x)/Q_m(x) \equiv 1$ is evaluated in the previous subsection. Consequently we consider only the case $m > n$ and $P_n(x)/Q_m(x)$ is even, that is,

$m \geq n+2$. Repeating the computations done for formula (8.2.8) we obtain

$$I_q^p = \frac{\pi i}{a} \sum_k \left(2 \operatorname*{Res}_{z=z_k} + \operatorname*{Res}_{z=a_k} \right) \left[\frac{P_n(z)}{Q_m(z)} \, \frac{1}{(a\, e^{iz} - iz) \cos z} \right]$$
$$:= S_5 + S_6, \tag{8.2.20}$$

where z_k are the zeros of $Q_m(z)$ in the upper half-plane, and S_5 and S_6 are defined in an obvious way. The series S_5 will appear in (8.2.27). We show that S_6 can be evaluated in closed form. We have

$$\begin{aligned} S_6 &= \frac{\pi i}{a} \sum_{k=-\infty}^{\infty} \operatorname*{Res}_{z=a_k} \left[\frac{P_n(z)}{Q_m(z)} \, \frac{1}{(a\, e^{iz} - iz) \cos z} \right] \\ &= \frac{\pi}{a} \sum_{k=-\infty}^{\infty} \frac{P_n(a_k)}{Q_m(a_k)} \, \frac{(-1)^{k+1}}{a(-1)^k - a_k} \\ &= \frac{\pi}{a} \sum_{k=-\infty}^{\infty} \frac{P_n(a_k)}{Q_m(a_k)} \, \frac{a_k(-1)^k + a}{a_k^2 - a^2}. \end{aligned} \tag{8.2.21}$$

To evaluate S_6 in closed form, we use the formulae (see [**21**], pp. 296–297)

$$\sum_{n=-\infty}^{\infty} f(n) = -\pi \sum_k \operatorname*{Res}_{z=\zeta_k} [f(z) \cot \pi z], \tag{8.2.22}$$

and

$$\sum_{n=-\infty}^{\infty} (-1)^n f(n) = -\pi \sum_k \operatorname*{Res}_{z=\zeta_k} \left[\frac{f(z)}{\sin \pi z} \right], \tag{8.2.23}$$

where $f(z)$ is a rational function such that $f(z) = O(1/|z|^2)$ as $z \to \infty$ with noninteger poles, ζ_k. We shall derive (8.2.22) and (8.2.23) in Chapter 10. Note that formulae (8.2.9) and (8.2.10) can be derived from (8.2.22) and (8.2.23).

Using (8.2.21)–(8.2.23) we obtain

$$\begin{aligned} S_6 = -\frac{\pi^2}{a} \sum_k \Bigg\{ &\operatorname*{Res}_{z=\tilde{z}_k} \left[\frac{P_n(\tilde{z})}{Q_m(\tilde{z})} \, \frac{a}{\tilde{z}^2 - a^2} \cot \pi z \right] \\ &+ \operatorname*{Res}_{z=\tilde{z}_k} \left[\frac{P_n(\tilde{z})}{Q_m(\tilde{z})} \, \frac{\tilde{z}}{(\tilde{z}^2 - a^2) \sin \pi z} \right] \Bigg\}, \end{aligned} \tag{8.2.24}$$

where $\hat{z}_k$ are the zeros of $Q_m(\tilde{z})(\tilde{z}^2 - a^2)$ and $\tilde{z} = (2z+1)\pi/2$. It follows from (8.2.24) that

$$S_6 = -\frac{\pi^2}{a} \sum_k \operatorname*{Res}_{z=\hat{z}_k} \left[\frac{P_n(\tilde{z})}{Q_m(\tilde{z})} \, \frac{a \cos \pi z + \tilde{z}}{(\tilde{z}^2 - a^2) \sin \pi z} \right]. \tag{8.2.25}$$

Let us compute separately the sum of the residues in (8.2.25) at the points $\tilde{z} = \pm a$, that is, at the points $\hat{z}_1 = a/\pi - 1/2$ and $\hat{z}_2 = -(a/\pi + 1/2)$:

$$\begin{aligned}
&\left(\operatorname*{Res}_{z=\hat{z}_1} + \operatorname*{Res}_{z=\hat{z}_2} \right) \left[\frac{P_n(\tilde{z})}{Q_m(\tilde{z})} \, \frac{a \cos \pi z + \tilde{z}}{(\tilde{z}^2 - a^2) \sin \pi z} \right] \\
&\qquad = \frac{P_n(a)}{Q_m(a)} \, \frac{a \cos \pi \hat{z}_1 + a}{2a\pi \sin \pi \hat{z}_1} + \frac{P_n(-a)}{Q_m(-a)} \, \frac{a \cos \pi \hat{z}_2 - a}{(-2a\pi) \sin \pi \hat{z}_2} \\
&\qquad\qquad \text{(since } P_n(x)/Q_m(x) \text{ is even)} \\
&\qquad = \frac{P_n(a)}{Q_m(a)} \left[\frac{\cos(a - \pi/2) + 1}{2\pi \sin(a - \pi/2)} + \frac{\cos(a + \pi/2) - 1}{2\pi \sin(a + \pi/2)} \right] \\
&\qquad = -\frac{P_n(a)}{Q_m(a)} \, \frac{1 + \sin a}{\pi \cos a}.
\end{aligned}$$

Then (8.2.25) can be written as

$$\begin{aligned}
S_6 = -\frac{\pi^2}{a} \sum_k \operatorname*{Res}_{z=\tilde{z}_k} \left[\frac{P_n(\tilde{z})}{Q_m(\tilde{z})} \, \frac{a \cos \pi z + \tilde{z}}{(\tilde{z}^2 - a^2) \sin \pi z} \right] \\
+ \frac{\pi}{a} \, \frac{P_n(a)}{Q_m(a)} \, \frac{1 + \sin a}{\cos a},
\end{aligned} \tag{8.2.26}$$

where $\tilde{z}_k$ are all the zeros of $Q_m(\tilde{z})$ and $\tilde{z} = (2z+1)\pi/2$. Upon substitution of (8.2.26) into (8.2.20) we obtain the formula for the evaluation of the integral I_q^p:

$$\begin{aligned}
I_q^p &= \int_{-\infty}^{\infty} \frac{P_n(x)}{Q_m(x)} \, \frac{dx}{x^2 - 2ax \sin x + a^2} \\
&= \frac{2\pi i}{a} \sum_k \operatorname*{Res}_{z=z_k} \left[\frac{P_n(z)}{Q_m(z) \, (a \, e^{iz} - iz) \cos z} \right] \\
&\quad - \frac{\pi^2}{a} \sum_k \operatorname*{Res}_{z=\tilde{z}_k} \left[\frac{P_n(\tilde{z})}{Q_m(\tilde{z})} \, \frac{a \cos \pi z + \tilde{z}}{(\tilde{z}^2 - a^2) \sin \pi z} \right] \\
&\quad + \frac{\pi}{a} \, \frac{P_n(a)}{Q_m(a)} \, \frac{1 + \sin a}{\cos a},
\end{aligned} \tag{8.2.27}$$

where z_k are the zeros of $Q_m(z)$ in the upper half-plane and $\tilde{z}_k$ are all the zeros of $Q_m(\tilde{z})$ with $\tilde{z} = (2z+1)\pi/2$.

Note 8.2.1. Although formula (8.2.27) is derived under the condition that $m \ge n+2$, in fact it is still valid if $P_n(x)/Q_m(x) \equiv 1$. In this case $\hat{z}_k$ are the zeros of $\tilde{z}^2 - a^2$, that is, $\hat{z}_1 = a/\pi - 1/2$, $\hat{z}_2 = -(a/\pi + 1/2)$ and we obtain

$$I_q^p = \int_{-\infty}^{\infty} \frac{dx}{x^2 - 2ax\sin x + a^2} = \frac{\pi}{a}\,\frac{1+\sin a}{\cos a},$$

which is formula (8.2.2).

EXAMPLE 8.2.1. *Evaluate the integral*

$$A = \int_{-\infty}^{\infty} \frac{dx}{(x^2+1)(x^2 - 2ax\sin x + a^2)}, \qquad 0 < a < \frac{\pi}{2}.$$

SOLUTION. Here, $P_n(z) = 1$, $Q_m(z) = z^2 + 1$ and the only zero of $Q_m(z)$ in the upper half-plane is $z_1 = i$. Then

$$Q_m(\tilde{z}) = 1 + \frac{(2z+1)^2\pi^2}{4},$$

which vanishes at $\tilde{z}_{1,2} = \pm i/\pi - 1/2$. By (8.2.27) we have

$$\begin{aligned} A = {} & \frac{2\pi i}{a} \operatorname*{Res}_{z=i}\left[\frac{1}{(z^2+1)\,(a\,e^{iz} - iz)\cos z}\right] \\ & - \frac{\pi^2}{a}\left(\operatorname*{Res}_{z=\tilde{z}_1} + \operatorname*{Res}_{z=\tilde{z}_2}\right)\left[\frac{1}{1+(2z+1)^2\pi^2/4}\,\frac{a\cos\pi z + (2z+1)\pi/2}{[(2z+1)^2\pi^2/4 - a^2]\sin\pi z}\right] \\ & + \frac{\pi}{a}\,\frac{1}{a^2+1}\,\frac{1+\sin a}{\cos a}. \end{aligned}$$

The sum of the residues at $\tilde{z}_1$ and $\tilde{z}_2$ is

$$\begin{aligned} & \left(\operatorname*{Res}_{z=\tilde{z}_1} + \operatorname*{Res}_{z=\tilde{z}_2}\right)\left[\frac{1}{(2z+1)^2\pi^2/4+1}\,\frac{a\cos\pi z + (2z+1)\pi/2}{[(2z+1)^2\pi^2/4 - a^2]\sin\pi z}\right] \\ & \qquad = \frac{1}{2\pi i}\,\frac{a\cos\pi\tilde{z}_1 + i}{(-1-a^2)(-\cos i)} + \frac{1}{2\pi i}\,\frac{a\cos\pi\tilde{z}_2 - i}{(-1-a^2)\cos i} \\ & \qquad = \frac{1}{2\pi}\left[\frac{a\sinh 1 + 1}{-(1+a^2)(-\cosh 1)} + \frac{a\sinh 1 + 1}{(1+a^2)\cosh 1}\right] \\ & \qquad = \frac{a\sinh 1 + 1}{\pi}\,\frac{1}{(1+a^2)\cosh 1}. \end{aligned}$$

Thus,

$$A = \frac{\pi}{a}\,\frac{1}{(a\,e^{-1}+1)\cosh 1} - \frac{\pi}{a}\,\frac{a\sinh 1 + 1}{(1+a^2)\cosh 1} + \frac{\pi}{a}\,\frac{1}{1+a^2}\,\frac{1+\sin a}{\cos a}. \quad \square$$

8.3. Forms containing $(h \sin ax + x \cos ax)^{-1}$

This section presents results obtained in [**3**]. We consider integrals of the form

$$I_\varphi = \text{p. v.} \int_{-\infty}^{\infty} \frac{P_n(x)}{Q_m(x)} \frac{dx}{\varphi(x)}, \tag{8.3.1}$$

where, in general, $\varphi(z)$ is the entire function $\varphi(z) = h \sin az + z \cos az$. We also consider similar integrals, I_φ^c and I_φ^s, where dx is replaced by $\cos bx\, dx$ and $\sin bx\, dx$, respectively.

The entire functions $\varphi(z)$ considered in this section generally come from the solution of Sturm–Liouville differential equations:

$$\frac{d}{dx}\left[k(x)\frac{du}{dx}\right] - q(x)u = -\lambda^2 \rho(x)u, \tag{8.3.2}$$

over the interval $0 < x < L$ or $0 < x < \infty$, with appropriate boundary conditions, where $k(x) > 0$, $\rho(x) > 0$, $q(x) \geq 0$ are given functions, which are continuous on the closed interval $[0, L]$, and $k'(x)$ is continuous on the open interval $(0, L)$. If the interval is semi-infinite, $0 \leq x < \infty$, then $|u|$ is bounded at infinity.

For given boundary conditions, the eigenvalues of (8.3.2) on a finite interval are the roots of some equation

$$\varphi(\lambda) = 0. \tag{8.3.3}$$

It is known [**7**] that, under these conditions, equation (8.3.3) has only simple real roots. In this case, the entire function, $\varphi(z)$, in the integral (8.3.1) has only real zeros.

On the infinite interval $0 \leq x < \infty$, one considers equation (8.3.2) with boundary conditions of the first, second or third kind at $x = 0$ and the condition $|u(x)| < M$, $M = \text{constant}$, as $x \to \infty$. Depending upon the behavior of the functions $k(x)$ and $\rho(x)$ in (8.3.2) as $x \to \infty$, this problem may have a discrete or a continuous spectrum (see [**5**]).

If (8.3.2) has only one regular singular point in the finite complex plane, that is, the functions $k(z)$, $q(z)$ and $\rho(z)$ are analytic, $k'(z)/k(z)$ has one simple pole, and $q(z)/k(z)$ and $\rho(z)/k(z)$ have only one pole whose order is not higher than two, then the eigenfunctions are (apart from a factor z^α) entire functions with simple real zeros.

The function $\varphi(z)$ was taken to be $\sin az$ and $\cos az$ in Section 8.1 and $a\, e^{iz} - iz$ (without zeros in the upper half-plane) in Section 8.2. In the present section, $\varphi(z)$ will be taken to be $h \sin az + z \cos az$, with $h > 0$ and $a > 0$. In Section 8.4, $\varphi(z)$ will be taken to be $J_p(az)/z^p$ and

$J_{p+\nu}(az)/z^{p-l}J_{l+\nu}(bz)$, where $J_p(z)$ is the Bessel function of the first kind of order p for $p = 0, 1, \dots$.

In the following two subsections we consider integrals of the form

$$\begin{aligned} I_\varphi &= \text{p. v.} \int_{-\infty}^{\infty} \frac{P_n(x)}{Q_m(x)} \frac{dx}{\varphi(x)}, \\ I_\varphi^{\mathrm{c}} &= \text{p. v.} \int_{-\infty}^{\infty} \frac{P_n(x)}{Q_m(x)} \frac{\cos bx}{\varphi(x)}\, dx, \\ I_\varphi^{\mathrm{s}} &= \text{p. v.} \int_{-\infty}^{\infty} \frac{P_n(x)}{Q_m(x)} \frac{\sin bx}{\varphi(x)}\, dx, \end{aligned} \tag{8.3.4}$$

where $\varphi(x) = h\sin ax + x\cos ax$, $h > 0$, $a > 0$, $P_n(x)$ and $Q_m(x)$ are polynomials of degrees n and m, respectively, and $m \ge n+1$. Using the methods of Section 8.1, we express these integrals as finite sums of residues at the zeros of the polynomial $Q_m(x)$.

8.3.1. The integral I_φ. Consider the integral

$$I_\varphi = \text{p. v.} \int_{-\infty}^{\infty} \frac{P_n(x)}{Q_m(x)} \frac{dx}{h\sin ax + x\cos ax}, \tag{8.3.5}$$

where the degrees of P_n and Q_m satisfy $m \ge n+1$ and $h > 0$.

The transcendental equation

$$h\sin a\lambda + \lambda\cos a\lambda = 0 \tag{8.3.6}$$

arises in the determination of the eigenvalues, λ_n, of the following Sturm–Liouville boundary value problem:

$$\frac{d^2u}{dx^2} + \lambda^2 u = 0, \qquad 0 < x < a, \tag{8.3.7}$$

$$u|_{x=0} = 0, \qquad \left(\frac{du}{dx} + hu\right)\bigg|_{x=a} = 0. \tag{8.3.8}$$

Since (8.3.7) is a particular case of (8.3.2), the entire function

$$\varphi(z) = h\sin az + z\cos az \tag{8.3.9}$$

has only simple real zeros.

First, we consider the case where $Q_m(x) \ne 0$ for real x and $P_n(x)/Q_m(x)$ is an odd function so that the integrand in (8.3.5) is even. If $P_n(x)/Q_m(x)$ is neither even nor odd, then it can be represented as the sum of an even and an odd function. Moreover, the integral of the quotient of an even function with the odd function $\varphi(x)$ is zero.

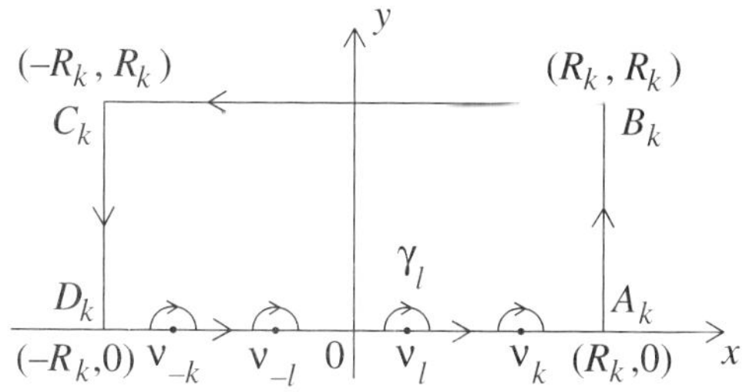

FIGURE 8.3. The path of integration in Subsection 8.3.1.

Set

$$f(z) = \frac{P_n(z)}{Q_m(z)} \frac{1}{h \sin az + z \cos az}, \qquad z \in \mathbb{C}. \tag{8.3.10}$$

Recalling that $h > 0$, we may rewrite (8.3.6) for real x in the form

$$h \tan ax + x = 0 \tag{8.3.11}$$

and consider the real roots, ν_k, of this last equation. It is easily seen by examining the graphs of $y = \tan ax$ and $y = -x/h$ that ν_k satisfy the inequality

$$\frac{k\pi}{a} < \nu_k < \frac{(k+1)\pi}{a}, \qquad k = 0, 1, 2, \dots ,$$

and that $-\nu_k$ also is a root.

Let C be a closed path consisting of the segment $[-R_k, R_k]$ of the real axis, where $aR_k = k\pi$ for $k = 1, 2, 3, \dots$ and the zeros, ν_l, $l = 0, \pm 1, \dots , \pm k$, of equation (8.3.11) are bypassed along the semicircles γ_l of radii δ in the upper half-plane, and the sides A_kB_k, B_kC_k and C_kD_k of the rectangle $A_kB_kC_kD_k$, with vertices $A_k = (R_k, 0)$, $B_k = (R_k, R_k)$, $C_k = (-R_k, R_k)$, $D_k = (-R_k, 0)$, shown in Fig 8.3.

By the residue theorem, we obtain

$$\left(\int_{-R_k}^{R_k} + \sum_{l=-k}^{k} \int_{\gamma_l} + \int_{\eta_k} \right) \left[\frac{P_n(z)}{Q_m(z)(h \sin az + z \cos az)} \right] dz$$
$$= 2\pi i \sum_k \operatorname*{Res}_{z=z_k} \left[\frac{P_n(z)}{Q_m(z)(h \sin az + z \cos az)} \right], \tag{8.3.12}$$

where η_k is the polygonal line $A_kB_k \cup B_kC_k \cup C_kD_k$, z_k are the zeros of $Q_m(z)$ lying inside C and the integral from $-R_k$ to R_k is evaluated along the line segments of the x-axis with the exclusion of the diameters of the semicircles γ_l.

We show that the integral I_{η_k} along the polygonal line η_k approaches zero as $R_k \to \infty$. Since

$$|\cos az| = \sqrt{\sinh^2 ay + \cos^2 ax}, \tag{8.3.13}$$

we have

$$\begin{aligned}
|I_{\eta_k}| &\le \left| \int_{\eta_k} \frac{P_n(z)}{Q_m(z)} \frac{dz}{h \sin az + z \cos az} \right| \\
&\le \int_{\eta_k} \left| \frac{P_n(z)}{Q_m(z)} \right| \frac{|dz|}{|\cos az|\,|h \tan az + z|} \\
&\qquad \text{(and, since } m \ge n+1,\text{)} \\
&\le \frac{C}{R_k} \int_{\eta_k} \frac{|dz|}{|\cos az|\,|h \tan az + z|},
\end{aligned}$$

that is,

$$|I_{\eta_k}| \le \frac{C}{R_k} \left(\int_{A_k B_k} + \int_{B_k C_k} + \int_{C_k D_k} \right) \left[\frac{|dz|}{|\cos az|\,|h \tan az + z|} \right]. \tag{8.3.14}$$

On the segment $A_k B_k$, $z = R_k + iy$ and $|dz| = dy$; thus we have

$$|\cos az|\Big|_{z \in A_k B_k} = \sqrt{\sinh^2 ay + 1} = \cosh ay > \frac{1}{2}\, e^{ay},$$

and

$$\begin{aligned}
|h \tan az + z|\Big|_{z \in A_k B_k} &= |h \tan(k\pi + iay) + R_k + iy| \\
&= |ih \tanh ay + R_k + iy| \\
&\ge R_k = \frac{1}{a}\, k\pi.
\end{aligned}$$

Therefore,

$$\begin{aligned}
\int_{A_k B_k} \frac{|dz|}{|\cos az|\,|h \tan az + z|} &\le \frac{2a}{k\pi} \int_0^{R_k} \frac{dy}{e^{ay}} \\
&= \frac{2}{k\pi} \left(1 - e^{-aR_k}\right) \\
&\to 0, \qquad \text{as} \quad R_k \to \infty.
\end{aligned} \tag{8.3.15}$$

Similarly, the integral along C_kD_k approaches zero as $R_k \to \infty$. On the segment B_kC_k, $z = x + iR_k$ and $|dz| = dx$; thus we have

$$\begin{aligned} |h \tan az + z|_{z=x+iR_k} &= \left| ih \frac{1 - e^{2iaz}}{1 + e^{2iaz}} + z \right|_{z=x+iR_k} \\ &= \left| ih \frac{1 - e^{-2aR_k}\, e^{2iax}}{1 + e^{-2aR_k}\, e^{2iax}} + x + iR_k \right| := g_k(x) \\ &\to \infty, \qquad \text{as } R_k \to \infty. \end{aligned}$$

Hence, we have

$$\begin{aligned} \int_{B_kC_k} \frac{|dz|}{|\cos az|\,|h \tan az + z|} &= \int_{R_k}^{-R_k} \frac{dx}{g_k(x)\sqrt{\sinh^2 k\pi + \cos^2 ax}} \\ &\qquad \text{(and putting } x = R_k t\text{)} \\ &= R_k \int_1^{-1} \frac{dt}{g_k(R_k t)\sqrt{\sinh^2 k\pi + \cos^2(aR_k t)}} \\ &\to 0, \qquad \text{as } R_k \to \infty. \end{aligned} \tag{8.3.16}$$

Using (8.3.14)–(8.3.16), we obtain that $I_{\eta_k} \to 0$ as $R_k \to \infty$.

Since ν_k is a simple pole of $f(z)$, by expanding f in a Laurent series in a neighborhood of the point ν_k, as $\delta \to 0$ we obtain

$$\lim_{\delta \to 0} \int_{\gamma_k} f(z)\, dz = -\pi i \operatorname*{Res}_{z=\nu_k} f(z). \tag{8.3.17}$$

(For a similar computation, see formula (6.1.9) in Subsection 6.1.2.) As $\delta \to 0$ for any fixed R_k, the second sum on the left-hand side of (8.3.12),

$$\begin{aligned} S_l &= \pi i \sum_{k=-l}^{l} \operatorname*{Res}_{z=\nu_k} \left[\frac{P_n(z)}{Q_m(z)} \frac{1}{h \sin az + z \cos az} \right] \\ &= \pi i \sum_{k=-l}^{l} \frac{P_n(\nu_k)}{Q_m(\nu_k)} \left[\frac{1}{(h \sin az + z \cos az)'} \right] \Bigg|_{z=\nu_k} \\ &= \pi i \frac{P_n(0)}{Q_m(0)} \frac{1}{h(a+1)} = 0, \end{aligned} \tag{8.3.18}$$

is zero since

$$\frac{P_n(-\nu_k)}{Q_m(-\nu_k)} = -\frac{P_n(\nu_k)}{Q_m(\nu_k)},$$

the derivative of the odd function $h\sin az + z\cos az$ is even, and $P_n(0) = 0$. (For a similar computation, see formula (8.1.9) in Subsection 8.1.2.) Hence S_l does not contribute to the right-hand side of (8.3.12).

Therefore, taking the limit in (8.3.12) as $R_k \to \infty$, $\delta \to 0$, we obtain

$$\begin{aligned}\text{p. v.} \int_{-\infty}^{\infty} \frac{P_n(x)}{Q_m(x)} \frac{dx}{h\sin ax + x\cos ax} \\ = 2\pi i \sum_k \operatorname*{Res}_{z=z_k} \left[\frac{P_n(z)}{Q_m(z)} \frac{1}{h\sin az + z\cos az}\right], \qquad (8.3.19)\end{aligned}$$

where $m \geq n+1$, $Q_m(x) \neq 0$ for real x and $\Im z_k > 0$. To the authors' knowledge this formula is absent from the literature, even in the form of examples.

EXAMPLE 8.3.1. *Compute the integral*

$$I = \text{p. v.} \int_{-\infty}^{\infty} \frac{x}{x^2+1} \frac{dx}{h\sin ax + x\cos ax}.$$

SOLUTION. Since the conditions of formula (8.3.19) are satisfied, we have

$$\begin{aligned} I &= 2\pi i \operatorname*{Res}_{z=i} \left[\frac{z}{z^2+1} \frac{1}{h\sin az + z\cos az}\right] \\ &= 2\pi i \frac{i}{2i} \frac{1}{h\sin ai + i\cos ai} \\ &= \frac{\pi}{h\sinh a + \cosh a}. \quad \square \end{aligned}$$

Second, we consider the case where $Q_m(x)$ has the simple real zeros $x = a_k$ for $k = 1, 2, \ldots, l$, and $a_k \neq \nu_k$. By bypassing the zeros a_k on the segment $[-R_k, R_k]$ along semicircles, δ_k, in the upper half-plane, we find that the term

$$A = \pi i \sum_{k=1}^{l} \operatorname*{Res}_{z=a_k} \left[\frac{P_n(z)}{Q_m(z)} \frac{1}{h\sin az + z\cos az}\right] \qquad (8.3.20)$$

has to be added to the right-hand side of (8.3.19). In the following lemma we show that A is zero.

LEMMA 8.3.1. *The finite sum* (8.3.20) *is equal to zero.*

PROOF. The proof of this lemma coincides almost literally with the proof of Lemma 8.1.2 in Subsection 8.1.2. We note that the even function $Q_m(x)$ satisfies the condition $Q_m(0) \neq 0$ because it follows from the condition $Q_m(0) = 0$ that, at least, $Q_m(x) = x^2 \widetilde{Q}_{m-2}(x)$, where $\widetilde{Q}_{m-2}(x)$ is an even polynomial in x and $\widetilde{Q}_{m-2} \neq 0$. However, in this case the integral in (8.3.19) is divergent in a neighborhood of $x = 0$. Therefore, we must have $Q_m(x) \neq 0$. □

It follows from Lemma 8.3.1 that if the real zeros, $x = a_k$, of $Q_m(x)$ are simple and $a_k \neq \nu_k$, then the integral (8.3.5) can be evaluated by (8.3.19).

8.3.2. The integrals I_φ^c and I_φ^s. We consider integrals of the form

$$\begin{aligned} I_\varphi^c &= \text{p. v.} \int_{-\infty}^{\infty} \frac{P_n(x)}{Q_m(x)} \frac{\cos bx}{h \sin ax + x \cos ax} \, dx, \\ I_\varphi^s &= \text{p. v.} \int_{-\infty}^{\infty} \frac{P_n(x)}{Q_m(x)} \frac{\sin bx}{h \sin ax + x \cos ax} \, dx, \end{aligned} \tag{8.3.21}$$

under the same conditions as in the previous subsection. In contrast to Subsections 8.1.2–8.1.4, we restrict ourselves to the case $|b| \leq a$. We assume that the function $P_n(x)/Q_m(x)$ is odd and $m \geq n+1$ in I_φ^c, while it is even and $m \geq n+2$ in I_φ^s.

Let z_k be the zeros of $Q_m(z)$ in the upper half-plane and let a_k be the simple real zeros of $Q_m(z)$. Moreover, we assume that $a_k \neq \nu_k$ where ν_k are the real zeros of the function $\varphi(x) = h \sin ax + x \cos ax$ (note that $\varphi(x)$ does not have other zeros). If these conditions are satisfied then, by a computation similar to the one in the previous subsection, the integrals (8.3.21) are expressed by means of a finite sum of residues at the zeros of $Q_m(z)$:

$$\begin{aligned} \text{p. v.} &\int_{-\infty}^{\infty} \frac{P_n(x)}{Q_m(x)} \frac{\cos bx}{h \sin ax + x \cos ax} \, dx \\ &= 2\pi i \sum_k \operatorname*{Res}_{z=z_k} \left[\frac{P_n(z)}{Q_m(z)} \frac{\cos bz}{h \sin az + z \cos az} \right] \end{aligned} \tag{8.3.22}$$

and

$$\begin{aligned} \text{p. v.} &\int_{-\infty}^{\infty} \frac{P_n(x)}{Q_m(x)} \frac{\sin bx}{h \sin ax + x \cos ax} \, dx \\ &= 2\pi i \sum_k \operatorname*{Res}_{z=z_k} \left[\frac{P_n(z)}{Q_m(z)} \frac{\sin bz}{h \sin az + z \cos az} \right]. \end{aligned} \tag{8.3.23}$$

The sum of residues at the real zeros, a_k, is equal to zero by virtue of Lemma 8.3.1.

One can, however, compute the integrals (8.3.21) also in the case where $|b| \leq |a|$ (as in Subsections 8.1.2–8.1.4) by using Jordan's Lemma and letting $\cos bx = \Re\, e^{ibx}$ and $\sin bx = \Im\, e^{ibx}$. In this case, the symmetry is lost and the answer is given by the sum of two terms: a finite sum of residues at the zeros of the polynomial $Q_m(z)$ and an infinite series (that is, the sum of the residues at the zeros, ν_k, of the function $\varphi(x) = h\sin ax + x\cos ax$). Equating these answers to the right-hand sides in (8.3.22) and (8.3.23), one can get equations for the determination of the sum of these series in the following form (for a similar detailed computation, see Subsection 8.1.2):

$$\Re D = \Re\left\{2\pi i \sum_k \operatorname*{Res}_{z=z_k}\left[\frac{P_n(z)}{Q_m(z)}\,\frac{\cos bz}{\varphi(z)}\right]\right\} \tag{8.3.24}$$

and

$$\Im D = \Im\left\{2\pi i \sum_k \operatorname*{Res}_{z=z_k}\left[\frac{P_n(z)}{Q_m(z)}\,\frac{\sin bz}{\varphi(z)}\right]\right\}, \tag{8.3.25}$$

where

$$D = 2\pi i \sum_k \operatorname*{Res}_{z=z_k}\left[\frac{P_n(z)}{Q_m(z)}\,\frac{e^{ibz}}{\varphi(z)}\right] + 2\pi i \sum_{k=1}^{\infty} \frac{P_n(\nu_k)}{Q_m(\nu_k)}\,\frac{e^{ib\nu_k}}{\varphi'(\nu_k)},$$

and $\varphi(z) = h\sin az + z\cos az$.

New closed-form expressions of these two series are obtained from (8.3.24) and (8.3.25) in the form

$$\begin{aligned} S_7 &= \sum_{k=1}^{\infty} \frac{P_n(\nu_k)}{Q_m(\nu_k)}\,\frac{\sin\nu_k}{\varphi'(\nu_k)} \\ &= \Re\left\{i \sum_k \operatorname*{Res}_{z=z_k}\left[\frac{P_n(z)}{Q_m(z)}\,\frac{e^{ibz} - \cos bz}{\varphi(z)}\right]\right\} \end{aligned} \tag{8.3.26}$$

and

$$\begin{aligned} S_8 &= \sum_{k=1}^{\infty} \frac{P_n(\nu_k)}{Q_m(\nu_k)}\,\frac{\cos\nu_k}{\varphi'(\nu_k)} \\ &= \Im\left\{i \sum_k \operatorname*{Res}_{z=z_k}\left[\frac{P_n(z)}{Q_m(z)}\,\frac{\sin bz - e^{ibz}}{\varphi(z)}\right]\right\}, \end{aligned} \tag{8.3.27}$$

where, using $\tan a\nu_k = -\nu_k/h$, we have

$$\begin{aligned} \varphi'(\nu_k) &= -[(1+ah)h/\nu_k + a\nu_k]\sin\nu_k \\ &= \frac{(1+ah)h + a\nu_k^2}{\sqrt{\nu_k^2 + h^2}}. \end{aligned}$$

8.4. Forms containing Bessel functions

We consider integrals of the form

$$\begin{aligned}
&\text{p. v.} \int_{-\infty}^{\infty} \frac{P_n(x)}{Q_m(x)} \frac{x^p}{J_p(ax)}\,dx, \\
&\text{p. v.} \int_{-\infty}^{\infty} \frac{P_n(x)}{Q_m(x)} \frac{x^{p-l} J_{l+\nu}(bx)}{J_{p+\nu}(ax)}\,dx,
\end{aligned} \tag{8.4.1}$$

where $p = 0, 1, 2, \dots$, and $J_\nu(z)$ is the Bessel function of the first kind of order ν, which can be represented by the following series:

$$J_\nu(z) = \sum_{k=0}^{\infty} \frac{(-1)^k}{k!\Gamma(k+\nu+1)} \left(\frac{z}{2}\right)^{2k+\nu}, \qquad |z| < \infty. \tag{8.4.2}$$

Here $\Gamma(\zeta)$ is Euler's gamma function given by the integral

$$\Gamma(\zeta) = \int_0^{\infty} e^{-t} t^{\zeta-1}\,dt, \qquad \Re\zeta > 0. \tag{8.4.3}$$

If $\zeta = p+1$ for $p = 0, 1, 2, \dots$, then it can easily be shown from (8.4.3) that $\Gamma(p+1) = p!$. In this case the series in (8.4.2) can be written in the form

$$J_p(z) = \sum_{k=0}^{\infty} \frac{(-1)^k}{k!(k+p)!} \left(\frac{z}{2}\right)^{2k+p}, \qquad |z| < \infty. \tag{8.4.4}$$

Using the same methods as in the previous subsections, we express the integrals (8.4.1) by means of a finite sum of residues at the zeros of $Q_m(z)$.

8.4.1. Integrals containing $x^p/J_p(ax)$. We consider integrals of the form

$$\text{p. v.} \int_{-\infty}^{\infty} \frac{P_n(x)}{Q_m(x)} \frac{x^p}{J_p(ax)}\,dx, \tag{8.4.5}$$

where $P_n(x)$ and $Q_m(x)$ are polynomials of degrees n and m, respectively, $m \ge n+p+1$, and the function $P_n(x)/Q_m(x)$ is even if p is odd. We first consider the case where $Q_m(x) \neq 0$ for real x. The equation

$$J_p(a\lambda) = 0 \tag{8.4.6}$$

appears in the determination of the eigenvalues of the following Sturm–Liouville boundary value problem:

$$\frac{d}{dx}\left(x\frac{du}{dx}\right) - \frac{p^2}{x}u = -\lambda^2 xu, \qquad 0 < x < a, \tag{8.4.7}$$

$$\left| u\big|_{x=0} \right| < M, \qquad u\big|_{x=a} = 0. \tag{8.4.8}$$

Comparing (8.4.7) with (8.3.2) we see that, in this case, $k(x) = x$, $\rho(x) = x$ and $q(x) = p^2/x$. Since $k(0) = 0$ and $\rho(0) = 0$, we have a singular case (see [**5**]); therefore, we assume, in (8.4.8), that the solution is bounded at $x = 0$. Hence

$$\varphi(z) = J_p(az) \tag{8.4.9}$$

is an entire function with only simple zeros (except the zero of order p at $z = 0$).

Set

$$f(z) = \frac{P_n(z)}{Q_m(z)} \frac{z^p}{J_p(az)}, \qquad z \in \mathbb{C}, \tag{8.4.10}$$

and let α_k be the nonzero roots of the equation

$$J_p(ax) = 0. \tag{8.4.11}$$

We use the following expansion of $J_p(z)$ as $|z| \to \infty$ (see [**17**], Vol. II, p. 85):

$$J_p(z) \sim \sqrt{\frac{2}{\pi z}} \cos\left(z - \frac{\pi}{4} - \frac{\pi}{2}p\right) \left[1 + O\left(\frac{1}{z}\right)\right], \tag{8.4.12}$$

as $z \to \infty$ and $-\pi < \arg z < \pi$. Thus, for sufficiently large N, the zeros, α_k, of $J_p(z)$ have the form

$$\alpha_k - \frac{\pi}{4} - \frac{\pi}{2}p \approx k\pi + \frac{\pi}{2}, \qquad \text{that is,} \quad \alpha_k \approx \left(k + \frac{p}{2} + \frac{3}{4}\right)\pi,$$

for $k = \pm N, \pm(N+1), \dots$. This means that, starting with some sufficiently large N, the roots of (8.4.11) satisfy the inequality

$$\left(k + \frac{p}{2}\right)\pi < a\alpha_k < \left(k + 1 + \frac{p}{2}\right)\pi. \tag{8.4.13}$$

Let C be a closed path consisting of the segment $[-R_k, R_k]$ of the real axis, and the sides A_kB_k, B_kC_k, C_kD_k of a rectangle (see Fig 8.3), where $aR_k = (k+1/2)p$, $k = N, N+1, \dots$, for N sufficiently large. The zeros, α_l, of equation (8.4.11) on $[-R_k, R_k]$ are bypassed along semicircles γ_l in the upper half-plane, and the points A_k, B_k, C_k and D_k have the following coordinates: $A_k = (R_k, 0)$, $B_k = (R_k, R_k)$, $C_k = (-R_k, R_k)$, $D_k = (-R_k, 0)$.

As can be seen from (8.4.4) the function $x^p/J_p(ax)$ tends to $p!2^p$ as $x \to 0$. By the residue theorem we obtain

$$\left[\int_{-R_k}^{R_k} + \sum_{l=-k}^{k} \int_{\gamma_l} + \int_{\eta_k}\right] \frac{P_n(z)}{Q_m(z)} \frac{z^p}{J_p(az)}\, dz$$

$$= 2\pi i \sum_k \operatorname*{Res}_{z=z_k} \left[\frac{P_n(z)}{Q_m(z)} \frac{z^p}{J_p(az)} \right], \tag{8.4.14}$$

where $\eta_k = A_k B_k \cup B_k C_k \cup C_k D_k$, z_k are the zeros of $Q_m(z)$ that lie inside C and the integral I_{η_k} from $-R_k$ to R_k is evaluated along the line segments of the x-axis excluding the arcs γ_l. We show that the integral along the polygonal line η_k approaches zero as $R_k \to \infty$. Using (8.4.12) we have

$$\begin{aligned}
|I_{\eta_k}| &\le \left| \int_{\eta_k} \frac{P_n(z)}{Q_m(z)} \frac{z^p}{J_p(az)}\, dz \right| \\
&\le \int_{\eta_k} \left| \frac{P_n(z) z^p}{Q_m(z)} \right| \frac{\sqrt{\pi |az|}\, |dz|}{\sqrt{2}\, |\cos(az - \pi/4 - \pi/2p)|} \\
&\le \int_{\eta_k} \frac{D}{\sqrt{|z|}} \frac{|dz|}{|\cos(az - \pi/4 - \pi/2p)|} \\
&\le \frac{D}{\sqrt{R_k}} \int_{\eta_k} \frac{|dz|}{|\cos(az - \pi/4 - \pi/2p)|}.
\end{aligned} \tag{8.4.15}$$

Since formula (8.4.15) is similar to (8.2.14), where one replaces $(l + 1/2)\pi$ by R_k, the rest of the proof is given after formula (8.2.14). Hence,

$$\lim_{R_k \to \infty} \int_{\eta_k} f(z)\, dz = 0. \tag{8.4.16}$$

Since α_k is a simple pole of $f(z)$ then, by using a Laurent series expansion in a neighborhood of the point α_k, we obtain

$$\lim_{\delta \to 0} \int_{\gamma_k} f(z)\, dz = -\pi i \operatorname*{Res}_{z=\alpha_k} f(z) \tag{8.4.17}$$

(for a similar detailed computation, see formula (6.1.9) in Subsection 6.1.2). Therefore, as $\delta \to 0$, and for each fixed R_k, the term

$$\begin{aligned}
S_l &= \pi i \sum_{\substack{k=-l \\ k \ne 0}}^{l} \operatorname*{Res}_{z=\alpha_k} \left[\frac{P_n(z)}{Q_m(z)} \frac{z^p}{J_p(az)} \right] \\
&= \frac{\pi i}{a} \sum_{\substack{k=-l \\ k \ne 0}}^{l} \frac{P_n(\alpha_k)}{Q_m(\alpha_k)} \frac{\alpha_k^p}{J_p'(a\alpha_k)}
\end{aligned} \tag{8.4.18}$$

is added to the right-hand side of (8.4.14). Here $J_p'(s)$ denotes the derivative of $J_p(s)$ with respect to s. Using formula (8.4.4) we obtain

$$
\begin{aligned}
A_k &= \frac{\alpha_k^p}{J_p'(a\alpha_k)} = \alpha_k^p \left[\frac{1}{2} \sum_{k=0}^{\infty} \frac{(-1)^k (2k+p)}{k!(k+p)!} \left(\frac{z}{2}\right)^{2k+p-1} \bigg|_{z=a\alpha_k} \right]^{-1} \\
&= \left[\frac{1}{2} \sum_{k=0}^{\infty} \frac{(-1)^k (2k+p)(a/2)^{2k+p-1} \alpha_k^{2k-1}}{k!(k+p)!} \right]^{-1},
\end{aligned}
$$

that is, A_k is an odd function of α_k.

Since, by assumption, $P_n(\alpha_k)/Q_m(\alpha_k)$ is even, then the function under the summation sign in (8.4.18) is an odd function of α_k, the term with $k=0$ being absent. Hence, $S_l = 0$. Therefore, considering the limit in (8.4.14) as $R_k \to \infty$, $\delta \to 0$, we obtain a formula for evaluating the integral (8.4.5) in the form

$$
\text{p. v.} \int_{-\infty}^{\infty} \frac{P_n(x)}{Q_m(x)} \frac{x^p}{J_p(ax)}\, dx = 2\pi i \sum_k \operatorname*{Res}_{z=z_k} \left[\frac{P_n(z)}{Q_m(z)} \frac{z^p}{J_p(az)} \right], \tag{8.4.19}
$$

where $Q_m(x) \neq 0$ for real x, z_k are the zeros of $Q_m(z)$ in the upper half-plane, $m \geq n+p+1$, and the function $P_n(x)/Q_m(x)$ is even. Formula (8.4.19) has not been found in the literature even in the form of examples.

EXAMPLE 8.4.1. *Compute the integral*

$$
I = \text{p. v.} \int_{-\infty}^{\infty} \frac{1}{x^2+1} \frac{x}{J_1(ax)}\, dx.
$$

SOLUTION. The conditions $P_n(x) = 1$, $Q_m(x) = x^2+1 \neq 0$ for real x, $p=1$, and $m = 2 = p+1$ are such that formula (8.4.19) is true. Therefore, I is equal to $2\pi i$ times the only residue at the point $z = i$ (this is the zero of z^2+1 in the upper half-plane),

$$
\begin{aligned}
I &= 2\pi i \operatorname*{Res}_{z=i} \left[\frac{z}{(z^2+1) J_1(az)} \right] \\
&= 2\pi i \frac{i}{2i J_1(ai)} = \pi i \frac{1}{i I_1(a)} \\
&= \frac{\pi}{I_1(a)},
\end{aligned}
$$

where (see formula (8.4.4))

$$
I_1(a) = \sum_{k=0}^{\infty} \frac{1}{k!(k+1)!} \left(\frac{a}{2}\right)^{2k+1}
$$

is the modified Bessel function of the first kind of order 1. □

Secondly, we consider the case $Q_m(x) = 0$ for real x at the points $x = a_k$, $k = 1, \dots, l$, where all the zeros a_k are simple and $a_k \neq \alpha_k$. By bypassing the zeros a_k on the segment $[-R_k, R_k]$ along semicircles δ_k in the upper half-plane, we find that the term

$$A = \pi i \sum_{k=1}^{l} \operatorname*{Res}_{z=a_k} \left[\frac{P_n(z)}{Q_m(z)} \frac{z^p}{J_p(az)} \right] \tag{8.4.20}$$

has to be added to the right-hand side of (8.4.19). In the following lemma we show that A is zero.

LEMMA 8.4.1. *The finite sum* (8.4.20) *is equal to zero.*

PROOF. The proof of this lemma coincides almost completely with the proof of Lemma 8.1.1 in Subsection 8.1.2. It should only be noted that in both lemmas, the even function $Q_m(x)$ satisfies the condition $Q_m(0) \neq 0$. The reason is the following: if $Q_m(0) = 0$ then we have, at least, that $Q_m(x) = x^2 \widetilde{Q}_{m-2}(x)$ (in general, we may have $Q_m(x) = x^4 \widetilde{\widetilde{Q}}_{m-4}(x)$, etc.), where $\widetilde{Q}_{m-2}(x)$ is an even polynomial in x and $\widetilde{Q}_{m-2}(0) \neq 0$. In this case, however, the integral in (8.4.5) is divergent in any neighborhood of $x = 0$. Hence, we must have $Q_m(0) \neq 0$. □

It follows from the previous Lemma 8.4.1 that, if $x = a_k$ are simple real zeros of $Q_m(x)$ and $a_k \neq \nu_k$, then integral (8.4.5) can be evaluated by formula (8.4.19).

8.4.2. Integrals containing ratios of Bessel functions. We consider integrals of the form

$$\text{p. v.} \int_{-\infty}^{\infty} \frac{P_n(x)}{Q_m(x)} \frac{x^{p-l} J_{l+\nu}(bx)}{J_{p+\nu}(ax)}\, dx, \qquad l, p = 0, 1, 2, \dots, \tag{8.4.21}$$

where $P_n(x)$ and $Q_m(x)$ are polynomials of degrees n and m, respectively, $P_n(x)/Q_m(x)$ is odd and $m \geq n + p - l + 2$. Using (8.4.4) it can easily be shown that the ratio

$$\frac{z^{p-l} J_{l+\nu}(bz)}{J_{p+\nu}(az)} = \frac{b^\nu \sum_{k=0}^{\infty} (-1)^k [k!\Gamma(k+l+\nu+1)]^{-1} (bz/2)^{2k}}{a^\nu \sum_{k=0}^{\infty} (-1)^k [k!\Gamma(k+p+\nu+1)]^{-1} (az/2)^{2k}} \tag{8.4.22}$$

is even. Although each of the functions $J_{l+\nu}(bz)$ and $J_{p+\nu}(az)$ has a branch point at $z = 0$ (and $z = \infty$) for non-integer ν, their ratio, as one can see from (8.4.22), has no branch point and is a meromorphic function.

In contrast with Subsections 8.1.2–8.1.4, we restrict ourselves to the case $|b| \leq |a|$. Let z_k be the zeros of $Q_m(z)$ in the upper half-plane, and let a_k be the simple real zeros of $Q_m(z)$ such that $a_k \neq \alpha_k$, where $\alpha_k \neq 0$ are real zeros of $J_p(az)$ (the function $J_p(az)$ does not have other zeros). In this case a computation similar to the one in the previous subsection leads to a formula expressing the integral (8.4.21) through a finite sum of the residues at the zeros of $Q_m(z)$:

$$\text{p. v.} \int_{-\infty}^{\infty} \frac{P_n(x)}{Q_m(x)} \frac{x^{p-l} J_{l+\nu}(bx)}{J_{p+\nu}(ax)} \, dx = 2\pi i \sum_k \operatorname*{Res}_{z=z_k} \left[\frac{P_n(z)}{Q_m(z)} \frac{z^{p-l} J_{l+\nu}(bz)}{J_{p+\nu}(az)} \right]. \quad (8.4.23)$$

By Lemma 8.4.1, the sum of the residues at the simple zeros a_k is equal to zero.

Exercises for Chapter 8

Evaluate the following integrals.

1. $\text{p. v.} \displaystyle\int_{-\infty}^{\infty} \frac{x}{x^4+1} \frac{dx}{\sin ax}.$

2. $\text{p. v.} \displaystyle\int_{-\infty}^{\infty} \frac{x}{(x^2+4)(x^2+1)} \frac{dx}{\sin ax}.$

3. $\text{p. v.} \displaystyle\int_{-\infty}^{\infty} \frac{x}{x^4+13x^2+36} \frac{dx}{\sin ax}.$

4. $\text{p. v.} \displaystyle\int_{-\infty}^{\infty} \frac{1}{(x^2+\alpha^2)(x^2+\beta^2)} \frac{dx}{\cos ax}, \qquad \alpha > 0, \quad \beta > 0.$

5. $\text{p. v.} \displaystyle\int_{-\infty}^{\infty} \frac{1}{x^2+4x+8} \frac{dx}{\cos ax}.$

6. $\text{p. v.} \displaystyle\int_{-\infty}^{\infty} \frac{1}{x^4+10x^2+9} \frac{dx}{\cos ax}.$

7. $\text{p. v.} \displaystyle\int_{-\infty}^{\infty} \frac{1}{(x^2+\alpha^2)(x^2+\beta^2)} \frac{\sin bx}{\sin ax} \, dx,$
   $0 < b < a, \quad \alpha > 0, \quad \beta > 0.$

8. $\text{p. v.} \displaystyle\int_{-\infty}^{\infty} \frac{x}{(x^2+\alpha^2)(x^2+\beta^2)} \frac{\cos bx}{\sin ax} \, dx,$
   $-a < b < a, \quad \alpha > 0, \quad \beta > 0.$

9. $\text{p. v.} \displaystyle\int_{-\infty}^{\infty} \frac{x \tan ax}{(x^2+\alpha^2)(x^2+\beta^2)} \, dx, \qquad a > 0, \quad \alpha > 0, \quad \beta > 0.$

10. $\displaystyle\int_{-\infty}^{\infty} \frac{dx}{(x^2+\beta^2)(x^2-2ax\sin x+a^2)}, \qquad 0 < a < \frac{\pi}{2}, \quad \beta > 0.$

11. p. v. $\displaystyle\int_{-\infty}^{\infty} \frac{x}{(x^2+1)(x^2+4)} \frac{dx}{b\sin ax + x\cos ax}$.

12. p. v. $\displaystyle\int_{-\infty}^{\infty} \frac{x}{x^2+16} \frac{dx}{b\sin ax + x\cos ax}$.

13. p. v. $\displaystyle\int_{-\infty}^{\infty} \frac{1}{x^2+\beta^2} \frac{x^2}{J_2(ax)}\, dx, \qquad \beta > 0.$

14. p. v. $\displaystyle\int_{-\infty}^{\infty} \frac{1}{(x^2+1)(x^2+4)} \frac{x}{J_1(ax)}\, dx.$

CHAPTER 9

Further Applications of the Theory of Residues

9.1. Counting zeros and poles of meromorphic functions

Let $f(z)$ be a meromorphic function which has a finite number of poles, $z_1, z_2, \ldots, z_m$, and a finite number of zeros, $\tilde{z}_1, \tilde{z}_2, \ldots, \tilde{z}_l$, in a simply connected domain D bounded by a closed path C. We assume that $f(z)$ is analytic on C and has no zeros or poles on C.

In this chapter, a zero of order n is counted n times and a pole of order p is counted p times. For short, we shall say "counting orders."

DEFINITION 9.1.1. The function

$$\varphi(z) = \frac{d}{dz}[\log f(z)] = \frac{f'(z)}{f(z)} \tag{9.1.1}$$

is called the *logarithmic derivative* of $f(z)$.

THEOREM 9.1.1. *If $\tilde{z}_k$ is a zero of order n_k and z_k is a pole of order p_k of $f(z)$, where n_k and p_k are positive integers, then $\tilde{z}_k$ and z_k are simple poles of $\varphi(z) = f'(z)/f(z)$ and the residues of $\varphi(z)$ at these points are*

$$\operatorname*{Res}_{z=\tilde{z}_k} \varphi(z) = n_k \quad \textit{and} \quad \operatorname*{Res}_{z=z_k} \varphi(z) = -p_k, \tag{9.1.2}$$

respectively.

PROOF. The proof is in two parts.

(1) Let $\tilde{z}_k$ be a zero of multiplicity n_k of $f(z)$. Then $f(z)$ can be represented in the form

$$f(z) = (z - \tilde{z}_k)^{n_k} f_1(z), \tag{9.1.3}$$

where $f_1(z)$ is analytic at $\tilde{z}_k$, and $f_1(z) \neq 0$, $f_1(z) \neq \infty$ in some neighborhood of $\tilde{z}_k$. Taking the logarithm of $f(z)$,

$$\log f(z) = n_k \log(z - \tilde{z}_k) + \log f_1(z), \tag{9.1.4}$$

and differentiating the result, we obtain

$$\varphi(z) := \frac{f'(z)}{f(z)} = \frac{n_k}{z - \tilde{z}_k} + \frac{f_1'(z)}{f_1(z)}. \tag{9.1.5}$$

Since $f_1(\tilde{z}_k) \neq 0$ and $f_1(\tilde{z}_k) \neq \infty$, then $f_1'(z)/f_1(z)$ can be expanded in a Taylor's series about $\tilde{z}_k$. Thus the function $f_1'(z)/f_1(z)$ is the regular part of the Laurent series for $\varphi(z)$ at $\tilde{z}_k$ and $n_k/(z - \tilde{z}_k)$ is its principal part. It is seen that $\tilde{z}_k$ is a simple pole of $\varphi(z)$ and

$$\operatorname*{Res}_{z=\tilde{z}_k} \varphi(z) = n_k,$$

so the first part of (9.1.2) is proven.

(2) Let z_k be a pole of order p_k of $f(z)$. Then $f(z)$ can be represented in the form

$$f(z) = \frac{f_2(z)}{(z - z_k)^{p_k}}, \tag{9.1.6}$$

where $f_2(z)$ is analytic at z_k and in some neighborhood of z_k; moreover, $f_2(z_k) \neq 0$ and $f_2(z_k) \neq \infty$. Taking the logarithm of $f(z)$,

$$\log f(z) = \log f_2(z) - p_k \log(z - z_k), \tag{9.1.7}$$

and differentiating the result, we obtain

$$\varphi(z) := \frac{f'(z)}{f(z)} = \frac{f_2'(z)}{f_2(z)} + \frac{(-p_k)}{z - z_k}. \tag{9.1.8}$$

Since $f_2(z_k) \neq 0$ and $f_2(z_k) \neq \infty$, then $f_2'(z)/f_2(z)$ can be expanded in a Taylor series centered at z_k. Thus $f_2'(z)/f_2(z)$ is the regular part of the Laurent series of $\varphi(z)$ while $-p_k/(z - z_k)$ is its principal part. It is seen that z_k is a simple pole of $\varphi(z)$, and

$$\operatorname*{Res}_{z=z_k} \varphi(z) = -p_k.$$

This completes the proof of the second formula in (9.1.2). □

Formulae (9.1.2) allow one to prove the following important theorem.

THEOREM 9.1.2. *Let $f(z)$ be a meromorphic function in a simply connected domain D bounded by the positively oriented simple closed path C. Suppose that $f(z)$ has no zeros or poles on C. Then the difference between the number, Z_f, of zeros and the number, P_f, of poles of $f(z)$ in D, counting orders, is given by the integral*

$$Z_f - P_f = \frac{1}{2\pi i} \oint_C \frac{f'(z)}{f(z)}\, dz. \tag{9.1.9}$$

PROOF. By Theorem 9.1.1, the singular points of $f'(z)/f(z)$ in D are the zeros, $\tilde{z}_k$ of order n_k, of $f(z)$ and the poles, z_k of order p_k, of $f(z)$. Then, by the Residue Theorem 5.2.2 and (9.1.2), we have

$$\begin{aligned}\frac{1}{2\pi i}\oint_C \frac{f'(z)}{f(z)}\,dz &= \sum_k \operatorname*{Res}_{z=\tilde{z}_k} \frac{f'(z)}{f(z)} + \sum_k \operatorname*{Res}_{z=z_k} \frac{f'(z)}{f(z)} \\ &= \sum_k n_k - \sum_k p_k = Z_f - P_f. \quad \square\end{aligned}$$

9.2. The argument principle

In this section, we give a geometric interpretation of formula (9.1.9). For this purpose, we define the variation of the argument of $f(z)$ as z traverses a simple closed path.

DEFINITION 9.2.1. Let C be a simple closed path in the z-plane and γ its image in the w-plane under the mapping $z \mapsto w = f(z)$. The change or *variation of the argument of* $f(z)$ as C is traversed once in the positive direction is denoted by $\operatorname{Var}_C \arg f(z)$ and is equal to the number, M_+, of times the point $w = 0$ is encircled by γ traversed in the positive direction minus the number, M_-, of times it is encircled by γ traversed in the negative direction, multiplied by 2π, that is,

$$\frac{1}{2\pi}\operatorname{Var}_C \arg f(z) = M_+ - M_- =: M. \tag{9.2.1}$$

Geometrically, formula (9.1.9) is equivalent to the argument principle.

THEOREM 9.2.1 (ARGUMENT PRINCIPLE). *Let $f(z)$ be a meromorphic function in a simply connected domain D bounded by the simple closed path C. Suppose that $f(z)$ has no zeros nor poles on C. Then the difference between the number, Z_f, of zeros and the number, P_f, of poles of $f(z)$ in D, counting orders, is given by the formula*

$$Z_f - P_f = \frac{1}{2\pi}\operatorname{Var}_C \arg f(z), \tag{9.2.2}$$

known as the argument principle.

PROOF. We rewrite (9.1.9) in the form

$$\begin{aligned}Z_f - P_f &= \frac{1}{2\pi i}\oint_C \frac{f'(z)}{f(z)}\,dz \\ &= \frac{1}{2\pi i}\oint_C d\log f(z)\end{aligned}$$

$$= \frac{1}{2\pi i} \oint_C d \ln |f(z)| + \frac{1}{2\pi i} \oint_C d\big(i \arg f(z)\big), \tag{9.2.3}$$

since

$$\log f(z) = \ln |f(z)| + i \arg f(z).$$

The simple closed path C encloses a simply connected domain D. Since $\ln |f(z)|$ is a real valued function of two variables and the integral of a total differential along the closed curve C is zero, (9.2.3) reduces to

$$Z_f - P_f = \frac{1}{2\pi} \operatorname{Var}_C \arg f(z). \quad \square$$

We illustrate two cases.

Case 1. The function $w = f(z)$ maps the closed path C, in the z-plane as shown in Fig 9.1(a), into the closed path γ, in the w-plane, not enclosing the point $w = 0$, as shown in Fig 9.1(b). Suppose that the point $z_0 \in C$ is mapped to the point $w_0 \in \gamma$. As z_0 traverses C once in the positive direction, w_0 traverses γ an integer number of times in the positive or negative direction. However, the number

$$\arg f(z)|_{z=z_0} = \arg w_0$$

does not change as z_0 goes once or several times along C. In fact, $\arg w_0$ increases (up to the point A) then decreases and when w_0 returns to its initial position, $\arg w_0$ returns to its initial value. In this case (9.2.2) gives

$$Z_f - P_f = \frac{1}{2\pi} \operatorname{Var}_C \arg f(z) = 0. \tag{9.2.4}$$

If $f(z)$ has no poles in D, then $P_f = 0$ and thus, by (9.2.4), $Z_f = 0$, that is, $f(z)$ has no zeros in D, if C is mapped onto γ as shown in Fig 9.1(b).

Case 2. The function $f(z)$ maps the closed path C into the closed path γ which encloses the point $w = 0$, as shown in Fig 9.1(c). In this case, $\operatorname{Var}_C \arg f(z)$ increases by 2π every time w_0 traverses γ in the positive direction and decreases by 2π every time w_0 traverses γ in the negative direction. Hence, we have

$$\operatorname{Var}_C \arg f(z) = 2\pi M, \tag{9.2.5}$$

where the number $M = M_+ - M_-$ is as defined in (9.2.1). It follows from (9.2.2) and (9.2.5) that

$$Z_f - P_f = M, \tag{9.2.6}$$

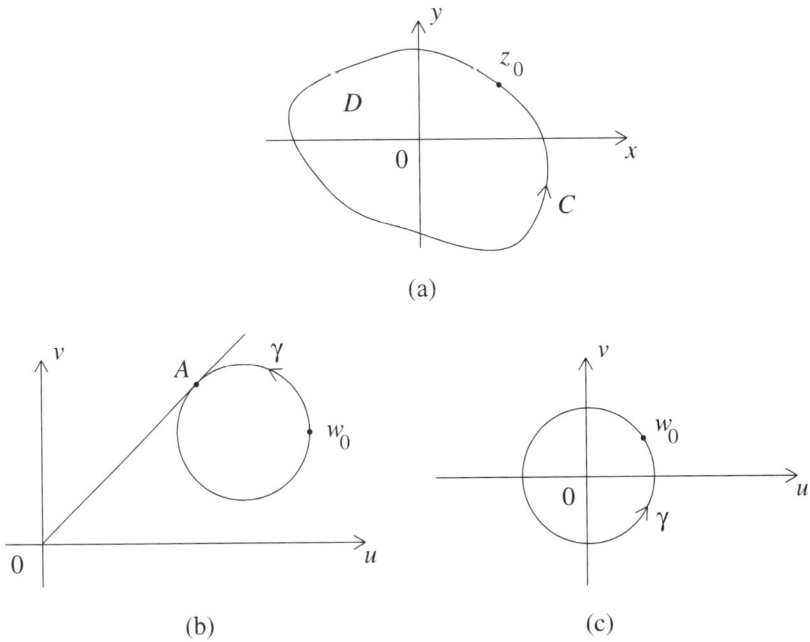

FIGURE 9.1. Geometric interpretation of formula (9.2.2). (a) The path C bounding the domain D in the z-plane; (b) the path γ does not encircle the point $w = 0$; (c) the path γ encircles the point $w = 0$.

that is, the difference between the number of zeros and the number of poles of $f(z)$, counting orders, in a simply connected domain D bounded by the path C (on which $f(z)$ has no zeros nor poles) is equal to the number of times γ is traversed as C is traversed once in the positive direction.

If the function $w = f(z)$ has no poles in D, then $P_f = 0$ and formula (9.2.6) reduces to

$$Z_f = M. \tag{9.2.7}$$

DEFINITION 9.2.2. The *index* of a point z_0 with respect to a closed curve C in the z-plane is the integer defined by the equation

$$n(C, z_0) = \frac{1}{2\pi i} \oint_C \frac{dz}{z - z_0}. \tag{9.2.8}$$

The index is also called the *winding number* of C with respect to z_0.

One can see that our definition of M is $n(\gamma, 0)$ in the w-plane, where γ is the image of C under the mapping $w = f(z)$.

Note 9.2.1. The fact that the path γ is traversed more than once in the w-plane as C is traversed once in the z-plane means that the w-plane is considered as a Riemann surface with a corresponding cut. The first time around γ is made on the first sheet of this surface; the second time around γ is made on the second sheet, and so on.

EXAMPLE 9.2.1. *Find the number of zeros of $w = z^2 - 0.5$ in the disk $D : |z| \le 1$ bounded by the path $C : |z| = 1$.*

SOLUTION. Since the equation of C is $z = e^{i\theta}$, then the image of C is

$$\gamma : w = u + iv = e^{2i\theta} - 0.5,$$

that is,

$$u = \cos(2\theta) - 0.5, \qquad v = \sin 2\theta, \qquad 0 \le 2\theta \le 2\pi. \tag{9.2.9}$$

Eliminating θ from (9.2.9) we obtain $(u+0.5)^2 + v^2 = 1$, which is the equation of a circle of radius 1 centered at $(u, v) = (-0.5, 0)$. Hence, the origin of the coordinate system lies inside the disk bounded by γ and therefore the function $w = z^2 + 0.5$ has zeros in D.

The number of zeros is equal to the number of times γ is traversed as C is traversed once. In this case, γ is traversed twice, that is, $M = 2$ (the first time as θ goes from 0 to π, and the second time as θ goes from π to 2π, since $u(0) = u(\pi) = u(2\pi) = 0.5$ and $v(0) = v(\pi) = v(2\pi) = 0$). However, in this simple example one can find directly the two zeros of $f(z)$ in D, namely, $w_{1,2} = \pm i\sqrt{0.5}$. □

9.3. Rouché's Theorem

Another method of counting zeros of analytic functions in a given region is by means of the following theorem due to Rouché.

THEOREM 9.3.1 (ROUCHÉ'S THEOREM). *Let $f(z)$ and $g(z)$ be analytic in a simply connected domain D and on its boundary, C, and suppose that the following inequality is satisfied for all $z \in C$:*

$$|f(z)| > |g(z)|. \tag{9.3.1}$$

Then $f(z)$ and $F(z) = f(z) + g(z)$ have the same number of zeros in D.

PROOF. Note, first, that (9.3.1) implies that, on C,

$$|f(z)| > 0 \quad \text{and} \quad |f(z) + g(z)| \ge |f(z)| - |g(z)| > 0.$$

Therefore, $f(z)$ and $f(z)+g(z)$ are not equal to zero on C and the argument principle (9.2.2) can be used for these functions with $P_f = P_F = 0$ since

$f(z)$ and $F(z)$ are analytic in D. Thus, if Z_f and Z_F denote the number of zeros of $f(z)$ and $F(z) = f(z) + g(z)$, respectively, in D, counting orders, we obtain

$$\begin{aligned} Z_F - Z_f &= \frac{1}{2\pi} \operatorname{Var}_C \arg[f(z) + g(z)] - \frac{1}{2\pi} \operatorname{Var}_C \arg f(z) \\ &= \frac{1}{2\pi} \operatorname{Var}_C \Big\{ \arg[f(z) + g(z)] - \arg f(z) \Big\}. \end{aligned} \tag{9.3.2}$$

Using the formula

$$\arg(z_1) - \arg(z_2) = \arg \frac{z_1}{z_2}$$

(see (1.1.32)), we have

$$\arg[f(z) + g(z)] - \arg f(z) = \arg \frac{f(z) + g(z)}{f(z)} = \arg \left[1 + \frac{g(z)}{f(z)}\right],$$

so that (9.3.2) can be written in the form

$$Z_F - Z_f = \frac{1}{2\pi} \operatorname{Var}_C \arg \left[1 + \frac{g(z)}{f(z)}\right]. \tag{9.3.3}$$

To show that the term on the right-hand side of (9.3.3) is equal to zero, we consider the function

$$w(z) = 1 + \frac{g(z)}{f(z)}. \tag{9.3.4}$$

Since, by assumption, $|f(z)| > |g(z)|$ for $z \in C$, it follows from (9.3.4) that

$$|w(z) - 1| = \left|\frac{g(z)}{f(z)}\right| \le \rho_0 < 1, \qquad z \in C. \tag{9.3.5}$$

Inequality (9.3.5) implies that $w(z)$ maps the path C onto the path γ, which lies entirely inside the disk $|w - 1| \le \rho_0 < 1$ (see Fig 9.2).

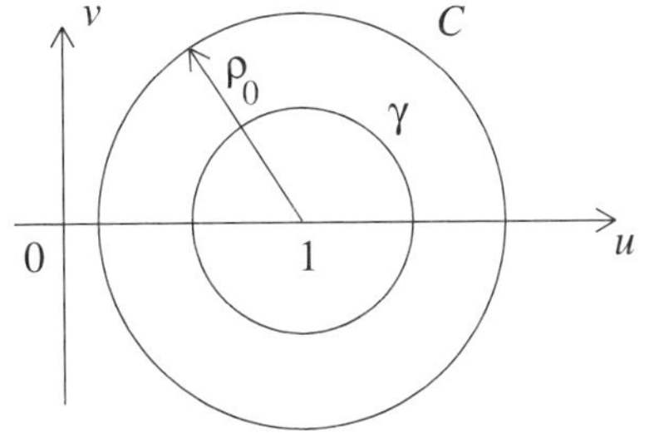

FIGURE 9.2. The path γ in the w-plane.

Therefore, γ does not enclose the point $w = 0$. Thus (see Fig 9.1(b))

$$\operatorname{Var}_C \arg w = \operatorname{Var}_C \arg \left[1 + \frac{g(z)}{f(z)}\right] = 0,$$

and formula (9.3.3) becomes

$$Z_F = Z_f. \quad \square$$

EXAMPLE 9.3.1. *Find the number of zeros of the polynomial*

$$F(z) = z^{10} - 7z^6 - 2z + 1$$

inside the unit disk $D : |z| \leq 1$.

SOLUTION. Let $F(z) = f(z) + g(z)$, where

$$f(z) = -7z^6 + 1 \quad \text{and} \quad g(z) = z^{10} - 2z.$$

Then, for every z on the unit circle $C : |z| = 1$

$$|f(z)| = |-7z^6 + 1| \geq |-7z^6| - 1 = 7 - 1 = 6,$$

and

$$|g(z)| = |z^{10} - 2z| \leq |z^{10}| + |2z| = 1 + 2 = 3.$$

Hence

$$|f(z)| > |g(z)| > 0, \qquad z \in C.$$

Therefore, by Rouché's Theorem, the number of zeros of $F(z)$ inside the unit disk, $D : |z| \leq 1$, is equal to the number of zeros of $f(z) = -7z^6 + 1$ in D. Solving the equation $f(z) = 0$, we obtain

$$z = 7^{-1/6} e^{2k\pi i/6}, \qquad k = 0, 1, \ldots, 5.$$

Therefore, $F(z)$ has six zeros in D. $\square$

The fundamental theorem of algebra (see Exercise 20, Section 3.4) follows simply from Rouché's Theorem, as shown in the following example.

EXAMPLE 9.3.2. *Use Rouché's Theorem to prove that a polynomial of degree* n,

$$p(z) = z^n + a_1 z^{n-1} + a_2 z^{n-2} + \cdots + a_n,$$

has exactly n *zeros.*

SOLUTION. Let

$$f(z) = z^n, \qquad g(z) = a_1 z^{n-1} + a_2 z^{n-2} + \cdots + a_n,$$

and consider the path $C_R : |z| = R$.

We have

$$|f(z)| = R^n, \qquad z \in C_R,$$

and

$$\begin{aligned}|g(z)| &= |a_1 z^{n-1} + a_2 z^{n-2} + \cdots + a_n| \\ &\leq |a_1| R^{n-1} + |a_2| R^{n-2} + \cdots + |a_n| \\ &=: \tilde{g}(R).\end{aligned}$$

Since

$$\lim_{R\to\infty} \frac{R^n}{\tilde{g}(R)} = +\infty,$$

then there exists R_0 such that

$$|f(z)| > |g(z)|, \qquad z \in C_R,$$

for all $R \geq R_0$. Hence, by Rouché's Theorem, the number of zeros of $p(z)$ in the disk $|z| \leq R$ is equal to the number of zeros, counting orders, of $f(z) = z^n$ in the same region. Since $z = 0$ is a zero of order n of z^n, then $p(z)$ has exactly n zeros in the disk $|z| \leq R$. □

EXAMPLE 9.3.3. *Find the number of roots of the equation*

$$z^{10} - a\,e^z = 0, \qquad 0 < a < e^{-1} \tag{9.3.6}$$

in the open unit disk $|z| < 1$.

SOLUTION. Let $f(z) = z^{10}$ and $g(z) = -a\,e^z$. On the circle $|z| = 1$,

$$|f(z)| = |z^{10}| = 1$$

and

$$\begin{aligned}|g(z)| &= |-a\,e^z| = a|e^{x+iy}| \\ &= a\,e^x\big|_{x=\cos\theta} < \frac{1}{e}\,e^{\cos\theta} \\ &= e^{-(1-\cos\theta)} \leq 1.\end{aligned}$$

Then, on the circle $|z| = 1$, we have

$$|f(z)| = 1, \qquad |g(z)| < 1.$$

Therefore, by Rouché's Theorem, equation (9.3.6) has the same number of zeros, counting orders, inside the unit disk as the equation $z^{10} = 0$, that is, 10 zeros. □

Note 9.3.1. The function $F(x) = x^{10} - a\,e^x$ is continuous on the real segment $-1 \leq x \leq 1$ and

$$\begin{aligned}F(-1) &= 1 - a\,e^{-1} > 1 - \frac{1}{e^2} > 0, \\ F(0) &= -a < 0,\end{aligned}$$

$$F(1) = 1 - ae > 0.$$

Since $F(x)$ is positive at $x = \pm 1$ and negative at $x = 0$, it has at least two real zeros on the segment $(-1, 1)$ and, hence, at most eight complex zeros inside the unit disk.

Exercises for Sections 9.2 and 9.3

Determine the number of zeros of the following polynomials in the indicated regions.

1. $z^6 - 5z^4 + z^3 - 2z$, in $|z| < 1$.
2. $2z^4 - 2z^3 + 2z^2 - 2z + 9$, in $|z| < 1$.
3. $2z^5 - 6z^2 + z + 1$, in $1 \leq |z| < 2$.
4. $z^7 - 2z^5 + 6z^3 - z + 1$, in $|z| < 1$.
   (Hint: Look for the biggest term when $|z| = 1$ and apply Rouché's Theorem.)
5. $z^4 - 6z + 3$, in $1 < |z| < 2$.
6. $z^4 + 8z^3 + 3z^2 + 8z + 3$, in $\Re z > 0$.
   (Hint: Sketch the image of the imaginary axis under the mapping by the given polynomial and apply the argument principle to a large half-disk.)

7. Prove the following form of Rouché's Theorem: Suppose $f(z)$ and $g(z)$ are meromorphic in a neighborhood of the closed disk $|z - a| \leq R$ with no zeros or poles on the circle $C : |z - a| = R$. If Z_f, Z_g (P_f, P_g) are the number of zeros (poles) of $f(z)$ and $g(z)$, respectively, inside C, counting orders, and if

$$|f(z) + g(z)| < |f(z)| + |g(z)|$$

on C, then

$$Z_f - P_f = Z_g - P_g.$$

(Hint: The inequality $\left|\dfrac{f(z)}{g(z)} + 1\right| < \left|\dfrac{f(z)}{g(z)}\right| + 1$ does not hold on C if $\dfrac{f(z)}{g(z)}$ is real. Then define a branch of $\log \dfrac{f(z)}{g(z)}$ as a well-defined primitive for $\dfrac{[f(z)/g(z)]'}{f(z)/g(z)}$.)

9.4. Simple-pole expansion of meromorphic functions

9.4.1. A theorem of Cauchy. A particular case of a theorem of Cauchy [**40**], p. 305, asserts that if all the poles z_n of a meromorphic function $f(z)$, which is analytic at $z = 0$, are simple and have increasing moduli, $|z_1| < |z_2| < |z_3| < \dots$, $n = 1, 2, 3, \dots$, and if $f(z)$ is bounded,

$$|f(z)| \leq M \qquad \forall z \in C_n, \qquad n = 1, 2, 3, \dots, \tag{9.4.1}$$

for some $M > 0$ on some regular system of paths C_n (to be defined later), then the following formula holds:

$$f(z) = f(0) + \sum_{n=1}^{\infty} \left(\frac{1}{z - z_n} + \frac{1}{z_n} \right) \operatorname*{Res}_{z=z_n} f(z) \tag{9.4.2}$$

(see, for example, [**44**], p. 175, [**42**], p. 266, [**33**], p. 430). This decomposition falls under the general theorem of Mittag–Leffler.

Partial fraction expansions of elementary meromorphic functions of a complex variable, such as

$$\frac{1}{\sin z}, \quad \frac{1}{\cos z}, \quad \frac{\sin az}{\sin z}, \quad \frac{\sin az}{\cos z}, \quad \frac{\cos az}{\cos z}, \quad \frac{\cos az}{\sin z}, \quad \tan z, \quad \cot z, \tag{9.4.3}$$

where $|a| < 1$, and the corresponding hyperbolic functions are derived in numerous text and reference books by means of (9.4.2). If the function $f(z)$ has a pole at $z = 0$ (for example, $1/\sin z$, $\cot z$), to apply formula (9.4.2) one has to consider the difference $f(z) - g(z; 0)$ instead of $f(z)$, where $g(z; 0)$ is the principal part of the Laurent series expansion of $f(z)$ with center $z = 0$. For example, instead of $1/\sin z$ and $\cot z$ one has to consider $1/\sin z - 1/z$ and $\cot z - 1/z$, respectively.

In this section, condition (9.4.1) is replaced by the condition

$$\lim_{n\to\infty} \oint_{C_n} \frac{f(\zeta)}{\zeta - z}\, d\zeta = 0 \tag{9.4.4}$$

and, instead of (9.4.2), the simpler formula,

$$f(z) = \sum_{n=1}^{\infty} \frac{1}{z - z_n} \operatorname*{Res}_{z=z_n} f(z), \tag{9.4.5}$$

is derived.

It is proved that the functions listed in (9.4.3) satisfy condition (9.4.4) and therefore they can be expanded in partial fractions by means of (9.4.5). The expansions obtained by this procedure coincide with the expansions produced by the less simple formula (9.4.2). The advantage of (9.4.5) over

(9.4.2) is that even if $f(z)$ has a simple pole at $z = 0$, there is no need to construct an auxiliary function which is regular at $z = 0$.

DEFINITION 9.4.1. A system of closed paths C_n ($n = 1, 2, 3, \dots$) is called *regular* if the following three conditions are satisfied:

(a) The path C_1 contains the point $z = 0$ and each path C_n lies inside the region bounded by the path C_{n+1}.

(b) The distance, d_n, from C_n to the origin increases without bound as n increases.

(c) The quotient of the length, l_n, of C_n to the distance d_n remains bounded:

$$\frac{l_n}{d_n} \leq A = \text{constant} > 0.$$

We note that the quotient in (c) is equal to 2π for a circle $|z| = R$, and $\sqrt{2}/2$ for a square centered at the origin.

9.4.2. Partial fraction expansion theorem. We prove the following theorem, which is a particular case of a theorem proved in [**41**], p. 219.

THEOREM 9.4.1 (PARTIAL FRACTION EXPANSION). *Suppose that a meromorphic function $f(z)$ satisfies the condition*

$$\lim_{n\to\infty} \oint_{C_n} \frac{f(\zeta)}{\zeta - z}\, d\zeta = 0 \tag{9.4.6}$$

on some regular system of paths C_n. Moreover, suppose that the poles z_k of $f(z)$ are simple and have strictly increasing moduli,

$$|z_1| < |z_2| < |z_3| < \cdots < |z_k| < \dots .$$

Then the partial fraction expansion formula (9.4.5) *holds for any z such that $z \neq z_k$ ($k = 1, 2, 3, \dots$) and $z \notin C_n$, $n = 1, 2, \dots$.*

PROOF. Consider the integral

$$\frac{1}{2\pi i} \oint_{C_n} \frac{f(\zeta)}{\zeta - z}\, d\zeta, \tag{9.4.7}$$

where z is an arbitrary but fixed point lying inside the closed path C_n and distinct from any poles z_k of $f(\zeta)$. The integrand in (9.4.7) has simple poles at $\zeta = z$ and at $\zeta = z_k$ inside the region G_n bounded by C_n. Therefore, by the residue theorem we have

$$\frac{1}{2\pi i} \oint_{C_n} \frac{f(\zeta)}{\zeta - z}\, d\zeta = f(z) + \sum_{z_k \in G_n} \operatorname*{Res}_{\zeta = z_k} \left[\frac{f(\zeta)}{\zeta - z}\right]. \tag{9.4.8}$$

However,

$$\begin{aligned}\operatorname*{Res}_{\zeta=z_k} \frac{f(\zeta)}{\zeta - z} &= \lim_{\zeta\to z_k} \left[(\zeta - z_k)\frac{f(\zeta)}{\zeta - z}\right] \\ &= \frac{1}{(z_k - z)} \lim_{\zeta\to z_k} [(\zeta - z_k)f(\zeta)] \\ &= \frac{1}{z_k - z} \operatorname*{Res}_{\zeta=z_k} f(\zeta) \\ &= \frac{1}{z_k - z} \operatorname*{Res}_{z=z_k} f(z).\end{aligned}$$

Then (9.4.8) can be written in the form

$$\frac{1}{2\pi i}\oint_{C_n} \frac{f(\zeta)}{\zeta - z}\, d\zeta = f(z) + \sum_{z_k\in G_n} \frac{1}{z_k - z} \operatorname*{Res}_{z=z_k} f(z). \tag{9.4.9}$$

Taking the limit in (9.4.9) as $n \to \infty$ and using (9.4.4), we obtain (9.4.5). Moreover, since the left-hand side of (9.4.9) tends to zero as $n \to \infty$, then the right-hand side of (9.4.9) also tends to zero; this fact, in turn, guarantees the convergence of the series in (9.4.5). □

Note 9.4.1. The series (9.4.5) should be understood in the following sense:

$$\sum_{k=1}^{\infty} \frac{1}{z - z_k} \operatorname*{Res}_{z=z_k} f(z) = \lim_{n\to\infty} \sum_{z_k\in G_n} \frac{1}{z - z_k} \operatorname*{Res}_{z=z_k} f(z).$$

That is, one first computes the terms related to the poles inside C_1; then one adds to the partial sum the terms related to the poles lying between C_1 and C_2, and so on.

9.4.3. Examples. We present two simple examples of partial fraction expansion of meromorphic functions.

EXAMPLE 9.4.1. *Expand the meromorphic function*

$$f(z) = \frac{1}{\sin z} \tag{9.4.10}$$

in partial fractions.

SOLUTION. To use (9.4.5) one has to show that (9.4.10) satisfies (9.4.6). For C_n we take the square $P_nQ_nR_nS_n$ with vertical sides through the points $\pm\gamma_n = \pm(2n+1)\pi/2$ (see Fig 9.3). Letting $\zeta = \xi + i\eta$, we have

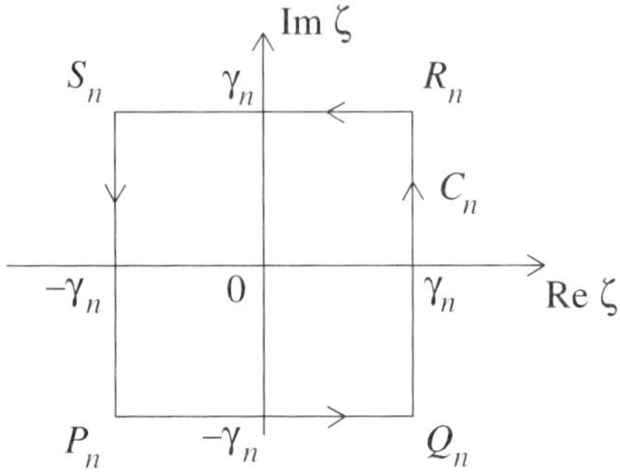

FIGURE 9.3. The path C_n for csc z in Example 9.4.1.

$$\begin{aligned} |I_n| := \left| \oint_{C_n} \frac{d\zeta}{(\zeta - z)\sin\zeta} \right| \\ &\leq \oint_{C_n} \frac{|d\zeta|}{|\zeta - z|\,|\sin\zeta|} \\ &\leq \frac{1}{d_n} \oint_{C_n} \frac{|d\zeta|}{|\sin\zeta|}, \end{aligned} \tag{9.4.11}$$

where

$$d_n = \min_{C_n} |\zeta - z|, \qquad |\sin\zeta| = \sqrt{\sinh^2\eta + \sin^2\xi}\,, \tag{9.4.12}$$

and $d_n \to \infty$ as $n \to \infty$ since z is fixed. Hence

$$|I_n| \leq \frac{1}{d_n}\left(\int_{P_nQ_n} + \int_{Q_nR_n} + \int_{R_nS_n} + \int_{S_nP_n} \right) \left[\frac{|d\zeta|}{|\sin\zeta|} \right]. \tag{9.4.13}$$

On the segment P_nQ_n, $\zeta = \xi - i\gamma_n$ and $|d\zeta| = d\xi$. Thus, letting $\xi = \gamma_n t$, we have

$$\begin{aligned} \int_{P_nQ_n} \frac{|d\zeta|}{|\sin\zeta|} &= \int_{-\gamma_n}^{\gamma_n} \frac{d\xi}{\sqrt{\sinh^2\gamma_n + \sin^2\xi}} \\ &= \gamma_n \int_{-1}^{1} \frac{dt}{\sqrt{\sinh^2\gamma_n + \sin^2\gamma_n t}} \\ &\to 0 \qquad \text{as } n \to \infty. \end{aligned} \tag{9.4.14}$$

Similarly, the integral along R_nS_n approaches 0 as $n \to \infty$. On the segment Q_nR_n, $\zeta = \gamma_n + i\eta$ and $|d\zeta| = d\eta$; thus we have

$$\begin{aligned}\int_{Q_nR_n} \frac{|d\zeta|}{|\sin\zeta|} &= \int_{-\gamma_n}^{\gamma_n} \frac{d\eta}{\sqrt{\sinh^2\eta + \sin^2\gamma_n}} \\ &= 2\int_0^{\gamma_n} \frac{d\eta}{\cosh\eta} = 4\arctan e^{\eta}\Big|_0^{\gamma_n} \\ &= 4\left(\arctan e^{\gamma_n} - \frac{\pi}{4}\right) \\ &\to \pi \qquad \text{as } n \to \infty. \end{aligned} \tag{9.4.15}$$

Similarly, one can show that the integral along S_nP_n approaches $-\pi$ as $n \to \infty$. It follows from (9.4.13)–(9.4.15) that $I_n \to 0$ as $n \to \infty$. Hence, by (9.4.5) we have

$$\begin{aligned}\frac{1}{\sin z} &= \sum_{n=-\infty}^{\infty} \frac{1}{z - n\pi} \operatorname*{Res}_{z=n\pi} \frac{1}{\sin z} \\ &= \sum_{n=-\infty}^{\infty} \frac{(-1)^n}{z - n\pi} \\ &= \sum_{n=0}^{\infty} \frac{(-1)^n}{z - n\pi} + \sum_{n=1}^{\infty} \frac{(-1)^n}{z + n\pi} \\ &= \frac{1}{z} + \sum_{n=1}^{\infty} (-1)^n \left[\frac{1}{z - n\pi} + \frac{1}{z + n\pi}\right] \\ &= \frac{1}{z} + 2z\sum_{n=1}^{\infty} \frac{(-1)^n}{z^2 - n^2\pi^2}. \quad \square \end{aligned} \tag{9.4.16}$$

Thus we have derived the well-known expansion (9.4.16) by means of formula (9.4.5).

Note 9.4.2. According to Note 9.4.1, the summation in (9.4.16) has to be taken by grouping terms as follows:

$$\frac{1}{z} - \left(\frac{1}{z-\pi} + \frac{1}{z+\pi}\right) + \left(\frac{1}{z-2\pi} + \frac{1}{z+2\pi}\right) - \cdots.$$

The summation in Example 9.4.2 will be done analogously.

It can similarly be proven that condition (9.4.4) holds for the remaining functions in (9.4.3) (except for $\tan z$ and $\cot z$). In the case of functions containing $\cos z$ in the denominator one has to take for the path C_n a square with vertical sides through the points $\pm n\pi$ for $n = 1, 2, 3, \ldots$.

The partial fraction expansions of the functions in (9.4.3) by means of (9.4.5) or (9.4.2) are identical.

We also note that it is not more complicated to prove (9.4.4) than to prove (9.4.1) for the functions in (9.4.3) (see, for example, the proof of (9.4.1) for $|f(z)| = |\cot z|$ in [**42**] on p. 268).

EXAMPLE 9.4.2. *Expand in partial fractions the meromorphic function*

$$f(z) = \cot z. \tag{9.4.17}$$

SOLUTION. We first show that condition (9.4.6) of Theorem 9.4.1 holds for the square $C_n = P_nQ_nR_nS_n$ shown in Fig 9.3. In this case it is convenient to carry out the proof by combining the integrals along the opposite sides of C_n. Using the identity

$$\begin{aligned}\cot\zeta &= \cot(\xi + i\eta)\\ &= \frac{\cos\xi\sin\xi - i\cosh\eta\sinh\eta}{\cosh^2\eta - \cos^2\xi},\end{aligned} \tag{9.4.18}$$

we have (see Fig 9.3 and Example 9.4.1)

$$\begin{aligned}&\left(\int_{P_nQ_n} + \int_{R_nS_n}\right)\left[\frac{\cot\zeta}{\zeta - z}\,d\zeta\right]\\
&\quad = \int_{-\gamma_n}^{\gamma_n}\left[\frac{\cot(\xi - i\gamma_n)}{\xi - i\gamma_n - z} - \frac{\cot(\xi + i\gamma_n)}{\xi + i\gamma_n - z}\right]d\xi\\
&\quad = \int_{-\gamma_n}^{\gamma_n}\frac{1}{(\xi - z)^2 + \gamma_n^2}\Big\{(\xi - z)\big[\cot(\xi - i\gamma_n) - \cot(\xi + i\gamma_n)\big]\\
&\qquad + i\gamma_n\big[\cot(\xi + i\gamma_n) + \cot(\xi - i\gamma_n)\big]\Big\}\,d\xi\\
&\quad = 2i\int_{-\gamma_n}^{\gamma_n}\frac{\xi - z}{(\xi - z)^2 + \gamma_n^2}\,\frac{\cosh\gamma_n\sinh\gamma_n}{\cosh^2\gamma_n - \cos^2\xi}\,d\xi\\
&\qquad + 2i\gamma_n\int_{-\gamma_n}^{\gamma_n}\frac{1}{(\xi - z)^2 + \gamma_n^2}\,\frac{\cos\xi\sin\xi}{\cosh^2\gamma_n - \cos^2\xi}\,d\xi\\
&\quad = 2i\int_{-1}^{1}\frac{t - z/\gamma_n}{(t - z/\gamma_n)^2 + 1}\,\frac{\cosh\gamma_n\sinh\gamma_n}{\cosh^2\gamma_n - \cos^2\gamma_n t}\,dt\\
&\qquad + 2i\int_{-1}^{1}\frac{1}{(t - z/\gamma_n)^2 + 1}\,\frac{\cos\gamma_n t\sin\gamma_n t}{\cosh^2\gamma_n - \cos^2\gamma_n t}\,dt.\end{aligned} \tag{9.4.19}$$

Since $\gamma_n = (2n+1)\pi/2$, then for all $t \in [-1, 1]$ we have

$$\lim_{n\to\infty}\frac{\cosh\gamma_n\sinh\gamma_n}{\cosh^2\gamma_n - \cos^2\gamma_n t} = 1, \qquad \lim_{n\to\infty}\frac{\cos\gamma_n t\sin\gamma_n t}{\cosh^2\gamma_n - \cos^2\gamma_n t} = 0.$$

Since the integrand in (9.4.19) is continuous on the interval $-1 \leq t \leq 1$ for $1 \leq n < \infty$, then the limit and the integration can be interchanged and we have

$$\lim_{n\to\infty}\left(\int_{P_nQ_n} + \int_{R_nS_n}\right)\left[\frac{\cot\zeta}{\zeta - z}\, d\zeta\right] = 2i\int_{-1}^{1}\frac{t}{t^2+1}\, dt$$
$$= 0. \tag{9.4.20}$$

Similarly,

$$\begin{aligned}
&\left(\int_{Q_nR_n} + \int_{S_nP_n}\right)\left[\frac{\cot\zeta}{\zeta - z}\, d\zeta\right] \\
&= i\int_{-\gamma_n}^{\gamma_n}\left[\frac{\cot(\gamma_n + i\eta)}{\gamma_n + i\eta - z} - \frac{\cot(\gamma_n - i\eta)}{\gamma_n - i\eta + z}\right] d\eta \\
&= i\int_{-\gamma_n}^{\gamma_n}\frac{1}{\gamma_n^2 + \eta^2 + 2iz\eta - z^2}\Big\{\gamma_n\left[\cot(\gamma_n + i\eta) - \cot(\gamma_n - i\eta)\right] \\
&\quad + (z - i\eta)\left[\cot(\gamma_n + i\eta) + \cot(\gamma_n - i\eta)\right]\Big\}\, d\eta \\
&= 2\gamma_n\int_{-\gamma_n}^{\gamma_n}\frac{1}{\gamma_n^2 + \eta^2 + 2iz\eta - z^2}\,\frac{\cosh\eta\sinh\eta}{\cosh^2\eta - \cos^2\gamma_n}\, d\eta \\
&\quad + 2\int_{-\gamma_n}^{\gamma_n}\frac{z - i\eta}{\gamma_n^2 + \eta^2 + 2iz\eta - z^2}\,\frac{\cos\gamma_n\sin\gamma_n}{\cosh^2\eta - \cos^2\gamma_n}\, d\eta \\
&\quad (\text{setting } \eta = \gamma_n t \text{ and noting that } \cos\gamma_n = 0) \\
&= 2\int_{-1}^{1}\frac{1}{t^2 + 1 + 2izt/\gamma_n - z^2/\gamma_n^2}\tanh\gamma_n t\, dt \\
&= 2\int_{-1}^{1}\frac{t^2 + 1 - z^2/\gamma_n^2 - 2izt/\gamma_n}{(t^2 + 1 - z^2/\gamma_n^2)^2 + 4z^2t^2/\gamma_n^2}\tanh\gamma_n t\, dt \\
&= 2\int_{-1}^{1}\frac{(t^2 + 1 - z^2/\gamma_n^2)\tanh\gamma_n t}{(t^2 + 1 - z^2/\gamma_n^2)^2 + 4z^2t^2/\gamma_n^2}\, dt \\
&\quad - 4\int_{-1}^{1}\frac{(izt/\gamma_n)\tanh\gamma_n t}{(t^2 + 1 - z^2/\gamma_n^2)^2 + 4z^2t^2/\gamma_n^2}\, dt \\
&\to 0,
\end{aligned} \tag{9.4.21}$$

as $n \to \infty$. The second last integral is zero since the integrand is odd and the limits of integration are symmetric. Thus (9.4.20) and (9.4.21) imply (9.4.6). Hence, (9.4.17) can be expanded in partial fractions by means of

(9.4.5) as follows:

$$\begin{aligned}\cot z &= \sum_{n=-\infty}^{\infty} \frac{1}{z-n\pi} \operatorname*{Res}_{z=n\pi} \cot z \\ &= \lim_{N\to\infty} \sum_{n=-N}^{N} \frac{1}{z-n\pi} \\ &= \lim_{N\to\infty} \left(\frac{1}{z} + \sum_{n=1}^{N} \frac{1}{z-n\pi} + \sum_{n=1}^{N} \frac{1}{z+n\pi} \right) \\ &= \frac{1}{z} + \lim_{N\to\infty} \sum_{n=1}^{N} \frac{2z}{z^2-n^2\pi^2} \\ &= \frac{1}{z} + 2z \sum_{n=1}^{\infty} \frac{1}{z^2-n^2\pi^2}. \quad \square \end{aligned} \tag{9.4.22}$$

We note that the series (9.4.22) and (9.4.16) are absolutely and uniformly convergent in any disk $|z| \le R$ with deleted rings $|z - n\pi| \le \delta$ $(n = 1, 2, 3, \dots)$ for arbitrary large R, since the series in (9.4.5) can be majorized by the convergent series of positive terms

$$\sum_{n=1}^{\infty} \frac{1}{|n^2\pi^2 - R|},$$

because

$$\frac{1}{|n^2\pi^2 - z|} \le \frac{1}{\big|n^2\pi^2 - |z|\big|} \le \frac{1}{|n^2\pi^2 - R|}.$$

The proof of (9.4.4) for the function $f(z) = \tan z$ can be done similarly; one has to take squares, $P_nQ_nR_nS_n$, with vertical sides through the points $\pm n\pi$ $(n = 1, 2, 3, \dots)$.

Remark 9.4.1. We note that a more general formula than (9.4.5) is given in Problem 27.02 on p. 262 in [**21**], namely,

$$\lim_{n\to\infty} \left[f(z) - \sum_{k=1}^{n} g(z; z_k) \right] = 0, \tag{9.4.23}$$

where $g(z; z_k)$ is the principal part of the Laurent series of a meromorphic function $f(z)$ with center $z = z_k$. If all the poles, z_k, of $f(z)$ are simple, then

$$g(z; z_k) = \frac{1}{z - z_k} \operatorname*{Res}_{z=z_k} f(z)$$

and formula (9.4.23) is transformed into (9.4.5). However, the derivation of (9.4.23) in [**21**] is done under more stringent conditions than for (9.4.4), namely,

$$\lim_{n\to\infty} \oint_{C_n} |f(z)| \frac{|dz|}{|z|+1} = 0. \tag{9.4.24}$$

It can be shown that $1/\sin z$ satisfies (9.4.24), but $\tan z$ and $\cot z$ do not satisfy this condition. Therefore, in [**42**], p. 268, one proves the boundedness of $|\cot z|$ on the paths C_n (that is, condition (9.4.2)), and then $f(z) = \cot z - 1/z$ is expanded in partial fractions by means of (9.4.2).

9.5. Infinite product expansion of entire functions

9.5.1. Infinite products. Consider a sequence

$$b_1,\ b_2,\ \dots,\ b_n, \dots,$$

where b_n is either a complex number or a complex-valued function of the complex variable z, such that $b_n \neq 0$ and $\lim_{n\to\infty} b_n \neq 0$ for all n. Denote partial products as follows:

$$P_1 = b_1, \qquad P_2 = b_1 b_2, \ \dots, \qquad P_n = b_1 b_2 \cdots b_n = \prod_{k=1}^{n} b_k, \ \dots . \tag{9.5.1}$$

DEFINITION 9.5.1. An expression of the form

$$\prod_{k=1}^{\infty} b_k \tag{9.5.2}$$

is called a *formal infinite product.*

DEFINITION 9.5.2. We say that the infinite product (9.5.2) is *convergent* and is equal to $P \neq 0$ if the limit

$$P = \lim_{n\to\infty} \prod_{k=1}^{n} b_k \tag{9.5.3}$$

exists, is finite and is not equal to zero. If (9.5.3) has no nonzero finite limit, then the infinite product (9.5.2) is said to be *divergent* and has no numerical value.

EXAMPLE 9.5.1. *Prove that*

$$\lim_{n\to\infty} \prod_{k=1}^{n} \left(1 + z^{2^{k-1}}\right) = \lim_{n\to\infty} (1+z)(1+z^2)\cdots\left(1+z^{2^{n-1}}\right)$$

$$= \frac{1}{1-z} \tag{9.5.4}$$

in the unit disk $|z| < 1$.

SOLUTION. We have

$$\begin{aligned}(1-z)\prod_{k=1}^{n}\left(1+z^{2^{k-1}}\right) &= (1-z)(1+z)(1+z^2)(1+z^4)\cdots\left(1+z^{2^{n-1}}\right)\\ &= (1-z^2)(1+z^2)(1+z^4)\cdots\left(1+z^{2^{n-1}}\right)\\ &= (1-z^4)(1+z^4)\cdots\left(1+z^{2^{n-1}}\right)\\ &= \left(1-z^{2^{n-1}}\right)\left(1+z^{2^{n-1}}\right)\\ &= 1-z^{2n}.\end{aligned}$$

Hence,

$$\begin{aligned}(1-z)\lim_{n\to\infty}\prod_{k=1}^{n}\left(1+z^{2^{k-1}}\right) &= \lim_{n\to\infty}\left(1-z^{2^n}\right)\\ &= 1, \end{aligned} \tag{9.5.5}$$

since $|z| < 1$. Formula (9.5.4) follows from (9.5.5). □

THEOREM 9.5.1 (NECESSARY CONDITION FOR CONVERGENCE). *If the infinite product* (9.5.2) *is convergent, then*

$$\lim_{n\to\infty} b_n = 1. \tag{9.5.6}$$

PROOF. Suppose that the limit

$$P = \lim_{n\to\infty}\prod_{k=1}^{n} b_k \neq 0 \tag{9.5.7}$$

is finite. Then the limit

$$P = \lim_{n\to\infty}\prod_{k=1}^{n-1} b_k \tag{9.5.8}$$

also exists and is finite. It follows from (9.5.7) and (9.5.8) that

$$\begin{aligned}\lim_{n\to\infty} b_n &= \lim_{n\to\infty}\frac{\prod_{k=1}^{n} b_k}{\prod_{k=1}^{n-1} b_k}\\ &= \frac{P}{P} = 1. \quad \square\end{aligned}$$

Note 9.5.1. The necessary condition (9.5.6) for the convergence of an infinite product is similar to the necessary condition of Theorem 4.1.1 for the convergence of an infinite series

$$\sum_{k=1}^{\infty} a_k, \tag{9.5.9}$$

that is, if the series converges, then $\lim_{n\to\infty} a_k = 0$.

However, the converse is not true. As in the case of series, the test (9.5.6) is only necessary. There exist divergent infinite products satisfying (9.5.6), as can be seen from the following example.

EXAMPLE 9.5.2. *Prove that*

$$\lim_{n\to\infty} \prod_{k=1}^{n} \left(1 + \frac{1}{k}\right) = \infty, \tag{9.5.10}$$

although

$$\lim_{k\to\infty} \left(1 + \frac{1}{k}\right) = 1.$$

SOLUTION. Considering the identity

$$\begin{aligned}\prod_{k=1}^{n} \left(1 + \frac{1}{k}\right) &= \exp\left(\ln \prod_{k=1}^{n} \left[1 + \frac{1}{k}\right]\right) \\ &= \exp\left(\sum_{k=1}^{n} \ln\left(1 + \frac{1}{k}\right)\right),\end{aligned} \tag{9.5.11}$$

we see that the infinite product on the left-hand side is divergent if the series on the right-hand side is divergent. Since

$$\ln(1 + 1/k) \sim 1/k, \qquad \text{as} \quad k \to \infty,$$

and the series $\sum_{k=1}^{\infty} 1/k$ is divergent, then the series $\sum_{k=1}^{\infty} \ln(1 + 1/k)$ is divergent. Therefore, (9.5.10) follows from (9.5.11). □

In the previous solution, we have used the following analog of Bertrand's test for the convergence of improper integrals (see [**50**], p. 71).

THEOREM 9.5.2. *If $a_k = O(1/k^\alpha)$ as $k \to \infty$, then the series $\sum_{k=1}^{\infty} a_k$ is convergent if $\alpha > 1$ and divergent if $\alpha \le 1$.*

From Theorem 9.5.2 it can easily be shown that

$$\prod_{k=1}^{\infty} \left(1 + \frac{1}{k^\alpha}\right) \tag{9.5.12}$$

is convergent for $\alpha > 1$ and divergent for $\alpha \le 1$.

If we let $b_k = 1+a_k$, then the necessary condition (9.5.6) for convergence of an infinite product,

$$\prod_{k=1}^{\infty}(1+a_k), \tag{9.5.13}$$

has the form

$$\lim_{k\to\infty} a_k = 0. \tag{9.5.14}$$

Furthermore, we have

$$\prod_{k=1}^{n}(1+a_k) = \exp\left(\sum_{k=1}^{n}\log(1+a_k)\right), \qquad k = 0, \pm 1, \pm 2, \dots, \tag{9.5.15}$$

where

$$\log(1+a_k) = \operatorname{Log}(1+a_k) + 2m\pi i \tag{9.5.16}$$

and

$$\operatorname{Log}(1+a_k) = \ln|1+a_k| + i\operatorname{Arg}(1+a_k),$$
$$-\pi < \operatorname{Arg}(1+a_k) \le \pi. \tag{9.5.17}$$

For finite n, one can take any branch of $\log(1+a_k)$ in (9.5.15), that is, any fixed value of m in (9.5.16), since $e^{2m\pi i} = 1$. For instance, taking $m = 1$, we have

$$\log(1+a_k) = \ln|1+a_k| + i\operatorname{Arg}(1+a_k) + 2\pi i. \tag{9.5.18}$$

However, it should be taken into account that the necessary condition (9.5.14) for convergence must be satisfied. Therefore

$$\begin{aligned}\lim_{n\to\infty}\operatorname{Arg}(1+a_k) &= \operatorname{Arg} 1\\ &= 0.\end{aligned}$$

If the series

$$\sum_{k=1}^{\infty}\operatorname{Log}(1+a_k) = \sum_{k=1}^{\infty}\left[\ln|1+a_k| + i\operatorname{Arg}(1+a_k)\right] \tag{9.5.19}$$

is convergent, then the series, with the kth term given by (9.5.18), is divergent, since

$$2\pi + 2\pi + 2\pi + \cdots + 2\pi + \ldots \to \infty, \qquad \text{as} \quad n\to\infty.$$

Hence, in this case, one should take

$$\prod_{k=1}^{n}(1+a_k) = \exp\left(\sum_{k=1}^{n}\operatorname{Log}(1+a_k)\right) \tag{9.5.20}$$

instead of (9.5.15). But it is irrelevant that the argument

$$A = \sum_{k=1}^{\infty}\operatorname{Arg}(1+a_k)$$

of the series satisfies the inequality $-\pi < A \leq \pi$, provided it converges. If, for instance, $A = 4\pi/3$, then (9.5.19) becomes

$$\begin{aligned}\sum_{k=1}^{\infty}\operatorname{Log}(1+a_k) &= \sum_{k=1}^{\infty}\ln|1+a_k| + \frac{4\pi}{3}\,i \\ &= B + \frac{4\pi}{3}\,i.\end{aligned}$$

It then follows from (9.5.20) that the limit

$$\lim_{n\to\infty}\prod_{k=1}^{n}(1+a_k) = e^{B+4\pi i/3}$$

exists. Conversely, if the limit

$$Q = \lim_{n\to\infty}\prod_{k=1}^{n}(1+a_k)$$

is finite and distinct from zero, it follows from (9.5.20) that

$$\begin{aligned}Q &= \lim_{n\to\infty}\exp\left(\sum_{k=1}^{n}\log(1+a_k)\right) \\ &= \exp\left(\lim_{n\to\infty}\sum_{k=1}^{n}\log(1+a_k)\right),\end{aligned}$$

that is,

$$\log Q = \lim_{n\to\infty}\sum_{k=1}^{\infty}\log(1+a_k).$$

Hence, the convergence of the series (9.5.20) is necessary and sufficient for the convergence of the infinite product (9.5.13). Since $a_k \to 0$ as $k \to \infty$, then

$$\log(1+a_k) \sim a_k, \qquad \text{as} \quad k \to \infty.$$

Thus, both series $\sum_{k=1}^{\infty} \log(1+a_k)$ and $\sum_{k=1}^{\infty} a_k$ either diverge or converge. Therefore, we have proved the following theorem.

THEOREM 9.5.3. *A necessary and sufficient condition for the convergence of the infinite product* (9.5.13) *is the convergence of the series* $\sum_{k=1}^{\infty} a_k$.

As in the case of a series, the concept of absolute convergence is introduced for an infinite product. It follows from (9.5.20) that a permutation of the factors in the infinite product $\prod_{k=1}^{\infty}(1+a_k)$ corresponds to a permutation of the terms of the series $\sum_{k=1}^{\infty} \log(1+a_k)$, which is either convergent or divergent together with the series $\sum_{k=1}^{\infty} a_k$. It is known that a permutation of the terms of the series $\sum_{k=1}^{\infty} a_k$ does not change its sum only if it is absolutely convergent. In this case, by (9.5.20), the infinite product also does not change value. Therefore it is natural to have the following definition.

DEFINITION 9.5.3. An infinite product $\prod_{k=1}^{\infty}(1+a_k)$ is said to be *absolutely convergent* if the series $\sum_{k=1}^{\infty} a_k$ is absolutely convergent.

The following theorem follows from this definition and Theorem 9.5.3.

THEOREM 9.5.4. *A necessary and sufficient condition for the absolute convergence of the infinite product* $\prod_{k=1}^{\infty}(1+a_k)$ *is the absolute convergence of the series* $\sum_{k=1}^{\infty} a_k$.

9.5.2. Infinite product expansion of entire functions. The expansion of an entire function in the form of an infinite product is a natural generalization of the expansion of a polynomial $P_n(z)$ into its factors.

DEFINITION 9.5.4. The infinite product

$$\prod_{k=1}^{\infty}[1+f_k(z)], \tag{9.5.21}$$

whose factors are not equal to zero in a domain D is said to be *uniformly convergent* in that domain if the sequence of functions

$$F_n(z) = \prod_{k=1}^{n}[1+f_k(z)], \qquad n = 1, 2, 3, \dots, \tag{9.5.22}$$

is uniformly convergent in D.

Using Theorem 9.4.1 for expanding a meromorphic function with simple poles into partial fractions, one obtains the following theorem for the expansion of an entire function in the form of infinite product.

THEOREM 9.5.5. *Let* $f(z)$ *be an entire function with zeros* z_k *of order*

n_k. *Suppose that the meromorphic function*

$$F(z) = \frac{f'(z)}{f(z)}$$

satisfies condition (9.4.6) *of Theorem* 9.4.1, *namely,*

$$\lim_{n\to\infty} \oint_{C_n} \frac{F(\zeta)}{\zeta - z}\, d\zeta = 0, \tag{9.5.23}$$

where the integral approaches zero uniformly in any disk $|z| \le R$ *not containing the disks* $|z - z_k| \le \delta$. *Supppose, moreover, that* $f(0) \neq 0$. *Then* $f(z)$ *has the infinite product representation*

$$f(z) = f(0) \prod_{k=1}^{\infty} \left(1 - \frac{z}{z_k}\right)^{n_k}, \tag{9.5.24}$$

which is uniformly convergent in any bounded region of the complex plane.

PROOF. It follows from Theorem 9.1.1 that the logarithmic derivative $F(z) = f'(z)/f(z)$ has simple poles at the zeros z_k of the entire function $f(z)$ and does not have any other poles. Since the order of the zero z_k is n_k, then, by (9.1.2),

$$\operatorname*{Res}_{z=z_k} F(z) = n_k. \tag{9.5.25}$$

Substituting (9.5.25) into (9.4.5), we obtain

$$F(z) = \sum_{k=1}^{\infty} \frac{n_k}{z - z_k} = \frac{d}{dz} \log f(z), \tag{9.5.26}$$

and integrating $F(z)$ along any arbitrary path joining the the origin to any point z and not passing through any zeros of $f(z)$, we obtain

$$\begin{aligned} \log f(z) - \log f(0) &= \sum_{k=1}^{\infty} n_k \log(z - z_k)\Big|_{z=0}^{z=z} \\ &= \sum_{k=1}^{\infty} n_k \log\left(1 - \frac{z}{z_k}\right). \end{aligned} \tag{9.5.27}$$

Then (9.5.24) follows by taking the exponential of (9.5.27). □

Note 9.5.2. Formula (9.5.24) has not been found in the literature, where one uses (9.4.2) instead of (9.4.5) for expanding $F(z)$ into partial fractions in the form

$$f(z) = f(0)\, e^{zf'(0)/f(0)} \prod_{k=1}^{\infty} \left(1 - \frac{z}{z_k}\right) e^{z/z_k}. \tag{9.5.28}$$

In this expression, each factor

$$\left(1 - \frac{z}{z_k}\right) e^{z/z_k}$$

is repeated n_k times, where n_k is the order of the zero z_k.

EXAMPLE 9.5.3. *Expand in an infinite product the function*

$$f(z) = \begin{cases} (\sin z)/z, & z \neq 0, \\ 1, & z = 0. \end{cases} \tag{9.5.29}$$

SOLUTION. The logarithmic derivative of $f(z)$ is

$$\begin{aligned} F(z) &= \left(\frac{\sin z}{z}\right)' \Big/ \left(\frac{\sin z}{z}\right) \\ &= \left(\frac{\cos z}{z} - \frac{\sin z}{z^2}\right) \Big/ \left(\frac{\sin z}{z}\right) \\ &= \cot z - \frac{1}{z}. \end{aligned}$$

It is proved in Example 9.4.2 that $\cot z$ satisfies condition (9.5.23), and one can easily show that $1/z$ also satisfies (9.5.23). Moreover, $f(0) = 1 \neq 0$, so that all the conditions of Theorem 9.5.5 are satisfied and one can use (9.5.24). In this case the zeros of $(\sin z)/z$ are

$$z_k = k\pi, \qquad k = \pm 1, \pm 2, \dots ,$$

with order $n_k = 1$. Hence, it follows from (9.5.24) that

$$\begin{aligned} \frac{\sin z}{z} &= \prod_{\substack{k=-\infty \\ k\neq 0}}^{\infty} \left(1 - \frac{z}{k\pi}\right) \\ &= \lim_{n\to\infty} \prod_{k=1}^{n} \left(1 - \frac{z}{k\pi}\right) \prod_{k=1}^{n} \left(1 + \frac{z}{k\pi}\right) \\ &= \prod_{k=1}^{\infty} \left(1 - \frac{z^2}{k^2\pi^2}\right). \end{aligned}$$

Therefore

$$\frac{\sin z}{z} = \prod_{k=1}^{\infty} \left(1 - \frac{z^2}{k^2\pi^2}\right). \tag{9.5.30}$$

This well-known formula is usually derived by means of the more complicated formula (9.5.28). □

EXAMPLE 9.5.4. *Expand in an infinite product the function*

$$f(z) = \begin{cases} [(\sin z)/z]^m, & z \neq 0, \\ 1, & z = 0, \end{cases} \tag{9.5.31}$$

where m is an arbitrary positive integer.

SOLUTION. The logarithmic derivative of $f(z)$ is

$$\begin{aligned} F(z) &= \left[\left(\frac{\sin z}{z}\right)^m\right]' \left[\left(\frac{\sin z}{z}\right)\right]^{-m} \\ &= \left(\frac{m \sin^{m-1} z \cos z}{z^m} - \frac{m \sin^m z}{z^{m+1}}\right) \frac{z^m}{\sin^m z} \\ &= m \cot z - \frac{m}{z}. \end{aligned}$$

Since this function is similar to the one in the previous example, condition (9.5.23) is satisfied and one can use (9.5.24). Thus, we have

$$\begin{aligned} \left(\frac{\sin z}{z}\right)^m &= \prod_{\substack{k=-\infty \\ k\neq 0}}^{\infty} \left(1 - \frac{z}{k\pi}\right)^m \\ &= \lim_{n\to\infty} \prod_{k=1}^{n} \left(1 - \frac{z}{k\pi}\right)^m \prod_{k=1}^{n} \left(1 + \frac{z}{k\pi}\right)^m \\ &= \prod_{k=1}^{\infty} \left(1 - \frac{z^2}{k^2\pi^2}\right)^m, \end{aligned}$$

that is,

$$\left(\frac{\sin z}{z}\right)^m = \prod_{k=1}^{\infty} \left(1 - \frac{z^2}{k^2\pi^2}\right)^m. \tag{9.5.32}$$

This formula can be found by raising (9.5.30) to the power m since it can easily be proved that

$$\left[\prod_{k=1}^{\infty}\left(1-\frac{z^2}{k^2\pi^2}\right)\right]^m=\prod_{k=1}^{\infty}\left(1-\frac{z^2}{k^2\pi^2}\right)^m.$$

We leave the proof to the reader. □

EXAMPLE 9.5.5. *Expand in an infinite product the function*

$$f(z)=\cos^m z, \qquad m\in\mathbb{N}. \tag{9.5.33}$$

SOLUTION. The logarithmic derivative of $f(z)$ is

$$\begin{aligned} F(z)&=(\cos^m z)'\cos^{-m} z\\ &=-m\cos^{m-1} z\,\sin z\,\cos^{-m} z\\ &=-m\tan z. \end{aligned}$$

One easily verifies that $\tan z$ satisfies condition (9.5.23), as in Example 9.4.2. Moreover, since the zeros of $f(z)$,

$$z_k=(2k-1)\frac{\pi}{2}, \qquad k=0,\pm1,\pm2,\dots$$

are of order $n_k=m$, then by (9.5.24) we have

$$\begin{aligned} \cos^m z&=\prod_{\substack{k=-\infty\\k\neq0}}^{\infty}\left[1-\frac{z}{(2k-1)\pi/2}\right]^m\\ &=\lim_{n\to\infty}\prod_{k=1}^{n}\left[1-\frac{z}{(2k-1)\pi/2}\right]^m\prod_{k=1}^{n}\left[1+\frac{z}{(2k-1)\pi/2}\right]^m\\ &=\prod_{k=1}^{\infty}\left[1-\frac{4z^2}{(2k-1)^2\pi^2}\right]^m. \end{aligned}$$

Thus,

$$\cos^m z=\prod_{k=1}^{\infty}\left[1-\frac{4z^2}{(2k-1)^2\pi^2}\right]^m. \quad \square \tag{9.5.34}$$

Exercises for Sections 9.4 and 9.5

Expand the following meromorphic functions in partial fractions.

1. $f(z) = \dfrac{1}{\sin z}$.

2. $f(z) = \pi \coth z$.

3. $f(z) = \tan \dfrac{\pi z}{2}$.

3. $f(z) = \sec \dfrac{\pi z}{2}$.

Evaluate the following infinite products.

5. $\displaystyle\prod_{k=1}^{\infty} \left[1 + \frac{1}{k(k+2)}\right]$.

6. $\displaystyle\prod_{k=2}^{\infty} \left[1 - \frac{2}{k(k+1)}\right]$.

7. Where does the infinite product $\prod_{k=1}^{\infty}\left(1 - z^k\right)$ converge absolutely?
8. Prove that the infinite product $\prod_{k=1}^{\infty}\left(1 + a_k\right)$ is absolutely convergent if and only if the infinite product $\prod_{k=1}^{\infty}\left(1 + |a_k|\right)$ is convergent.

9. Prove that $\sinh z = \prod_{k=1}^{\infty}\left(1 + \frac{z^2}{k^2\pi^2}\right)$.

CHAPTER 10

Series Summation by Residues

10.1. Type of series considered

In this chapter, we consider the summation of series of the form

$$S_1 = \sum_{k=-\infty}^{\infty} f(k), \qquad S_2 = \sum_{k=-\infty}^{\infty} (-1)^k f(k),$$
$$S_3 = \sum_{k=-\infty}^{\infty} (-1)^k f(k)\, e^{iak}, \quad S_4 = \sum_{k=-\infty}^{\infty} f(k)\, e^{iak}, \qquad (10.1.1)$$
$$S_5 = \sum_{k=1}^{\infty} f(k), \qquad S_6 = \sum_{k=1}^{\infty} (-1)^k f(k),$$

where $f(z) = P_n(z)/Q_m(z)$, $P_n(z)$ and $Q_m(z)$ are polynomials of degrees n and m, respectively, $m \geq n+2$ (also $m \geq n+1$ for the series S_3 and S_4). We also consider the summation of series of the form

$$S_7 = \sum_{k=-\infty}^{\infty} f(\gamma_k), \quad S_8 = \sum_{k=-\infty}^{\infty} f(\gamma_k)\, e^{i\gamma_k a}, \qquad (10.1.2)$$

where γ_k are the real roots of some transcendental equation (for instance, the zeros of an entire function). Finally we consider the summation of S_7 where γ_k are the complex roots of a transcendental equation (for instance, the roots of the equation $\sinh z \pm z = 0$).

Series of the form (10.1.1) are considered in the literature (see, for example, [**21**], pp. 241–247, [**44**], pp. 188–191, [**51**]), but a systematic study of the series (10.1.2) seems not to have been done. There are two examples in [**21**] for the case where γ_k are the roots of the equation $\tan x = x$. The summation of the series (10.1.1) and (10.1.2) is done by means of one common method based on the following theorem.

THEOREM 10.1.1. *Let $P_n(z)/Q_m(z)$ be a proper rational function, that is, $m > n$, and let $F(z)$ be an entire function such that the poles, γ_k, of $F'(z)/F(z)$ tend to infinity as $k \to \infty$. Also let C_k be a regular system of paths (see Definition 9.4.1). If*

$$\lim_{k\to\infty} \oint_{C_k} \frac{P_n(z)}{Q_m(z)} \frac{F'(z)}{F(z)}\, dz = 0, \tag{10.1.3}$$

then

$$\sum_k \operatorname*{Res}_{z=\gamma_k} \left[\frac{P_n(z)}{Q_m(z)} \frac{F'(z)}{F(z)}\right] = -\sum_k \operatorname*{Res}_{z=z_k} \left[\frac{P_n(z)}{Q_m(z)} \frac{F'(z)}{F(z)}\right], \tag{10.1.4}$$

where z_k are the zeros of the polynomial $Q_m(z)$ and $z_k \neq \gamma_l$ for all k and l.

PROOF. By the residue theorem,

$$\oint_{C_k} \frac{P_n(z)}{Q_m(z)} \frac{F'(z)}{F(z)}\, dz$$
$$= 2\pi i\left(\sum_k \operatorname*{Res}_{z=z_k} + \sum_k \operatorname*{Res}_{z=\gamma_k}\right) \left[\frac{P_n(z)}{Q_m(z)} \frac{F'(z)}{F(z)}\right], \tag{10.1.5}$$

where γ_k are the poles of $F'(z)/F(z)$ and z_k are the zeros of $Q_m(z)$ inside the path C_k. Considering the limit of (10.1.5) as $k \to \infty$ and using (10.1.3), we obtain (10.1.4). □

By choosing $F(z)$ properly, one can find formulae for evaluating the sums S_1 to S_6.

10.2. Summation of S_1

We obtain a formula for the summation of series of the form

$$S_1 = \sum_{k=-\infty}^{\infty} \frac{P_n(k)}{Q_m(k)}, \qquad m \geq n+2, \tag{10.2.1}$$

by taking the entire function

$$F(z) = \sin \pi z$$

whose zeros are $\gamma_k = k$. In this case, the path C_k is conveniently chosen to be a square with vertices A_k, B_k, D_k, E_k at the points

$$(\pm(2k+1)/2, \pm(2k+1)/2)$$

(see Fig 10.1). Here the function

$$\frac{F'(z)}{F(z)} = \pi \cot \pi z$$

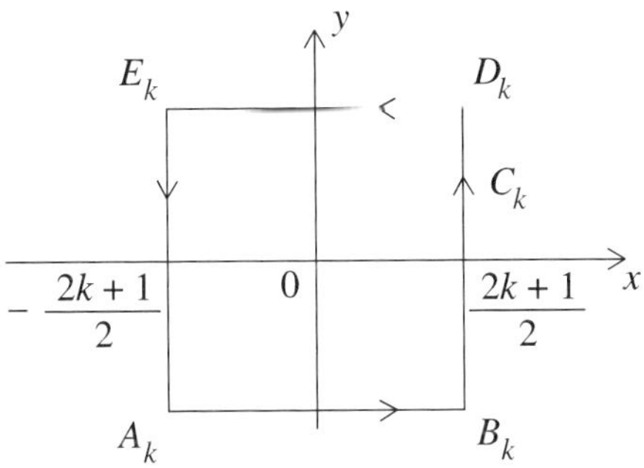

FIGURE 10.1. The square path C_k.

has simple poles at $\gamma_k = k$. We need to prove that

$$\lim_{k\to\infty} \oint_{C_k} \frac{P_n(z)}{Q_m(z)} \cot \pi z \, dz = 0. \tag{10.2.2}$$

Since $m \geq n+2$, (10.2.2) can be proven similarly to (9.4.23). But the boundedness of $|\cot \pi z|$ on the path $A_kB_kD_kE_k$, namely $|\cot \pi z| < M$ for all $k = 1, 2, \ldots$, follows in a simpler way than in (9.4.23) (see [**42**], p. 268). Hence, using Theorem 9.4.1 and substituting $F(z) = \sin \pi z$ in (10.1.4), we obtain the following formula for evaluating S_1:

$$\sum_{k=-\infty}^{\infty} \frac{P_n(k)}{Q_m(k)} = -\pi \sum_k \operatorname*{Res}_{z=z_k} \left[\frac{P_n(z)}{Q_m(z)} \cot \pi z\right], \qquad m \geq n+2, \tag{10.2.3}$$

where z_k are the zeros of $Q_m(z)$ and no z_k is equal to an integer.

Note 10.2.1. If, for some $k = \kappa$, $z_\kappa = N$ where N is a positive or negative integer, then (10.2.3) reduces to

$$\sum_{\substack{k=-\infty \\ k\neq\kappa}}^{\infty} \frac{P_n(k)}{Q_m(k)} = -\pi \sum_k \operatorname*{Res}_{z=z_k} \left[\frac{P_n(z)}{Q_m(z)} \cot \pi z\right]. \tag{10.2.4}$$

We note that the residue at the point $z_k = N$ is included in the right-hand side.

Note 10.2.2. If $P_n(k)/Q_m(k)$ is an even function, then (10.2.3) can be written in the form

$$\frac{1}{2}\frac{P_n(0)}{Q_m(0)} + \sum_{k=1}^{\infty} \frac{P_n(k)}{Q_m(k)} = -\frac{\pi}{2} \sum_k \operatorname*{Res}_{z=z_k} \left[\frac{P_n(z)}{Q_m(z)} \cot \pi z\right]. \tag{10.2.5}$$

EXAMPLE 10.2.1. *Sum the series*

$$\sum_{k=1}^{\infty} \frac{1}{k^2 + a^2}, \qquad a > 0. \tag{10.2.6}$$

SOLUTION. The conditions $P_n(z) = 1$, $Q_m(z) = z^2 + a^2$, $n = 0$, and $m = 2$, are such that (10.2.5) is true. The roots of the equation $z^2 + a^2 = 0$ are $z_1 = ai$ and $z_2 = -ai$. Therefore by (10.2.5) we have

$$\begin{aligned}\sum_{k=1}^{\infty} \frac{1}{k^2 + a^2} &= -\frac{1}{2a^2} - \frac{\pi}{2} \left(\operatorname*{Res}_{z=ai} + \operatorname*{Res}_{z=-ai} \right) \left[\frac{1}{z^2 + a^2} \cot \pi z \right] \\ &= -\frac{1}{2a^2} - \frac{\pi}{2} \left[\frac{\cot \pi ai}{2ai} + \frac{\cot \pi ai}{2ai} \right] \\ &= -\frac{1}{2a^2} + \frac{\pi}{2a} \coth \pi a,\end{aligned}$$

since $\cot \pi ai = -i \coth \pi a$. Hence,

$$\sum_{k=1}^{\infty} \frac{1}{k^2 + a^2} = \frac{1}{2a} \left(\pi \coth \pi a - \frac{1}{a} \right). \quad \sqcup \tag{10.2.7}$$

Note 10.2.3. To obtain the formula

$$\sum_{k=1}^{\infty} \frac{1}{k^2} = \frac{\pi^2}{6}$$

it is sufficient to consider the limit in (10.2.7) as $a \to 0$ (or use formula (10.2.3) as the reader may check):

$$\begin{aligned}\sum_{k=1}^{\infty} \frac{1}{k^2} &= \lim_{a \to 0} \frac{1}{2a^2} (\pi a \coth \pi a - 1) \\ &= \frac{1}{2} \lim_{a \to 0} \frac{\pi a \cosh \pi a - \sinh \pi a}{a^2 \sinh \pi a} \\ &= \frac{1}{2} \lim_{a \to 0} \frac{1}{\pi a^3} \left\{ \pi a \left[1 + \frac{(\pi a)^2}{2!} + \dots \right] - \left[\pi a + \frac{(\pi a)^3}{3!} + \dots \right] \right\} \\ &= \frac{\pi^2}{6}.\end{aligned}$$

Note 10.2.4. To sum the series

$$\sum_{k=1}^{\infty} \frac{1}{(k^2 + a^2)^2}$$

it suffices to assume that $a^2 = \alpha$ in (10.2.7) and differentiate with respect to the parameter α:

$$\sum_{k=1}^{\infty} \frac{1}{(k^2+a^2)^2} = -\frac{d}{d\alpha}\left[\frac{\pi \coth \pi\sqrt{\alpha}}{2\sqrt{\alpha}} - \frac{1}{2\alpha}\right].$$

10.3. Summation of S_2

To evaluate the sum

$$S_2 = \sum_{k=-\infty}^{\infty} (-1)^k \frac{P_n(k)}{Q_m(k)}, \qquad m \geq n+2, \tag{10.3.1}$$

one need only take

$$\frac{F'(z)}{F(z)} = \frac{\pi}{\sin \pi z}, \tag{10.3.2}$$

since then

$$\left.\frac{\pi}{(\sin \pi z)'}\right|_{z=k} = (-1)^k.$$

Let us use the same system of paths shown in Fig 10.1. To use Theorem 10.1.1 one has to prove that

$$\lim_{k\to\infty} \oint_{C_k} \frac{P_n(z)}{Q_m(z)} \frac{dz}{\sin \pi z} = 0. \tag{10.3.3}$$

Since $m \geq n+2$, formula (10.3.3) can be proven as was done in Example 9.4.1. Hence, all the conditions of Theorem 10.1.1 are satisfied, and, substituting (10.3.2) into (10.1.4), we obtain the following formula for evaluating S_2:

$$\sum_{k=-\infty}^{\infty} (-1)^k \frac{P_n(k)}{Q_m(k)} = -\pi \sum_k \operatorname*{Res}_{z=z_k} \left[\frac{P_n(z)}{Q_m(z)} \frac{1}{\sin \pi z}\right],$$
$$m \geq n+2, \qquad z_k \notin \mathbb{Z}. \tag{10.3.4}$$

Note 10.3.1. If, for $k = k_1, k_2, \ldots, k_l$, z_k coincide with the integers $N_1, N_2, \ldots, N_l$, respectively, then one has to drop the terms of the series (10.3.4) with $k_1, k_2, \ldots, k_l$ from the left-hand side, but keep the residues on the right-hand side at the points $z_k = N_1, N_2, \ldots, N_l$.

If the function $P_n(k)/Q_m(k)$ is even, it follows from (10.3.4) that

$$\frac{1}{2}\frac{P_n(0)}{Q_m(0)} + \sum_{k=1}^{\infty}(-1)^k \frac{P_n(k)}{Q_m(k)} = -\frac{\pi}{2} \sum_k \operatorname*{Res}_{z=z_k} \left[\frac{P_n(z)}{Q_m(z)} \frac{1}{\sin \pi z}\right]. \tag{10.3.5}$$

EXAMPLE 10.3.1. *Sum the series*

$$\sum_{k=1}^{\infty} \frac{(-1)^k}{k^2 + a^2}, \qquad a > 0. \tag{10.3.6}$$

SOLUTION. In this case

$$f(z) = \frac{1}{z^2 + a^2},$$

so that the singular points of $f(z)$, $z = \pm ai$, are simple poles. Using (10.3.5) we obtain

$$\begin{aligned}
\sum_{k=1}^{\infty} \frac{(-1)^k}{k^2 + a^2} &= -\frac{1}{2a^2} - \frac{\pi}{2}\left(\operatorname*{Res}_{z=ai} + \operatorname*{Res}_{z=-ai}\right)\left[\frac{1}{z^2 + a^2}\,\frac{1}{\sin \pi z}\right] \\
&= -\frac{1}{2a^2} - \frac{\pi}{2}\left(\frac{1}{2ai \sin \pi ai} + \frac{1}{2ai \sin \pi ai}\right) \\
&= -\frac{1}{2a^2} + \frac{\pi}{2}\,\frac{1}{a \sinh \pi a}.
\end{aligned}$$

Therefore

$$\sum_{k=1}^{\infty} \frac{(-1)^k}{k^2 + a^2} = -\frac{1}{2a^2} + \frac{\pi}{2a \sinh \pi a}. \quad \square \tag{10.3.7}$$

EXAMPLE 10.3.2. *Evaluate the series*

$$\sum_{k=1}^{\infty} \frac{(-1)^k}{(k^2 + a^2)^2}. \tag{10.3.8}$$

SOLUTION. It suffices to differentiate (10.3.7) with respect to $\alpha = a^2$:

$$\begin{aligned}
\sum_{k=1}^{\infty} \frac{(-1)^k}{(k^2 + a^2)^2} &= \frac{d}{d\alpha}\left[\frac{1}{2\alpha} - \frac{\pi}{2\sqrt{\alpha}\, \sinh(\pi\sqrt{\alpha}\,)}\right] \\
&= \frac{1}{2a}\left[-\frac{1}{a^3} + \frac{\pi}{2a^2 \sinh \pi a} + \frac{\pi^2 \cosh \pi a}{2a \sinh^2 \pi a}\right]. \quad \square
\end{aligned} \tag{10.3.9}$$

EXAMPLE 10.3.3. *Evaluate the series*

$$\sum_{k=1}^{\infty} \frac{(-1)^k}{k^4}. \tag{10.3.10}$$

SOLUTION. It is sufficient to consider the limit as $a \to 0$ in (10.3.9) or use Note 10.3.1:

$$\begin{aligned}
\sum_{k=1}^{\infty} \frac{(-1)^k}{k^4} &= \lim_{a\to 0} \frac{1}{4a^4}\left[\frac{\pi^2 a^2 \cosh \pi a}{\sinh^2 \pi a} + \frac{\pi a}{\sinh \pi a} - 2\right] \\
&= \lim_{a\to 0} \frac{\pi^2 a^2 \cosh \pi a + \pi a \sinh \pi a + 1 - \cosh 2\pi a}{4a^4 \sinh^2 \pi a} \\
&= \lim_{a\to 0}\left[\pi^2 a^2\left(1 + \frac{(\pi a)^2}{2!} + \frac{(\pi a)^4}{4!} + \dots\right)\right. \\
&\quad + \pi a\left(\pi a + \frac{(\pi a)^3}{3!} + \frac{(\pi a)^5}{5!} + \dots\right) \\
&\quad \left. - \frac{(2\pi a)^2}{2!} - \frac{(2\pi a)^4}{4!} - \frac{(2\pi a)^6}{6!} - \dots\right]\frac{1}{4\pi^2 a^6} \\
&= \lim_{a\to 0} \frac{\pi^6 a^6}{4\pi^2 a^6}\left(\frac{1}{4!} + \frac{1}{5!} - \frac{2^6}{6!}\right) \\
&= -\frac{7\pi^4}{720}.
\end{aligned} \tag{10.3.11}$$

To derive (10.3.11) one uses the identity $2\sinh^2 \pi a = \cosh 2\pi a - 1$ and Maclaurin's series for the functions $\cosh y$ and $\sinh y$. □

In [**21**], in Problem 30.03(8) on p. 297, instead of $\sum_{k=1}^{\infty}(-1)^n/n^3$ one should read $\sum_{n=1}^{\infty}(-1)^n/n^4$ since the given answer (apart from the sign) coincides with (10.3.11). The series $\sum_{n=1}^{\infty} 1/n^3$ and $\sum_{n=1}^{\infty}(-1)^n/n^3$ cannot be evaluated in closed form by means of (10.3.5). However, these can be evaluated by means of a partial fraction expansion of the logarithmic derivative of the gamma function (see the hint for Problem 30.10 on p. 299 in [**21**]).

EXAMPLE 10.3.4. *Evaluate the series*

$$\sum_{k=1}^{\infty} \frac{(-1)^k}{k^4} \tag{10.3.12}$$

in closed form by means of the formula

$$\sum_{\substack{k=-\infty \\ k\neq 0}}^{\infty} \frac{(-1)^k}{k^4} = -\pi \operatorname*{Res}_{z=0} \frac{1}{z^4 \sin \pi z}. \tag{10.3.13}$$

SOLUTION. Since $z = 0$ is a pole of order 5 of the function $1/(z^4 \sin \pi z)$, it follows from (10.3.13) that

$$2\sum_{k=1}^{\infty} \frac{(-1)^k}{k^4} = -\pi \frac{1}{4!} \lim_{z\to 0} \left(\frac{z^5}{z^4 \sin \pi z}\right)^{(4)}$$

$$= -\frac{\pi}{4!} \lim_{z\to 0} \frac{d^4}{dz^4} \left(\frac{z}{\sin \pi z}\right)$$

$$= -\frac{\pi^4}{4!} \lim_{z_1\to 0} \frac{d^4}{dz_1^4} \left(\frac{z_1}{\sin z_1}\right), \tag{10.3.14}$$

where we have set $\pi z = z_1$.

A direct computation of the fourth derivative at $z_1 = 0$ is tedious. We use the trick which allows us to do it faster not only in the case of the fourth derivative, but also in the case of the sixth derivative, which is needed in the next example. Replacing z_1 by z, we have

$$\frac{d^4}{dz^4} \left(\frac{z}{\sin z}\right)\Bigg|_{z=0} = \frac{d^4}{dz^4} \frac{z}{z - \frac{z^3}{3!} + \frac{z^5}{5!} - \dots}$$

$$= \frac{d^4}{dz^4} \frac{1}{1 - \frac{z^2}{3!} + \frac{z^4}{5!} - \dots}\Bigg|_{z=0} \tag{10.3.15}$$

(we have kept only the terms up to z^4 since higher powers of z, after differentiation, will disappear as $z \to 0$). Expanding the function on the right-hand side of (10.3.15) in a Maclaurin series (in even powers of z since the function is even), we have

$$f(z) = \frac{1}{1 - \frac{z^2}{6} + \frac{z^4}{120} - \dots}$$

$$= 1 + a_2 z^2 + a_4 z^4 + \dots, \tag{10.3.16}$$

where

$$a_2 = \frac{1}{2!} f''(0), \qquad a_4 = \frac{1}{4!} f^{(4)}(0).$$

Our aim is to compute a_4. It follows from (10.3.16) that

$$1 = \left(1 + a_2 z^2 + a_4 z^4 + \dots\right)\left(1 - \frac{z^2}{6} + \frac{z^4}{120} - \dots\right). \tag{10.3.17}$$

Equating the coefficients of z^2 and z^4 in (10.3.17), we obtain

$$0 = a_2 - \frac{1}{6}, \qquad \text{that is,} \quad a_2 = \frac{1}{6},$$

$$0 = a_4 - \frac{a_2}{6} + \frac{1}{120}, \quad \text{that is,} \quad a_4 = \frac{7}{360}.$$

Thus,

$$f^{(4)}(0) = \frac{7 \cdot 4!}{360} = \frac{d^4}{dz_1^4}\left(\frac{z_1}{\sin z_1}\right)\Big|_{z=0}. \tag{10.3.18}$$

Substituting (10.3.18) into (10.3.14) we obtain, as in the previous example, that

$$\sum_{k=1}^{\infty} \frac{(-1)^k}{k^4} = -\frac{7\pi^4}{720}. \quad \square$$

EXAMPLE 10.3.5. *Sum the series*

$$\sum_{k=1}^{\infty} \frac{(-1)^k}{k^6}. \tag{10.3.19}$$

SOLUTION. Using the formula

$$\sum_{\substack{k=-\infty \\ k\neq 0}}^{\infty} \frac{(-1)^k}{k^6} = -\pi \operatorname*{Res}_{z=0} \frac{1}{z^6 \sin \pi z}, \tag{10.3.20}$$

we have

$$\begin{aligned} 2\sum_{k=1}^{\infty} \frac{(-1)^k}{k^6} &= -\pi \frac{1}{6!} \lim_{z\to 0} \left(\frac{z}{\sin \pi z}\right)^{(6)} \\ &= -\frac{\pi^6}{6!} \lim_{z\to 0} \left(\frac{z}{\sin z}\right)^{(6)} \\ &= -\frac{\pi^6}{6!}\left[\frac{d^6}{dz^6}\left(\frac{1}{1 - \frac{z^2}{3!} + \frac{z^4}{5!} - \frac{z^6}{7!} + \dots}\right)\right]\Bigg|_{z=0} \\ &= -\frac{\pi^6}{6!} f^{(6)}(0). \end{aligned} \tag{10.3.21}$$

To derive (10.3.21) one uses the substitution $\pi z = z_1$ and replaces z_1 by z again. Thus we have

$$\begin{aligned} f(z) &= \frac{1}{1 - \frac{z^2}{3!} + \frac{z^4}{5!} - \frac{z^6}{7!} + \dots} \\ &= 1 + a_2 z^2 + a_4 z^4 + a_6 z^6 + \dots . \end{aligned}$$

Hence

$$1 = \left(1 + a_2 z^2 + a_4 z^4 + a_6 z^6 + \dots\right)\left(1 - \frac{z^2}{3!} + \frac{z^4}{5!} - \frac{z^6}{7!} + \dots\right).$$

Equating the coefficients of z^2, z^4 and z^6, we have the system

$$\begin{aligned} 0 &= a_2 - \frac{1}{3!} \\ 0 &= a_4 - \frac{a_2}{3!} + \frac{1}{5!} \\ 0 &= a_6 - \frac{a_4}{3!} + \frac{a_2}{5!} - \frac{1}{7!}, \end{aligned}$$

whose solution is

$$a_2 = \frac{1}{6}, \qquad a_4 = \frac{7}{360}, \qquad a_6 = \frac{31}{15120} = \frac{f^{(6)}(0)}{6!}.$$

Substituting $f^{(6)}(0)$ into (10.3.21) we obtain

$$\sum_{k=1}^{\infty} \frac{(-1)^k}{k^6} = -\frac{31\pi^6}{30240},$$

which coincides with formula 5.1.2(3) (for $s = 6$) on p. 652 in [**38**]. □

10.4. Summation of S_3 and S_4

To evaluate the series

$$S_3 = \sum_{k=1}^{\infty} \frac{(-1)^k P_n(k)}{Q_m(k)} e^{ika}, \qquad \begin{cases} |a| \le \pi, \text{ if } m \ge n+2, \\ |a| < \pi, \text{ if } m = n+1, \end{cases} \tag{10.4.1}$$

and

$$S_4 = \sum_{k=1}^{\infty} \frac{P_n(k)}{Q_m(k)} e^{ikb}, \qquad \begin{cases} 0 \le b \le 2\pi, \text{ if } m \ge n+2 \\ 0 < b < 2\pi, \text{ if } m = n+1, \end{cases} \tag{10.4.2}$$

it is sufficient to assume that

$$\frac{F'(z)}{F(z)} = \frac{e^{iaz}}{\sin \pi z} \tag{10.4.3}$$

in (10.1.3) and (10.1.4) since

$$\operatorname*{Res}_{z=k} \frac{P_n(z)}{Q_m(z)} \frac{e^{iaz}}{\sin \pi z} = \frac{(-1)^k}{\pi} \frac{P_n(k)}{Q_m(k)} e^{iak}.$$

One can solve (10.4.3) for $F(z)$ by quadrature, but this is unnecessary. In order to use formula (10.1.4) one need only prove that

$$\lim_{k\to\infty} \oint_{C_k} \frac{P_n(z)}{Q_m(z)} \frac{e^{iaz}}{\sin \pi z} = 0, \tag{10.4.4}$$

where $|a| \leq \pi$ if $m \geq n+2$ and $|a| < \pi$ if $m = n+1$.

The modulus of the integral on the left-hand side of (10.4.4) has the bound

$$\begin{aligned} |J_k| &:= \left| \oint_{C_k} \frac{P_n(z)}{Q_m(z)} \frac{e^{iaz}}{\sin \pi z} \, dz \right| \\ &\leq \oint_{C_k} \left| \frac{P_n(z)}{Q_m(z)} \right| \frac{|e^{iaz}|}{|\sin \pi z|} \, |dz| \\ &\leq |A| \oint_{C_k} \frac{1}{|z|^p} \frac{e^{-ay}}{|\sin \pi z|} \, |dz| \\ &= |A| \int_{A_k B_k \cup B_k D_k \cup D_k E_k \cup E_k A_k} \left[\frac{1}{|z|^p} \frac{e^{-ay}}{|\sin \pi z|} \, |dz| \right], \end{aligned} \tag{10.4.5}$$

where $|A| = \text{constant} > 0$, $p = 2$ if $m \geq n+2$ and $p = 1$ if $m = n+1$. Since the functions $|\sin \pi z|$ and $|e^{iaz}|$ grow exponentially everywhere in the complex plane as $z \to \infty$, except along the real axis, then one can assume that $|a| \leq \pi$ in (10.4.5) if $m \geq n+2$ (that is, for the case $p = 2$). The detailed proof of the fact that the function $|e^{i\pi z}/ \sin \pi z| = |\cot \pi z + i|$ is bounded on the system of paths $A_k B_k D_k E_k$ in Fig 10.1 is given in [**42**]. Hence

$$|\cot \pi z| \Big|_{z \in A_k B_k D_k E_k} < M$$

for all $k = 1, 2, \ldots$. Therefore, in the case $p = 2$, it follows from (10.4.5) that

$$|J_k| \leq |A| M \oint_{C_k} \frac{|dz|}{|z|^2} \to 0, \qquad \text{as} \quad k \to \infty.$$

In the case $m = n+1$, that is, $p = 1$ in (10.4.5), to prove that the integral in (10.4.5) approaches zero as $k \to \infty$, one has to satisfy the inequality $|a| < \pi$. Since on the edge $A_k B_k$ in Fig 10.1 we have

$$z = x - i\gamma_k, \qquad |z| = \sqrt{x^2 + \gamma_k^2}, \qquad |dz| = dx, \qquad \gamma_k = \frac{2k+1}{2},$$

then

$$\begin{aligned} \int_{A_k B_k} \frac{e^{-ay}}{|z \sin \pi z|} \, |dz| &= \int_{-\gamma_k}^{\gamma_k} \frac{e^{a\gamma_k}}{\sqrt{x^2 + \gamma_k^2} \sqrt{\sinh^2 \pi \gamma_k + \sin^2 \pi x}} \, dx \\ &\quad \text{(and setting } x = \gamma_k t) \\ &= \int_{-1}^{1} \frac{e^{a\gamma_k} \, dt}{\sqrt{t^2 + 1} \sqrt{\sinh^2 \pi \gamma_k + \sin^2 (\pi \gamma_k t)}} \\ &\to 0 \end{aligned}$$

as $k \to \infty$ if $|a| < \pi$. Similarly, it can be shown that the integral along D_kE_k approaches zero as $k \to \infty$ if $|a| < \pi$.

Since on the side B_kD_k we have

$$z = \gamma_k + iy, \qquad |dz| = dy,$$

then

$$\begin{aligned}
\int_{B_kD_k} \frac{e^{-ay}\,|dz|}{|z||\sin \pi z|} &= \int_{-\gamma_k}^{\gamma_k} \frac{e^{-ay}\,dy}{\sqrt{\gamma_k^2 + y^2}\,\sqrt{\sinh^2 \pi y + \sin^2 \pi\gamma_k}} \\
&= \int_{-\gamma_k}^{\gamma_k} \frac{e^{-ay}\,dy}{\sqrt{\gamma_k^2 + y^2}\,\cosh \pi y} \\
&= \int_{-1}^{1} \frac{e^{-a\gamma_k t}\,dt}{\sqrt{t^2+1}\,\cosh \pi\gamma_k t} \\
&\to 0,
\end{aligned}$$

as $k \to \infty$ if $|a| < \pi$. In deriving the last formula we have used the identities

$$\sin^2 \pi\gamma_k = 1, \qquad \sinh^2 \pi y + 1 = \cosh^2 \pi y,$$

and the substitution $y = \gamma_k t$. Similarly, one can show that the integral along E_kA_k approaches zero as $k \to \infty$ if $|a| < \pi$. This completes the proof of (10.4.4) and also stresses the importance of the strict inequality $|a| < \pi$ if $m = n + 1$. Therefore, substituting

$$\frac{F'(z)}{F(z)} = \frac{e^{iaz}}{\sin \pi z}$$

in (10.1.4), we obtain the following formula for the evaluation of series S_3:

$$\sum_{k=-\infty}^{\infty} (-1)^k \frac{P_n(k)}{Q_m(k)}\, e^{iak} = -\pi \sum_k \operatorname*{Res}_{z=z_k} \left[\frac{P_n(z)}{Q_m(z)}\, \frac{e^{iaz}}{\sin \pi z} \right], \tag{10.4.6}$$

where $|a| \le \pi$ if $m \ge n + 2$ and $|a| < \pi$ if $m = n + 1$, $z_k \notin \mathbb{Z}$.

If some $z_k \in \mathbb{Z}$, one has to use Note 10.3.1.

Substituting $(-1)^k = e^{ik\pi}$ in (10.4.6) we have

$$\sum_{k=-\infty}^{\infty} \frac{P_n(k)}{Q_m(k)}\, e^{ibk} = -\pi \sum_k \operatorname*{Res}_{z=z_k} \left[\frac{P_n(z)}{Q_m(z)}\, \frac{e^{i(b-\pi)z}}{\sin \pi z} \right], \tag{10.4.7}$$

where $b = a + \pi$, that is, $0 \le b \le 2\pi$ if $m \ge n + 2$ and $0 < b < 2\pi$ if $m = n + 1$.

Separating the real and imaginary parts in (10.4.6) we obtain

$$\sum_{k=-\infty}^{\infty} (-1)^k \frac{P_n(k)}{Q_m(k)} \sin ak = -\pi \Im \sum_k \operatorname*{Res}_{z=z_k} \left[\frac{P_n(z)}{Q_m(z)} \frac{e^{iaz}}{\sin \pi z} \right] \tag{10.4.8}$$

and

$$\sum_{k=-\infty}^{\infty} (-1)^k \frac{P_n(k)}{Q_m(k)} \cos ak = -\pi \Re \sum_k \operatorname*{Res}_{z=z_k} \left[\frac{P_n(z)}{Q_m(z)} \frac{e^{iaz}}{\sin \pi z} \right], \tag{10.4.9}$$

where $|a| \leq \pi$ if $m \geq n+2$ and $|a| < \pi$ if $m = n+1$.

The series on the left-hand sides of (10.4.8) and (10.4.9) coincide with the series (8.1.60) and (8.1.45) that have been obtained in Chapter 8 as a by-product of the evaluation of integrals. However, the present formulae (10.4.8) and (10.4.9) are computationally more convenient than the former ones.

It can easily be shown that, for given values of the polynomials $P_n(k)$ and $Q_m(k)$, formulae (10.4.8) and (10.4.9), on the one hand, and (8.1.60) and (8.1.45), on the other hand, lead to the same results.

Similarly, equating the real and imaginary parts in (10.4.7), we obtain

$$\sum_{k=-\infty}^{\infty} \frac{P_n(k)}{Q_m(k)} \sin bk = -\pi \Im \sum_k \operatorname*{Res}_{z=z_k} \left[\frac{P_n(z)}{Q_m(z)} \frac{e^{i(b-\pi)z}}{\sin \pi z} \right], \tag{10.4.10}$$

and

$$\sum_{k=-\infty}^{\infty} \frac{P_n(k)}{Q_m(k)} \cos bk = -\pi \Re \sum_k \operatorname*{Res}_{z=z_k} \left[\frac{P_n(z)}{Q_m(z)} \frac{e^{i(b-\pi)z}}{\sin \pi z} \right], \tag{10.4.11}$$

where $0 \leq b \leq 2\pi$ if $m \geq n+2$ and $0 < b < 2\pi$ if $m = n+1$.

If $P_n(k)/Q_m(k)$ is odd in (10.4.8) and (10.4.10), and even in (10.4.9) and (10.4.11), then it follows from (10.4.8)–(10.4.11) that

$$\sum_{k=1}^{\infty} (-1)^k \frac{P_n(k)}{Q_m(k)} \sin ak = -\frac{\pi}{2} \Im \sum_k \operatorname*{Res}_{z=z_k} \left[\frac{P_n(z)}{Q_m(z)} \frac{e^{iaz}}{\sin \pi z} \right], \tag{10.4.12}$$

and

$$\frac{P_n(0)}{2Q_m(0)} + \sum_{k=1}^{\infty} (-1)^k \frac{P_n(k)}{Q_m(k)} \cos ak$$
$$= -\frac{\pi}{2} \Re \sum_k \operatorname*{Res}_{z=z_k} \left[\frac{P_n(z)}{Q_m(z)} \frac{e^{iaz}}{\sin \pi z} \right], \tag{10.4.13}$$

where $|a| \le \pi$ if $m \ge n+2$ and $|a| < \pi$ if $m = n+1$, and $P_n(k)/Q_m(k)$ is odd in (10.4.12) and even in (10.4.13).

Similarly, we have

$$\sum_{k=1}^{\infty} \frac{P_n(k)}{Q_m(k)} \sin bk = -\frac{\pi}{2} \Im \sum_k \operatorname*{Res}_{z=z_k} \left[\frac{P_n(z)}{Q_m(z)} \frac{e^{i(b-\pi)z}}{\sin \pi z} \right], \tag{10.4.14}$$

and

$$\frac{P_n(0)}{2Q_m(0)} + \sum_{k=1}^{\infty} \frac{P_n(k)}{Q_m(k)} \cos bk = -\frac{\pi}{2} \Re \sum_k \operatorname*{Res}_{z=z_k} \left[\frac{P_n(z)}{Q_m(z)} \frac{e^{i(b-\pi)z}}{\sin \pi z} \right], \tag{10.4.15}$$

where $0 \le b \le 2\pi$ if $m \ge n+2$ and $0 < b < 2\pi$ if $m = n+1$, and $P_n(k)/Q_m(k)$ is odd in (10.4.14) and even in (10.4.15).

EXAMPLE 10.4.1. *Evaluate the series*

$$\sum_{k=1}^{\infty} \frac{k \sin kx}{k^2 + \alpha^2}, \qquad \alpha > 0. \tag{10.4.16}$$

SOLUTION. Since

$$k/(k^2 + \alpha^2) = O(1/k),$$

that is, $m = n+1$, we can use formula (10.4.14) on the interval $0 < x < 2\pi$:

$$\begin{aligned} \sum_{k=1}^{\infty} \frac{k \sin kx}{k^2 + \alpha^2} &= -\frac{\pi}{2} \Im \left(\operatorname*{Res}_{z=i\alpha} + \operatorname*{Res}_{z=-i\alpha} \right) \left[\frac{z}{z^2 + \alpha^2} \frac{e^{i(x-\pi)z}}{\sin \pi z} \right] \\ &= -\frac{\pi}{2} \Im \left[\frac{1}{2} \frac{e^{i(x-\pi)i\alpha}}{\sin \pi i\alpha} - \frac{1}{2} \frac{e^{-i(x-\pi)i\alpha}}{\sin \pi i\alpha} \right] \\ &= -\frac{\pi}{4} \frac{1}{\sinh \pi\alpha} \Im \left\{ \frac{1}{i} \left[e^{(\pi-x)\alpha} - e^{-(\pi-x)\alpha} \right] \right\} \\ &= \frac{\pi}{2} \frac{\sinh \alpha(\pi - x)}{\sinh \pi\alpha}, \qquad 0 < x < 2\pi. \end{aligned} \tag{10.4.17}$$

This result coincides with Formula 1.445(1) in [**23**], p. 40. □

EXAMPLE 10.4.2. *Evaluate the series*

$$\sum_{k=1}^{\infty} \frac{\cos kx}{k^2 + \alpha^2}, \qquad \alpha > 0. \tag{10.4.18}$$

SOLUTION. Since

$$1/(k^2+\alpha^2) = O(1/k^2),$$

that is, $m = n+2$, we can use formula (10.4.15) on the interval $0 \le x \le 2\pi$:

$$\begin{aligned}
\sum_{k=1}^{\infty} \frac{\cos kx}{k^2+\alpha^2} &= -\frac{1}{2\alpha^2} - \frac{\pi}{2}\Re\left(\operatorname*{Res}_{z=i\alpha} + \operatorname*{Res}_{z=-i\alpha}\right)\left[\frac{1}{z^2+\alpha^2}\,\frac{e^{i(x-\pi)z}}{\sin \pi z}\right] \\
&= -\frac{1}{2\alpha^2} - \frac{\pi}{2}\Re\left[\frac{e^{i(x-\pi)i\alpha}}{2i\alpha \sin \pi i\alpha} + \frac{e^{-i(x-\pi)i\alpha}}{2i\alpha \sin \pi i\alpha}\right] \\
&= -\frac{1}{2\alpha^2} + \frac{\pi}{4\alpha \sinh \pi\alpha}\left[e^{\alpha(\pi-x)} + e^{-\alpha(\pi-x)}\right] \\
&= \frac{\pi}{2\alpha}\,\frac{\cosh \alpha(\pi-x)}{\sinh \pi\alpha} - \frac{1}{2\alpha^2}, \qquad 0 \le x \le 2\pi.
\end{aligned}$$

This result coincides with Formula 5.4.5(1) in [**38**], p. 730. Note that for Formula 1.445(2) in [**23**], p. 40, the open interval $0 < x < 2\pi$ can be replaced by the closed interval $0 \le x \le 2\pi$. □

EXAMPLE 10.4.3. *Evaluate the series*

$$\sum_{k=1}^{\infty} \frac{(-1)^k \cos kx}{k^2+\alpha^2}, \qquad \alpha > 0. \tag{10.4.19}$$

SOLUTION. Since in this example $m = 2$, $n = 0$, that is, $m = n+2$, we can use formula (10.4.13) on the interval $-\pi \le x \le \pi$:

$$\begin{aligned}
\sum_{k=1}^{\infty} \frac{(-1)^k \cos kx}{k^2+\alpha^2} &= -\frac{1}{2\alpha^2} - \frac{\pi}{2}\Re\left(\operatorname*{Res}_{z=i\alpha} + \operatorname*{Res}_{z=-i\alpha}\right)\left[\frac{1}{z^2+\alpha^2}\,\frac{e^{ixz}}{\sin \pi z}\right] \\
&= -\frac{1}{2\alpha^2} - \frac{\pi}{2}\Re\left[\frac{e^{ix\alpha i}}{2\alpha i \sin \pi\alpha i} + \frac{e^{-ix\alpha i}}{2\alpha i \sin \pi\alpha i}\right] \\
&= -\frac{1}{2\alpha^2} + \frac{\pi}{4\alpha \sinh \pi\alpha}\left(e^{-\alpha x} + e^{\alpha x}\right) \\
&= \frac{\pi}{2\alpha}\,\frac{\cosh \alpha x}{\sinh \alpha\pi} - \frac{1}{2\alpha^2}, \qquad -\pi \le x \le \pi.
\end{aligned}$$

This formula coincides with Formula 1.445(3) in [**23**], p. 40. □

EXAMPLE 10.4.4. *Evaluate the series*

$$\sum_{k=1}^{\infty} \frac{(-1)^k k \sin kx}{k^2+\alpha^2}, \qquad \alpha > 0. \tag{10.4.20}$$

SOLUTION. Since, in this example, $m = n+1$, we use formula (10.4.12) on the interval $-\pi < x < \pi$:

$$\begin{aligned}
\sum_{k=1}^{\infty} \frac{(-1)^k k \sin kx}{k^2+\alpha^2} &= -\frac{\pi}{2} \Im \left(\operatorname*{Res}_{z=i\alpha} + \operatorname*{Res}_{z=-i\alpha} \right) \left[\frac{z}{z^2+\alpha^2} \frac{e^{ixz}}{\sin \pi z} \right] \\
&= -\frac{\pi}{2} \Im \left[\frac{e^{ix\alpha i}}{2 \sin \pi \alpha i} - \frac{e^{-ix\alpha i}}{2 \sin \pi \alpha i} \right] \\
&= -\frac{\pi}{4\alpha \sinh \pi \alpha} \Im \left[-i \left(e^{-\alpha x} - e^{\alpha x} \right) \right] \\
&= -\frac{\pi}{2\alpha} \frac{\sinh \alpha x}{\sinh \alpha \pi}, \qquad -\pi < x < \pi.
\end{aligned}$$

This result coincides with Formula 1.445(4) in [**23**], p. 40. □

Similarly, one can evaluate series of the form (10.4.8)–(10.4.11) whose nth term depends on $2k+1$ but not on k. For this purpose it is sufficient to use the integral

$$\oint_{C_k} \frac{P_n(z)}{Q_m(z)} \frac{e^{iaz}}{\cos(\pi z/2)} \, dz,$$

where $|a| \leq \pi/2$ if $m \geq n+2$ and $|a| < \pi/2$ if $m = n+1$, and the path C_k is taken to be the boundary of the square whose vertices are the points $\pm k\pi \pm k\pi i$.

We present the final result (leaving its derivation as an exercise for the reader)

$$\begin{aligned}
&\sum_{k=-\infty}^{\infty} (-1)^k \frac{P_n(2k+1)}{Q_m(2k+1)} e^{ia(2k+1)} \\
&\qquad = \frac{\pi}{2} \sum_k \operatorname*{Res}_{z=z_k} \left[\frac{P_n(z)}{Q_m(z)} \frac{e^{iaz}}{\cos \pi z/2} \right], \qquad (10.4.21)
\end{aligned}$$

where $|a| \leq \pi/2$ if $m \geq n+2$ and $|a| < \pi/2$ if $m = n+1$, $Q_m(z_k) = 0$. Substituting

$$(-1)^k = \frac{1}{i} e^{i(2k+1)\pi/2},$$

in (10.4.21) we have

$$\sum_{k=-\infty}^{\infty} \frac{P_n(2k+1)}{Q_m(2k+1)} e^{ib(2k+1)} = \frac{\pi i}{2} \sum_k \operatorname*{Res}_{z=z_k} \left[\frac{P_n(z)}{Q_m(z)} \frac{e^{i(b-\pi/2)z}}{\cos \pi z/2} \right], \qquad (10.4.22)$$

where $b = a + \pi/2$, that is, $0 \leq b \leq \pi$ if $m \geq n+2$ and $0 < b < \pi$ if $m = n+1$.

Finally, separating the real and imaginary parts of formulae (10.4.21) and (10.4.22), we obtain four formulae similar to (10.4.8)–(10.4.11), where the argument k in the functions under the summation sign is replaced by $2k+1$,

$$\sum_{k=-\infty}^{\infty} (-1)^k \frac{P_n(2k+1)}{Q_m(2k+1)} \sin(2k+1)a$$

$$= \frac{\pi}{2} \Im \left\{ \sum_k \operatorname*{Res}_{z=z_k} \left[\frac{P_n(z)}{Q_m(z)} \frac{e^{iaz}}{\cos \pi z/2} \right] \right\}, \quad (10.4.23)$$

and

$$\sum_{k=-\infty}^{\infty} (-1)^k \frac{P_n(2k+1)}{Q_m(2k+1)} \cos(2k+1)a$$

$$= \frac{\pi}{2} \Re \left\{ \sum_k \operatorname*{Res}_{z=z_k} \left[\frac{P_n(z)}{Q_m(z)} \frac{e^{iaz}}{\cos \pi z/2} \right] \right\}, \quad (10.4.24)$$

where $|a| \le \pi/2$ if $m \ge n+2$ and $|a| < \pi/2$ if $m = n+1$.

Similarly,

$$\sum_{k=-\infty}^{\infty} \frac{P_n(2k+1)}{Q_m(2k+1)} \sin(2k+1)b$$

$$= \frac{\pi}{2} \Im \left\{ i \sum_k \operatorname*{Res}_{z=z_k} \left[\frac{P_n(z)}{Q_m(z)} \frac{e^{i(b-\pi/2)z}}{\cos \pi z/2} \right] \right\}, \quad (10.4.25)$$

and

$$\sum_{k=-\infty}^{\infty} \frac{P_n(2k+1)}{Q_m(2k+1)} \cos(2k+1)b$$

$$= \frac{\pi}{2} \Re \left\{ i \sum_k \operatorname*{Res}_{z=z_k} \left[\frac{P_n(z)}{Q_m(z)} \frac{e^{i(b-\pi/2)z}}{\cos \pi z/2} \right] \right\}, \quad (10.4.26)$$

where $0 \le b \le \pi$ if $m \ge n+2$ and $0 < b < \pi$ if $m = n+1$.

The series on the left-hand sides in (10.4.23) and (10.4.24) coincide with the series (8.1.68) and (8.1.69), respectively, already derived in Chapter 8 as a by-product for the evaluation of integrals. However, the present formulae (10.4.23) and (10.4.24) are computationally more convenient than (8.1.68) and (8.1.69)). It can be shown that for given values of $P_n(2k+1)$ and $Q_m(2k+1)$, formulae (8.1.68) and (8.1.69) (for the case $2a = \pi$) and formulae (10.4.23) and (10.4.24) give the same results.

If the functions under the summation sign in (10.4.23) and (10.4.25) are odd and those under the summation sign in (10.4.24) and (10.4.26) are even, then the series in (10.4.23)–(10.4.26) can be transformed so that the summation index will go from 0 to $+\infty$ by means of some trick.

If $f(x)$ is even, then

$$\sum_{k=-\infty}^{\infty} f(2k+1) = \sum_{k=-\infty}^{-1} f(2k+1) + \sum_{k=0}^{\infty} f(2k+1)$$

(and subtituting $k=-l-1$ in the first sum)

$$= \sum_{l=\infty}^{0} f(-(2l+1)) + \sum_{k=0}^{\infty} f(2k+1)$$

$$= 2\sum_{k=0}^{\infty} f(2k+1). \tag{10.4.27}$$

Similarly, if $f(x)$ is odd, then

$$\sum_{k=-\infty}^{\infty} (-1)^k f(2k+1) = \sum_{k=-\infty}^{-1} (-1)^k f(2k+1) + \sum_{k=0}^{\infty} (-1)^k f(2k+1)$$

$$= \sum_{l=\infty}^{0} (-1)^{l-1} f(-2l-1) + \sum_{k=0}^{\infty} (-1)^k f(2k+1)$$

$$= 2\sum_{k=0}^{\infty} (-1)^k f(2k+1). \tag{10.4.28}$$

Using formulae (10.4.27) and (10.4.28) we can transform (10.4.23)–(10.4.26) to the form

$$\sum_{k=0}^{\infty} (-1)^k \frac{P_n(2k+1)}{Q_m(2k+1)} \sin(2k+1)a$$

$$= \frac{\pi}{4}\Im\left\{\sum_k \operatorname*{Res}_{z=z_k}\left[\frac{P_n(z)}{Q_m(z)}\frac{e^{iaz}}{\cos\pi z/2}\right]\right\} \tag{10.4.29}$$

and

$$\sum_{k=0}^{\infty} (-1)^k \frac{P_n(2k+1)}{Q_m(2k+1)} \cos(2k+1)a$$

$$= \frac{\pi}{4}\Re\left\{\sum_k \operatorname*{Res}_{z=z_k}\left[\frac{P_n(z)}{Q_m(z)}\frac{e^{iaz}}{\cos\pi z/2}\right]\right\}, \tag{10.4.30}$$

where $|a| \le \pi/2$ if $m \ge n+2$ and $|a| < \pi/2$ if $m = n+1$. Similarly,

$$\sum_{k=0}^{\infty} \frac{P_n(2k+1)}{Q_m(2k+1)} \sin{(2k+1)b} = \frac{\pi}{4}\Im\left\{ i \sum_k \operatorname*{Res}_{z=z_k} \left[\frac{P_n(z)}{Q_m(z)} \frac{e^{i(b-\pi/2)z}}{\cos \pi z/2} \right] \right\} \tag{10.4.31}$$

and

$$\sum_{k=0}^{\infty} \frac{P_n(2k+1)}{Q_m(2k+1)} \cos{(2k+1)b} = \frac{\pi}{4}\Re\left\{ i \sum_k \operatorname*{Res}_{z=z_k} \left[\frac{P_n(z)}{Q_m(z)} \frac{e^{i(b-\pi/2)z}}{\cos \pi z/2} \right] \right\}, \tag{10.4.32}$$

where $0 \le b \le \pi$ if $m \ge n+2$ and $0 < b < \pi$ if $m = n+1$.

10.5. Series with neither even nor odd terms

We evaluate the series

$$S_5 = \sum_{k=1}^{\infty} \frac{P_n(k)}{Q_m(k)}, \qquad m \ge n+2, \tag{10.5.1}$$

and

$$S_6 = \sum_{k=1}^{\infty} (-1)^k \frac{P_n(k)}{Q_m(k)}, \qquad m \ge n+1, \tag{10.5.2}$$

where $P_n(x)/Q_m(x)$ is neither even nor odd.

In [**1**] Sect. 6.8, p. 264, rational series are summed by means of polygamma functions which are defined as follows. The logarithmic derivative of the gamma function

$$\psi(z) = \frac{d[\log \Gamma(z)]}{dz} = \frac{\Gamma'(z)}{\Gamma(z)}$$

is called the ψ or *digamma function.* The nth derivatives of the digamma functions for $n = 0, 1, 2, \dots$ are called *polygamma functions.* The expansion of the digamma function in partial fractions is given in [**1**], p. 259, formula (6.3.16)):

$$\psi(z+1) = -\gamma + \sum_{n=1}^{\infty} \frac{z}{n(n+z)}$$

$$= -\gamma + \sum_{n=1}^{\infty} \left(\frac{1}{n} - \frac{1}{n+z} \right), \tag{10.5.3}$$

for $z \neq -1, -2, -3, \dots$, where

$$\gamma = -\psi(1) = 0.577215665\dots \tag{10.5.4}$$

is Euler's constant. Differentiating (10.5.3) we have

$$\psi'(z+1) = \sum_{n=1}^{\infty} \frac{1}{(n+z)^2}, \qquad \psi''(z+1) = -2\sum_{n=1}^{\infty} \frac{1}{(n+z)^3}, \tag{10.5.5}$$

and so on. In [**1**], values of the polygamma functions $\psi^{(n)}(z)$ for real z and $n = 0, 1, 2, 3$ are listed in Tables 6.1 and 6.2, pp. 267–271, and values of the digamma function $\psi(z)$ for complex values of z are listed in Table 6.8, pp. 288–293. We shall restrict ourselves to an example, similar to the one given in [**1**], p. 264.

Suppose one has to evaluate the series

$$\sum_{n=1}^{\infty} u(n) = \sum_{n=1}^{\infty} \frac{A(n)}{B_1(n)B_2(n)}, \tag{10.5.6}$$

where

$$\begin{aligned} B_1(n) &= (n+\alpha_1)(n+\alpha_2)\cdots(n+\alpha_m), \\ B_2(n) &= (n+\beta_1)^2(n+\beta_2)^2\cdots(n+\beta_r)^2, \end{aligned} \tag{10.5.7}$$

and $A(n)$ is a polynomial in n whose degree does not exceed $m+2r-2$ and the constants α_i and β_i are distinct. Expanding $u(n)$ in partial fractions, we obtain

$$u(n) = \sum_{k=1}^{m} \frac{a_k}{n+\alpha_k} + \sum_{k=1}^{r} \frac{b_{1k}}{n+\beta_k} + \sum_{k=1}^{r} \frac{b_{2k}}{(n+\beta_k)^2}, \tag{10.5.8}$$

where

$$\sum_{k=1}^{m} a_k + \sum_{k=1}^{r} b_{1k} = 0, \tag{10.5.9}$$

since the sum of residues of the analytic function $u(z)$ in $\mathbb{C}$ is equal to zero. Substituting (10.5.8) into (10.5.6), we have

$$\sum_{n=1}^{\infty} u(n) = \sum_{k=1}^{m} a_k \sum_{n=1}^{\infty} \left(\frac{1}{n+\alpha_k} - \frac{1}{n} + \frac{1}{n} \right) + \sum_{k=1}^{r} b_{1k} \sum_{n=1}^{\infty} \left(\frac{1}{n+\beta_k} - \frac{1}{n} + \frac{1}{n} \right) + \sum_{k=1}^{r} b_{2k} \sum_{n=1}^{\infty} \frac{1}{(n+\beta_k)^2}, \tag{10.5.10}$$

and by (10.5.9) we obtain

$$\lim_{N\to\infty} \sum_{n=1}^{N} \frac{1}{n} \left(\sum_{k=1}^{m} a_k + \sum_{k=1}^{r} b_{1k} \right) = \lim_{N\to\infty} \sum_{n=1}^{N} 0 = 0.$$

Therefore, by (10.5.3) and (10.5.5) we obtain from (10.5.10) that

$$\sum_{n=1}^{\infty} u(n) = -\sum_{k=1}^{m} a_k[\psi(1+\alpha_k)+\gamma] - \sum_{k=1}^{r} b_{1k}[\psi(1+\beta_k)+\gamma] + \sum_{k=1}^{r} b_{2k}\psi'(1+\beta_k),$$

or, taking (10.5.9) into account,

$$\sum_{n=1}^{\infty} u(n) = -\sum_{k=1}^{m} a_k \psi(1+\alpha_k) - \sum_{k=1}^{r} [b_{1k}\psi(1+\beta_k) + b_{2k}\psi'(1+\beta_k)], \tag{10.5.11}$$

that is, the series (10.5.6) is evaluated in closed form.

To evaluate

$$S_6 = \sum_{k=0}^{\infty} (-1)^k \frac{P_n(k)}{Q_m(k)}$$

we use the formulae 8.370 and 8.372(1) on pp. 947 in [**23**]:

$$\beta(x) = \frac{1}{2} \left[\psi\left(\frac{x+1}{2}\right) - \psi\left(\frac{x}{2}\right) \right], \tag{10.5.12}$$

$$\beta(x) = \sum_{k=0}^{\infty} \frac{(-1)^k}{x+k}. \tag{10.5.13}$$

It can be shown that the series (10.5.13) is uniformly convergent for all $x > 0$ (see, for example, [**32**], p. 819). Hence we have

$$\beta'(x) = \sum_{k=0}^{\infty} \frac{(-1)^{k+1}}{(x+k)^2}, \qquad \beta''(x) = 2\sum_{k=0}^{\infty} \frac{(-1)^k}{(x+k)^3}. \tag{10.5.14}$$

Note that by means of (10.5.12) and the tables for $\psi(x)$ in [**1**], one can evaluate $\beta(x)$, $\beta'(x)$ and $\beta''(x)$.

As an example, we evaluate the following series in closed form:

$$\sum_{n=0}^{\infty}(-1)^n u(n) = \sum_{n=0}^{\infty}(-1)^n \frac{A(n)}{B_1(n)B_2(n)}, \tag{10.5.15}$$

where $A(n)$, $B_1(n)$ and $B_2(n)$ are the same as in (10.5.7), but here the degree of the polynomial $A(n)$ is at most $m+2r-1$. The expansion of the rational function $u(n)$ in partial fractions is similar to (10.5.8):

$$\begin{aligned}\sum_{n=0}^{\infty}(-1)^n u(n) = \sum_{k=1}^{m} a_k \sum_{n=0}^{\infty} \frac{(-1)^n}{n+\alpha_k} + \sum_{k=1}^{r} b_{1k} \sum_{n=0}^{\infty} \frac{(-1)^n}{n+\beta_k} \\ + \sum_{k=1}^{r} b_{2k} \sum_{n=0}^{\infty} \frac{(-1)^n}{(n+\beta_k)^2}.\end{aligned} \tag{10.5.16}$$

Using formulae (10.5.13), (10.5.14) and (10.5.16) we obtain

$$\sum_{n=0}^{\infty}(-1)^n u(n) = \sum_{k=1}^{m} \beta(\alpha_k) + \sum_{k=1}^{r} b_{1k}\beta(\beta_k) + \sum_{k=1}^{r} b_{2k}\beta'(\beta_k). \tag{10.5.17}$$

Hence, the series (10.5.15) is evaluated in closed form.

10.6. Series involving real zeros of entire functions

We consider series of the form

$$S_7 = \sum_{k=-\infty}^{\infty} f(\gamma_k) \tag{10.6.1}$$

and

$$S_8 = \sum_{k=-\infty}^{\infty} f(\gamma_k)\, e^{i\gamma_k a}, \tag{10.6.2}$$

where $f(z) = P_n(z)/Q_m(z)$, $P_n(z)$ and $Q_m(z)$ are polynomials of degrees n and m, respectively, $m \geq n+2$, and γ_k are the zeros of an entire function.

The formulae of this section have not been found in the literature. We shall consider two cases.

Case 1. We consider the case where γ_k are the roots of the equation

$$\tan x = -Cx, \qquad C = \text{constant}, \quad C \geq -1. \tag{10.6.3}$$

This last equation is a particular case of the equation

$$\cot \lambda l = \frac{\lambda^2 - h_1 h_2}{\lambda(h_1 + h_2)}, \qquad h_1 \geq 0, \quad h_2 \geq 0. \tag{10.6.4}$$

Equation (10.6.4) is of the form

$$\cot \lambda l = -\frac{h_2}{\lambda}, \quad \text{that is,} \quad \tan \lambda l = -\frac{\lambda}{h_2},$$

if $h_1 \to \infty$ and coincides with (10.6.3) if $C = 1/h_2$ and $h_1 = 0$. Equation (10.6.4) has only real roots $\lambda = \lambda_n$, since these roots are the eigenvalues of the following self-adjoint Sturm–Liouville problem (see [**14**], p. 256, Problem 112):

$$X''(x) + \lambda X = 0, \qquad 0 < x < l, \tag{10.6.5}$$

$$X'(0) - h_1 X(0) = 0, \qquad X'(l) + h_2 X(l) = 0. \tag{10.6.6}$$

In a note to the table of the first seven roots of (10.6.3) in [**14**], p. 684, it is stated that "all the roots of (10.6.3) are real if $C \geq -1$." Negative values of C occur in Sturm–Liouville problems for spheres.

To sum the series

$$S_7 = \sum_{k=-\infty}^{\infty} f(\gamma_k),$$

where γ_k are the roots of equation (10.6.3), we let

$$F(z) = \sin z + Cz \cos z, \qquad C \geq -1, \tag{10.6.7}$$

in formula (10.1.4) of Theorem 10.1.1. Then

$$\frac{F'(z)}{F(z)} = \frac{\cos z + C(\cos z - z \sin z)}{\sin z + Cz \cos z} \tag{10.6.8}$$

and in order to use (10.1.4) one has to prove that

$$\lim_{k \to \infty} \oint_{C_k} \frac{P_n(z)}{Q_m(z)} \frac{\cos z + C(\cos z - z \sin z)}{\sin z + Cz \cos z} \, dz = 0, \tag{10.6.9}$$

where C_k is the system of paths shown in Fig 10.1. Since $m \geq n+2$, the proof of formula (10.6.9) is similar to the one used for the summation of S_3. We have

$$\begin{aligned}\left|\frac{F'(z)}{F(z)}\right|_{z\in C_k} &= \left|\frac{1+C(1-z\tan z)}{\tan z + Cz}\right|_{z\in C_k}\\ &\leq \left.\frac{|C|\,|\tan z| + (1+|C|)/|z|}{|C| - |\tan z|/|z|}\right|_{z\in C_k}\\ &< M_1 = \text{constant} > 0\end{aligned}$$

as $k \to \infty$ since

$$|\tan z|\Big|_{z\in C_k} < M = \text{constant} > 0, \qquad \text{for all } k.$$

Hence, substituting (10.6.8) into (10.1.4) and taking the fact into account that $z=0$ is also a pole of $F'(z)/F(z)$, S_7 is evaluated in closed form,

$$\begin{aligned}&\sum_{k=-\infty}^{\infty} \frac{P_n(\gamma_k)}{Q_m(\gamma_k)}\\ &= \left(-\operatorname*{Res}_{z=0} - \sum_k \operatorname*{Res}_{z=z_k}\right)\left[\frac{P_n(z)}{Q_m(z)}\,\frac{\cos z + C(\cos z - z\sin z)}{\sin z + Cz\cos z}\right],\end{aligned} \tag{10.6.10}$$

where $m \geq n+2$, $\tan\gamma_k + C\gamma_k = 0$ and $z_k \neq 0$.

EXAMPLE 10.6.1. *Sum the series*

$$\sum_{k=-\infty}^{\infty} \frac{1}{\gamma_k^2 + a^2}, \qquad a > 0, \tag{10.6.11}$$

where γ_k are the roots of the equation $\tan x = x$.

SOLUTION. We see that $C = -1$ in (10.6.3). Hence we have

$$\begin{aligned}\sum_{k=-\infty}^{\infty} \frac{1}{\gamma_k^2 + a^2} &= -\left(\operatorname*{Res}_{z=0} + \operatorname*{Res}_{z=ai} + \operatorname*{Res}_{z=-ai}\right)\frac{1}{z^2+a^2}\,\frac{z\sin z}{\sin z - z\cos z}\\ &= -\lim_{z\to 0}\frac{1}{z^2+a^2}\,\frac{z^2\left(z - \frac{z^3}{3!} + \dots\right)}{z - \frac{z^3}{3!} + \dots - z\left(1 - \frac{z^2}{2!} + \dots\right)}\\ &\quad - \frac{1}{2}\,\frac{\sin ai}{\sin ai - ai\cos ai} - \frac{1}{2}\,\frac{\sin ai}{\sin ai - ai\cos ai}\\ &= \frac{\sinh a}{a\cosh a - \sinh a} - \frac{3}{a^2}.\end{aligned}$$

This answer coincides with [**21**], Problem 30.09(2), if, changing the lower limit, in [**21**] one takes $\sum_{k=-\infty}^{\infty}$ instead of $\sum_{k=1}^{\infty}$. □

EXAMPLE 10.6.2. *Sum the series*

$$\sum_{k=-\infty}^{\infty} \frac{1}{\gamma_k^2}, \tag{10.6.12}$$

where $\tan\gamma_k = \gamma_k$, $\gamma_k \neq 0$.

SOLUTION. One can find the sum of the above series by letting $a \to 0$ in the formula of Example 10.6.1. However, the computation of the limit is not simple. Therefore we use formula (10.6.10) directly with $C = -1$,

$$\begin{aligned}
\sum_{k=-\infty}^{\infty} \frac{1}{\gamma_k^2} &= -\operatorname*{Res}_{z=0} \frac{1}{z^2} \frac{z\sin z}{\sin z - z\cos z} \\
&= -\operatorname*{Res}_{z=0} \frac{\sin z}{z(\sin z - z\cos z)} \\
&= -\frac{1}{2!}\lim_{z\to 0}\left(\frac{z^2\sin z}{\sin z - z\cos z}\right)'' \\
&= -\frac{1}{2}\left[\frac{z^2\left(z - \frac{z^3}{3!} + \frac{z^5}{5!} - \dots\right)}{z - \frac{z^3}{3!} + \frac{z^5}{5!} - \dots - z\left(1 - \frac{z^2}{2!} + \frac{z^4}{4!} - \dots\right)}\right]''\Bigg|_{z=0} \\
&= -\frac{3}{2}\left[\frac{1 - \frac{z^2}{3!} + \frac{z^4}{5!} - \dots}{1 - \frac{z^2}{10} + \dots}\right]''\Bigg|_{z=0} \\
&= -\frac{3}{2}\frac{d^2 f(z)}{dz^2}\bigg|_{z=0},
\end{aligned} \tag{10.6.13}$$

where

$$f(z) = \frac{1 - \frac{z^2}{3!} + \frac{z^4}{5!} - \dots}{1 - \frac{z^2}{10} + \dots} = 1 + a_2 z^2 + \dots, \tag{10.6.14}$$

and $a_2 = f''(0)/2$. It follows from (10.6.14) that

$$1 - \frac{z^2}{6} + \frac{z^4}{120} - \dots = \left(1 + a_2 z^2 + \dots\right)\left(1 - \frac{z^2}{10} + \dots\right). \tag{10.6.15}$$

Equating the coefficients of z^2 in (10.6.15), we have

$$-\frac{1}{6} = a_2 - \frac{1}{10}, \qquad \text{that is,} \qquad a_2 = -\frac{1}{15} = \frac{f''(0)}{2}. \tag{10.6.16}$$

Substituting (10.6.16) into (10.6.13) we obtain

$$\sum_{k=-\infty}^{\infty} \frac{1}{\gamma_k^2} = \frac{1}{5}. \tag{10.6.17}$$

Since $\tan z$ is an odd function, one sees that $\gamma_{-k} = -\gamma_k < 0$ is a root of $\tan z = z$ if γ_k is a root. Hence (10.6.17) reduces to

$$\sum_{k=1}^{\infty} \frac{1}{\gamma_k^2} = \frac{1}{10}, \qquad \gamma_k > 0. \quad \square \tag{10.6.18}$$

Note 10.6.1. It is interesting to compare the sum (10.6.18) with the sum of the asymptotic values of the roots $\gamma_k > 0$ evaluated graphically (γ_k are the abscissas of the points of intersection of the curves $y = \tan x$ and $y = x$; see Fig 10.2).

It follows from the graph that

$$\lim_{k \to \infty} (\gamma_{k+1} - \gamma_k) = \pi$$

and

$$\gamma_1 \approx 4.49 \approx 3\pi/2, \qquad \gamma_2 \approx 7.73 \approx 5\pi/2, \ldots, \qquad \gamma_k \approx (2k+1)\pi/2, \ldots.$$

Hence,

$$\begin{aligned} \sum_{k=1}^{\infty} \frac{1}{\gamma_k^2} &\approx \sum_{k=1}^{\infty} \frac{1}{(2k+1)^2\pi^2/4} \\ &= \frac{4}{\pi^2}\left[\sum_{k=0}^{\infty} \frac{1}{(2k+1)^2} - 1\right]. \end{aligned}$$

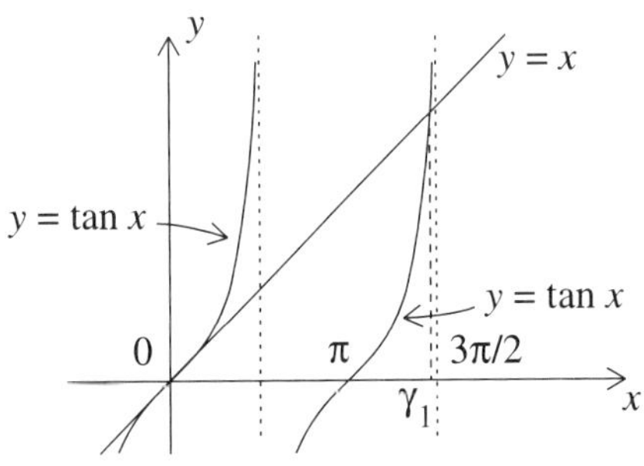

FIGURE 10.2. Positive roots of $\tan x = x$.

The last series can easily be computed by means of (10.4.32) if $P_n(z) = 1$, $Q_m(z) = z^2$ and $b = 0$. In this case

$$\begin{aligned}
\sum_{k=0}^{\infty} \frac{1}{(2k+1)^2} &= \frac{\pi}{4}\Re\left[i \operatorname*{Res}_{z=0} \frac{1}{z^2}\, \frac{e^{-i\pi z/2}}{\cos(\pi z/2)}\right] \\
&= \frac{\pi}{4}\Re \lim_{z\to 0}\left[i\, \frac{e^{-i\pi z/2}}{\cos(\pi z/2)}\right]' \\
&= \frac{\pi^2}{8}\Re\left[i \lim_{z\to 0} \frac{-i\, e^{-i\pi z/2}\cos(\pi z/2) + \sin(\pi z/2)\, e^{-i\pi z/2}}{\cos^2(\pi z/2)}\right] \\
&= \frac{\pi^2}{8}.
\end{aligned}$$

This coincides with Formula 5.1.4(1) in [**38**], p. 653. Then

$$\begin{aligned}
\sum_{k=1}^{\infty} \frac{1}{\gamma_k^2} &\approx \frac{4}{\pi^2}\left[\frac{\pi^2}{8} - 1\right] \\
&= \frac{1}{2} - \frac{4}{\pi^2} = 0.0947\dots,
\end{aligned}$$

which differs by 5% from the exact answer 0.1. This approach may be useful for approximating the values of series which cannot be evaluated in closed form provided the series formed by the asymptotics of the nth terms of the series can be computed exactly.

EXAMPLE 10.6.3. *Sum the series*

$$\sum_{k=-\infty}^{\infty} \frac{1}{\gamma_k^4}, \tag{10.6.19}$$

where $\tan\gamma_k = \gamma_k$, $\gamma_k \neq 0$.

SOLUTION. Using formula (10.6.10) with $C = -1$, we have

$$\begin{aligned}
\sum_{k=-\infty}^{\infty} \frac{1}{\gamma_k^4} &= -\operatorname*{Res}_{z=0} \frac{1}{z^4}\, \frac{z\sin z}{\sin z - z\cos z} \\
&= -\frac{1}{4!}\lim_{z\to 0}\left(\frac{z^2\sin z}{\sin z - z\cos z}\right)^{(4)} \\
&= -\frac{1}{4!}\lim_{z\to 0}\left[\frac{z^2\left(z - \frac{z^3}{3!} + \frac{z^5}{5!} - \dots\right)}{z - \frac{z^3}{3!} + \frac{z^5}{5!} - \dots - z\left(1 - \frac{z^2}{2!} + \frac{z^4}{4!} - \dots\right)}\right]^{(4)}
\end{aligned}$$

$$= -\frac{1}{4!}\lim_{z\to 0}\left[\frac{z^3\left(1-\frac{z^2}{3!}+\frac{z^4}{5!}-\dots\right)}{\frac{z^3}{3}-\frac{z^5}{30}+\frac{z^7}{840}-\dots}\right]^{(4)}$$

$$= -\frac{1}{4!}\lim_{z\to 0}\left[\frac{1-\frac{z^2}{6}+\frac{z^4}{120}-\dots}{\frac{1}{3}-\frac{z^2}{30}+\frac{z^4}{840}-\dots}\right]^{(4)}$$

$$= -\frac{1}{4!}\left.\frac{d^4 f(z)}{dz^4}\right|_{z=0}, \tag{10.6.20}$$

where

$$f(z) = \frac{1-\frac{z^2}{6}+\frac{z^4}{120}-\dots}{\frac{1}{3}-\frac{z^2}{30}+\frac{z^4}{840}-\dots} = 3 + a_2 z^2 + a_4 z^4 + \dots \tag{10.6.21}$$

and

$$a_2 = \frac{1}{2!}f''(0), \qquad a_4 = \frac{1}{4!}f^{(4)}(0). \tag{10.6.22}$$

It follows from (10.6.21) that

$$1 - \frac{z^2}{6} + \frac{z^4}{120} - \dots$$
$$= \left(3 + a_2 z^2 + a_4 z^4 + \dots\right)\left(\frac{1}{3} - \frac{z^2}{30} + \frac{z^4}{840} - \dots\right). \tag{10.6.23}$$

In (10.6.23), equating the coefficients of z^2 and z^4, respectively, for a_2 and a_4 we have

$$\begin{aligned} -\frac{1}{6} &= -\frac{1}{10} + \frac{a_2}{3}, &&\text{that is,}\quad a_2 = -\frac{1}{5},\\ \frac{1}{120} &= \frac{3}{840} - \frac{a_2}{30} + \frac{a_4}{3}, &&\text{that is,}\quad a_4 = -\frac{1}{175}. \end{aligned} \tag{10.6.24}$$

Substituting (10.6.22) and (10.6.24) into (10.6.20) we obtain

$$\sum_{k=-\infty}^{\infty} \frac{1}{\gamma_k^4} = \frac{1}{175},$$

that is,

$$\sum_{k=1}^{\infty} \frac{1}{\gamma_k^4} = \frac{1}{350} = 0.002857\dots. \tag{10.6.25}$$

The approximate sum

$$\begin{aligned}\sum_{k=1}^{\infty} \frac{1}{\gamma_k^4} &\approx \frac{16}{\pi^4} \sum_{k=1}^{\infty} \frac{1}{(2k+1)^4} \\ &= \frac{16}{\pi^4}\left[\sum_{k=0}^{\infty} \frac{1}{(2k+1)^4} - 1\right] \\ &= \frac{16}{\pi^4}\left(\frac{\pi^4}{96} - 1\right) \\ &= 0.0024109\ldots \end{aligned} \tag{10.6.26}$$

is derived by means of Formula 5.1.4(1) on p. 653 in [**38**],

$$\sum_{k=0}^{\infty} \frac{1}{(2k+1)^4} = \frac{\pi^4}{96}.$$

This result can also be easily obtained by means of formula (10.6.10). The difference between (10.6.26) and (10.6.25) is about 15%, and it reduces almost to zero if one takes the sum of the first four terms of the following series (see, for example, Table 5 on p. 757 in [**14**]):

$$\begin{aligned}\sum_{k=1}^{\infty} \frac{1}{\gamma_k^4} &\approx \sum_{k=1}^{4} \frac{1}{\gamma_k^4} \\ &= \frac{1}{(4.4943)^4} + \frac{1}{(7.7253)^4} + \frac{1}{(10.9410)^4} + \frac{1}{(14.0662)^4} \\ &= 0.0028327\ldots.\end{aligned}$$

This result differs from the exact value 0.002857 only by 1%. This difference can be further decreased if we add the sum of the terms of the asymptotics for the nth term, starting with $n = 5$:

$$\begin{aligned}\sum_{k=1}^{\infty} \frac{1}{\gamma_k^4} &= 0.002827 + \frac{16}{\pi^4} \sum_{k=5}^{\infty} \frac{1}{(2k+1)^4} \\ &= 0.002827 + \frac{16}{\pi^4}\left(\frac{\pi^4}{96} - 1 - \frac{1}{3^4} - \frac{1}{5^4} - \frac{1}{7^4} - \frac{1}{9^4}\right) \\ &= 0.0028539. \quad \square\end{aligned}$$

To evaluate the series

$$S_8 = \sum_{k=-\infty}^{\infty} \frac{P_n(\gamma_k)}{Q_m(\gamma_k)}\, e^{ia\gamma_k}, \tag{10.6.27}$$

where γ_k are the poles of $F'(z)/F(z)$,

$$F(z) = \sin z + Cz\cos z, \qquad C \geq -1, \tag{10.6.28}$$

$|a| \leq 1$ if $m \geq n+2$, and $|a| < 1$ if $m = n+1$, we replace $F'(z)/F(z)$ in (10.1.3) and (10.1.4) with

$$\frac{F'(z)}{F(z)}\, e^{iaz}.$$

Since the functions $|\sin z|$ and $|\cos z|$ grow exponentially, condition (10.1.3) is satisfied for the system of paths shown in Fig 10.1. Hence, one can use formula (10.1.4),

$$\sum_{k=-\infty}^{\infty} \frac{P_n(\gamma_k)}{Q_m(\gamma_k)}\, e^{ia\gamma_k} = -\sum_k \operatorname*{Res}_{z=z_k} \left[\frac{P_n(z)}{Q_m(z)}\, \frac{e^{iaz}F'(z)}{F(z)}\right], \tag{10.6.29}$$

where $|a| \leq 1$ if $m \geq n+2$ and $|a| < 1$ if $m = n+1$, and z_k are the zeros of $Q_m(z)$. Separating the real and imaginary parts of (10.6.29), we obtain

$$\sum_{k=-\infty}^{\infty} \frac{P_n(\gamma_k)}{Q_m(\gamma_k)} \cos a\gamma_k = -\Re \sum_k \operatorname*{Res}_{z=z_k} \left[\frac{P_n(z)}{Q_m(z)}\, \frac{e^{iaz}F'(z)}{F(z)}\right] \tag{10.6.30}$$

and

$$\sum_{k=-\infty}^{\infty} \frac{P_n(\gamma_k)}{Q_m(\gamma_k)} \sin a\gamma_k = -\Im \sum_k \operatorname*{Res}_{z=z_k} \left[\frac{P_n(z)}{Q_m(z)}\, \frac{e^{iaz}F'(z)}{F(z)}\right], \tag{10.6.31}$$

with the same conditions as in (10.6.29).

Case 2. Let γ_k be the zeros of the Bessel function of the first kind of order ν:

$$J_\nu(\gamma_k) = 0, \qquad \gamma_k \neq 0,$$

and suppose that $F(z) = J_\nu(z)$ in (10.1.3) and (10.1.4). Since

$$J_\nu(z) \sim \sqrt{\frac{2}{\pi z}} \left[\cos\left(z - \frac{\nu\pi}{2} - \frac{\pi}{4}\right) + O\left(\frac{1}{z}\right)\right]$$

as $z \to \infty$ and $m \geq n+2$, condition (10.1.3) is satisfied and one can use formula (10.1.4),

$$\sum_{k=-\infty}^{\infty} \frac{P_n(\gamma_k)}{Q_m(\gamma_k)} = -\left(\operatorname*{Res}_{z=0} + \sum_k \operatorname*{Res}_{z=z_k}\right) \left[\frac{P_n(z)}{Q_m(z)}\, \frac{J'_\nu(z)}{J_\nu(z)}\right]. \tag{10.6.32}$$

Note that the function $J_\nu'(z)/J_\nu(z)$ has no branch points since

$$\begin{aligned}\frac{J_\nu'(z)}{J_\nu(z)} &= \sum_{k=0}^{\infty}(-1)^k \frac{(k+\nu/2)(z/2)^{2k+\nu-1}}{k!\Gamma(k+\nu+1)} \Big/ \sum_{k=0}^{\infty}(-1)^k \frac{(z/2)^{2k+\nu}}{k!\Gamma(k+\nu+1)} \\ &= \sum_{k=0}^{\infty}(-1)^k \frac{(k+\nu/2)(z/2)^{2k-1}}{k!\Gamma(k+\nu+1)} \Big/ \sum_{k=0}^{\infty}(-1)^k \frac{(z/2)^{2k}}{k!\Gamma(k+\nu+1)}\end{aligned}$$

for $|z| < \infty$, that is, $J_\nu'(z)/J_\nu(z)$ is a ratio of two entire functions.

EXAMPLE 10.6.4. *Show that*

$$\sum_{k=-\infty}^{\infty} \frac{1}{\beta_k^2} = \frac{1}{6}, \tag{10.6.33}$$

where $J_2(\beta_k) = 0$.

Formula (10.6.33) occurs in hydrodynamical problems (see [**46**], p. 245, Formula 6.52).

SOLUTION. By formula (10.6.32), we have

$$\begin{aligned}\sum_{k=-\infty}^{\infty} \frac{1}{\beta_k^2} &= -\operatorname*{Res}_{z=0}\left[\frac{1}{z^2}\frac{J_2'(z)}{J_2(z)}\right] \\ &= -\frac{1}{2!}\lim_{z\to 0}\left[z\sum_{k=0}^{\infty}(-1)^k \frac{(k+1)(z/2)^{2k+1}}{k!(k+2)!} \Big/ \sum_{k=0}^{\infty}(-1)^k \frac{(z/2)^{2k+2}}{k!(k+2)!}\right]'' \\ &= -\frac{1}{2!}\lim_{z\to 0}\left[\frac{1-\frac{4}{3!}\left(\frac{z}{2}\right)^2+\dots}{\frac{1}{2}-\frac{1}{3!}\left(\frac{z}{2}\right)^2+\dots}\right]'' \\ &= -\frac{1}{2}f''(0),\end{aligned} \tag{10.6.34}$$

where

$$\begin{aligned}f(z) &= \frac{1-\frac{4}{3!}\left(\frac{z}{2}\right)^2+\dots}{\frac{1}{2}-\frac{1}{3!}\left(\frac{z}{2}\right)^2+\dots} \\ &= 2 + a_2z^2 + \dots,\end{aligned} \tag{10.6.35}$$

and $a_2 = f''(0)/2$. It follows from (10.6.35) that

$$1-\frac{4}{3!}\left(\frac{z}{2}\right)^2+\dots = \left[\frac{1}{2}-\frac{1}{3!}\left(\frac{z}{2}\right)^2+\dots\right]\left[2+a_2z^2+\dots\right]. \tag{10.6.36}$$

Equating the coefficients of z^2 in (10.6.36) we obtain

$$-\frac{4}{6}\cdot\frac{1}{4}=\frac{a_2}{2}-\frac{2}{3!}\cdot\frac{1}{4}, \qquad \text{that is,} \qquad a_2=-\frac{1}{6}.$$

Then from (10.6.34) we obtain

$$\sum_{k=-\infty}^{\infty}\frac{1}{\beta_k^2}=\frac{1}{6}, \qquad \text{or} \qquad \sum_{k=1}^{\infty}\frac{1}{\beta_k^2}=\frac{1}{12}. \quad \square$$

EXAMPLE 10.6.5. *Sum the series*

$$\sum_{k=-\infty}^{\infty}\frac{1}{1+\gamma_k^2}, \tag{10.6.37}$$

where $J_0(\gamma_k)=0$.

SOLUTION. By formula (10.6.32), we have

$$\begin{aligned}\sum_{k=-\infty}^{\infty}\frac{1}{1+\gamma_k^2} &= -\left(\operatorname*{Res}_{z=i}+\operatorname*{Res}_{z=-i}\right)\left[\frac{1}{1+z^2}\,\frac{J_0'(z)}{J_0(z)}\right]\\ &=\frac{1}{2i}\,\frac{J_1(i)}{J_0(i)}-\frac{1}{2i}\,\frac{J_1(-i)}{J_0(-i)}\\ &=\frac{1}{i}\,\frac{J_1(i)}{J_0(i)}=\frac{I_1(1)}{I_0(1)},\end{aligned}$$

where

$$I_l(z)=\sum_{k=0}^{\infty}\frac{(z/2)^{2k+l}}{k!\Gamma(k+l+1)}, \qquad l=0,1,$$

is the modified Bessel function of the first kind of order l. Hence,

$$\sum_{k=-\infty}^{\infty}\frac{1}{1+\gamma_k^2}=\frac{I_1(1)}{I_0(1)}, \qquad J_0(\gamma_k)=0. \quad \square$$

EXAMPLE 10.6.6. *Sum the series*

$$\sum_{k=-\infty}^{\infty}\frac{1}{\gamma_k^4}, \qquad J_0(\gamma_k)=0. \tag{10.6.38}$$

SOLUTION. By (10.6.32) we obtain

$$\begin{aligned}\sum_{k=-\infty}^{\infty} \frac{1}{\gamma_k^4} &= -\operatorname*{Res}_{z=0}\left[\frac{J_0'(z)}{z^4 J_0(z)}\right] \\ &= \operatorname*{Res}_{z=0} \frac{J_1(z)}{z^4 J_0(z)} \\ &= \frac{1}{2}\lim_{z\to 0}\left[\frac{J_1(z)}{zJ_0(z)}\right]'' \\ &= \frac{1}{2}\lim_{z\to 0}\left[\sum_{k=0}^{\infty}(-1)^k\frac{(z/2)^{2k+1}}{k!(k+1)!}\Big/ z\sum_{k=0}^{\infty}(-1)^k\frac{(z/2)^{2k}}{(k!)^2}\right]'' \\ &= \frac{1}{4}\lim_{z\to 0}\left(\frac{1-\frac{z^2}{8}+\dots}{1-\frac{z^2}{4}+\dots}\right)'' \\ &= \frac{1}{4}f''(0), \end{aligned} \tag{10.6.39}$$

where

$$\begin{aligned} f(z) &= \frac{1-\frac{z^2}{8}+\dots}{1-\frac{z^2}{4}+\dots} \\ &= 1 + a_2 z^2 + \dots, \end{aligned} \tag{10.6.40}$$

and $a_2 = f''(0)/2$. It then follows from (10.6.40) that

$$1 - \frac{z^2}{8} + \dots = \left(1 + a_2 z^2 + \dots\right)\left(1 - \frac{z^2}{4} + \dots\right).$$

Equating the coefficients of z^2, we obtain

$$-\frac{1}{8} = a_2 - \frac{1}{4}, \qquad \text{that is,} \qquad a_2 = \frac{1}{8}, \quad f''(0) = \frac{1}{4}.$$

Thus,

$$\sum_{k=-\infty}^{\infty} \frac{1}{\gamma_k^4} = \frac{1}{16} \quad \text{and} \quad \sum_{k=1}^{\infty} \frac{1}{\gamma_k^4} = \frac{1}{32},$$

where $J_0(\gamma_k) = 0$. □

10.7. Series involving complex zeros of entire functions

We consider series of the form

$$S_9 = \sum_{k=-\infty}^{\infty} f(\gamma_k), \tag{10.7.1}$$

where $f(z) = P_n(z)/Q_m(z)$, $P_n(z)$ and $Q_m(z)$ are polynomials of degrees n and m, respectively, $m \geq n+2$, and γ_k are the complex roots of the equations

$$\sinh z + z = 0, \tag{10.7.2}$$

$$\sinh z - z = 0. \tag{10.7.3}$$

The roots of (10.7.2) and (10.7.3) appear in the solution of the biharmonic equation

$$\Delta\Delta u = 0, \tag{10.7.4}$$

where

$$\Delta = \frac{\partial^2}{\partial x^2} + \frac{\partial^2}{\partial y^2} \tag{10.7.5}$$

is the Laplacian, for the case of an infinite strip in elasticity problems (see, for example, [**43**], pp. 330–333, and [**48**], pp. 26–37). Complex roots of $z \tan z = c$ occur in dielectric spectroscopy (see [**22**]). A quasi-global selective method of solution of elementary transcendental equations based on the iteration theory of Fatou and Julia can be found in [**16**] and [**39**] and will be covered in Chapter 11.

The roots of equations (10.7.2) and (10.7.3), except $z = 0$, are complex. It is easily seen that if $\gamma_k = a_k + ib_k$ is a root of one of these equations, then $a_k - ib_k$, $-a_k + ib_k$, $-a_k - ib_k$ are also roots. Thus, the latter roots are located at the vertices $\pm a_k \pm ib_k$ of rectangles in the complex plane. Let, for example, $z = a + bi$ be a root of (10.7.3), that is,

$$\sinh(a + bi) = a + bi.$$

Separating the real and imaginary parts in the above equation we obtain

$$\sinh a \cos b = a \quad \text{and} \quad \cosh a \sin b = b.$$

Since these equations do not change form if a is replaced by $-a$ or b by $-b$, then $-a \pm bi$ and $a - bi$ are also roots of these equations.

To sum the series (10.7.1) in closed form, we assume that the function $F(z)$ in (10.1.3) is of the form

$$F(z) = \sinh z \pm z. \tag{10.7.6}$$

Then

$$\frac{F'(z)}{F(z)} = \frac{\cosh z \pm 1}{\sinh z \pm z}, \tag{10.7.7}$$

and for the validity of (10.1.4) one has to prove that

$$\lim_{k\to\infty} \oint_{C_k} \frac{P_n(z)}{Q_m(z)} \frac{\cosh z \pm 1}{\sinh z \pm z}\, dz = 0, \tag{10.7.8}$$

where C_k is the system of paths shown in Fig 10.1. Since $m \geq n+2$, the proof of (10.7.8) is similar to the one for S_3. Hence, substituting (10.7.7) into (10.1.4), we obtain

$$\sum_{k=-\infty}^{\infty} \frac{P_n(\gamma_k)}{Q_m(\gamma_k)} = -\left(\operatorname*{Res}_{z=0} + \sum_k \operatorname*{Res}_{z=z_k}\right) \left[\frac{P_n(z)}{Q_m(z)} \frac{\cosh z \pm 1}{\sinh z \pm z}\right], \tag{10.7.9}$$

where $m \geq n+2$ and $\sinh \gamma_k \pm \gamma_k = 0$.

EXAMPLE 10.7.1. *Sum the series*

$$\sum_{k=-\infty}^{\infty} \frac{1}{\gamma_k^2}, \tag{10.7.10}$$

where γ_k are roots of

$$\sinh z + z = 0.$$

SOLUTION. Since $P_n(z)/Q_m(z) = 1/z^2$, it follows from (10.7.9) that

$$\begin{aligned}
\sum_{k=-\infty}^{\infty} \frac{1}{\gamma_k^2} &= -\operatorname*{Res}_{z=0} \left[\frac{\cosh z + 1}{z^2(\sinh z + z)}\right] \\
&= \frac{1}{2} \lim_{z\to 0} \left[\frac{z(1+\cosh z)}{\sinh z + z}\right]'' \\
&= \frac{1}{2} \lim_{z\to 0} \left[\frac{z\left(2 + \frac{z^2}{2} + \dots\right)}{2z + \frac{z^3}{6} + \dots}\right]'' \\
&= \frac{1}{2} \lim_{z\to 0} \left(\frac{2 + \frac{z^2}{2} + \dots}{2 + \frac{z^2}{6} + \dots}\right)'' \\
&= \frac{1}{2} \lim_{z\to 0} \left[\frac{\frac{4}{3}z + \dots}{\left(2 + \frac{z^2}{6} + \dots\right)^2}\right]' \\
&= \frac{1}{2} \cdot \frac{4 \cdot 2^2}{3 \cdot 2^4} = \frac{1}{6}.
\end{aligned}$$

If $\gamma_k = a_k + ib_k$ is a root of the equation $\sinh z + z = 0$, then $-a_k \pm ib_k$ and $a_k - ib_k$ are also roots of the same equation. Hence,

$$\sum_{k=-\infty}^{\infty} \left[\frac{1}{(a_k + ib_k)^2} + \frac{1}{(a_k - ib_k)^2} + \frac{1}{(-a_k - ib_k)^2} + \frac{1}{(-a_k + ib_k)^2} \right] = \frac{1}{6},$$

that is,

$$4\Re \left[\sum_{k=-\infty}^{\infty} \frac{1}{(a_k + ib_k)^2} \right] = \frac{1}{6}.$$

Finally,

$$\sum_{k=-\infty}^{\infty} \frac{a_k^2 - b_k^2}{(a_k^2 + b_k^2)^2} = \frac{1}{24},$$

where $\gamma_k = a_k + ib_k$, $a_k > 0$, $b_k > 0$, and $\sinh \gamma_k + \gamma_k = 0$. □

Exercises for Chapter 10

Evaluate the following series.

1. $\displaystyle\sum_1^\infty \frac{1}{k^4 + a^4}$.

2. $\displaystyle\sum_1^\infty \frac{1}{k^4}$.

3. $\displaystyle\sum_1^\infty \frac{(-1)^k}{k^2}$.

4. $\displaystyle\sum_1^\infty \frac{(-1)^k}{k^4 + a^4}$.

5. $\displaystyle\sum_1^\infty \frac{\sin kx}{k^3}$, $\quad 0 \le x \le 2\pi$.

6. $\displaystyle\sum_1^\infty \frac{\cos kx}{k^4}$, $\quad 0 \le x \le 2\pi$.

7. $\displaystyle\sum_{-\infty}^\infty \frac{1}{\gamma_k^6}$, $\quad$ where $\tan \gamma_k = \gamma_k$, $\gamma_k \ne 0$.

8. $\displaystyle\sum_{-\infty}^\infty \frac{1}{\alpha_k^2}$, $\quad$ where $J_1(\alpha_k) = 0$, $k \ne 0$.

CHAPTER 11

Numerical Solutions of Transcendental Equations

11.1. Introduction

In interactive or automatic scientific computation, one looks for adapted methods to solve specific problems that occur in the applications. In this chapter, which follows [**16**], [**39**] and the references therein, we present a combination of global and local iterative methods to find selective roots of elementary transcendental equations,

$$F(z,c) - z = 0, \qquad c \in \mathbb{C}.$$

Such equations occur in two-point boundary value problems, which could be called complex Sturm–Liouville eigenvalue problems, after separation of variables in initial-boundary value problems in physics and engineering. Examples of such equations are found in dielectric spectroscopy, scattering problems for metallic grooves, and orbit determination.

It will be shown that the iteration functions in question,

$$z_{n+1} = F(z_n, c),$$

have very few attractive fixed points, $z = F(z,c)$, and very few critical values. Geometric considerations will identify bounded regions of the plane which contain the attractive fixed points and the critical values of F. Moreover, the attractive fixed points of all but a few branches of the inverse, F^{-1}, of F have relatively large basins of attraction. An application of the Fatou–Julia iteration theory for entire and meromorphic functions will ensure convergence to the specified roots, while attempting to avoid attractive cycles. In the presence of slow convergence near multiple zeros, Steffensen's procedure or an interpolation scheme will accelerate convergence. In cases where the specified roots are known to lie in convex regions, good starting values can be supplied for an efficient use of a fast convergent local method, such as Newton's method.

In Section 11.2, transcendental equations are derived from the boundary conditions of some Sturm–Liouville problems in the complex plane. Section 11.3 presents basic concepts of the Fatou–Julia theory which will be used here. Section 11.4 discusses drawbacks of local methods, such as Newton's method, in the context at hand. Section 11.5 presents almost global iteration schemes. In Section 11.6, Newton's method is used effectively to find roots in some cases. An interpolation procedure is described in Section 11.7. Application to Kepler's equation is done in Section 11.8. Finally, Section 11.9 presents a programming strategy.

11.2. Complex Sturm–Liouville problems

Several boundary conditions for the two second-order ordinary differential equations

$$y'' = \mp\lambda^2 y, \qquad a \le x \le b,$$

will be seen to lead to elementary transcendental equations.

First, the differential equation

$$y'' = -\lambda^2 y, \qquad a \le x \le b, \tag{11.2.1}$$

admits the general solution

$$y(x) = \alpha\cos\lambda x + \beta\sin\lambda x,$$

whose derivative is

$$y'(x) = -\alpha\lambda\sin\lambda x + \beta\lambda\cos\lambda x.$$

The boundary conditions

$$y(a) = y'(a); \quad y'(b) = 0,$$

may be written as the linear homogeneous linear system

$$\alpha(\cos\lambda a + \lambda\sin\lambda a) + \beta(\sin\lambda a - \lambda\cos\lambda a) = 0,$$

$$-\alpha\lambda\sin\lambda b + \beta\lambda\cos\lambda b = 0,$$

and so the associated boundary-value problem has a nontrivial solution if and only if the corresponding determinant vanishes,

$$\cos\lambda b(\cos\lambda a + \lambda\sin\lambda a) + \sin\lambda b(\sin\lambda a - \lambda\cos\lambda a) = 0,$$

that is,

$$\cos\lambda(b-a) = \lambda\sin\lambda(b-a)$$

or

$$\lambda = \cot\lambda(b-a),$$

Diff. eqs.	Boundary conditions	Transc. eqs.
$y'' = -\lambda^2 y$	$y(a) + y'(a) = 0,\ \ y'(b) - 0$	$\cot z = \mp z/c$
	$y(a) \pm y'(a) = 0,\ \ y(b) = 0$	$\tan z = \pm z/c$
	$y(a) = 0,\ \ y'(b) = 0$	$\cos z = 0$
	$y(a) \pm y'(b) = 0,\ \ y(b) = 0$	$\sin z = \pm z/c$
	$y(a) \pm \lambda y(b) = 0,\ \ y'(b) = 0$	$\cos z = \mp z/c$
$y'' = \lambda^2 y$	$y(a) \pm y'(a) = 0,\ \ y(b) = 0$	$\tanh z = \pm z/c$
	$y(a) \pm y'(b) = 0,\ \ y(b) = 0$	$\sinh z = \pm z/c$
	$y(a) \pm y'(b) = 0,\ \ \lambda y(b) + y'(b) = 0$	$e^z = \pm z/c$

TABLE 11.1. The table lists some boundary conditions and their corresponding transcendental equations for equations (11.2.1) and (11.2.2). Here $z = \lambda(b-a)$ and $c = b - a$.

so that, with $\lambda(b-a) = z$ and $b - a = c$, one obtains the transcendental equation

$$\cot z = \frac{z}{c}.$$

Other boundary conditions for equation (11.2.1) and the equivalent transcendental equations are listed in Table 11.1.

Secondly, the differential equation

$$y'' = \lambda^2 y, \qquad a \le x \le b, \tag{11.2.2}$$

admits the general solution

$$y(x) = \alpha\, e^{\lambda x} + \beta\, e^{-\lambda x}$$

whose derivative is

$$y'(x) = \alpha\lambda\, e^{\lambda x} - \beta\lambda\, e^{-\lambda x}.$$

Again, boundary conditions for equation (11.2.2) and the equivalent transcendental equations are listed in Table 11.1.

By introducing the complex variable z and complex parameters λ, a, b and c, the distinction between equations (11.2.1) and (11.2.2) disappears. After some transformations like $\tanh z = -i \tan iz$ and $c \mapsto 1/c$, the transcendental equations contained in Table 11.1 become

$$z = c\cot z, \quad z = c\tan z, \quad z = \frac{1}{c}\cos z, \quad z = \frac{1}{c}\sin z, \tag{11.2.3}$$

referred to as the four trigonometric equations in this chapter, and the exponential equation

$$z = \frac{1}{c}\, e^{z}. \tag{11.2.4}$$

These trigonometric and exponential transcendental equations have infinitely many roots, except possibly for at most two values of c, as follows from an extension to Picard's theorem [**36**], p. 75. The problem considered in this chapter is to find any specified roots of these equations.

In the following subsections, we present two examples of transcendental equations found in the applications.

11.2.1. Dielectric spectroscopy. We give an example of transcendental equations that are found in dielectric spectroscopy [**22**].

Coaxial transmission lines have been used as sample cells in dielectric measurements for many years. Reflection measurements with the sample terminating in an open circuit lead to solving the *permittivity equation*

$$z \tan z = c \tag{11.2.5}$$

for the unknown normalized propagation constant, $i\bar{z}$, for a set of experimentally obtained values of the complex normalized admittance, $-\bar{c}$. Similarly, a short circuit termination leads to the permittivity equation

$$z \cot z = c. \tag{11.2.6}$$

The length of the sample, its position and the impedance terminating the line were chosen, in the past, to provide the best accuracy at each frequency being used. However, over the past 25 years, commercially available automated network analyzers have been able to measure impedance over an increasing range of frequencies. For optimal use of this instrumentation, it is not practical to adjust the length of the sample or the termination to obtain the best performance at each frequency. Thus (11.2.5) and (11.2.6) are to be solved over a wide range of values of c, for some of which the roots z may come close to double roots of these equations.

11.2.2. Scattering problem for a metallic groove. The solution of the scattering problem for a groove in a metallic plane by the modal method leads to transcendental equations [**39**].

Modal methods are widely used to solve electromagnetic scattering problems for rough surfaces. These methods consist in expanding the electric and magnetic fields inside each groove in eigenfunctions that satisfy the boundary conditions. They are useful in providing explicit analytical representations of the fields inside the asperities of the surface. They also give a simple way of understanding the physical interpretation of the results.

For infinite gratings of simple geometries (rectangular, semicircular, etc.), the eigenfunctions are known simple functions. But, in the general case of a groove with arbitrary profile on a surface made of an isotropic material (dielectrics, metals, etc.), to find the modal functions is a very complicated process, making the use of the modal method inconvenient in such cases. However, an arbitrary profile can be approximated by layers of rectangular shape. In each layer the fields can be expanded in modal functions corresponding to a rectangular groove, these functions being combinations of sines and cosines. Then, the problem can be solved by matching the fields at the interfaces.

To solve the scattering problem for a metallic surface with a groove of arbitrary shape, the first step consists in finding the modal eigenfunctions of a rectangular groove in the metallic surface. This calculation leads to transcendental equations that must be solved numerically. One has to find the roots of the complex-valued transcendental equations appearing in the calculation of the modal functions of a rectangular groove in a metallic surface. These equations can be reduced to the transcendental equations

$$\cos z = cz \quad \text{and} \quad \sin z = cz, \tag{11.2.7}$$

where $z, c \in \mathbb{C}$.

11.3. Fatou–Julia iteration theory

A few results from the Fatou–Julia global iteration theory, as extended to the iteration of meromorphic functions [**26**] and their inverses, will now be listed. A general presentation of the iteration theory for rational functions, notation and references are found in the survey [**10**]. The complex plane and the extended complex plane, or Riemann sphere, will be denoted by $\mathbb{C}$ and $\overline{\mathbb{C}} = \mathbb{C} \cup \infty$, respectively.

DEFINITION 11.3.1. Let

$$\varphi : \mathbb{C} \to \overline{\mathbb{C}}$$

be a transcendental meromorphic function and consider the iteration

$$z_{n+1} = \varphi(z_n), \qquad n = 0, 1, 2, \dots. \tag{11.3.1}$$

A *fixed point* s of φ, $s = \varphi(s)$, is *attractive*, *repulsive* or *indifferent* as the absolute value of its *multiplier*, $\varphi'(s)$, satisfies $|\varphi'(s)| < 1$, > 1 or $= 1$, respectively.

The inverse, φ^{-1}, of the function φ may have two kinds of finite singularities or critical points, namely *algebraic critical points*, which are the zeros of $\varphi'(z)$, and *transcendental critical points*, which are the finite exceptional or asymptotic values of φ. The image, $\varphi(z)$, of a critical point, z, will be called, for short, a *critical value* of φ.

DEFINITION 11.3.2. Let

$$\varphi^n(z) = \varphi[\varphi^{n-1}(z)], \qquad \varphi^0(z) = z, \tag{11.3.2}$$

denote the nth iterate of z by φ. The *Julia set* of φ, $\mathcal{J}(\varphi)$, is the set of nonnormality of φ:

$$\mathcal{J}(\varphi) := \{z; \{\varphi^n(z)\}_{n=1}^{\infty} \text{ is not a normal family}\}. \tag{11.3.3}$$

The *Fatou set* or *set of normality* of φ, $\mathcal{F}(\varphi)$, is defined in a similar way:

$$\mathcal{F}(\varphi) := \{z; \{\varphi^n(z)\}_{n=1}^{\infty} \text{ is a normal family}\}. \tag{11.3.4}$$

The *Rådtröm set* of φ, $\mathcal{R}(\varphi)$, is the set of predecessors of the essential singularities of φ:

$$\mathcal{R}(\varphi) := \{z; \varphi^n(z) \text{ is not defined for some } n \in \mathbb{N}\}. \tag{11.3.5}$$

We see that that the sets $\mathcal{F}$, $\mathcal{J}$ and $\mathcal{R}$ satisfy the relation

$$\mathcal{F}(\varphi) = \overline{\mathbb{C}} \setminus (\mathcal{J}(\varphi) \cup \mathcal{R}(\varphi)). \tag{11.3.6}$$

Of course, for entire and rational functions $\mathcal{R}$ is empty. For the meromorphic functions considered here, $\mathcal{J}$ and $\mathcal{R}$ are nonempty sets without isolated points. Moreover, $\mathcal{J}$ and $\mathcal{F}$ are completely invariant with respect to φ, that is, invariant under φ and φ^{-1}.

DEFINITION 11.3.3. A *k-cycle* of φ is a set of k distinct points,

$$s_0, \quad s_1, \quad \dots, \quad s_{k-1},$$

satisfying the relations

$$s_1 = \varphi(s_0), \quad s_2 = \varphi^2(s_0), \quad \dots, \quad s_{k-1} = \varphi^{k-1}(s_0), \quad s_0 = \varphi^k(s_0).$$

The *multiplier* of a k-cycle is

$$(\varphi^k)'(s_m) = \varphi'(s_{k-1}) \cdots \varphi'(s_1)\varphi'(s_0). \tag{11.3.7}$$

The multiplier of a cycle is seen to be the same at every point s_m, $m = 0, 1, \dots, k-1$, of the cycle.

DEFINITION 11.3.4. A k-cycle is *attractive*, *repulsive* or *indifferent* as

$$|(\varphi^k)'(s_m)| < 1, \quad > 1 \quad \text{or} \quad = 1.$$

respectively. A *fixed point* is a 1-cycle.

Any element s_m of a k-cycle is a fixed point of φ^k. The attractive cycles of φ are the repulsive cycles of φ^{-1} and conversely, since

$$\varphi'(z)\big(\varphi^{-1})'(\varphi(z)\big) = 1.$$

DEFINITION 11.3.5. The *immediate basin of attraction* of an attractive fixed point s is the largest connected open set Ω such that

$$z_n = \varphi^n(z_0) \to s, \quad \text{as} \quad n \to \infty \quad \text{for all} \quad z_0 \in \Omega.$$

The immediate basin of attraction of a k-cycle is the union of the immediate basins of attraction of the elements s_m considered as fixed points of φ^k. Attractive fixed points and attractive cycles are in $\mathcal{F}(\varphi)$.

Figures depicting the basin of attraction of the function $\cos(z)/c$, for different values of $c \in \mathbb{C}$, can be found in [**39**].

The following result, derived in [**26**], will be needed.

THEOREM 11.3.1. *The immediate basin of attraction of every attractive fixed point or cycle of $\varphi = c \cot z$, respectively, $\varphi = c \tan z$, contains at least one critical point of φ^{-1}.*

Remark 11.3.1. Ttheorem Thm11.3.1 was known to be true for rational [**10**] and entire [**8**] iteration functions.

One also remarks that if a critical point, z, is in the immediate basin of attraction of an attractive fixed point or cycle, then the critical value $\varphi(z)$ is also in the same basin.

11.4. Local iteration methods

In this section, drawbacks of local iterative methods, whose convergence to a specified root relies on close starting values, will be illustrated with an application of Newton's method for the solution to the transcendental equation

$$c - z \tan z = 0. \tag{11.4.1}$$

DEFINITION 11.4.1. Newton's iteration function for the equation $f(z) = 0$ is

$$z_{n+1} = z_n - \frac{f(z_n)}{f'(z_n)} := N(z_n). \tag{11.4.2}$$

In the present case one has

$$z_{j+1} = \frac{z_j^2 + c\cos^2 z_j}{z_j + \sin z_j \cos z_j}, \tag{11.4.3}$$

where $j = 0, 1, 2, 3, \ldots$, and z_0 is chosen arbitrarily. This iteration is constructed in such a way that solutions of (11.4.1) are fixed points of the Newton iteration

$$N(z, c) = \frac{z^2 + c\cos^2 z}{z + \sin z \cos z} \tag{11.4.4}$$

in the sense that $c - z\tan z = 0$ implies $N(z, c) = z$. It is obvious by inspection that the zeros of $\cos z$ are also fixed points of $N(z, c)$. Since $N(z, c)$ is the Newton iteration for the solution of (11.4.1), it is known that the roots of (11.4.1) are attractive fixed points of (11.4.4), regardless of the multiplicity of these solutions ($N'(z, c) = 0$ in the case of simple solutions); here $' = d/dz$. From the expression

$$N'(z, c) = \frac{2(z\sin z - c\cos z)(z\sin z + \cos z)}{(z + \sin z\cos z)^2} \tag{11.4.5}$$

it is seen that zeros of $\cos z$ are repulsive fixed points ($N'(z, c) = 2$).

Writing the iteration (11.4.4) in the form

$$N(z, c) - \frac{c + c^2(\tan^2 z + 1)}{\tan z + z(\tan^2 z + 1)}$$

and recalling that $\tan(\pm iy) \to \pm i$ as $y \to \infty$ in such a way that $y^2[\tan^2(iy) + 1] \to 0$, it is seen that $N(z, c) \to -ic$ as $z \to \infty$ along the positive imaginary axis and $N(z, c) \to ic$ as $z \to \infty$ along the negative imaginary axis. Moreover, it is shown [**27**], by an argument of Julia [**30**], pp. 92–94, that these are the only asymptotic values of $N(z, c)$. Thus, the set of points at which the inverse function is singular includes the transcendental critical points defined by the asymptotic values $\pm ic$ and the algebraic critical points z_ν determined by the equations $N'(z_\nu, c) = 0$.

Theorem 11.3.1 shows that all the attractive fixed points and cycles of $N(z, c)$ can be discovered by constructing iteration sequences starting from the respective critical points of $N(z, c)$. According to (11.4.5), the set of such values includes the successors of the simple roots of (11.4.1), which are themselves attractive fixed points, and the successors of the solutions of the transcendental equation

$$1 + z\tan z = 0. \tag{11.4.6}$$

This, surprisingly, corresponds to the case $c = -1$ of (11.4.1). The roots of (11.4.6) are double zeros of $N'(z, -1)$ given by (11.4.5). It follows that the algebraic critical points of $N(z, -1)$ are themselves superattractive fixed points. The transcendental critical points of $N(z, -1)$ are $\pm i$ which are found to lie in the basins of attraction of the roots of smallest modulus of (11.4.6), namely $\pm 1.20i$. By Theorem 11.3.1, for $c = -1$, $N(z, -1)$ has no attractive cycle of order bigger than 1. Figure 11.1 shows the immediate basins of attraction to the tenth, and part of the ninth, roots of (11.4.6).

The Julia set of N is the boundary of the components of the basins of attraction to the different roots. The numbers in Fig 11.1 sample the basins of attraction to the various roots. Thus an iteration started at the point $28 + 2i$ near the tenth root, 28.20, converges to the first root, $-1.20i$.

Another drawback with a local method, such as Newton's method, is the presence of many attractive cycles. Figure 11.2 in the c-plane describes part of the Mandelbrot bifurcation set giving rise to attractive cycles in the forward orbits of the algebraic critical point $w_2 = 2.8$ for the corresponding values of the parameter c. At each point, c, of the c-plane, an integer n, $-9 \leq n \leq 20$, respectively, $21 \leq n \leq 99$, indicates the rank, -9 to 20, of the root of (11.2.5), respectively, the order plus 20 of the attractive cycle which lies in the forward orbit of the algebraic critical point $w_2 = 2.8$. The positive semi-axes, $\Re c$ and $\Im c$, point downwards and to the right, respectively. The origin is at the top left corner and the step size in $y = \Im z$ is 0.2 per two-character column.

Finally, convergence to parasitic solutions or even to strange attractors may happen with higher order methods, as will be mentioned and defined later in the elliptic case of Kepler's equation.

11.5. Almost global iteration functions

We consider almost global iteration methods for the solution of the transcendental equations

$$z = c \cot z, \quad z = c \tan z, \quad z = \frac{1}{c} \cos z, \quad z = \frac{1}{c} \sin z, \quad z = \frac{1}{c}\, e^{z}.$$

As the treatment of the first four equations involving trigonometric functions can be done by the same general approach, it will be dealt with first. The last equation involving the exponential function will be considered last.

Generally, for a direct iteration method

$$z_{n+1} = F(z_n), \tag{11.5.1}$$

the *region of attractivity* will be defined as

$$\mathcal{A} = \{z;\ |F'(z)| < 1\} \tag{11.5.2}$$

and its image will be denoted by

$$\mathcal{B} = \{F(z);\ z \in \mathcal{A}\} =: F(\mathcal{A}). \tag{11.5.3}$$

```
The endpoints for X are          :     0.230000D+02,   0.300000D+02
The endpoints for Y are          :     0.000000D+00,   0.500000D+01
The step size for X is           :     0.100000D+00
The step size for Y is           :     0.100000D+00

25.0  9 9 9 9 9 9 9 9 9 9 9-1 3 1 1 1-1-1-1-1-1-1-1-1-1-1-1-1-1-1-1-1-1-1-1-1-1-1-1-1 1 1 1
25.1  9 9 9 9 9 9 9 9 9 9 9-1 1 1 1-2-1-1-1-1-1-1-1-1-1-1-1-1-1-1-1-1-1-1-1-1-1-1-1-2 1 1 1
25.2  9 9 9 9 9 9 9 9 9 9 9 910-1-1-1-1-1-1-1-1-1-1-1-1-1-1-1-1-1-1-1-1-1-1-1-1-1-1 1 1 1 1
25.3  9 9 9 9 9 9 9 9 9 9 9 9-1-1-1 1 1-1-1-1-1-1-1-1-1-1-1-1-1-1-1-1-1-1-1-1-1-1-1 1 1 1 1
25.4  9 9 9 9 9 9 9 9 9 9 9 9 1 1 2 3 1-1-1-1-1-1-1-1-1-1-1-1-1-1-1-1-1 1-1-1-1-1 1 1 1 1 1
25.5  9 9 9 9 9 9 9 9 9 9 9 9-1 7 8-1 1-1 1 1-1 2-1-1-1-1 2-1-1 1-1 1 2 2 2 2-1 1 1 1 1 1 1
25.6  9 9 9 9 9 9 9 9 9 9 9 9 8 8 8 8 8-1 6-1-1 1-1 1 1-1 1-1-1 3 1 2 2 2 2 2 1 1 1 1 1 1 1
25.7  9 9 9 9 9 9 9 9 9 9 9-1 8 8 8 8 8 7 7 7-1 6 7 5 5 4 4 4 3 3 3 2 2 2 2 2-1 1 1 1 1 1 1
25.8  9 9 9 9 9 9 9 9 9 9-1 1 1 8 8 8-1 1 7 7 5 6-1 5 5 3 4-1 1 3 1 1 1 2 2-1 1 1 1 1 1 1 1
25.9  9 9 9 9 9 9 9 9 9 9-1-1-1 1 1-1 1 1-1 1 1-1 1 3-1 1-1 1 1 1 1 1 1 1-1-1 1 1 1 1 1 1 1
26.0  9 9 9 9 9 9 9 9 910-1 1 1 1-1-1 1 1-1-1 1-1 1 1 1 1 1 1 1 1 1 1 1 1 1 1 1 1 1 1 1 1 1
26.1  9 9 9 9 9 9 9 9 1 0 1 1 1 1 1-1-2 1 1 1 1 1 1 1 1 1 1 1 1 1 1 1 1 1 1 1 1 1 1 1 1 1 1
26.2  9 9 9 9 9 9 9 2-1-1-1-1 1 1 1 1 1 1 1 1 1 1 1 1 1 1 1 1 1 1 1 1 1 1 1 1 1 1 1 1 1 1 1
26.3  9 9 9 9 9 9-1 8 1-1-1-1-1 1 1 1 1 1 1 1 1 1 1 1 1 1 1 1 1 1 1 1 1 1 1 1 1 1 1 1 1 1 1
26.4  9 9 9 9 9 1-1 8 7-1-1-1-1 1 1 1 1 1 1 1 1 1 1 1 1 1 1 1 1 1 1 1 1 1 1 1 1 1 1 1 1 1 1
26.5  9 9 9 9-1-1 1-1-1 1 4-1-1 1 1 1 1 1 1 1 1 1 1 1 1 1 1 1 1 1 1 1 1 1 1 1 1 1 1 1 1 1 1
26.6  9 9-1 1-1-1 1 1 1 1 1 1 1 1 1 1 1 1 1 1 1 1 1 1 1 1 1 1 1 1 1 1 1 1 1 1 1 1 1 1 1 1 1
26.7  9 1 1 1 1 1 1 1 1 1 1 1 1 1 1 1 1 1 1 1 1 1 1 1 1 1 1 1 1 1 1 1 1 1 1 1 1 1 1 1 1 1 1
26.8 1010-1-1 1-1 1 1 1 1 1 1 1 1 1 1 1 1 1 1 1 1 1 1 1 1 1 1 1 1 1 1 1 1 1 1 1 1 1 1 1 1 1
26.9 10101011-1 3-4 1 1 0-1-1-1 1 1 1 1 1 1 1 1 1 1 1 1 1 1 1 1 1 1 1 1 1 1 1 1 1 1 1 1 1 1
27.0 1010101010-11112-1-1-1-1-1 1 1 1 1 1 1 1 1 1 1 1 1 1 1 1 1 1 1 1 1 1 1 1 1 1 1 1 1 1 1
27.1 101010101010-1-1-1-1-1-1 1 1 1 1 1 1 1 1 1 1 1 1 1 1 1 1 1 1 1 1 1 1 1 1 1 1 1 1 1 1 1
27.2 10101010101010 1-1-1 7 1 1 1 1 1 1 1 1 1 1 1 1 1 1 1 1 1 1 1 1 1 1 1 1 1 1 1 1 1 1 1 1
27.3 1010101010101010-1 9 1 1 1-1-1 1 1 1 1 1 1 1 1 0 0 1 1 1 1 1 1 1-1 1 1 1-1 1 1 1 1 1 1
27.4 101010101010101010 1 1 1 1-1 1 1 4-1 1 1 1 1 1 0 0 1 1 1-1 1 1 1-3 1-2-2-2 1 1 1 1 1 1
27.5 10101010101010101010-1 1 111 1 1-1131416-1-1-1 0 0-1-1-1-1-1-3-1-3-1-2-2-2-2 1 1 1 1 1
27.6 1010101010101010101010-11111111212-1-1-1-1-1-1 0 0-1-1-1-1-1-1-1 1-1-2-2-2-2 1 1 1 1 1
27.7 1010101010101010101010-1111111-1 1-1-1-1-1-1-1-1-1-1-1-1-1-1-1-1 1 1 1 1 1-1-1 1 1 1 1
27.8 1010101010101010101010 1-1-1 1 1-1-1-1-1-1-1-1-1-1-1-1-1-1-1-1-1-1-1-1-1-1-1-1 1 1 1 1
27.9 1010101010101010101010 1 1-1 0 1-1-1-1-1-1-1-1-1-1-1-1-1-1-1-1-1-1-1-1-1-1-1-1-1 1 1 1
28.0 1010101010101010101010 7 1-1-1-1-1-1-1-1-1-1-1-1-1-1-1-1-1-1-1-1-1-1-1-1-1-1-1-1-1 1 1
28.1 101010101010101010101010 9-1 1 2-1-1-1-1-1-1-1-1-1-1-1-1-1-1-1-1-1-1-1-1-1-1-1-1-1 1 1
28.2 1010101010101010101010-3 1 1 1 1-1-1-1-1-1-1-1-1-1-1-1-1-1-1-1-1-1-1-1-1-1-1-1-1 1 1 1
28.3 101010101010101010101010-1 1-1-1-1-1-1-1-1-1-1-1-1-1-1-1-1-1-1-1-1-1-1-1-1-1-1-1-1 1 1
28.4 101010101010101010101010 1-1-1-1 2-1-1-1-1-1-1-1-1-1-1-1-1-1-1-1-1-1-1-1-1-1-1-1 1 1 1
28.5 101010101010101010101010 1-1-1 1 1-2-1-1-1-1-1-1-1-1-1-1-1-1-1-1-1-1 1-1-1-1-1-1 1 1 1
28.6 101010101010101010101010-1-1-1-1 1-1-1 1-1-1-1-1-1-1-1-1-1-1-1-1-1-1 2 2 2-1 1 1 1 1 1
28.7 101010101010101010101010-1 9 9 912-1 6 1-1 1-1 1-1 1-1 2-1-1 1-1-1 2 2 2 2 2 1 1 1 1 1
28.8 101010101010101010101010 9 9 9 9 9-1 8-1-1 1-1 1-1 1-1 1-1 2 3 3 1 2 2 2 2-1 1 1 1 1 1
28.9 1010101010101010101010 1 1 9 9 9 9 8 8 8 7 7 6 6 6 5 5 4 4 3 3 3 1 2 2 2 2 1 1 1 1 1 1
29.0 10101010101010101010 1 1 2 1 7-1 1 1-1-1-1-1 1-3 1-1 1 3-1 1 2-1 1-1 1-1-1 1 1 1 1 1 1
29.1 101010101010101010 9 1 1 1-3-1-1 1 1-1 1 1-1 1-1 1-1 1 1-1 1 1-1 1 1 1-2-1 1 1 1 1 1 1
29.2 1010101010101010-1-1 1 1 1 1 5-1 1 1 2 1 1 1 1 1 1 1 1 1 1 1 1 1 1 1 1 1 1 1 1 1 1 1 1
29.3 10101010101010-1 1-1-1 1 1 1 1-1-1 1 1 1 1 1 1 1 1 1 1 1 1 1 1 1 1 1 1 1 1 1 1 1 1 1 1
29.4 10101010101010 4-1-1-1-1 1 1 1 1 1 1 1 1 1 1 1 1 1 1 1 1 1 1 1 1 1 1 1 1 1 1 1 1 1 1 1
29.5 1010101010 1-1 9-1-1-1-1-1-1 1 1 1 1 1 1 1 1 1 1 1 1 1 1 1 1 1 1 1 1 1 1 1 1 1 1 1 1 1
29.6 10101010 1-1 1-1 7 7 1-1-1 1 1 1 1 1 1 1 1 1 1 1 1 1 1 1 1 1 1 1 1 1 1 1 1 1 1 1 1 1 1
29.7 101010-1-1-1 1 1 1 1 1 1 1 1 1 1 1 1 1 1 1 1 1 1 1 1 1 1 1 1 1 1 1 1 1 1 1 1 1 1 1 1 1
29.8 10 2-1 1 1 1 1 1 1 1 1 1 1 1 1 1 1 1 1 1 1 1 1 1 1 1 1 1 1 1 1 1 1 1 1 1 1 1 1 1 1 1 1
29.9 11-1-1 1 1 1 1 1 1 1 1 1 1 1 1 1 1 1 1 1 1 1 1 1 1 1 1 1 1 1 1 1 1 1 1 1 1 1 1 1 1 1 1
30.0 11111112-1-1 1 1 1 0 1-3 1 1 1 1 1 1 1 1 1 1 1 1 1 1 1 1 1 1 1 1 1 1 1 1 1 1 1 1 1 1 1
```

FIGURE 11.1. The upper part of the immediate basin of attraction to the tenth, and part of that to the ninth, superattractive fixed points of $N(z, -1)$. The positive semi-axes, $\Re z$ and $\Im z$, point downwards and to the right, respectively. The step size in $y = \Im z$ is 0.1 per two-character column, starting at $y = 0$. A point z sampled by, say, 10 is in the basin of attraction of the tenth root, 28.20.

```
The maximum number of iterations is      :   10000
The iteration levels are                 :     500   6000   7000  10000
The maximum length of cycle detected is :      79
The endpoints for Re c are               :      0.000000D+00,   0.140000D+02
The endpoints for Im c are               :      0.000000D+00,   0.800000D+01
The step size for Re c is                :      0.200000D+00
The step size for Im c is                :      0.200000D+00

 0.0    2 2 2 2 2 2 2 2 2 2 2 2 2 2 2 2 2 2 2 2 2 2 2 2 2 3-2 3 3 3 3 3 3 3 3 3 3 3 3 3
 0.2    2 2 2 2 2 2 2 2 2 2 2 2 2 2 2 2 2 2 2 2 2 2 2 2 2 3 3 3 3 3 3 3 3 3 3 3 3 3 3 3
 0.4    2 2 2 2 2 2 2 2 2 2 2 2 2 2 2 2 2 2 2 2 2 2 2 2 2-3 3 3 3 3 3 3 3 3 3 3 3 3 3 3
 0.6    2 2 2 2 2 2 2 2 2 2 2 2 2 2 2 2 2 2 2 2 2 2 2 2 2-3 0 3 3 3 3 3 3 3 3 3 3 3 326
 0.8    2 2 2 2 2 2 2 2 2 2 2 2 2 2 2 2 2 2 2 2 2 2 2 2 2-2 3 3 3 3 3 3 3 3 3 3 3 3 3 3
 1.0    2 2 2 2 2 2 2 2 2 2 2 2 2 2 2 2 2 2 2 2 2 2 2 2 224 3 3 3 3 3 3 3 3 3 4 3 32432
 1.2    2 2 2 2 2 2 2 2 2 2 2 2 2 2 2 2 2 2 2 2 2 2 2 2 2 3 3 3 3 3 3 3 3 3 3-3 3 32424
 1.4    2 2 2 2 2 2 2 2 2 2 2 2 2 2 2 2 2 2 2 2 2 2 2 2-22424 3 3 3 3 3 3 326 2 2242424
 1.6    2 2 2 2 2 2 2 2 2 2 2 2 2 2 2 2 2 2 2 2 2 2 2 2242424 02222 422222428 3-3322424
 1.8    2 2 2 2 2 2 2 2 2 2 2 2 2 2 2 2 2 2 2 2 2 2 232242425222222 5222211-3 330 3-324
 2.0    2 2 2 2 2 2 2 2 2 2 2 2 2 2 2 2 2 2 2 2 2 2 2 6-2-4-5222222222222 2-4 326 43224
 2.2    2 2 2 2 2 2 2 2 2 2 2 2 2 2 2 2 2 2 2 2 2 2 2 3-3-1 0222222222222-3 4-4 2 324-2
 2.4    2 2 2 2 2 2 2 2 2 2 2 2 2 2 2 2 2 2 2 2 2 2-2 3-6-2-32222222222222224 3 3 4 4 3 3
 2.6    2 2 2 2 2 2 2 2 2 2 2 2 2 2 2 2 2 2 2 2 3 32326 126 53222222222226 2 423 3 3 3-3
 2.8    2 2 2 2 2 2 2 2 2 2 2 2 2 2 2 2 2 2 2 224 2 3 326-422-32222-422 333 42626-3-3 4
 3.0    2 2 2 2 2 2 2 2 2 2 2 2 2 2 2 2 2 2 22222-324 022222222-3-322222222262638-1 3 0
 3.2    2 2 2 2 2 2 2 2 2 2 2 2 2 2 2 2 2 2 22222 6222222222222222222222222222232-424 3
 3.4    2 2 2 2 2 2 2 2 2 2 2 2 2 2 2 22222222222222222222222222222222222222222 0-22892
 3.6    2 2 2 2 2 2 2 2 2 2 2 2 222 322222222 2222222222222222222222222222222222230 322
 3.8    2 2 2 2 2 2 2 2 2 2 2 22222 3 3222222 22222222222222222222222222222222222222222
 4.0    2 2 2 2 2 2 2 2 2 22222222222 3222222222222222222222222222222222222222222222222
 4.2    2 2 2 2 2 22222 322222222 22222222222 1 422222222222222222222222222222222222222
 4.4    3222222222222 3 222222222222222222222222222222222222222222222222222222222222222
 4.6    3222222222222222222222222222222222222222222222222222222222222222222222222222222
 4.8    3 322222222222222222222222222222222222222-1222222222222222222222222222222222222
 5.0    3 3 422222222222222222222222222222222222222222222222222222222222222222222222222
 5.2    3 3 322222222222222222222222222222222222222222222222222222222222222222222222222
 5.4    3 3 3222222222222222222222222222222222222 2222222222222222222222222222222222222
 5.6    3 3 32222 4 522222222222222222222222222222222222222222222222222222222222222222
 5.8    3 3 3 3 4 422222222222222222222222222222222222222222222222222222222222222222222
 6.0    3 3 3 3222222222222222222222222222222222222222222222222222222222222222222222222
 6.2    3 3 3 3 3 2222222222222222222222222222222 0 02222222222222222222222222222222222
 6.4    3 3 3 3 32222222222222222222222222222222222222222222222222222222222222222222222
 6.6    3 3 3 3 3 3 4222222222222222222222222222222222222222222222222222222222222222222
 6.8    3 3 3 3 3 3 3 3 222222222222222222222222222222222222222222222222222222222222222
 7.0    3 3 3 3 3 3 322222222222222222222222222 722222222222222222222222222222222222222
 7.2    3 3 3 3 3 322222222222222222222222222 6 62222222222222222222222222222222222222
 7.4    3 3 3 3 3 3222222222222222222222222 5221122222222222222222222222222222222222222
 7.6    3 3 3 3 3 3 32222222222 5 4 4 422 5 5 42222222222222222222222222222222222222222
 7.8    3 3 3 3 3 3222222222222 4 4 4 4 422 6222222222222222222222222222222222222222222
 8.0    3 3 3 3 3 3222222 5 4 4 4 4 4 4 42222 02222222222222222222222222222222222222222
 8.2    3 3 3 3 3 3222222 4 4 4 4 4 4 4 42222-42222222222222222222222222222222222222222
 8.4    3 3 3 3 3 32222 4 4 4 4 4 4 422222222222222222222222222222222222222222222222222
 8.6    3 3 3 3 3 3 4 4 4 4 4 4 4222222222222222222222222222222222222222222222222222222
 8.8    3 3 3 3 3 322222222 5222222 522222222222222222222222222222222222222222222222222
 9.0    3 3 3 3 3 322222222222222222222222222222222222222222222222222222222222222222222
 9.2    3 3 3 3 3 3 3222222222222222222222222222222222222222222222222222222222222222222
 9.4    3 3 3 3 3 3 3 22222222222222222**2222222222222222222222222222222222222222222222
 9.6    3 3 3 3 3 3 3 3 322222222222222-42222222222222222222222222222222222222222222222
 9.8    3 3 3 3 3 3 3 32222222222222222222222222222222222222222222222222222222222222222
10.0    3 3 3 3 3 3 3 3 322222222222222222222222222222222222222222222222222222222222222
10.2    3 3 3 3 3 3 3 3 32222 3 2222222222222222222222222222222222222222222222222222222
10.4    3 3 3 3 3 3 3 3 3 3222222222222222222222222222222222222222222222222222222222222
10.6    3 3 3 3 3 3 3 3 3 322 422222222222222222222222222222222222222222222222222222222
10.8    3 3 3 3 3 3 3 3 3 3222222222222222222222222222222222222222222222222222222222222
11.0    3 3 3 3 3 3 3 3 3 3222222222222222222222222222222222222222222222222222222222222
```

FIGURE 11.2. The positive semi-axes, $\Re c$ and $\Im c$, point downwards and to the right, respectively. At each point, c, the integer n indicates that the nth root of (11.2.5), if $-9 \leq n \leq 20$, or the $n-20$ cycle, if $21 \leq n \leq 99$, lies in the forward orbit of the algebraic critical point $w_2 = 2.8$.

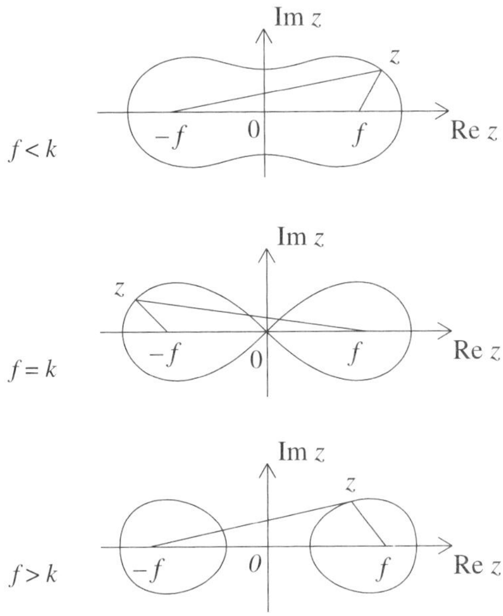

FIGURE 11.3. The figure shows ovals of Cassini $\{z; |z-f|\,|z+f| = k^2\}$ with foci $f > 0$ and $-f$; the product of the distances r_1 and r_2 from the point z to the foci is constant: $r_1 r_2 = k^2$. Top, $f < k$; center, $f = k$; bottom, $f > k$.

11.5.1. The four trigonometric equations. We first need the following definition.

DEFINITION 11.5.1. An *oval of Cassini* is a closed curve defined in standard position by the relation

$$\mathcal{O} = \{z;\ |z-f|\,|z+f| = k^2\} \tag{11.5.4}$$

where the points $\pm f$ are the foci of the oval and the constant k^2 is the product of the distances of the point z describing the oval to the two foci as shown in Fig 11.3.

For the four trigonometric equations,

$$z = c\cot z, \quad z = c\tan z, \quad z = \frac{1}{c}\cos z, \quad z = \frac{1}{c}\sin z,$$

it will turn out that the set $\mathcal{B}$ defined in (11.5.3) is the region bounded by some oval of Cassini.

$T(z,c)$	$\mathcal{A}$	f	k^2	T.C.P.	Alg.crit.pts.	A.C.V.
$c\cot z$	$\lvert\sin z\rvert > \sqrt{\lvert c\rvert}$	ic	$\lvert c\rvert$	$\pm ic$	None	None
$c\tan z$	$\lvert\cos z\rvert > \sqrt{\lvert c\rvert}$	ic	$\lvert c\rvert$	$\pm ic$	None	None
$(1/c)\cos z$	$\lvert\sin z\rvert < \lvert c\rvert$	$1/c$	1	None	$\pm k\pi$	$\pm 1/c$
$(1/c)\sin z$	$\lvert\cos z\rvert < \lvert c\rvert$	$1/c$	1	None	$\pm(2k+1)\pi/2$	$\pm 1/c$

TABLE 11.2. For the four trigonometric equations, the table lists the regions $\mathcal{A}$, the parameters f and k^2 of the oval $\mathcal{O}$ which is the boundary of the region $\mathcal{B}$, the finite transcendental critical points (T.C.P.) and the algebraic critical values (A.C.V.).

Now for the trigonometric equations, the iteration function F will be denoted by

$$T(z,c) := c\cot z, \quad c\tan z, \quad (1/c)\cos z, \quad \text{and} \quad (1/c)\sin z,$$

respectively. It can then be seen that the regions $\mathcal{A}$ and the parameters of the oval $\mathcal{O}$, which defines the region $\mathcal{B}$ for the four trigonometric equations, are as listed in Table 11.2. Moreover the only finite critical points of $c\cot z$ and $c\tan z$ are the transcendental critical points $\pm ic$, and the only critical points of $(1/c)\cos z$ and $(1/c)\sin z$ are the algebraic critical points $k\pi$ and $(2k+1)\pi/2$, respectively, with algebraic critical values $\pm 1/c$.

The level curves, $|\sin z| = \text{constant}$ and $|\cos z| = \text{constant}$, bounding $\mathcal{A}$ are shown in Fig 11.4.

With these considerations we have the following theorem.

THEOREM 11.5.1. *For any given value $c \neq 0$ the iteration function $z_{n+1} = T(z_n, c)$ has two attractive fixed points, and these are in $\mathcal{B}$, if $\mathcal{B} \subset \mathcal{A}$, and only if $\mathcal{B} \cap \mathcal{A} \neq \emptyset$, where $\mathcal{A}$ is the region of attractivity of T and $\mathcal{B} = T(\mathcal{A})$. The two fixed points, if any, are in the forward orbits of the two transcendental critical points of T for* cot *and* tan, *and of the two algebraic critical values of T for* cos *and* sin, *respectively.*

The proof of the first part follows from the fact that the mapping $T : \mathcal{A} \to \mathcal{A} \cap \mathcal{B}$ is contracting. The proof of the last part follows from Theorem 11.3.1 and Remark 11.3.1.

11.5.2. The inverse iteration function. The repulsive fixed points of T are attractive fixed points of properly chosen branches of the multiple-valued inverse iteration

$$z_{n+1} = T^{-1}(z_n). \tag{11.5.5}$$

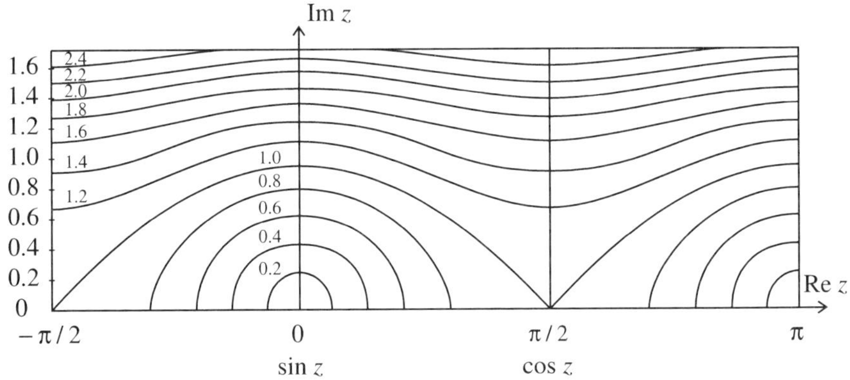

FIGURE 11.4. The figure shows the level curves $|\sin z| = k$ of $\sin z$, respectively, $|\cos z| = k$ of $\cos z$, in the upper half-plane, by taking the origin at $0 + 0i$, respectively, at $\pi/2 + 0i$.

Thesc fixed points will be seen to be in the forward orbit of almost any point on the chosen branches.

In order to have a clear view of the branches of T^{-1} one needs to locate the double roots of $z - T(z, c) = 0$. For the four trigonometric equations, these roots turn out to be roots of equations which are independent of c as is easily seen.

The equations for the double roots of the equations $z = c \cot z$ and $z = \tan z$ are

$$z + \sin z \cos z = 0 \quad \text{and} \quad z - \sin z \cos z = 0, \tag{11.5.6}$$

respectively. Table 11.3 lists the first seven double roots of these equations and the values of c for which double roots exist.

Similarly, the equations for the double roots of the equations $cz = \cos z$ and $cz = \sin z$ are

$$z \tan s = -1 \quad \text{and} \quad z \cot z = 1, \tag{11.5.7}$$

respectively. Tables 11.4 and 11.5 lists the first seven double roots of these equations and the values of c for which double roots exist.

It is to be noted that there are no roots of multiplicity higher than 2, except for $\sin z = cz$, $c = 1$ where the origin is a root of multiplicity 3; but in this case only two roots bifurcate as c moves away from the value 1 since the third one remains at the origin.

By drawing the images of the real c-axis and branch cuts joining the branch points c_j and $\bar{c}_j$ through infinity on the Riemann c-sphere for each

	$z + \sin z \cos z = 0$	$c = z \tan z$
j	$+p_j, \pm\bar{p}_j$	$c_j, \bar{c}_j$
1	0	0
2	$2.106 + 1.125i$	$-1.651 + 2.060i$
3	$5.356 + 1.552i$	$-2.058 + 5.335i$
4	$8.537 + 1.776i$	$-2.278 + 8.523i$
5	$11.699 + 1.929i$	$-2.431 + 11.689i$
6	$14.854 + 2.047i$	$-2.548 + 14.846i$
7	$18.005 + 2.142i$	$-2.643 + 17.998i$

	$z - \sin z \cos z = 0$	$c = z \cot z$
j	$\pm q_j, \pm\bar{q}_j$	$c_j, \bar{c}_j$
1	0	1
2	$3.749 + 1.384i$	$1.895 - 3.719i$
3	$6.950 + 1.676i$	$2.180 - 6.933i$
4	$10.119 + 1.858i$	$2.361 - 10.107i$
5	$13.277 + 1.992i$	$2.493 - 13.268i$
6	$16.430 + 2.097i$	$2.598 - 16.422i$
7	$19.579 + 2.183i$	$2.684 - 19.573i$

TABLE 11.3. For the equations $z = c \cot z$ and $z = c \tan z$ the table lists the equations for the double roots, the corresponding values of c and the numerical values of the first seven double roots, p_j and q_j, respectively, and the corresponding values of the complex number c_j.

of the four multiple-valued mappings

$$c \to \{z : z \tan z = c\}, \qquad c \to \{z : z \cot z = c\}, \tag{11.5.8}$$

and

$$c \to \{z : \cos z = cz\}, \qquad c \to \{z : \sin z = cz\}, \tag{11.5.9}$$

one obtains rough graphs of the images of the four quadrants, I, II, III and IV, of the c-plane into the four regions I_k, II_k, III_k and IV_k as shown in Figs. 11.5 to 11.8, respectively. For each $k = 1, 2, \dots$, the union $I_k \cup II_k \cup III_k \cup IV_k$ forms a fundamental region of the functions $c = z \tan z$, $c = z \cot z$, $c = (1/z) \cos z$, and $c = (1/z) \sin z$. In Figs. 11.5 and 11.6 one obtains the corresponding regions in the second, third and fourth quadrants of the z-plane by reflection through the origin and reflection in the real axis

	$z \tan z = -1$	$c = (\cos z)/z$	Foci
j	$\pm\xi_j$	$\pm c_j$	$\pm 1/c_j$
1	$1.199\,678\,6\,i$	$-1.508\,880\,i$	$0.662\,743\,i$
2	2.798 386 5	−0.336 508	− 2.971 698
3	6.121 250 5	0.161 228	6.202 395
4	9.317 866 5	−0.106 708	− 9.371 373
5	12.486 454 4	0.079 831	12.526 434
6	15.644 128 4	−0.063 792	−15.676 056
7	18.796 404 4	0.053 126	18.822 986

TABLE 11.4. For the equation $cz = \cos z$ the table lists the equations for the double roots, the corresponding values of c and the numerical values of the first seven double roots, ξ_j, the corresponding values c_j, and the foci $\pm 1/c_j$ of the ovals $\mathcal{O}$.

	$z \cot z = 1$	$c = (\sin z)/z$	Foci
j	$\pm\eta_j$	c_j	$\pm 1/c_j$
1	0	0	∞
2	4.493 409 5	−0.217 234	− 4.603 339
3	7.725 251 8	0.128 375	7.789 706
4	10.904 121 7	−0.091 325	−10.949 880
5	14.066 193 9	0.070 913	14.101 695
6	17.220 755 3	−0.057 972	−17.249 766
7	20.371 303 0	0.049 030	20.395 833

TABLE 11.5. For the equation $cz = \sin z$ the table lists the equations for the double roots, the corresponding values of c and the numerical values of the first seven double roots, η_j, the corresponding values c_j, and the foci $\pm 1/c_j$ of the ovals $\mathcal{O}$.

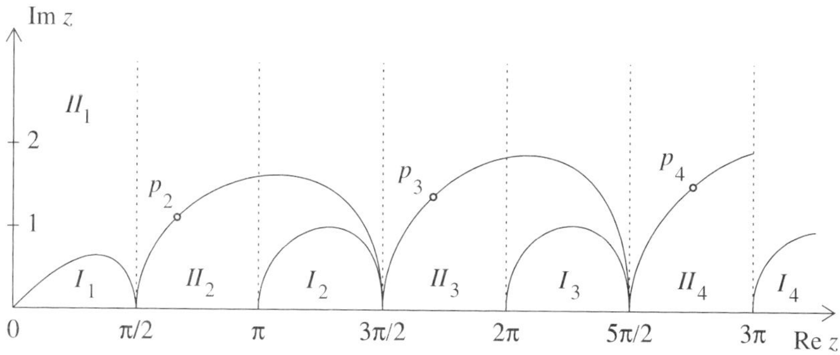

FIGURE 11.5. The figure shows the regions I_k and II_k, $k \geq 1$, which are images of the upper half of the c-plane into the first quadrant of the z-plane by the multiple-valued mapping $c \to \{z; z\tan z = c\}$. The points $p_1 = 0, p_2, p_3, \ldots,$ are the double roots of the equation $z\tan z = c$. The region II_1 is unbounded.

since $z\tan z$ and $z\cot z$ are even and real functions, that is,

$$-z\tan(-z) = z\tan z$$

and $x\tan x$ is real for real x, and similarly for the second function; this fact will also be used later in Table 11.8.

Now with

$$t = (ic - z)(ic + z) \quad \text{and} \quad t = cz \pm \sqrt{(cz)^2 - 1}, \tag{11.5.10}$$

the branches $T^{-1}(t, c) = T^{-1}(z, c)$ of the inverses of $T(z, c) = c\cot z$ and $T(z, c) = c\tan z$ are

$$\frac{1}{2i}\log t + k\pi \quad \text{and} \quad \frac{1}{2i}\log\left(-\frac{1}{t}\right), \tag{11.5.11}$$

respectively. Similarly, the branches of the inverses of $T(z, c) = (1/c)\cos z$ and $T(z, c) = (1/c)\sin z$ are

$$-\log t + 2k\pi \quad \text{and} \quad -i\log(it) + 2k\pi, \tag{11.5.12}$$

respectively.

Tables 11.6 and 11.7 give the vertical strips containing the values of $T^{-1}(z, c, k)$ for the four trigonometric equations. By choosing appropriate branch cuts in the t-plane and corresponding branch cuts in the z-plane, one

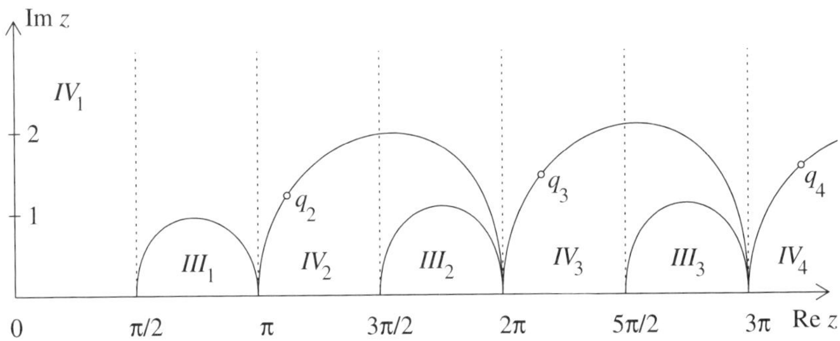

FIGURE 11.6. The figure shows the regions III_k and IV_k, $k \geq 1$, which are images of the lower half of the c-plane into the first quadrant of the z-plane by the multiple-valued mapping $c \to \{z; z\cot z = c\}$. The points $q_1 = 0, q_2, q_3, \dots,$ are the double roots of the equation $z \cot z = c$. The region IV_1 is unbounded.

$T(z,c)$	$T^{-1}(t,c,k)$	t-Cut	z-Cut	$\Re T^{-1}(z,c,k)$
$c\cot z$	$\frac{1}{2i}\log t + k\pi$	$[-\infty, 0]$	$[-ic, ic] \ni 0$	$(-\frac{\pi}{2} + k\pi, \frac{\pi}{2} + k\pi]$
		$[0, +\infty]$	$[-ic, ic] \ni \infty$	$(k\pi, \pi + k\pi]$
$c\tan z$	$\frac{1}{2i}\log(-\frac{1}{t}) + k\pi$	$[-\infty, 0]$	$[-ic, ic] \ni \infty$	$(-\frac{\pi}{2} + k\pi, \frac{\pi}{2} + k\pi]$
		$[0, +\infty]$	$[-ic, ic] \ni 0$	$(k\pi, \pi + k\pi]$

TABLE 11.6. Given $t = (ic - z)/(ic + z)$ and a branch cut in the t-plane, the table lists the corresponding branch cut in the z-plane and the vertical strip containing the values of $T^{-1}(z, c, k)$ for the first two trigonometric equations.

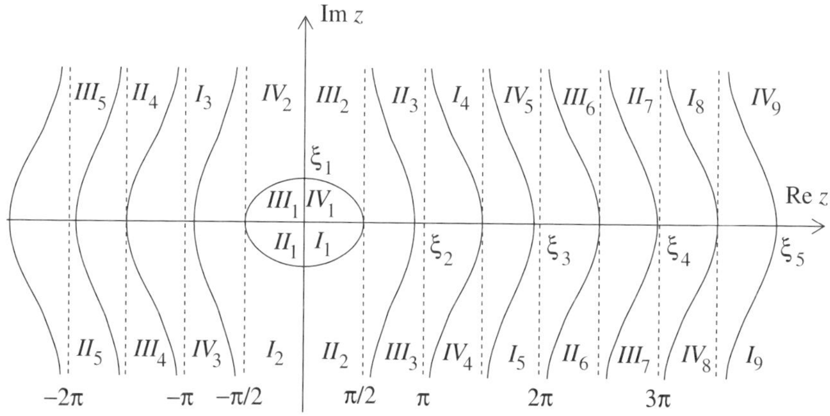

FIGURE 11.7. The regions I_k, II_k, III_k and IV_k of the z-plane are the images by the multiple-valued mapping $c \to \{z; \cos z = cz\}$ of the quadrants I, II, III and IV of the c-plane, respectively. The points ξ_j are the double roots of $\cos z = cz$. The dotted vertical lines are the asymptotes $\Re z = \pm n\pi/2$. The boundary of the regions cuts the real axis successively in $\pm\pi/2, \pm\xi_2, \pm 3\pi/2, \pm\xi_3$, etc., and the imaginary axis in $\pm\xi_1 = \pm 1.2i$. The central "ellipse" contains the first root of $\cos z = cz$.

obtains vertical strips which contain the real part of $T^{-1}(z, c)$ as indicated in these tables.

If a specified root $s^{(k)}$ for a given value of c lies in a vertical strip S_l which does not intersect the oval $\mathcal{O}$, by choosing the branch cut which does not intersect S_l, the iteration

$$z_{n+1} = T^{-1}(z_n, c, l) \tag{11.5.13}$$

will converge to $s^{(k)}$ for any $z_0 \in S_l \setminus \mathcal{A}$, provided $s^{(k)}$ is the unique root in S_l. This follows from the fact that T^{-1} is contracting in $S_l \setminus \mathcal{A}$.

The uniqueness can been seen from Figs. 11.5 to 11.8, except in the following cases. In Fig 11.5, when $c \in II$, the roots $s^{(1)}$ and $s^{(k)}$, $k > 1$, could possibly lie in the same vertical strip of width 2π, and similarly for $s^{(1)}$ and $s^{(k)}$, $k > 1$, in Fig 11.6, when $c \in IV$.

A similar situation occurs for $s^{(1)}$ in Fig 11.7; but in this case, this difficulty will be resolved by means of Newton's iteration in the next section.

Any double root of the given equations lies on the boundary of the region $\mathcal{A}$, and the oval $\mathcal{O}$ is tangent to that boundary at that point. At

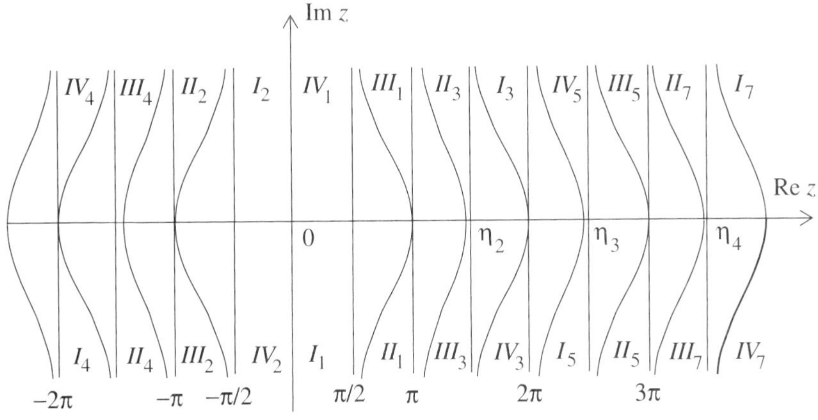

FIGURE 11.8. The regions I_k, II_k, III_k and IV_k of the z-plane are the images by the multiple-valued mapping $c \to \{z; \sin z = cz\}$ of the quadrants I, II, III and IV of the c-plane, respectively. The points η_j are the double roots of $\sin z = cz$. The dotted vertical lines are the asymptotes $\Re z = \pm n\pi/2$. The boundary of the regions cuts the real axis successively in $\eta_1 = 0, \pm\pi, \pm\eta_2, \pm 2\pi, \pm\eta_3$, etc.

a double root the multiplier, T', of T is equal to 1, which is a rational number. Hence the double root is an indifferent fixed point of T which lies in the Julia set of T, and by the Flower Theorem [**10**], it can be reached by both (11.5.1) and (11.5.13). In this case, an already, but slowly, convergent orbit may need to be accelerated by means of *Steffensen's procedure*, which is defined as follows.

DEFINITION 11.5.2. Steffensen's procedure for three iterates z_n, z_{n+1}, and z_{n+2} is

$$z'_n = z_n - \frac{(z_{n+1} - z_n)^2}{z_{n+2} - 2z_{n+1} + z_n}. \tag{11.5.14}$$

Moreover the direct and inverse iteration functions distinguish between the pair of roots after these have bifurcated from a double root if one root lies inside the oval and the other lies outside.

The only difficult case is when both roots are outside the region of attractivity $\mathcal{A}$ but near a double root. In this case the immediate basin of attraction of the specified root for T^{-1} could be relatively small and the iteration (11.5.13) may lead to attractive cycles. Thus, one may have to try different starting values or interpolation as explained in Section 11.7.

$T(z,c)$	$T^{-1}(t,c,k)$	t-Cut	Loc. of c	$\Re T^{-1}(z,c,k)$
$\frac{1}{c}\cos z$	$-\log t + 2k\pi$	$[0,+\infty]$	$III \cup II$	$[2k\pi, 2\pi + 2k\pi)$
		$[-\infty,0]$	$I \cup IV$	$[-\pi + 2\pi, \pi + 2\pi)$
$\frac{1}{c}\sin z$	$-i\log(it) + 2k\pi$	$[0,+\infty]$	$IV \cup III$	$[2k\pi, 2\pi + 2k\pi)$
		$[-\infty,0]$	$II \cup I$	$[-\pi + 2k\pi, \pi + 2k\pi)$

TABLE 11.7. Given $t = cz \pm \sqrt{(cz)^2 - 1}$, a branch cut in the t-plane and the quadrant containing c, the table gives the vertical strip containing the values of $T^{-1}(z,c,k)$ for the last two trigonometric equations.

It is to be noticed that indifferent fixed points lie on closed curves, each one going through a double root. We illustrate this situation for the equation

$$\cos z = cz.$$

Since, in this case, $|F_c'(z^*)| = 1$ and

$$z^* = F_c(z^*), \tag{11.5.15}$$

it follows that

$$-\frac{\sin z^*}{c} = e^{i2\pi\alpha}. \tag{11.5.16}$$

Eliminating c from (11.5.15) and (11.5.16), we see that z^* is a solution of

$$z \tan z = -e^{i2\pi\alpha}. \tag{11.5.17}$$

The solutions of (11.5.17) are plotted in Fig 11.9. These solutions lie on different ovals. Remember that each indifferent fixed point corresponds to a specific value of c, and only for some values of c does there correspond at most one indifferent fixed point. These points lie in the Julia set of F_c for the corresponding c.

11.5.3. The exponential equation. We turn now to the exponential equation

$$z = \frac{1}{c}\, e^z =: E(z,c). \tag{11.5.18}$$

Here the region of attractivity $\mathcal{A}$ is the unit disk in the z-plane for c lying outside the region $G \cup \Gamma$ shown in the right-hand part of Fig 11.10, where

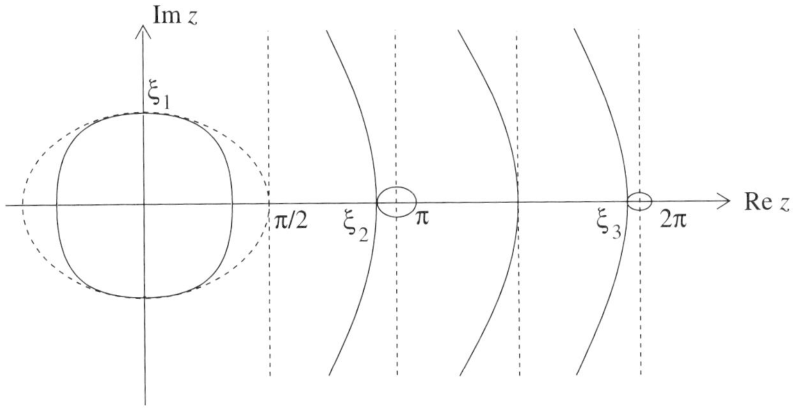

FIGURE 11.9. Three ovals of indifferent fixed points of $F_c(z) = (\cos z)/c$. The points ξ_1, ξ_2 and ξ_3 are double roots of $\cos z = cz$.

the boundary curve Γ is defined by the relation

$$\Gamma = \left\{ c; c = \frac{1}{\xi} e^{\xi}, \ |\xi| = 1 \right\}.$$

One sees that the oval $\mathcal{O}$ has reduced in this case to the unit circle.

Since $E(z, c)$ has only one transcendental critical point, $z = 0$, and no algebraic critical points, the iteration

$$z_{n+1} = E(z_n, c)$$

can have at most one attractive fixed point. Hence for any $c \notin G \cup \Gamma$, the iteration started at $z_0 = 0$, will converge to the first root, $s^{(1)}$, of (11.5.18) inside the unit disk.

The only double root of (11.5.18) is $z = 1$, and this occurs only when $c = e$. By considering the images of the real and imaginary axes, $a = \Re c$ and $b = \Im c$, respectively, under the multiple-valued mapping $c \to \{z; e^z = cz\}$, one obtains Fig 11.11.

The inverse E^{-1} of E is given by the logarithm

$$E^{-1}(z, c, k) = \ln z + \ln c + 2k\pi i, \qquad k \in \mathbb{Z}, \tag{11.5.19}$$

where the branch cut of $\ln z$ and $\ln c$ are taken along the negative real axes in the z-plane and c-plane, respectively. For c in a given quadrant, one can choose a value of k such that the values of (11.5.19) will lie in a horizontal strip of width 2π which covers only one image of the given quadrant, except for the first images I_1, II_1, III_1 and IV_1.

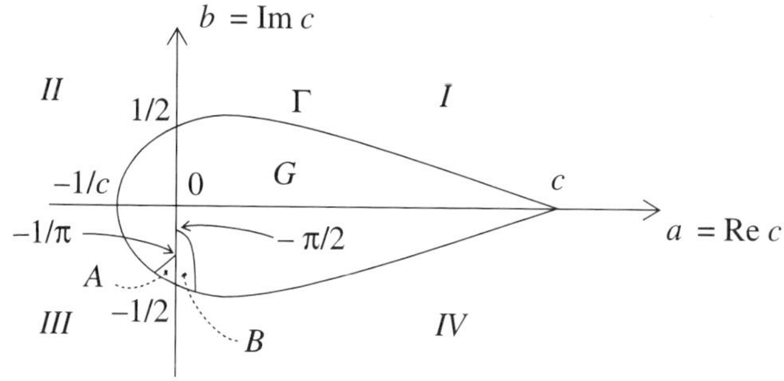

FIGURE 11.10. Images of the four quadrants of the c-plane into the z-plane under the multiple-valued mapping $c \to \{z; e^z = cz\}$. Region G of the c-plane mapped into $\{z^{(1)}; |z^{(1)}| > 1\}$. The curve Γ is the boundary of G.

In case of the first images, for example, for $c \in III$ in the small region $A \subset G$, $E^{-1}(z, c, k)$ has two attractive fixed points, $s^{(1)} \in III_1 \cap A'$ and $s^{(2)} \in III_2$. In such case, to have convergence to $s^{(1)}$ one needs a good starting value, say, $z_0 \in A'$. The same holds for $c \in B \subset G$ and for $c \in II$ and $c \in I$ inside regions which are symmetric to A and B with respect to the axis $\Re c$.

Otherwise, for $c \in G \setminus A \subset III$, (11.5.19) will converge to $s^{(1)}$ for $z_0 \in III_1 \setminus A'$, as is illustrated in Fig 11.11. The same holds for $c \in G$ and in the other three quadrants.

11.6. Effective use of Newton's methods

When a root to $f(z) = 0$ is known to lie in a convex region, one can produce good starting values for Newton's method. This situation occurs for any root $s^{(k)}$ to the first two trigonometric equations, rewritten in the form

$$f(z) := z \sin z - c \cos z = 0, \qquad c \in I, \tag{11.6.1}$$
$$g(z) := z \cos z - c \sin z = 0, \qquad c \in III, \tag{11.6.2}$$

and similarly for the first root $s^{(1)}$ of the third trigonometric equation

$$h(z) := \cos z - cz = 0. \tag{11.6.3}$$

The starting values, z_0, shown in Table 11.8 for (11.6.1) and (11.6.2) are obtained by truncated continued fractions. Those of (11.6.3) are obtained by a rational approximation of the first degree.

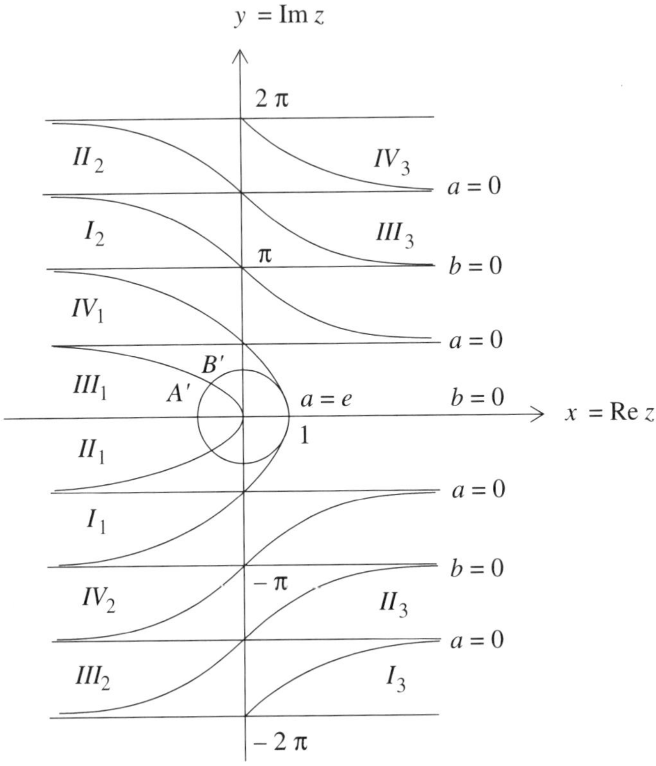

FIGURE 11.11. Images of the four quadrants of the c-plane into the z-plane under the multiple-valued mapping $c \to \{z; e^z = cz\}$.

In dielectric spectroscopy, where typical values of c lie in the annulus $10^{-2} \leq |c| \leq 10^2$, Newton's method for (11.6.1) and (11.6.2) converges very rapidly.

11.7. Interpolation near a double root

When one is looking for a solution of

$$\cos z = cz, \qquad c \in \mathbb{C}, \tag{11.7.1}$$

which lies near a double root, high precision is difficult to achieve. The double root bifurcates into two roots that are close to each other. There can be endless iterations which hardly move closer to the root. Even Steffensen's method may fail to improve the estimate, may converge to another root or may be divergent.

Equations	Quadrants	Starting values z_0	kth roots
$z\sin z -$ $c\cos z = 0$	$c \in I \cup IV$	$\pi\sqrt{\frac{c}{\pi^2+4c}}$ $(k-1)\pi + \frac{\pi}{2}\frac{2}{c+2k-1}$	$s^{(1)} \in I_1 \cup IV_1$ $s^{(k)} \in I_k \cup IV_k$ $k > 1$
$z\cos z -$ $c\sin z = 0$	$c \in III \cup II$	$\pi\sqrt{\frac{1-c}{4-c}}$ $(k-\frac{1}{2})\pi + \frac{\pi}{2}\frac{c}{c-2k+1}$	$s^{(1)} \in III_1 \cup II_1$ $s^{(k)} \in III_k \cup II_k$ $k > 1$
$\cos z - cz = 0$	$c \in I$ $c \in II$ $c \in III$ $c \in IV$	$\frac{\pi}{2}\frac{1}{\bar{\alpha}c+1}$ $\frac{\pi}{2}\frac{1}{\alpha c-1}$ $\frac{\pi}{2}\frac{1}{\bar{\alpha}c-1}$ $\frac{\pi}{2}\frac{1}{\alpha c+1}$	$s^{(1)} \in I_1$ $s^{(1)} \in II_1$ $s^{(1)} \in III_1$ $s^{(1)} \in IV_1$

TABLE 11.8. The table lists the starting values, z_0, to obtain the kth roots, $s^{(k)}$, with Newton's method for the given equations and values of c in appropriate quadrants; here $\alpha = 7/8 - (2/3)i$.

Here, we explain the instability as the iterates get closer to a double root. Let us suppose that $\tilde{z}$ is an *indifferent fixed point* of $F_{\tilde{c}}(z)$, that is,

$$F_{\tilde{c}}(\tilde{z}) = \tilde{z} \quad \text{and} \quad |F'_{\tilde{c}}(\tilde{z})| = 1,$$

implying that

$$\cos\tilde{z} = \tilde{c}\tilde{z} \quad \text{and} \quad \sin\tilde{z} = -\tilde{c}\, e^{i\varphi}. \tag{11.7.2}$$

For any angle $\varphi \in [-\pi, \pi)$, $\tilde{z}$ is on one of the ovals of Fig 11.9. Notice that $\tilde{z}$ tends to the double root ξ_i in the oval as $\varphi \to 0$.

Let $c \approx \tilde{c}$. We want to find a root z of $E_c(z) = 0$ and write

$$e = c - \tilde{c}, \qquad u = z - \tilde{z}.$$

Thus,

$$\begin{aligned} E_c(z) &= \cos z - cz \\ &= \cos(\tilde{z}+u) - (\tilde{c}+e)(\tilde{z}+u). \end{aligned} \tag{11.7.3}$$

Expanding $\cos(\tilde{z}+u)$ to second order around $\tilde{z}$ and using (11.7.2), we obtain

$$\begin{aligned} E_c(\tilde{z}+u) &= \tilde{c}\,(e^{i\varphi}-1)u - \frac{1}{2}\tilde{c}\tilde{z}u^2 + O(|u|^3) - e(\tilde{z}+u) \\ &= 0. \end{aligned}$$

Solving for e,

$$e = \frac{\tilde{c}\,(e^{i\varphi}-1)u - \frac{1}{2}\tilde{c}\tilde{z}u^2 + O(|u|^3)}{\tilde{z}+u},$$

and replacing $1/(\tilde{z}+u)$ by its Taylor expansion in powers of u ,

$$\frac{1}{\tilde{z}+u} = \frac{1}{\tilde{z}} - \frac{u}{\tilde{z}^2} + \frac{u^2}{\tilde{z}^3} + O(|u|^3),$$

we obtain

$$e = \frac{\tilde{c}\,(e^{i\varphi}-1)u}{\tilde{z}} - \tilde{c}\left(\frac{1}{2} + \frac{(e^{i\varphi}-1)}{\tilde{z}^2}\right)u^2 + O(|u|^3). \tag{11.7.4}$$

As $\tilde{z}$ approaches the double root, the first term tends to zero, and for $\tilde{z} = \xi_i$ and $\tilde{c} = c_i$ we have $e \approx -\tilde{c}u^2/2$, that is,

$$c - c_i \approx -\tilde{c}(z-\xi_i)^2/2.$$

This means that in a neighborhood of a double root, very small differences in c result in large differences in z, causing instability.

If greater precision than the one obtained by iterating F_c or G_c and improved by Steffensen's formula is desired, we can use **interpolation** for solving (11.7.1) near a double root.

The double root ξ_i and its corresponding c_i are known. Suppose we have $c \approx c_i$ and want to find the root z^* of $\cos z = cz$. We start with a very good estimate, z_0, of z^*, and choose four points around z_0, namely,

$$z_1 = z_0 + \delta, \quad z_2 = z_0 - \delta, \quad z_3 = z_0 + i\delta, \quad z_4 = z_0 - i\delta,$$

for a small value of δ. We use (11) to compute $c_1, \ldots, c_4$ and construct an interpolating polynomial (with complex coefficients) that verifies

$$P(c_i) = z_i, \qquad i = 0,1,2,3,4, \tag{11.7.5}$$

and interpolate for c,

$$\tilde{z}_0 = P(c).$$

We take $\tilde{z}_0$ as the new z_0 and repeat the procedure until $|\cos\tilde{z}_0 - c\tilde{z}_0|$ is sufficiently small or two consecutive values of $\tilde{z}_0$ are close enough.

It is crucial to start the process with a very good estimate for z^*. To compute the initial value z_0, we consider

$$\begin{aligned} E_c(z) &= \cos z - cz \\ &= \cos(\xi_i + u) - (c_i + e)(\xi_i + u). \end{aligned} \tag{11.7.6}$$

Replacing $\cos(\xi_i + u)$ by its third-order Taylor expansion,

$$\cos(\xi_i + u) = \cos\xi_i - (\sin\xi_i)u - \frac{\cos\xi_i}{2}u^2 + \frac{\sin\xi_i}{6}u^3 + O(|u|^4),$$

and recalling that

$$\cos \xi_i = c_i \xi_i \quad \text{and} \quad -\sin \xi_i = c_i,$$

we get

$$\begin{aligned} E_c(z) &= E_c(\xi_i + u) \\ &= -\frac{c_i}{6}\,u^3 - \frac{c_i}{2}\,\xi_i\,u^2 - e\xi_i - eu + O(|u|^4), \end{aligned} \tag{11.7.7}$$

and write

$$Q_c(u) = -\frac{c_i}{6}\,u^3 - \frac{c_i}{2}\,\xi_i\,u^2 - eu - e\xi_i. \tag{11.7.8}$$

To find an estimate of z^* we calculate the three roots u_1, u_2 and u_3 of the polynomial $Q_c(u)$. Suppose that the two roots that give the two smallest values for $|E_c(\xi_i + u_j)|$ are u_1 and u_2. Then we have the two starting values $z_0 = \xi_i + u_1$ and $z_0 = \xi_i + u_2$ to begin successive interpolations. These interpolations will converge separately to each one of the roots z^* that are close to the double root ξ_i.

11.8. Kepler's equation

In the two-body problem [**19**], pp. 84–91, time and spatial position are related by *Kepler's equation*

$$M = E - \epsilon \sin E, \qquad 0 < \epsilon < 1, \tag{11.8.1}$$

in the elliptic case, and

$$M = \epsilon \sinh F - F, \qquad \epsilon > 1, \tag{11.8.2}$$

in the hyperbolic case, where M is the *mean anomaly*, ϵ is the *eccentricity*, E is the *eccentric anomaly* or reference area, and F is the hyperbolic reference area.

By means of a Fourier series expansion, the eccentric anomaly is given by

$$E = M + 2\sum_{m=1}^{\infty} \frac{1}{m} J_m(m\epsilon)\sin(mM), \tag{11.8.3}$$

where J_n is the Bessel function of the first kind of order m.

Local methods with a faster convergence rate, such as Chebyshev's formula, may have greater drawbacks. This formula is defined as follows.

DEFINITION 11.8.1. *Chebyshev's formula* for an equation $f(z) = 0$ is

$$z_{n+1} = z_n - \frac{f(z_n)}{f'(z_n)}\left[1 + \frac{1}{2}\frac{f(z_n)}{f'(z_n)}\frac{f''(z_n)}{f'(z_n)}\right], \tag{11.8.4}$$

with cubic convergence to simple roots.

Concerning the drawbacks of local methods, it is reported in [**12**] that iteration with Chebyshev's formula, when applied to the elliptic Kepler equation, may lead to divergence or to convergence to parasitic solutions, namely, attractive fixed points of the Chebyshev iteration function which are not solutions of Kepler's equation, or even to strange attractors in case of poorly chosen starting values. The expression *strange attractor* is taken to mean that there is neither convergence nor divergence but the endless iterations will generate almost random values, z_n, inside a bounded region. It appears that once a value of z_n is within this region, it gets trapped forever with no hope of converging to a fixed point, or of getting out of the trap.

Kepler's equation also appears in problems of class D [**28**] in the testing of Runge–Kutta, multistep and Runge–Kutta–Nyström methods, for the periodic solution of systems of ordinary differential equations. Here one attempts to solve the equation of two-body motion

$$\begin{aligned} \ddot{x} &= -x/r^3, \qquad x(0) = 1-\epsilon, \quad \dot{x}(0) = 0, \\ \ddot{y} &= -y/r^3, \qquad x(0) = 0, \qquad \dot{y}(0) = [(1+\epsilon)/(1-\epsilon)]^{1/2}, \end{aligned} \tag{11.8.5}$$

where $r^2 = x^2 + y^2$ and ϵ is the eccentricity. The analytical periodic solution of (11.8.5),

$$\begin{aligned} x &= \cos u - \epsilon, \qquad & y &= \sqrt{1-\epsilon^2}\sin u, \\ \dot{x} &= \frac{-\sin u}{a - \epsilon\cos u}, \qquad & \dot{y} &= \frac{\sqrt{1-\epsilon^2}\cos u}{1-\epsilon\cos u}, \end{aligned}$$

where u is given by Kepler's equation

$$u - \epsilon \sin u = t,$$

is used to determine the global error at all stages of the numerical computation.

To apply the theory developed in the present chapter to Kepler's equation one rewrites (11.8.2) and (11.8.2) in the form

$$z = \epsilon \sin z + d =: K(z). \tag{11.8.6}$$

One sees that the region of attractivity is

$$\mathcal{A} = \{z : |\cos z| < 1/|\epsilon|\}$$

and the region $\mathcal{B} = K(\mathcal{A})$ is bounded by the oval

$$\mathcal{O} = \{z; |z - d - \epsilon|\,|z - d + \epsilon| < 1\}.$$

In the elliptic case, that is, when $0 < \epsilon < 1$, the equation

$$f(x) = x - \epsilon \sin x - d, \qquad \epsilon, d \in \Re,$$

has only one real root, $x(\epsilon, d)$, for each given real values of ϵ and d, and no multiple roots. This root is an attractive fixed point of K and can be reached by the iteration

$$x_{n+1} = K(x_n, \epsilon),$$

from one of the only two critical values of K,

$$x_0 = \pm\epsilon + d,$$

which lies in the immediate basin of attraction of the root.

When $\epsilon > 1$, that is, in the hyperbolic case, the equation

$$g(iy) = iy - \epsilon \sin(iy) - id, \qquad \epsilon, d \in \Re,$$

has only one imaginary root $iy(\epsilon, d)$ for each given real values of ϵ and d, and no multiple roots. This root is an attractive fixed point of

$$iy_{n+1} = \arcsin\left(\frac{iy_n - id}{\epsilon}\right),$$

that is,

$$\begin{aligned} y &= \operatorname{arcsinh}\left(\frac{y-d}{\epsilon}\right) \\ &= \log\left(\frac{y-d}{\epsilon} + \sqrt{\left[\frac{y-d}{\epsilon}\right]^2 + 1}\right). \end{aligned}$$

Convergence can be accelerated by means of Steffensen's procedure (11.5.14), especially when the eccentricity, ϵ, is close to 1.

11.9. A programming strategy

The following programming strategy has worked quite well in interactive and automatic searches for specified roots of the transcendental equations considered in this chapter.

(a) If Table 11.8 applies, use Newtons's methods. Else
(b) Else, if Theorem 11.5.1 applies, use the direct iteration $z_{n+1} = F(z_n, c)$.

(c) Else, use the inverse iteration $z_{n+1} = F^{-1}(z_n, c, k)$ with the appropriate branch; in case of nonconvergence, try the other branch cut.

(d) If (c) does not produce convergence when c is near a branch point, use Steffensen's procedure or interpolation, or change starting values to avoid attractive cycles.

In dielectric spectroscopy, one has experimental values of c which lie on a smooth curve; hence one can use extrapolation to pick the next starting value, z_0, in terms of the next value of c. Near a double root, one can interpolate to a value of z for a value of c lying between two experimental values of c, one on each side of, and slightly away from, the branch point in order to get a starting value that will give convergence to the correct root.

This strategy is likely to converge to any specified root and avoid undesired attractive cycles. Finally, the recourse to the inverse iteration function may be useful in the solution of other transcendental equations.

Answers to Odd-Numbered Exercises

Answers to Odd Exercises for Section 1.1

Page 12 in the text.

1. $-12 - 23i$.

3. $9/13$.

5. $\pi - \operatorname{Arctan} 1/\sqrt{5}$.

7. $-15 + 4i$.

9. $(1 + 4k)\pi/2, \quad k = 0, \pm 1, \pm 2, \ldots$

11. -14.

13. $x = -2, \quad y = 3$.

15. $z = -6/5 - 8i/5$.

17. $|z| = 1, \quad \arg z = 2\pi/3 + 2\pi k, \quad k = 0, \pm 1, \pm 2, \ldots, \quad \operatorname{Arg} z = 2\pi/3$.

19. $1/2 - i/2$.

21. $6/5 - 2i/5$.

25. $z_1 = 0, \quad z_2 = 1, \quad z_{3,4} = -1/2 \pm i\sqrt{3}/2$.

27. If $z = x + iy$ then $|\bar{z}| = |x - iy| = \sqrt{x^2 + (-y)^2} = |z|$.

29. $\overline{z_1 z_2} = \overline{(x_1 + iy_1)(x_2 + iy_2)} = x_1x_2 - y_1y_2 - i(y_1x_2 + x_1y_2)$,
$\bar{z}_1\bar{z}_2 = (x_1 - iy_1)(x_2 - iy_2) = x_1x_2 - y_1y_2 - i(y_1x_2 + x_1y_2)$.

31. The three points, z_1, z_2 and z_3, lie on a straight line if and only if there exists a real number $k \neq 0$ such that $z_2 - z_1 = k(z_3 - z_2)$.

33. Let $z_1 = x_1 + iy_1$ and $z_2 = x_2 + iy_2$. Show that the left-hand side is equal to the right-hand side.

35. $z_4 = z_1 + z_3 - z_2$.

37. $z_1 = 3\sqrt{2} + i2\sqrt{2} = \sqrt{26}\, e^{i \operatorname{Arctan} 2/3}$, $z_2 = -5 + i = \sqrt{26}\, e^{i(\pi - \operatorname{Arctan} 1/5)}$. The angle $\alpha = \pi - \operatorname{Arctan} 1/5 - \operatorname{Arctan} 2/3$.

39. Let $z = x + iy$, $w = u + iv$. Then

$$|1 - \bar{z}w|^2 - |z - w|^2 = (1 - ux - vy)^2 + (yu - xv)^2 - (x - u)^2 - (y - v)^2.$$

Simplifying the above expression we obtain $(1 - |z|^2)(1 - |w|^2)$.

41. If $a = 0$ the proof is trivial. If $a \neq 0$ we have

$$(1 - a\bar{z})z = z - a\bar{z}z = z - a|z|^2 = z - a,$$

since $|z| = 1$. Therefore $z/(z - a) = 1/(1 - a\bar{z})$.

43. $|z_1| = |(z_1 + z_2) + (-z_2)| \leq |z_1 + z_2| + |z_2|$ by means of the triangle inequality. Hence $|z_1 + z_2| \geq |z_1| - |z_2|$. Similarly, $|z_2 + z_1| \geq |z_2| - |z_1|$. Hence, $|z_1 + z_2| \geq \big||z_1| - |z_2|\big|$. The equality holds if two points z_1 and z_2 lie on a straight line.

45. $2(\cos 2\pi/3 + i \sin 2\pi/3)$.

47. $1/2[\cos(-7\pi/12) + i \sin(-7\pi/12)]$.

49. $1/2[\cos \pi/4 + i \sin \pi/4]$.

51. $1/256(\cos 0 + i \sin 0)$.

53. $(\cos\alpha + i \sin\alpha)^n = \cos n\alpha + i \sin n\alpha = 1$, hence $\cos n\alpha = 1$, $\sin n\alpha = 0$. Then $(\cos\alpha - i \sin\alpha)^n = \cos n\alpha - i \sin n\alpha = 1$.

55. $(2 - z)/(2 + z)$ is pure imaginary if z is a point on the circle $x^2 + y^2 = 4$.

57. $3\,e^{i\pi/6}$, $3\,e^{i5\pi/6}$, $3\,e^{-i\pi/2}$.

59. $e^{-i7\pi/72}$, $e^{-i31\pi/72}$, $e^{-i55\pi/72}$, $e^{i17\pi/72}$, $e^{i41\pi/72}$, $e^{i65\pi/72}$.

Answers to Odd Exercises for Section 1.2

Page 23 in the text.

1. The set is open; its interior is doubly connected.

3. The set is closed; its interior is not connected.

5. The set is neither open nor closed; its interior is not connected.

7. The set is closed; its interior is simply connected.

9. Closed disk of radius 2 and center $2 - i$.

11. Domain below the line $y = x$ $(y < x)$.

13. Domain below the two branches of the hyperbola $y = 1/(2x)$.

15. Hyperbola $xy = -1$, the set is closed.

17. Lower part of the circle $x^2 + y^2 = 1 \quad (y < 0)$.

19. Upper part of the semicircle of radius r with center at z_0.

21. The part of the hyperbola $x^2 - y^2 = 1$ joining the points $(\cosh 1, -\sinh 1)$

and $(\cosh 1, \sinh 1)$. (Hint: Use the identity $\cosh^t - \sinh^2 t = 1$.)

23. $z(t) = t + i(2t+1), \quad 0 \le t \le 1$.

25. $z(t) = R\cos t + iR\sin t, \quad -\pi/2 \le t \le \pi/2$.

27. $z(t) = 1 + 4\cos t + i(-3 + 3\sin t), \quad 0 \le t \le 2\pi$.

29. The limit does not exist.

31. 1.

33. 0.

35. 1.

37. The relative positions of the images of $-z$ and $\bar{z}$ on the Riemann sphere with respect to the image of z are the points lying in the same plane parallel to the z-plane which are diametrically opposite to z and symmetric to z with respect to the real axis, respectively.

39. Let $z_n = x_n + iy_n$ and $\alpha = c + id$. Then

$$\lim_{n\to\infty} x_n = c \quad \text{and} \quad \lim_{n\to\infty} y_n = d.$$

Therefore,

$$\lim_{n\to\infty} = \alpha \implies \lim_{n\to\infty} |z_n| = |\alpha|.$$

However, if $\lim\limits_{n\to\infty} |z_n| = |\alpha|$, then it is possible, for example, that $\lim\limits_{n\to\infty} = -c$ and $\lim\limits_{n\to\infty} = -d$. In this case

$$\lim_{n\to\infty} z_n = \lim_{n\to\infty} (x_n + iy_n) = -c - id \neq \alpha.$$

41. $z_1 = -z_2$.

Answers to Odd Exercises for Section 1.3

Page 29 in the text.

1. $z \neq \pm 2i$.

3. $z \neq 0$.

5. $\Re f(z) = 3x^2 - 3y^2 + 2y, \quad \Im f(z) = 6xy - 2x$.

7. $\Re f(z) = x^3 - 3xy^2 + x + 2, \quad \Im f(z) = 3x^2y - y^3 + y$.

9. $\Re f(z) = x(1+2y), \quad \Im f(z) = -x^2 + y^2 - y$.

11. $-12i$.

13. $\sqrt{13}$.

15. The limit does not exist.

17. The limit does not exist.

19. 0 if $m < n$; a_m/b_n if $m = n$; ∞ if $m > n$.

21. $z_1 = 0, \quad z_{2,3} = \pm i$.

23. $z_1 = e^{i\pi/4}, \quad z_2 = e^{i3\pi/4}, \quad z_3 = e^{-i3\pi/4}, \quad z_4 = e^{-i\pi/4}$.

25. $f(z)$ is continuous everywhere except at n points which are the roots of the equation

$$b_n z^n + \cdots + b_1 z + b_0 = 0.$$

27. Let $z_0 = x_0 + iy_0$ be an arbitrary point in a complex plane. Then

$$\lim_{z\to 0} f(z) = \lim_{x\to 0, y\to 0} x = x_0 = f(z_0).$$

Hence, $f(z)$ is continuous.

29. Put $z = x + iy$ and show that $f(z + h + ik) - f(z) \to 0$ as $h + ik \to 0$.

Answers to Odd Exercises for Section 1.4

Page 39 in the text.

1. $f'(z) = -2/(3z+4)^2$.

3. $f'(z) = (-8z^4 - 36z^2 + 14z)/(2z^3 + 7)^3$.

5. $f(z) = u + iv = x$, $u_x = 1$, and $v_y = 0$. Therefore $u_x \neq v_y$. Hence $f(z)$ is nowhere differentiable.

7. $f(z) = u + iv = x^2 + y^2$ implies that $u = x^2 + y^2$, $v = 0$. Therefore,

$$u_x = 2x, \qquad u_y = 2y, \qquad v_x = 0, \qquad v_y = 0.$$

The Cauchy–Riemann equations are satisfied only at $z = 0$. It follows that $f(z) = |z|^2$ is differentiable at $z = 0$, but it is not analytic since the Cauchy–Riemann equations are satisfied at no other points.

9. $f(z)$ is analytic everywhere except at the points

$$z_1 = 1, \quad z_2 = e^{i\pi/3}, \quad z_3 = e^{i2\pi/3}, \quad z_4 = e^{i\pi}, \quad z_5 = e^{-i\pi/3}, \quad z_6 = e^{-i2\pi/3}.$$

11. Put $z = x + iy = r\cos\theta + i\sin\theta$ and use the chain rule:

$$u_r = u_x x_r + u_y y_r = u_x \cos\theta + u_y \sin\theta, \qquad \text{etc.}$$

13. The function $u(x, y)$ is harmonic. Hence $f(z) = 1/z + C$.

15. v is not harmonic since $v_{xx} + v_{yy} \neq 0$.

17. $f(z) = z^3 + C$.

19. Show that $\lim_{x\to 0, y\to 0} u_x$ and $\lim_{x\to 0, y\to 0} v_x$ do not exist, where $f(z) = u(x, y) + iv(x, y)$.

21. Let $f(z) = u(x, y) + iv(x, y)$. Since $f(z)$ is a polynomial, it is analytic, hence $u_x = v_y$ and $u_y = -v_x$. We have

$$\overline{f(\bar{z})} = u(x, -y) - iv(x, -y).$$

Show that the Cauchy–Riemann equations are satisfied for $g(z) = \overline{f(\bar{z})}$. Similarly,

$$h(z) = \overline{f(\bar{z})} = u(x, y) - iv(x, y).$$

Show that the Cauchy–Riemann equations for $h(z)$ give

$$f'(0) = (u_x + iv_x)|_{z=0} = 0.$$

23. Using the Cauchy–Riemann equations, we can write $f'(z)$ in the form

$$f'(z) = u_x - iu_y = v_y + iv_x$$

Since $f'(z) \equiv 0$ in D then

$$u_x \equiv 0, \qquad u_y \equiv 0.$$

Therefore, $u \equiv C_1$. Similarly, $v \equiv C_2$. Hence, $f(z) = u+iv = C_1+iC_2 = C$.

Answers to Odd Exercises for Section 1.5

Page 48 in the text.

1. $i\,e^3$.

3. $e^{\cos 1}[\cos(\sin 1) + i \sin(\sin 1)]$.

5. Put $z = x + iy$ and show that

$$e^z = e^x(\cos y + i \sin y) \neq 0$$

since $e^x \neq 0$ for all finite real x, and $\cos y + i \sin y \neq 0$ for all real y.

7. Hint: Consider the limit of $f(z)$ as $z \to 0$ along different rays.

9. $\mathrm{Log}(3i) = \ln 3 + i\pi/2$.

11. $\log(1 + i) = 1/2 \ln 2 + i(\pi/4 + 2\pi k), \quad k = 0, \pm 1, \pm 2, \ldots$.

13. $z = \ln 4 + i(\pi + 2\pi k), \quad k = 0, \pm 1, \pm 2, \ldots$.

15. $z = -i \ln 2 + 5\pi/6 + 2\pi k, \quad k = 0, \pm 1, \pm 2, \ldots$.

17. $\sin z = \sin x \cosh y + i \cos x \sinh y$.

19. $\cosh z = \cosh x \cos y + i \sinh x \sin y$.

25. Put $z = x + iy$. Then show that

$$\cos \bar{z} = \cos x \cosh y + i \sin x \sinh y = u + iv.$$

Show that $u_x \neq v_y$ unless $x = n\pi$. But the vertical lines $x = n\pi$ are not open sets. Hence $\cos \bar{z}$ is analytic nowhere in C. A similar proof holds for the functiion $\sin \bar{z}$.

27. $z \neq 0, \quad z \neq -1.$

29. $z \neq i(\pi + 2\pi k), \quad k = 0, \pm 1, \pm 2, \ldots .$

31. $z = \pi/2 + 2\pi k - i \ln\left(1 + \sqrt{2}\right), \quad k = 0, \pm 1, \ldots .$

33. $z = i\left[\operatorname{Arctan}\left(\sqrt{15}\right) + 2\pi k\right], \quad k = 0, \pm 1, \ldots .$

35. The zeros of $\cosh z$ and $\sinh z$ are $i(\pi/2 + \pi k), \quad k = 0, \pm 1, \ldots,$ and $i\pi n, \quad n = 0, \pm 1, \ldots,$ respectively.

37. Solve the equations in terms of logarithms and prove that the roots are real if $-1 \leq a \leq 1$.

39. $e^{-\pi/2}$, $e^{\pi/2}[\cos(\ln 2) + i \sin(\ln 2)]$.

41. $-i \ln\left(\sqrt{2} - 1\right) + 2\pi k, \quad k = 0, \pm 1, \ldots .$

43. $\pi/2 + 2\pi k - i \ln\left(2 + \sqrt{3}\right), \quad k = 0, \pm 1, \ldots .$

Answers to Odd Exercises for Sections 2.1 and 2.2

Page 57 in the text.

1. $z(t) = t + it^2$, $1 \leq t \leq 3$; $z'(t) = 1 + 2it$.

3. $z(t) = -2\cos t + i2\sin t$, $\pi/3 \leq t \leq \pi/2$; $z'(t) = -2\sin t + i2\cos t$.

5. $z(t) = t + i/t^2$, $1 \leq t \leq 4$; $z'(t) = 1 - 2i/t^3$.

7. $\arg w'(z_0) = \pi$, $|w'(z_0)| = 2$.

9. $\arg w'(z_0) = -\pi/2$, $|w'(z_0)| = 1/2$.

11. $z = \pi/2 + \pi k$, $k = 0, \pm 1, \ldots .$

13. $z = -2$.

15. $z_1 = 1$, $z_2 = -4$.

17. $z = -1$.

19. (a) The curves intersect at $z = 1 + i$ at an angle $\alpha = \pi/4$.

(b) The image of γ_1 under the mapping $w = z^2$ in the w-plane is the parabola

$$u = \frac{1}{4}v^2 - 1, \quad \text{for} \quad -1 \leq u \leq 0 \quad \text{and} \quad 0 \leq v \leq 2,$$

where $w = u + iv$; the image of γ_2 is the segment $0 \leq v \leq 2$ of the imaginary v-axis. The angle, β, between the images of γ_1 and γ_2 at the point of intersection is $\beta = \pi/4$. Hence, $\beta = \alpha$ since the mapping is conformal.

Answers to Odd Exercises for Section 2.3

Page 64 in the text.

1. Translation by means of the vector $-i$.

3. Rotation by the angle $-\pi/2$ around the origin.

5. Rotation by the angle $\pi/3$ around the origin.

7. $w = (7 - 6i)z/5 + (2 - 6i)/5$.

9. $w = (1 + 2i)z + 2 - 2i$.

11. $\widetilde{D} = \{w; \Im w \geq 0\}$.

13. $\widetilde{D}$ is the domain between the lines $v = -u$ and $v = -u + \sqrt{2}$ in the w-plane.

15. $\widetilde{D} = \{(u, v) \in \mathbb{R}^2;\ 3 \leq u \leq 6,\ 1 \leq v \leq 4\}$.

17. $\widetilde{D} = \{w; |w| < 3, -\pi/4 \leq \operatorname{Arg} w \leq 0\}$.

19. $w = z - 1$.

Hint. To solve exercises 21–25, put $z = x+iy$. Then $w = u+iv = 1/(x+iy)$, and $u = x/(x^2 + y^2)$, $v = -y/(x^2 + y^2)$. Use the equation of the curve in the z-plane and these two formulae for u and v to eliminate x and y.

21. $w = -1/2$.

23. $w = 1/4$.

25. $(u + 1/4)^2 + (v + 1/4)^2 = 1/8$.

27. Put $z = x + iy$. Find the images of each boundary of D. The image of the line $x = 0$ is the line $u = 0$ in the w-plane, and the image of the line $x = 2$ is the circle

$$(u - 1/4)^2 + v^2 = 1/16$$

in the w-plane. The image of D is the domain between the circle

$$(u - 1/4)^2 + v^2 = 1/16$$

and the straight line $u = 0$.

29. The boundary, Γ, of the image of D consists of the two rays

$$u = 0, \quad 0 \leq v < +\infty, \quad \text{and} \quad u = 0, \quad -\infty < v \leq -1,$$

and the semicircle

$$u^2 + (v + 1/2)^2 = 1/4, \qquad u > 0.$$

The image of D is the domain to the right of Γ.

31. Any linear transformation is a combination of translation, dilation and rotation. Show that the reflection cannot be represented as a combination of the above three transformations.

33. Let $z = x$ and $w = u + iv$, where $v = 0$. Then $u = ax + b$, where a and b are some complex constants. Let

$$a = a_1 + ia_2 \quad \text{and} \quad b = b_1 + ib_2.$$

Then

$$u = (a_1 + ia_2)x + b_1 + ib_2.$$

Hence $u = a_1x + b_1$ and $0 = a_2x + b_2$. Since $a_2x + b_2 = 0$, then $w = a_1x + b_1$, where a_1 and b_1 are real.

Answers to Odd Exercises for Section 2.4

Page 74 in the text.

1. $\Re w > 0$.

3. Put $z = x + iy$. Then

$$w = u + iv = \frac{x^2 + y^2 - 1}{(x-1)^2 + y^2} - i\,\frac{2y}{(x-1)^2 + y^2}.$$

Show that the semi-infinite ray $x = 0, -\infty < y < 0$ is mapped onto the semicircle $u^2 + v^2 = 1, v > 0$ in the w-plane and the semi-infinite ray $y = 0, 0 < x < +\infty$ is mapped onto the two rays: $-\infty < u < -1, v = 0$ and $1 < u < +\infty, v = 0$. The image of D is the domain

$$\widetilde{D} = \{(u, v) \in \mathbb{R}^2;\ v > 0, u^2 + v^2 > 1\}.$$

5. $(u - 4/3)^2 + v^2 = 4/9$.

7. Let $w = (az + b)/(cz + d)$. Since $z_2 = 0$ is mapped into $w_2 = \infty$, then $w(0) = b/d = \infty$ if $d = 0, b \neq 0$. Let $a/c = \alpha$, $b/c = \beta$. Then we have a system $\alpha + \beta = -1$, $\alpha - i\beta = 1$, whose solution is $\alpha = -i$, $\beta = -1 + i$. Hence, $w = (i - 1)/z - i$.

9. $w = [(6 + 7i)z + 1 - 13i]/(17z - 3 + 5i)$.

11. $z_1 = 0$, $z_2 = -1$.

13. $z_{1,2} = \pm 1$.

15. $z = b/(a - 1)$ if $a \neq 1$; if $a = 1$ then any $z \in \mathbb{C}$ is a fixed point if and only if $b = 0$. If $a = 1$, $b \neq 0$ there are no fixed points.

17. $w = (-z + i)/(z + i)$.

19. Suppose

$$z = 2 \mapsto w = \infty, \qquad z = 2i \mapsto w = 0, \qquad z = -2 \mapsto w = 1.$$

Then the mapping is given by the function $w = (1-i)(z-2i)/(z-2)$.

21. Suppose

$$z = 0 \mapsto w = \infty, \qquad z = -3 \mapsto w = 0, \qquad z = -1-2i \mapsto w = i.$$

Then $w = (3+i)/4 + (9+3i)/(4z)$.

23. $w = (az+b)/(cz+d)$, where a, b, c and d are real and $ad - bc > 0$.

Answers to Odd Exercises for Section 2.5

Page 79 in the text.

1. $z = -2 - i$.

3. $z = 1/5 - 3/5i$.

5. Suppose $w = k(z+a)/(z+b)$. Then the condition $w(0) = i$ gives $ka/b = i$. Using the symmetry principle we obtain $w(\infty) = -i$, that is, $k = -i$. Then $w = -i(z+a)/(z-a)$. Using the fact that the circle $|z| = 1$ is mapped onto the real u-axis, show that $|a| = 1$. Then the condition $\operatorname{Arg} w'(0) = \pi/4$ gives $\operatorname{Arg} a = \pi/4$. Hence,

$$w = -i\left(z + e^{i\pi/4}\right) \Big/ \left(z - e^{i\pi/4}\right).$$

7. $w = e^{i3\pi/4}(z+1-i)/(z+1+i)$.

9. $w = i(z-3i)/(z+3i)$.

11. $w = \left(7+4\sqrt{3}\right) e^{i\alpha} \left(z+8-4\sqrt{3}\right) / \left(z+8+4\sqrt{3}\right), \qquad R = 2+\sqrt{3}.$

Answers to Odd Exercises for Section 2.6

Page 91 in the text.

1. $\widetilde{D} = \{w; 0 \le \operatorname{Arg} w \le 2\pi/3\}$.

3. $\widetilde{D} = \{w; 1 < |w| < 8, 0 < \operatorname{Arg} w < 3\pi/4\}$.

5. $w = -iz^3$.

7. $w = z^{1/\alpha}$.

9. $w = e^{i\pi/3}(2z - \sqrt{3} + i)^2/(2z + \sqrt{3} + i)^2$.

11. $w = \sqrt{(z-1)/(2-z)}$.

13. $w = \sqrt{[2+(1+i)z]/[z(1-i)-2-2i]}$.

15. $w = \sqrt{i/(2i+z)}$.

Answers to Odd Exercises for Section 2.7

Page 99 in the text.

1. $\widetilde{D} = \{w; 1 < |w| < e, 0 < \operatorname{Arg} w < \pi\}$.

3. $\widetilde{D} = \{w; |w| < 1, 0 < \operatorname{Arg} w < \pi/4\}$.

5. $\widetilde{D} = \{w; |w| > 1, 0 < \operatorname{Arg} w < \pi\}$.

7. $\widetilde{D} = \{w; |w| < 1, 0 < \operatorname{Arg} w < \pi\}$.

9. $\widetilde{D} = \{w; |w| > 1, 0 < \operatorname{Arg} w < \pi/2\}$.

11. $w = -e^{-2z}$.

13. $w = \sqrt{e^{3iz}}$.

15. $\widetilde{D} = \{w; -\infty < \Re w < +\infty, 0 < \Im w < \pi/2\}$.

17. $\widetilde{D} = \{w; 0 < \Re w < \ln 2, 0 < \Im w < \pi\}$.

19. $\widetilde{D} = \{w; 2 < \Re w < 2 + \ln 2, 1 < \Im w < \pi/2 + 1\}$.

21. $\widetilde{D} = \{w; -\infty < \Re w < +\infty, -\pi/2 < \Im w < 0\}$.

23. $w = \operatorname{Log} z$.

25. $w = -(i/\pi) \operatorname{Log}\left(e^{i\pi/4} z + i\right)$.

Answers to Odd Exercises for Sections 2.8 and 2.9

Page 111 in the text.

1. The whole w-plane with a cut joining the points -1 and 1 along the real u-axis.

3. The upper half-plane $\Im w > 0$ with the cut $[1, +\infty) \cup (-\infty, -1]$ along the extended real axis. This is easily visualized on the Riemann sphere.

5. The domain between the segment $[-1/2(R+1/R), 1/2(R+1/R)]$ of the real u-axis and the lower part of the ellipse

$$\frac{4u^2}{(R+1/R)^2} + \frac{4v^2}{(R-1/R)^2} = 1.$$

7. $\widetilde{D} = \{w; \Re w > 0, \Im w > 0\}$.

9. The whole w-plane with the cut $[1, +\infty) \cup (-\infty, -1]$ along the extended real axis.

11. The domain bounded by the positive real axis, the negative imaginary axis and the ellipse

$$\frac{u^2}{\cosh^2 \pi/2} + \frac{v^2}{\sinh^2 \pi/2} = 1.$$

13. $w = \cos z$.

15. $w = -\cos\left[-i\pi(z-2)\right]$.

Answers to Odd Exercises for Section 3.2

Page 124 in the text.

1. $11/2 - i$.

3. $1 + 2i/3$.

5. $2\pi i$.

7. $64/15 + 20i/3$.

9. $32/3 + 8i/3$.

11. $2/5 + 4i/5$.

13. 0.

15. $2i/3$.

17. $-\pi R - 2Ri$.

19. $[(\ln 5)/2 + i\,\mathrm{Arctan}\,0.5]^3 = 0.0021448 + 0.801067i$.

21. 2π.

23. $\pi/\left(32\sqrt{2}\right)$. (Hint: Put $z = 2\,e^{i\theta}$.)

25. $2\pi M/R$, where M is a constant such that $|u(z)| < M$ for all $z \in \mathbb{C}$. Also, $\lim_{R\to\infty} \int_{C_R} u(z)/z^2\,dz = 0$.

Answers to Odd Exercises for Section 3.3

Page 135 in the text.

1. e^{z^2} is analytic inside C.

3. The integrand is not analytic at $z = 1$ and $z = \pi/2 + n\pi$, $n = 0, \pm 1, \ldots$. All these points lie outside C. Therefore Cauchy's Theorem is satisfied.

5. 0.

7. The integrand is not analytic at the three points which are the roots of the equation $z^3 + 0.125 = 0$. All these points lie inside C. Therefore one cannot conclude that the integral is equal to zero.

Answers to Odd Exercises for Section 3.4

Page 145 in the text.

1. $\pi i e/2$.

3. $2\pi i \cos 2$.

5. 0.

7. 0.

9. $\pi i \sin 2/3$.

11. Use the proof of Cauchy's integral formula.

13. Let C be denote the circle $|z - z_0| = r\, e^{i\varphi}$. Then

$$\begin{aligned} f^{(n)}(z_0) &= \frac{n!}{2\pi i} \oint_C \frac{f(z)}{(z-z_0)^{n+1}}\, dz \\ &= \frac{n!}{2\pi i} \int_0^{2\pi} \frac{f\left(z_0 + r\, e^{i\varphi}\right) ri\, e^{i\varphi}}{\left(r\, e^{i\varphi}\right)^{n+1}}\, d\varphi \\ &= \frac{n!}{2\pi i} \int_0^{2\pi} f\left(z_0 + r\, e^{i\varphi}\right)\, e^{-in\varphi}\, d\varphi. \end{aligned}$$

Therefore

$$\begin{aligned} |f^{(n)}(z_0)| &\le \frac{n!}{2\pi r^n} \max_{|z-z_0|=r} |f(z)| 2\pi \\ &= \frac{n!}{r^n} \max_{|z-z_0|=r} |f(z)|. \end{aligned}$$

15. Since $\Re f(z) \le c$, we have

$$\left| e^{f(z)} \right| = \left| e^{\Re f(z) + i \Im f(z)} \right| = e^{\Re f(z)} \le c.$$

Hence the function $e^{f(z)}$ is uniformly bounded in the whole complex plane, and by Liouville's Theorem 3.4.5, it is constant in $\mathbb{C}$. Therefore $f(z)$ is constant in $\mathbb{C}$.

17. Applying Cauchy's estimate for $n = 2, 3, \ldots$ at every point $z \in \mathbb{C}$, we get

$$|f^{(n)}(z)| < M/r^{n-1}, \qquad \text{for all } r > 0 \text{ and } n = 2, 3, \ldots .$$

Integrating the equation $f''(z) = 0$, we obtain that $f(z)$ is a polynomial of degree at most 1. Assuming that there exist positive constants M and R such that $|f(z)| \le M|z|^n$ if $|z| \ge R$, one can show that, in this case, $f(z)$ is a polynomial of degree at most m.

19. $u(x,y) = \Re f(z) = e^x \cos y$. If there is a maximum or a minimum at a point (x,y) inside R, then we have

$$u_x = e^x \cos y = 0 \quad \text{and} \quad u_y = -e^x \sin y = 0.$$

Since the system of equations $u_x = 0$, $u_y = 0$ has no real solutions, $u(x,y)$ has no maxima nor minima inside R. Finally,

$$u_{\max} = e \quad \text{at } x = 1,\ y = 0, \quad \text{and} \quad u_{\min} = -e \quad \text{at } x = 1,\ y = \pi.$$

21. Consider the function $g(z) = 1/f(z)$. Since $f(z) \neq 0$ for $|z| < r$, it follows from the maximum modulus principle that the maximum of $|g(z)|$, that is, the minimum of $|f(z)|$, is assumed on $|z| = r$. On the other hand, the maximum of $|f(z)|$ is assumed on $|z| = r$. Since $|f(z)| =$ constant on $|z| = r$, the maximum and the minimum of $|f(z)|$ coincide. Therefore $f(z) =$ constant.

23. It follows from Cauchy's estimate that

$$|f^{(n)}(z)| \le n!M/r^n, \quad \text{where} \quad M = \max_{|\zeta - z| = r} |f(\zeta)|.$$

Suppose $|f^{(n)}(z)| > n!n^n$. Then we should have $n!n^n < n!M/r^n$, that is, $n^n < Mr^{-n}$. The last inequality cannot be satisfied for all n since n^n grows faster than r^{-n} for any fixed $0 < r < 1$.

25. There is no a contradiction with Liouville's Theorem since $|\cos z|$ is not bounded in $\mathbb{C}$. Therefore Liouville's Theorem cannot be applied.

Answers to Odd Exercises for Section 4.1

Page 162 in the text.

1. $\lim_{n\to\infty} z_n = 1/2 - i/2$.

3. $\lim_{n\to\infty} z_n = 0$.

5. $\lim_{n\to\infty} z_n = 2$. (Hint: Use the formula $\sin(iz) = i \sinh z$.)

7. $\lim_{n\to\infty} z_n = 0$.

9. Convergent.

11. Convergent.

13. Divergent. (Hint: Use the Stirling formula $n! \sim \sqrt{2\pi n}\,(n/e)^n$.)

15. Divergent.

17. Use partial sums $A_k = \sum_{n=1}^{k} z_n$ and $B_k = \sum_{k=1}^{n} \zeta_n$ such that

$$A_k + B_k = \sum_{n=1}^{k} (z_n + \zeta_n).$$

Then show that $\lim_{k\to\infty} (A_k + B_k) = A + B$.

19. Put $z_n = x_n + iy_n$, $A = a + ib$. Then evaluate the limit

$$\lim_{n\to\infty} |z_n| = \lim_{n\to\infty} \sqrt{x_n^2 + y_n^2}$$

using the fact that

$$\lim_{n\to\infty} x_n = a, \qquad \lim_{n\to\infty} y_n = b.$$

21. Yes.

23. $z \in \mathbb{R}$.

25. $|z| \geq R$ where $R > 4$.

27. $|z+1| \geq R$ where $R > 1$.

29. $z \in \mathbb{C}$.

Answers to Odd Exercises for Section 4.2

Page 171 in the text.

1. $R = 1$, $|z| \leq 1$.

3. $R = 0$. The series converges only at $z = -1$.

5. $R = 1/e$, $|z+2| < 1/e$.

7. $R = 1$, $|z+i| < 1$.

9. $R \geq \min\{R_1, R_2\}$.

11. $R \geq R_1 R_2$.

13. R_1.

15. R_1.

17. No, since $|3+4i| = 5 > |-3+3i| = 3\sqrt{2}$.

19. $(z^2+z)/(1-z)^3$.

21. $-z\,\mathrm{Log}(1-z) + z + \mathrm{Log}(1-z)$.

Answers to Odd Exercises for Section 4.3

Page 180 in the text.

1. $\cos z = \sum_{n=1}^{\infty} \frac{(-1)^{n-1}(z+\pi/2)^{2n-1}}{(2n-1)!}, \qquad R = \infty.$

3. $\dfrac{1}{z} = -\sum_{n=0}^{\infty}(z+1)^n, \quad R = 1.$

5. $\cos^2 z = 1 - z^2 + \dfrac{z^4}{3} - \dfrac{2z^6}{45} + \dfrac{z^8}{315} - \dots, \quad R = \infty.$

7. $\dfrac{z}{z^2+4} = \sum_{n=0}^{\infty} \dfrac{(-1)^n z^{2n+1}}{2^{2n+2}}, \quad R = 2.$

9. $z^4+2z^3-z+1 = 31+55(z-2)+36(z-2)^2+10(z-2)^3+(z-2)^4, \quad R = \infty.$

11. $\dfrac{z-2}{(z+3)(z-1)} = \dfrac{3}{4} - \dfrac{z+1}{4} + \dfrac{3(z+1)^2}{16} - \dfrac{(z+1)^3}{16} + \dfrac{3(z+1)^4}{64} - \dots, \quad R = 2.$

17. $\cos(3z-2) = \cos 1 \sum_{n=0}^{\infty} \dfrac{(-1)^n 3^{2n}}{(2n)!}(z-1)^{2n}$

$+ \sin 1 \sum_{n=0}^{\infty} \dfrac{(-1)^{n+1} 3^{2n+1}}{(2n+1)!}(z-1)^{2n+1}, \quad R = \infty.$

19. $e^{z^2+2z} = \dfrac{1}{e} \sum_{n=0}^{\infty} \dfrac{(z+1)^{2n}}{n!}, \quad R = \infty.$

21. $\dfrac{\cos^2 z}{1+z^2} = 1 - 2z^2 + \dfrac{7z^4}{3} - \dots, \quad R = 1.$

23. $\operatorname{Log}(1+\cos z) = \operatorname{Log} 2 - \dfrac{z^2}{4} - \dfrac{z^4}{96} - \dots, \quad R = \pi.$

25. $e^{1/z} = e - e(z-1) + \dfrac{3e(z-1)^2}{2} + \dots, \quad R = 1.$

31. $-2J_2(x) - J_0(x) + C.$

Answers to Odd Exercises for Section 4.4

Page 190 in the text.

1. $\sum_{n=0}^{\infty} \dfrac{(-1)^n z^{2n-2}}{(2n+1)!}.$

3. $\sum_{n=2}^{\infty} \dfrac{z^{n-2}}{n!}.$

5. $\sum_{n=0}^{\infty} \dfrac{(-1)^n}{z^{2n-3}(2n+1)!}.$

7. We have

$$
\begin{aligned}
e^{z+1/z} &= e^z\, e^{1/z} \\
&= \left(1+\frac{z}{1!}+\frac{z^2}{2!}+\frac{z^3}{3!}+\cdots+\frac{z^n}{n!}+\dots\right) \\
&\quad\times\left(1+\frac{1}{1!z}+\frac{1}{2!z^2}+\frac{1}{3!z^3}+\cdots+\frac{1}{n!z^n}+\dots\right) \\
&= 1+\frac{z}{1!}+\frac{z^2}{2!}+\frac{z^3}{3!}+\cdots+\frac{z^n}{n!}+\dots \\
&\quad+\frac{1}{1!z}+\frac{1}{1!1!}+\frac{z}{2!}+\frac{z^2}{3!}+\cdots+\frac{z^{n-1}}{n!}+\dots \\
&\quad+\frac{1}{2!z^2}+\frac{1}{2!z}+\frac{1}{2!2!}+\frac{z}{2!3!}+\cdots+\frac{z^{n-2}}{2!n!}+\dots \\
&= \left(1+\frac{1}{1!1!}+\frac{1}{2!2!}+\cdots+\frac{1}{n!n!}+\dots\right) \\
&\quad+\left(1+\frac{1}{2!}+\frac{1}{2!3!}+\dots\right)z+\left(\frac{1}{2!}+\frac{1}{3!}+\frac{1}{2!4!}+\dots\right)z^2+\dots \\
&\quad+\left(\frac{1}{n!}+\frac{1}{n!(n+1)!}+\dots\right)z^n+\cdots+\left(1+\frac{1}{2!}+\frac{1}{2!3!}+\dots\right)\frac{1}{z}+\dots \\
&= \sum_{n=-\infty}^{\infty} I_n(2)z^n,
\end{aligned}
$$

where

$$
I_n(x)=\sum_{k=0}^{\infty}\frac{(x/2)^{2k+n}}{k!(k+n)!}
$$

is the modified Bessel function of the first kind of order n for $n\in\mathbb{N}$.

9. $\dfrac{z+2}{(z-3)^2}=\dfrac{z-3+5}{(z-3)^2}=\dfrac{5}{(z-3)^2}+\dfrac{1}{z-3}.$

11. $\displaystyle\sum_{n=0}^{\infty}\frac{(-1)^n}{(z-i)^{2n}(2n+1)!}+(2+i)\sum_{n=0}^{\infty}\frac{(-1)^n}{(z-i)^{2n+1}(2n+1)!}.$

13. We have

$$
\begin{aligned}
\frac{\cos z}{z+4} &= \frac{\cos(z+4-4)}{z+4}=\frac{\cos(z+4)\cos 4+\sin(z+4)\sin 4}{z+4} \\
&= \frac{\cos 4}{z+4}\sum_{n=0}^{\infty}\frac{(-1)^n(z+4)^{2n}}{(2n)!}+\frac{\sin 4}{z+4}\sum_{n=0}^{\infty}\frac{(-1)^n(z+4)^{2n+1}}{(2n+1)!} \\
&= \cos 4\sum_{n=0}^{\infty}\frac{(-1)^n(z+4)^{2n-1}}{(2n)!}+\sin 4\sum_{n=0}^{\infty}\frac{(-1)^n(z+4)^{2n}}{(2n+1)!}.
\end{aligned}
$$

15. We have

(a) $$\frac{1}{3}\sum_{n=0}^{\infty}(\ 1)^{n+1}z^n \qquad \frac{1}{3}\sum_{n=0}^{\infty}\frac{z^n}{2^{n+1}};$$

(b) $$\frac{1}{3}\sum_{n=0}^{\infty}\frac{(-1)^{n+1}}{z^{n+1}}-\frac{1}{3}\sum_{n=0}^{\infty}\frac{z^n}{2^{n+1}};$$

(c) $$\frac{1}{3}\sum_{n=0}^{\infty}\frac{(-1)^{n+1}}{z^{n+1}}+\frac{1}{3}\sum_{n=0}^{\infty}\frac{2^n}{z^{n+1}}.$$

17. We have

(a) $$-\frac{1}{2}\sum_{n=0}^{\infty}(-1)^n z^n+\frac{5}{2}\sum_{n=0}^{\infty}\frac{(-1)^n z^n}{3^{n+1}};$$

(b) $$-\frac{1}{2}\sum_{n=0}^{\infty}\frac{(-1)^n}{z^{n+1}}+\frac{5}{2}\sum_{n=0}^{\infty}\frac{(-1)^n z^n}{3^{n+1}};$$

(c) $$-\frac{1}{2}\sum_{n=0}^{\infty}\frac{(-1)^n}{z^{n+1}}+\frac{5}{2}\sum_{n=0}^{\infty}\frac{(-1)^n 3^n}{z^{n+1}}.$$

19. $$\sum_{n=0}^{\infty}\frac{(-1)^n 3^n}{(z-1)^{n+1}}-\sum_{n=0}^{\infty}\frac{(-1)^n(z-1)^n}{4^{n+1}}.$$

21. $$-\frac{1}{4(z+i)^2}+\frac{i}{4(z+i)}+\frac{3}{16}-\frac{i(z+i)}{8}-\frac{5(z+i)^2}{64}+\dots.$$

Answers to Odd Exercises for Section 5.1

Page 210 in the text.

1. $z=\pm 4i$ are zeros of order 1.

3. $z=2k\pi$, $k=0,\pm 1,\dots$, are zeros of order 2, and $z=\pm 3$ are zeros of order 3.

5. $z=0$ is a zero of order 5; $z=k\pi$, $k=\pm 1\pm 2,\dots$, are zeros of order 3.

7. $z=0$ is a zero of order 1; $z=k\pi$, $k=\pm 1\pm 2,\dots$, are zeros of order 2.

9. Zero of order 3.

11. Zero of order 6.

13. Zero of order 2.

15. $z=\pm 2i$ are simple poles; $z=1$ is a pole of order 2.

17. $z=(\pi+2k\pi)i$, $k=0,\pm 1,\dots$, are simple poles; the function has an essential singularity at infinity.

19. $z=2k\pi$, $k=\pm 1,\pm 2,\dots$, are poles of order 2, and $z=0$ is a pole of order 1.

21. $z=i$ is an essential singularity.

23. $z = k\pi$, $k = 0, \pm 1, \pm 2, \ldots$, are poles of order 2.

25. If $n = m$, then z_0 is either a removable singularity or a pole of order $\leq m$. If $n \neq m$, then z_0 is a pole of order $\max(n, m)$.

Answers to Odd Exercises for Section 5.2

Page 227 in the text.

1. The residues are

$$\operatorname*{Res}_{z=0} \frac{1}{z - z^3} = 1, \qquad \operatorname*{Res}_{z=1} \frac{1}{z - z^3} = -\frac{1}{2},$$
$$\operatorname*{Res}_{z=-1} \frac{1}{z - z^3} = -\frac{1}{2}, \qquad \operatorname*{Res}_{z=\infty} \frac{1}{z - z^3} = 0.$$

3. The residues are

$$\operatorname*{Res}_{z=0} \frac{z^2 + 4z + 1}{z^2(z+1)} = 3, \quad \operatorname*{Res}_{z=-1} \frac{z^2 + 4z + 1}{z^2(z+1)} = -2, \quad \operatorname*{Res}_{z=\infty} \frac{z^2 + 4z + 1}{z^2(z+1)} = -1.$$

5. The residues are

$$\operatorname*{Res}_{z=1} \frac{e^z}{(z-1)(z+3i)^2} = -\frac{2e}{25} - \frac{3e}{50} i,$$
$$\operatorname*{Res}_{z=-3i} \frac{e^z}{(z-1)(z+3i)^2} = -\frac{1}{50} \cos 3 + \frac{9}{25} \sin 3 + i \left(\frac{9}{25} \cos 3 + \frac{1}{50} \sin 3 \right),$$
$$\operatorname*{Res}_{z=\infty} \frac{e^z}{(z-1)(z+3i)^2} = \frac{2e}{25} + \frac{1}{50} \cos 3 - \frac{9}{25} \sin 3 + i \left(\frac{3e}{50} - \frac{9}{25} \cos 3 - \frac{1}{50} \sin 3 \right).$$

7. $\displaystyle \operatorname*{Res}_{z=2k\pi i} \frac{1}{e^z - 1} = 1, \qquad k = 0, \pm 1, \pm 2, \ldots .$

9. The residues are

$$\operatorname*{Res}_{z=0} \frac{1 - \cos z}{z^2 \sin z} = \frac{1}{2},$$
$$\operatorname*{Res}_{z=n\pi} \frac{1 - \cos z}{z^2 \sin z} = \begin{cases} 0, & n = 2k, \\ -2/(n^2\pi^2), & n = 2k - 1, \end{cases} \qquad k = \pm 1, \pm 2, \ldots .$$

11. We have

$$z^2 \sin\left(\frac{1}{z-1} \right) = [(z-1)^2 + 2z - 1] \sin\left(\frac{1}{z-1} \right)$$

$$= [(z-1)^2 + 2(z-1) + 1] \sin\left(\frac{1}{z-1}\right)$$

$$= [(z-1)^2 + 2(z-1) + 1]\left[\frac{1}{z-1} - \frac{1}{3!(z-1)^3} + \frac{1}{5!(z-1)^5} + \dots\right].$$

Hence, $\operatorname*{Res}_{z=1} z^2 \sin\left(\frac{1}{z-1}\right) = 1 - \frac{1}{3!} = \frac{5}{6}$.

13. $\operatorname*{Res}_{z=0}\left(\frac{\sin z}{z} + \frac{1}{z^3} + e^{1/z}\right) = 1.$

15. $e^{z/(z-1)} = e^{1+1/(z-1)} = e\left(1 + \frac{1}{1!(z+1)} + \frac{1}{2!(z+1)^2} + \dots\right).$

Hence, $\operatorname*{Res}_{z=1} e^{z/(z-1)} = e$.

17. A pole of order 2 at $z = z_0$. Since $\psi(z_0) = 0$, $\psi'(z_0) \neq 0$, one has

$$\psi(z) = \alpha(z)(z - z_0) = [\alpha_0 + \alpha_1(z - z_0) + \dots](z - z_0),$$

where $\alpha_0 = \psi'(z_0)$, $\alpha_1 = 0.5\psi''(z_0)$. Then

$$\begin{aligned}\operatorname*{Res}_{z=z_0} f(z) &= \lim_{z\to z_0}\left[\frac{\varphi(z)(z-z_0)^2}{[\psi(z)]^2}\right]' \\ &= \lim_{z\to z_0}\left[\frac{\varphi(z)}{[\psi'(z_0) + 0.5\psi''(z_0)(z-z_0) + \dots]^2}\right]' \\ &= \frac{\varphi'(z_0)[\psi'(z_0)]^2 - 2\psi'(z_0)0.5\psi''(z_0)\varphi(z_0)}{[\psi'(z_0)]^4} \\ &= \frac{\varphi'(z_0)\psi'(z_0) - \psi''(z_0)\varphi(z_0)}{[\psi'(z_0)]^3}.\end{aligned}$$

19. 0.

21. 0.

23. 0.

25. $-4\pi i/3$.

27. 0.

$$\operatorname*{Res}_{z=2i} \frac{e^{1/z}}{z^2+4} = -\frac{i}{4}\, e^{-i/2}, \qquad \operatorname*{Res}_{z=-2i} \frac{e^{1/z}}{z^2+4} = \frac{i}{4}\, e^{i/2}.$$

To compute the residue of $e^{1/z}/(z^2+4)$ at $z = 0$, we expand $e^{1/z}$ in a Laurent series about $z = 0$ and $1/(z^2+4)$ in a Taylor series about $z = 0$ and multiply the results:

$$\frac{1}{z^2+4} = \frac{1}{4(1+z^2/4)} = \sum_{n=0}^{\infty} \frac{(-1)^n z^{2n}}{2^{2n+2}},\ e^{1/z} = \sum_{n=0}^{\infty} \frac{1}{n! z^n}.$$

Hence

$$e^{1/z}\,\frac{1}{z^2+4} = \sum_{n=0}^{\infty} \frac{(-1)^n z^{2n}}{2^{2n+2}} \sum_{n=0}^{\infty} \frac{1}{n!z^n}.$$

Multiplying the series and collecting the terms containing $1/z$, we obtain

$$\operatorname*{Res}_{z=0} \frac{e^{1/z}}{z^2+4} = \frac{1}{2}\sum_{n=0}^{\infty} \frac{(-1)^n (1/2)^{2n+1}}{(2n+1)!} = \frac{1}{2}\sin\frac{1}{2}.$$

The sum of the residues at the three singular points, $z = 2i$, $z = -2i$ and $z = 0$, is equal to 0.

29. 0.

31. 0.

33. 0.

Answers to Odd Exercises for Chapter 6

Page 262 in the text.

1. $\pi/\sqrt{2}$.

3. $\pi/2$.

5. $7\pi/50$.

Answers to Odd Exercises for Chapter 7

Page 294 in the text.

1. $2\pi/(3\sqrt{3}\,)$.

3. $\pi\left(1+\sqrt{2}\right)/4$.

5. $\dfrac{\pi}{2(1-a^2b^2)}\left(\dfrac{1}{a}\ln a + b\ln b\right)$.

7. $\dfrac{a\pi}{2b(b^2-a^2)}\ln\left(\dfrac{b}{a}\right)$.

9. $16\pi^3/(81\sqrt{3}\,)$.

11. $\dfrac{\pi}{2a(\ln^2(a)+\pi^2/4)} - \dfrac{1}{1+a^2}$.

13. $-\pi/4$.

15. $-\pi/16$.

17. $\dfrac{\pi}{20}\ln 8 - \dfrac{\pi}{30}\ln 13$.

19. $-\dfrac{\pi^2}{8}\ln 26 + \dfrac{\pi}{6}\operatorname{Arctan}\left(\dfrac{1}{5}\right)\ln 2 + \dfrac{\pi^2}{6}\ln 29 - \dfrac{\pi}{6}\operatorname{Arctan}(2)\ln 29$
$-\dfrac{\pi}{6}\operatorname{Arctan}\left(\dfrac{2}{5}\right)\ln 5.$

21. $-\dfrac{19\pi}{377}\ln 2 + \dfrac{19\pi}{754}\ln 29 + \dfrac{21\pi}{754}\operatorname{Arctan}\left(\dfrac{2}{5}\right).$

Answers to Odd Exercises for Chapter 8

Page 346 in the text.

1. $\dfrac{\pi \sin\left(a/\sqrt{2}\right)\cosh\left(a/\sqrt{2}\right)}{\sinh^2\left(a/\sqrt{2}\right) + \sin^2\left(a/\sqrt{2}\right)}.$

3. $\dfrac{\pi}{5}\left(\dfrac{1}{\sinh 2a} - \dfrac{1}{\sinh 3a}\right).$

5. $\dfrac{\pi\cos 2a\cosh 2a}{2(\sinh^2 2a + \cos^2 2a)}.$

7. $\dfrac{\pi}{(\beta^2-\alpha^2)}\left(\dfrac{\sinh b\alpha}{\alpha\sinh a\alpha} - \dfrac{\sinh b\beta}{\beta\sinh a\beta}\right).$

9. $\dfrac{\pi}{(\beta^2-\alpha^2)}\left(\dfrac{e^{-a\alpha}}{\cosh a\alpha} - \dfrac{e^{-a\beta}}{\cosh a\beta}\right).$

11. $\dfrac{\pi}{3}\left(\dfrac{1}{b\sinh a + \cosh a} - \dfrac{1}{b\sinh 2a + 2\cosh 2a}\right).$

13. $\dfrac{\pi\beta}{I_2(a\beta)}$, where $I_2(z)$ is the modified Bessel function of the first kind of order 2.

Answers to Odd Exercises for Sections 9.2 and 9.3

Page 358 in the text.

1. Four zeros. Put $f(z) = -5z^4 + z^3$, $g(z) = z^6 - 2z$.

3. Three zeros. Use Rouché's Theorem in the domains $|z| < 1$ and $|z| < 2$.

5. Three zeros.

Answers to Odd Exercises for Sections 9.4 and 9.5

Page 377 in the text.

1. $\csc z = \dfrac{1}{z} - 2z\sum_{n=1}^{\infty} \dfrac{(-1)^{n-1}}{z^2 - n^2\pi^2}$.

3. $\tan \dfrac{\pi z}{2} = \dfrac{4}{\pi}\sum_{n=0}^{\infty} \dfrac{(-1)^n(2n+1)}{(2n+1)^2 - z^2}$.

5. We have

$$\prod_{k=1}^{\infty}\left[1 + \frac{1}{k(k+2)}\right] = \prod_{k=1}^{\infty}\frac{k^2+2k+1}{k(k+2)} = \prod_{k=1}^{\infty}\frac{(k+1)^2}{k(k+2)}.$$

Let $b_n = \prod_{k=1}^{n}\dfrac{(k+1)^2}{k(k+2)}$. Then $b_1 = 2 \times \dfrac{2}{3}$, $b_2 = 2 \times \dfrac{3}{4}$, $b_3 = 2 \times \dfrac{4}{5}$. One can prove by mathematical induction that $b_n = 2\,\dfrac{n+1}{n+2}$. Hence, $\lim_{n\to\infty} b_n = \lim_{n\to\infty} 2\,\dfrac{n+1}{n+2} = 2$.

7. $|z| < 1$.

Answers to Odd Exercises for Chapter 10

Page 414 in the text.

1. $-\dfrac{1}{2a^4} + \dfrac{\pi}{2\sqrt{2}\,a^3}\,\dfrac{\cos(\pi a/\sqrt{2})\sin(\pi a/\sqrt{2}) + \cosh(\pi a/\sqrt{2})\sinh(\pi a/\sqrt{2})}{\cosh^2(\pi a/\sqrt{2}) - \cos^2(\pi a/\sqrt{2})}$.

3. $-\pi^2/12$.

5. $(x-\pi)^3/12 - \pi^2(x-\pi)/12$.

7. $73/189000 = 0.000386\ldots$.

References

1. M. Abramowitz and I. A. Stegun (eds.), *Handbook of mathematical functions*, 10th printing, with corrections, National Bureau of Standards, Washington, D.C., 1972.
2. L. V. Ahlfors, *Complex analysis*, 3rd ed., McGraw-Hill, New York, 1979.
3. J. M. Antimirov, *A new class of integrals evaluated in closed form*, Proc. Acad. Sci. Latvia, Section B, Natural Sciences, **7/8** (1995), 120–123.
4. M. Ya. Antimirov, *On one contact thermal problem*, Latvian Mathematical Annals, Zinātne, Riga, 1965, pp. 197–226 (Russian).
5. M. Ya. Antimirov, A. A. Kolyshkin, and R. Vaillancourt, *Applied integral transforms*, CRM Monograph Series, no. 2, Amer. Math. Soc., Providence, RI, 1993.
6. M. Ya. Antimirov, A. A. Kolyshkin, and R. Vaillancourt, *Nouvelles classes d'intégrales définies*, C. R. Acad. Sci. Paris, **319**, Série I, (1994), 1089–1094.
7. V. Ya. Arsenin, *Mathematical physics. Basic equations and special functions*, Nauka, Moscow, 1966 (Russian).
8. I. N. Baker, *Limit functions and sets of non-normality in iteration theory*, Ann. Acad. Sci. Fenn. Ser. A I, no 467 (1970), 1–11.
9. D. Bierens de Haan, *Nouvelles tables d'intégrales définies*, edition of 1867 — corrected, Hafner Publishing Co., New York, 1957.
10. P. Blanchard, *Complex analytic dynamics on the Riemann sphere*, Bull. Amer. Math. Soc. (N.S.), **11** (1984), 85–141.
11. Y. S. Bougrov and S. M. Nikolski, *Cours de mathématiques supérieures*, translated from the Russian, Mir, Moscow, 1983.
12. R. Broucke, *On Kepler's equation and strange attractors*, J. Astron. Sci., **28** (1980), 255–265.
13. B. M. Budak and S. V. Fomin, *Multiple integrals and series*, Nauka, Moscow, 1965 (Russian).
14. B. M. Budak, A. A. Samarskii and A. N. Tikhonov, *A collection of problems on mathematical physics*, Macmillan, New York, 1964.
15. R. V. Churchill and J. W. Brown, *Complex variables and applications*, 5th ed., McGraw-Hill, New York, 1990.
16. N. Doual, J. L. Howland and R. Vaillancourt, *Selective solutions to transcendental equations*, Computers Math. Applic., **22** (1991), 61–76.

17. A. Erdélyi *et al.*, *Higher transcendental functions*, 3 vol., McGraw-Hill, New York, 1953–1955.
18. A. Erdélyi *et al.*, *Tables of integral transforms*, Vol I, McGraw-Hill, New York, 1954.
19. P. D. Escobal, *Methods of orbit determination*, published by John Wiley & Sons, New York, London, Sydney, in 1965, reprinted by Robert E. Krieger, Huntington, N. Y., 1975, pp. 84–91.
20. M. A. Evgrafov, *Analytic functions*, transl. by Scripta Mathematica, ed. by B. R. Gelbaum, W. B. Saunders, Philadelphia and London, 1966.
21. M. A. Evgrafov, J. V. Sidorov, M. V. Fedoryuk, M. I. Shabunin, and K. A. Bezanov, *Collection of problems on the theory of analytic functions*, Moscow, Nauka, 1969 (Russian).
22. S. Gélinas, V. N. Tran and R. Vaillancourt, *Global iterative solution of dielectric spectroscopy equation*, Trans. IEEE Instr. Meas. **IM-39** (1990), 615–620.
23. I. S. Gradshteyn and I. M. Ryzhik, *Tables of integrals, series and products*, 4th ed. prepared by Yu. V. Geronimus and M. Yu Tseytlin, transl. from the Russian by Scripta Technica, Academic Press, New York, 1965.
24. W. Gröbner and N. Hofreiter, *Integral Tafel. Zweiter Teil. Bestimmte Integrale*, Wien and Innsbruck, Springer-Verlag, 1950.
25. P. Henrici, *Applied and computational complex analysis*, 3 vol., Wiley, New York, 1974.
26. J. L. Howland and R. Vaillancourt, *Attractive cycles in the iteration of meromorphic functions*, Numer. Math., **46** (1985), 323–337.
27. J. L. Howland, A. Thompson and R. Vaillancourt, *On the dynamics of a meromorphic function*, Appl. Math. Notes, **15** (1990), 7–37.
28. T. E. Hull, W. H. Enright, B. M. Fellen and A. E. Sedgwick, *Comparing numerical methods for ordinary differential equations*, SIAM J. Numer. Anal., **9** (1972), 603–637.
29. V. A. Ilyin and E. G. Poznyak, *Basics of mathematical analysis*, Nauka, Moscow, 1973 (Russian).
30. G. Julia, *Leçons sur les fonctions uniformes à point singulier essentiel isolé*, Gauthier-Villars, Paris, 1924, par. 57, pp. 92–94.
31. M. L. Krasnov, A. I. Kissélev and G. I. Makarenko, *Fonctions d'une variable complexe. Calcul opérationnel. Théorie de la stabilité*, Mir, Moscow, 1985.
32. E. Kreyzsig, *Advanced engineering mathematics*, 7th ed., Wiley, New York, 1993.
33. M. A. Lavrentiev and B. V. Chabat, *Méthodes de la théorie des fonctions d'une variable complexe*, transl. from the Russian by Henri Damadian, Mir, Moscow, 1972.
34. A. I. Markushevich, *Theory of functions of a complex variable*, 3 vol., Prentice-Hall, Englewood Cliffs, NJ, 1965.
35. J. E. Marsden and A. J. Tromba, *Vector calculus*, 2nd ed., W. H. Freeman, New York, San Francisco, 1976.

36. R. Nevanlinna, *Le théorème de Picard–Borel et la théorie des fonctions méromorphes*, Gauthier-Villars, Paris, 1929, par. 39, p. 75.
37. R. Nevanlinna, *Analytic functions*, transl. from the 2nd German ed. by Phillip Emig, Springer-Verlag, New York, 1970.
38. A. P. Prudnikov, Y. A. Brychkov and O. I. Marichev, *Integrals and series*, translated from the Russian by N. M. Queen, Vol. 1, Gordon and Breach, New York, 1986.
39. A. M. C. Ruedin, D. C. Skigin and R. Vaillancourt, *Computer solution of the scattering problem for a groove in a metallic plane using the modal method*, Computers & Mathematics with Applications, 1998, to appear.
40. S. Saks and A. Zygmund, *Analytic functions*, 3rd ed., Elsevier, New York, 1971.
41. G. Sansone and J. Gerretsen, *Lectures on the theory of functions of a complex variable. I. Holomorphic functions*, P. Noordhoff, Gronigen, 1960.
42. Yu. Sidorov, M. Fedoriouk, and M. Chabounine, *Leçons sur la théorie des fonctions d'une variable complexe*, Nauka, Moscow, 1976.
43. I. N. Sneddon, *The use of integral transforms*, McGraw-Hill, New York, 1972.
44. M. R. Spiegel, *Theory and problems of complex variables*, Schaum's Outline Series, Schaum Publishing Co., New York, 1964.
45. A. Sveshnikov and A. Tikhonov, *The theory of functions of a complex variable*, Mir, Moscow, 1978. (Russian).
46. S. M. Targ, *Basic problems in the theory of laminar flows*, GTTI, Moscow–Leningrad, 1951 (Russian).
47. E. F. Tsarkov, *Random perturbations of functional differential equations*, Zinātne, Riga, 1989 (Russian).
48. Ya. S. Uflyand, *Integral transforms for elasticity problems*, Soviet Academy of Science Publ. House, Moscow–Leningrad, 1963 (Russian).
49. L. A. Vainstein, *Theory of diffraction and factorization method*, Sovetskoye Radio, Moscow, 1966 (Russian).
50. E. T. Whittaker and G. N. Watson, *A course of modern analysis*, 4th ed., University Press, Cambridge, 1963.
51. A. D. Wunsch, *Complex variables with applications*, Addison-Wesley, Reading, MA, 1994.

Index

About the Authors

M. Ya. Antimirov was born in 1937 in Grozny, Russia. He graduated from Kirgiz State University in Frunze (Kirgizstan) in theoretical physics in 1960. From 1960 to 1964, he was assistant and senior lecturer in the Department of Mathematics of the Grozny Petroleum Institute. In 1964–1965, he did his doctoral studies at the Computer Center of the Latvian State University in Riga, Latvia, where he held a Senior Research Fellowship from 1965 to 1968. He obtained his Ph.D. (Candidate in physics and mathematics) in 1966. From 1968 to 1991 he was docent in the Department of Engineering (formerly Applied) Mathematics of the Riga Technical University and is now professor in the same department. In 1991, he obtained the degree of Doctor of Mathematics and Physics at the Physics Institute of the Academy of Sciences of Latvia and was promoted to full professorship in his university. He published more than 75 research papers.

A. A. Kolyshkin was born in 1954 in Tallinn, Estonia. He graduated in applied mathematics from the Riga Technical University in Riga, Latvia, in 1976, and obtained his Ph.D. (Candidate in physics and mathematics) from Saint-Petersburg (then Leningrad) State University in 1981. In 1979–1980, he held a Research Fellowship in the Laboratory of Nondestructive Testing at the Riga Technical University. Since then he has been with the Department of Engineering (formerly Applied) Mathematics of the Riga Technical University, as assistant in 1980–1983, senior lecturer in 1983–1990 and associate professor (docent) from 1990. In 1987–1988, he visited the University of Ottawa under the auspices of the Canada-USSR Academic Exchange Program and in 1991–1992 as a visiting researcher, and almost every year thereafter, where he is now adjunct professor of mathematics.

Rémi Vaillancourt was born in 1934 in Maniwaki, Québec, Canada. He took several bachelor's and master's degrees at the University of Ottawa, Ontario, Canada, and his Ph.D. under the direction of P. D. Lax and K. O. Friedrichs at the Courant Institute of Mathematical Sciences at New York University in 1969. He held an Office of Naval Research Fellowship at the University of Chicago where he worked under the direction of A. P. Calderón in 1969–1970. He is known for the Yamanugi–Nogi–Vaillancourt pseudo-difference operators and the Calderón–Vaillancourt theorem for non-classical pseudodifferential operators. Since 1970 he has been at the University of Ottawa with the department of Mathematics and Statistics, and jointly with the department of Computer Science in 1974-1997 and with the School of Informatiton Technology and Engineering since 1997. He is full professor. He served as chairman of the department of Mathematics in 1972–1976, as vice-president, president and past president of the Canadian Mathematical Society in 1975–1981, as chairman of the Canadian National Committee for the International Mathematical Union in 1979–1988, and as president and past president of the Canadian Applied Mathematics Society in 1993–1997. He collaborated with late Hitoshi Kumano-go and Michihiro Nagase in translating and updating Kumano-go's monograph entitled *Pseudo-Differential Operators*, published by The MIT Press. He co-authored, with M. Ya. Antimirov and A. A. Kolyshkin, a monograph entitled *Applied Integral Transforms* in the CRM Monograph Series, published by AMS, and co-edited, with A. L. Smirnov, a survey entitled *Asymptotic Methods in Mechanics* in the CRM Proceedings & Lecture Notes, published by AMS. He co-authored, with Ryuichi Ashino, a MATLAB compendium called *Introduction to Matlab* published in Japanese by Kyoritsu Shuppan in Tokyo in July 1997. He co-authored, with M. Ya. Antimirov and A. A. Kolyshkin, a monograph entitled *Mathematical Models for Eddy Current Testing* to be published by Les Publications CRM in Montréal, later in 1997. His mathematical hobby is writing reviews for *Mathematical Reviews* and *Zentralblatt für Mathematik*.

90182
9 780120 595457
ISBN 0-12-059545-1